PEARSON ALWAYS LEARNING

Jorge Dorribo Camba • Jeffrey Otey • Matthew Whiteacre

Foundations of Graphics for Engineers

Third Edition

Cover Art: Courtesy of Photodisc/Getty Images, Art Explosion by Nova.

Pearson Learning Solutions, 501 Boylston Street, Suite 900, Boston, MA 02116
A Pearson Education Company
www.pearsoned.com

Printed in the United States of America

2 3 4 5 6 7 8 9 10 V092 17 16 15 14 13 12

000200010271288726

TS

PEARSON

ISBN 10: 1-256-80445-2
ISBN 13: 978-1-256-80445-1

Contents

1 CHAPTER

Introduction to Engineering Graphics

1. Introduction

Technical illustrations have been used throughout history to communicate ideas and designs. From cave drawings portraying prehistoric life, to architectural drawings in ancient Greece, to Leonardo da Vinci's famous designs during the Renaissance, to the development of descriptive geometry and computer graphics, drawings have been used for the sole purpose of communicating information. This information can be used to convince others of the validity of your design or to provide detailed instructions so your design can be constructed.

Drawings are graphical representations of real objects or ideas. They are used in many different areas, such as visual arts, artistic painting, illustrations, engineering and technical applications. Artistic drawings are concerned with aesthetics and beauty, creating compositions, shapes, and color combinations that make a drawing visually appealing. Engineering and technical drawings are used to capture and communicate accurately the technical detail required to bring a design to reality. This book will focus on the latter.

Engineering drawings appear in many different forms, depending on the engineering specialty. For example, civil engineers design buildings, bridges, and roads. Mechanical engineers create parts and assemblies that need to be manufactured. Aerospace engineers draw planes, rockets, and satellites. Electrical engineers draw circuit board layouts and schematics. Chemical engineers design piping systems, chemical process plants, valves, and vessels.

All these drawings are used as a tool to describe complex models, systems, and structures precisely. For this reason, certain rules and guidelines must be followed in engineering graphics to guarantee accuracy, precision, and successful performance of the final product while avoiding any potential misunderstanding or misinterpretation. This book is concerned with engineering graphics, commonly defined as a visual form of communication used to share, document, and analyze technical ideas for construction and manufacturing. It is hoped that students will see the vital role graphics fills in engineering while grasping the importance of following standard practices in order to convey the designer's intent.

2. Standards and Conventions

All technical communication systems rely on well defined sets of rules that guarantee that any individual using the system understands the same message. In other words, messages must be delivered clearly and unambiguously. The letters of the alphabet, for example, are the foundation of written communication, and grammar defines how a message should be properly constructed.

Rules can be divided in two categories: standards and conventions. A standard is a formal document that institutes a uniform method or process to perform an action or defines a criterion for a product. Standards are defined by organizations such as ANSI (American National Standards Institute) responsible for governing and regulating many engineering standards used in the United States. Other organizations include ISO (International Standards Organization) and JIS (Japanese Industrial Standards).

Conventions or conventional practices are methods and criteria used so extensively that they become dominant and generally accepted and adopted by the public. They are also known as "industry standards" or "de facto standards." There are no organizations or formal documents that define conventions. These conventions can differ between industries or even firms within the same industry.

Engineering graphics are regulated by both standards and conventions. How lines are represented or how a drawing is dimensioned are examples of clearly stipulated rules. In the United States, the Y series of the ANSI standards regulate engineering drawings.

3. The Engineering Design Process

Engineering design is a team activity, where individuals with different experience and skills work toward a common purpose. Depending on the complexity of the project, one person or a group may be assigned responsibility of a single specific function of the overall design.

The engineering design process is the sequence of steps involved in solving an engineering problem or developing new products or processes by applying scientific principles, mathematics and engineering science. The engineering design process is not universal, as it varies from company to company, or even from problem to problem. Some companies use traditional approaches, while others use more concurrent processes.

Although the number of steps may vary, a general linear engineering design process is comprised of the following stages (see Figure 1.1).

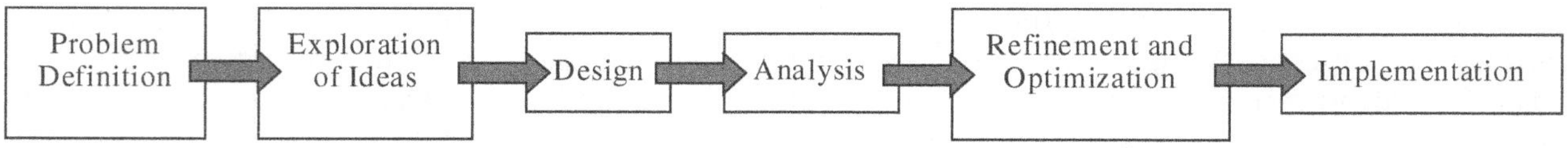

Figure 1.1 Linear engineering design process

3.1 Problem Definition

Problem definition is the first step of the design process where all the boundaries and constraints of the problem are established. It involves a clear definition of what objectives need to be met, under what conditions, and the limitations.

In this stage, a great amount of information is communicated through rough sketches. Most engineering problems can be visualized easier on paper than through words. Occasionally, these sketches may be combined with formal diagrams, dimensions, and charts to present the problem in a way that is easily understood (see Figure 1.2).

3.2 Exploration of Ideas

The second stage of the design process involves the identification of possible approaches or solutions to the defined problem. This process is also commonly referred to as brainstorming, where dif-

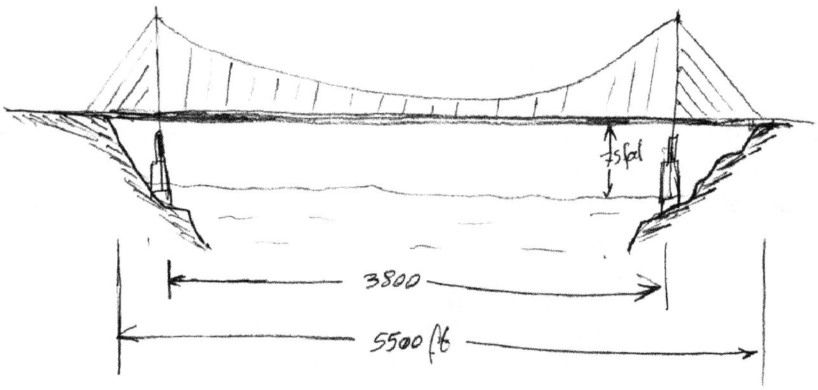

Figure 1.2 Sketch used for problem definition

ferent people in a group suggest potential solutions to the problem and share those ideas with the rest of the group. At this stage of the design process, no value judgments are made about individual ideas; the quantity of ideas, not quality of ideas is what is needed. When sufficient ideas have been recorded, they are discussed in further detail, evaluated, and narrowed down into the most plausible design solutions. In this stage, drawings and pictures of previous designs are used and multiple sketches and preliminary models are produced.

3.3 Design

In this stage, the selected ideas are studied thoroughly and developed. Rough sketches from previous steps are converted into accurate models and scaled drawings and illustrations (See Figure 1.3). Early prototypes may also be constructed at this stage, in order to clarify design intent and to increase visualization.

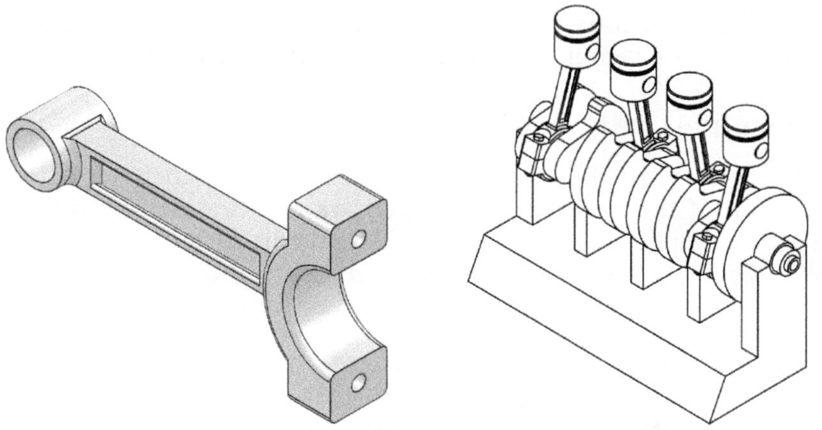

Figure 1.3 Design

3.4 Analysis

During the analysis process, the models are evaluated using mathematical and engineering principles to determine the suitability and validity of the designs with respect to its functionality, strength, safety, and cost (among other factors). Numerical techniques and classical methods such as finding areas, inclines, and forces are commonly used. Computer-based graphical analysis tools such as Finite Element Analysis (FEA), calculation of physical properties, or thermal, stress, and flow analysis tools are becoming more and more popular and reliable as technology advances (See Figure 1.4).

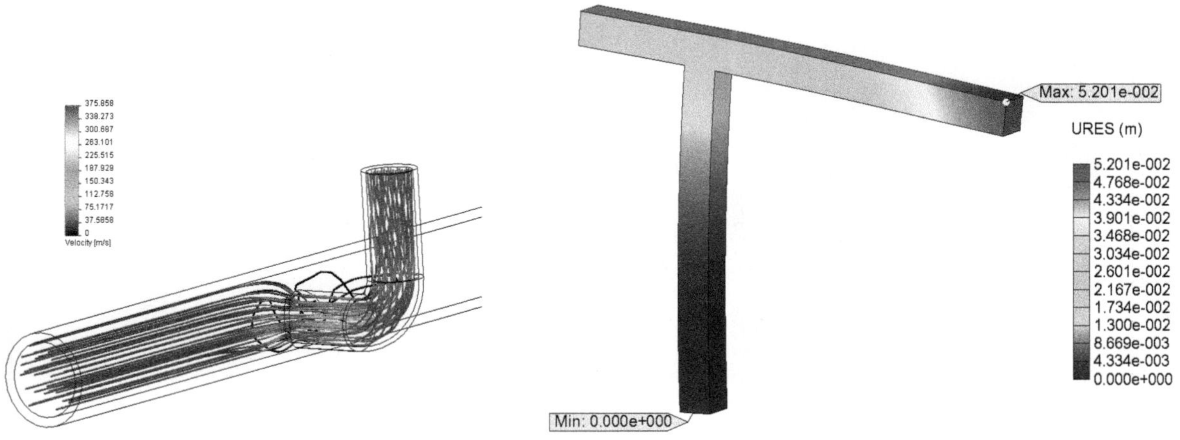

Figure 1.4 Flow analysis and FEA analysis

Figure 1.5 Prototype manufacturing

3.5 Refinement and Optimization

The results obtained in the analysis stage are used to make changes and/or adjustments to the design to make it more efficient. Many CAD packages assist in the refinement process. Revised models and drawings are produced at this stage. Ultimately, prototypes are manufactured and tested and shown to focus groups to study consumer reaction. Recent advances in computer aided machining and manufacturing allow the rapid creation of physical parts directly from 3D models (see Figure 1.5).

3.6 Implementation

The final version of the design becomes the product that will be manufactured based on the specifications defined in previous steps. To ensure proper fabrication and successful performance of the design, explicit and accurate detailed drawings and instructions must be provided to the machinist or manufacturing shop. Other information such as packaging, storage, and shipping may be included.

3.7 Documentation

Although not a specific stage of the engineering design process, documentation is a practice that complements each step of the sequence. Documentation includes recording and presenting the functionality and architecture of the design. These documents include mathematical models, analysis, results, and tests. Graphics are widely used in documents and reports. Charts, graphs, 3D views of a product, and even animations and simulations are effective data presentation and documentation tools.

4. Engineering Graphics Technology

With the advent of computer graphics and CAD, the role of graphics in the engineering design team has changed significantly. Historically, when drawings were created by hand using traditional instruments such as mechanical pencils, protractors, and triangles, the draftsperson was the individual in the design team relied upon responsible for producing technical drawings. This person was responsible for documenting the design from the rough sketches and information given by the engineer. The draftsperson was very proficient with technical instruments and was considered a "technical artist." With the introduction of computer technology, the time and effort required to produce high quality engineering drawings was drastically reduced. Today, every member of the design team is expected to have the knowledge and skills required to produce professional drawings and the ability to learn and use modern CAD and parametric modeling applications.

Lettering and Sketching

2
CHAPTER

1. Introduction

Engineering is a profession that places great importance on the ability to clearly and effectively communicate ideas. No matter how efficient or novel a design happens to be, the information needed to describe it must be conveyed in a methodical and clear manner. Lettering that is easily read and the ability to sketch accurately are tools required of today's engineer.

While computer programs are used for almost all engineering drawings nowadays (word processing and CAD programs), the ability to quickly illustrate a proposed design when computers are not available can be highly beneficial. While this chapter discusses only the basic concepts of lettering and sketching, only continued practice will ensure improvement.

2. Lettering

Lettering is essentially text used to provide additional information to a drawing. This information includes notes, labels, dimensions, titles, author's name, date of creation, and any information that needs to be communicated through text. The use of text in a drawing should be kept to a minimum. It should be short and concise.

There are many styles of lettering in existence, some easier to read than others. Lettering styles differ in appearance based on the function of the document. Diplomas and wedding announcements are artistic in nature, so they require a more visually appealing lettering style. Engineering drawings require a lettering style that is easy to read so potential mistakes in manufacturing are minimized. The function of lettering in engineering drawings is to clearly and without error provide information needed for manufacturing. The lettering style used most commonly in the engineering profession is Single Stroke Gothic, also referred to as Sans Serif. A serif is a curved line used to make a letter more aesthetically appealing. Lettering styles using serifs are often drawn with calligraphy pens and are considered an art form. Engineering lettering is accomplished with pencils. See Figure 2.1 for a comparison between a serif dominated lettering style and Single Stroke Gothic. Single Stroke Gothic lettering for the entire alphabet and the numerals from 0 to 9 is shown in Figure 2.2.

Figure 2.1 Comparison between Old English and Single Stroke Gothic lettering style

A B C D E F G H I J K L M N O P Q R S T U V W X Y Z
1 2 3 4 5 6 7 8 9 0

Figure 2.2 Single Stroke Gothic lettering style

Notice that each letter is simple, elegant, and easily read. Single Stroke Gothic lettering is aptly named because each letter can be drawn with a series of single strokes. Gothic is defined as black letters. In most cases (mm being just one exception), all letters are shown in upper case. Ideally, each stroke should advance the hand to the next letter in sequence. Ensure that the pencil is rotated as each stroke is constructed so letter width does not vary as the notation is put down on paper. Hold the pencil between the thumb, forefinger, and second finger without gripping it too tightly. The most ergonomically efficient position to hold the pencil is at a 60° angle with the paper, as shown in Figure 2.3.

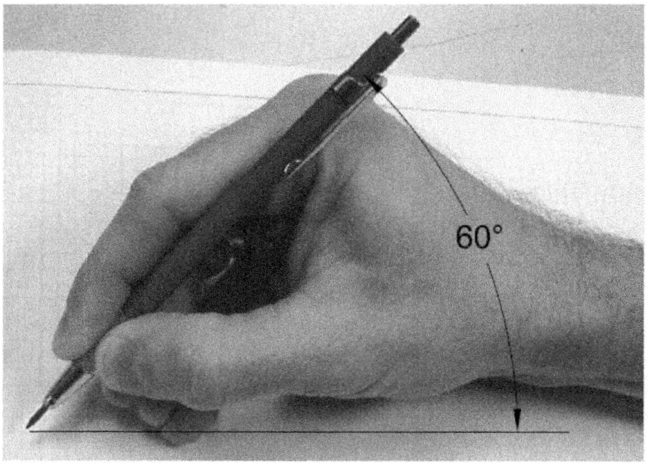

Figure 2.3 Proper pencil holding position, using a mechanical lead holder

Lettering size is customarily 1/8″ (3 mm) but could vary depending upon industry standards. Fractions take up twice the standard letter height, but the numerals are usually less than 1/8″ to allow space for the vinculum (horizontal line between the numerator and the denominator). It is also not unheard of for lettering size to depend on the sheet size that is used. The larger the paper size the larger the lettering, so all text can be easily read. For most commonly used paper sizes such as 8.5″ × 11″ or 11″ × 17″, lettering height of 1/8″ (3 mm) should be used for notes, dimensions, and any text in the drawing, and a height of 1/4″ (6 mm) should be used for the text in the title block. Text included in title blocks provides general information such as drawing title, author, date, version, etc. Guidelines are normally used as visual aids to estimate proper lettering height.

Some letters are wider than others, so using equal spacing between letters would make the text appear confusing. Spacing between letters should be balanced to make the text easy to read. Spacing between words is usually as wide as the letter "O". Spacing between lines is generally half the height of the letters. See Figure 2.4 for an example.

PROPER LETTERING SPACING.
CROWDED LETTERING SPACING.

Figure 2.4 Proper letter spacing

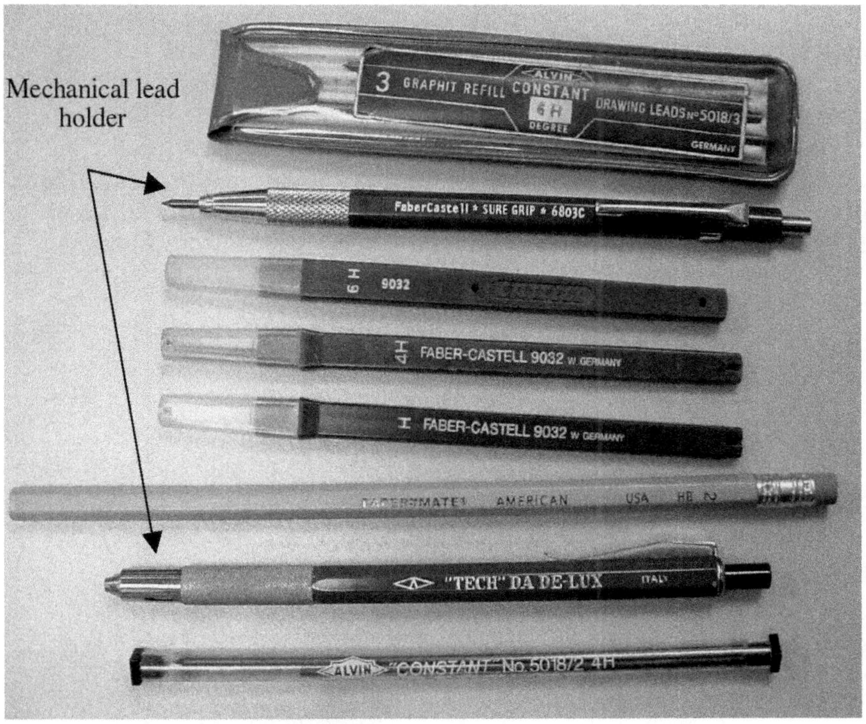

Figure 2.5 Commonly available lead types

For lettering and sketching purposes, it is most common for pencils to be used. There exists a lead hardness scale that is a gradation from hardest (lightest) to softest (darkest). Leads are manufactured to be either light or dark depending upon the amount of clay mixed with the graphite. Lighter leads have more clay mixed in with the graphite while darker leads have less clay. The scale goes in the following order, from lightest to darkest: 9H, 8H, 7H, 6H, 5H, 4H, 3H, 2H, H, F, HB, B, 2B, 3B, 4B, 5B, 6B, 7B, 8B, and 9B. "H" represents hardness and "B" represents blackness. The No. 2 pencil lead is actually HB lead. Hard leads, such as 6H are very light. They have the advantage of not smearing and are commonly used in surveying and for construction lines. HB leads provide for black lines although the lines smear over time. A small collection of the lead types commonly available is shown in Figure 2.5.

3. Sketching

Sketching is a quick, freehand method of drawing, but more than that, it is a form of communication. Very rarely will the person designing an assembly be the one manufacturing it. For this reason, adequate and effective information needs to be passed on to the person actually constructing the design. Sketching differs from drawing in that no mechanical instruments are used (compass, straight edge, protractor, etc.). These tools can slow down the sketching process or add a level of accuracy not needed in early stages of the design process. In many cases, mechanical instruments are not even available. For these reasons, traditional sketching remains as an important skill in engineering. Continued practice and the use of basic techniques allow for the creation of straight lines, perfect circles, and good sketches, entirely freehand, without relying on any drawing instruments.

Sketching can be done on any media, whether engineering paper, sketch pad, or even a paper napkin. In fact, many great ideas and designs have been devised on a paper napkin during a lunch meeting at a restaurant or during a short break at a coffee shop. Engineering sketches are combinations of simple geometric figures, with appropriate notes, so that design intent can be conveyed. A good sketch should be able to stand alone and not require any additional explanation or information from the designer (see Figure 2.6). In contrast with artistic sketches, engineering sketching is not concerned with shading, shadows, or other aesthetic flourishes. Engineering sketches are simple dark line drawings with enough information so that a CAD operator can draw the part or a machinist can construct a prototype. Engineering sketches are legal documents that can be used to show ownership of a design or as a proof in case of a lawsuit, so it is necessary to sign and date all sketches.

Sketching can also aid in solving problems or arranging the information in such a way that a solution presents itself. Sketching an idea will improve visualization and can be used to more easily illustrate design flaws or missing information that is needed for design refinement.

Figure 2.6 High quality sketch of a classic car (courtesy of Donivan Potter)

4. Sketching Techniques

Good sketching requires practice. A good sense of proportion and the ability to estimate sizes, distances, angles, and other geometric relationships are also useful (see Figure 2.7). The following techniques will give basic guidelines and tips to improve sketches.

- Light construction lines can be drawn to guide the permanent lines added later.
- Quick, confident strokes will always appear more professional, even if drawn at an inexact angle or location, than slowly drawn strokes, where hand shaking will negatively affect the lines.
- All sketched lines should be drawn with the eye focused on the endpoint, rather than following the pencil. If necessary, break long lines into shorter more manageable segments.
- Sketching perfect circles and ellipses is not an easy task. Begin with a box and lightly sketch the diagonals to find the center point. Mark the midpoints of the edges of the box. They are the contact points with the circle. Circles and ellipses can be sketched by tracing the different sements using the contact points. Rotate the paper if necessary (see Figure 2.8).
- After a preliminary sketch is completed, the desired lines can then be darkened and any stray marks can be erased.

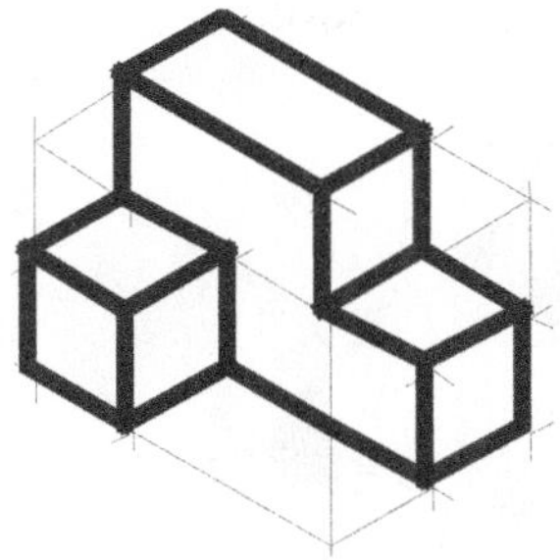

Good proportionate sketch

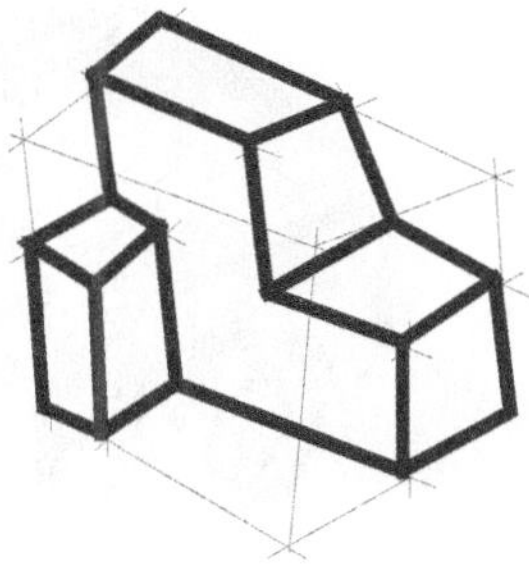

Bad sketch

Figure 2.7 Examples of proportion and estimation

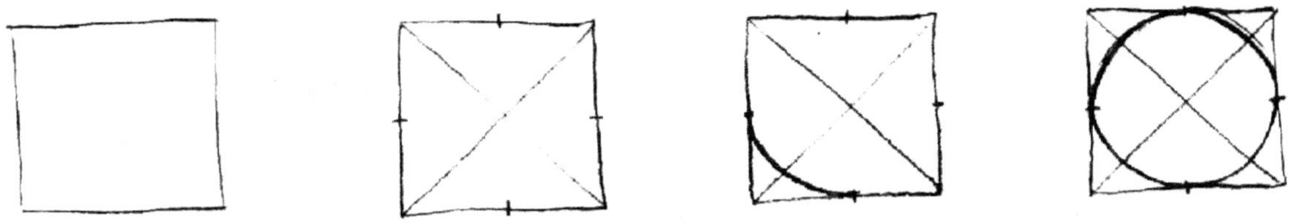

Figure 2.8 General steps to sketch circles

The general process of sketching a simple object is shown step by step in Figure 2.9.

Lightly sketch construction lines

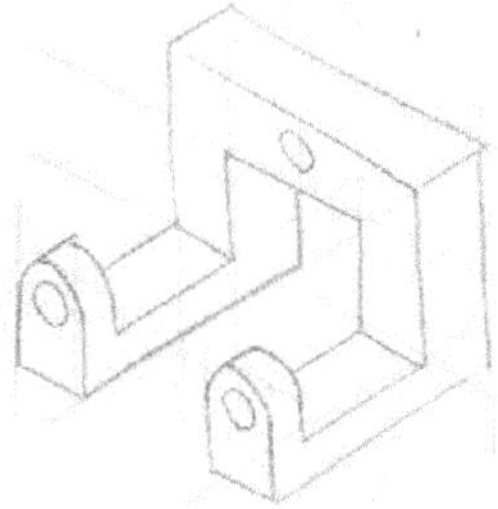

Using darker lines, add the edges of the object

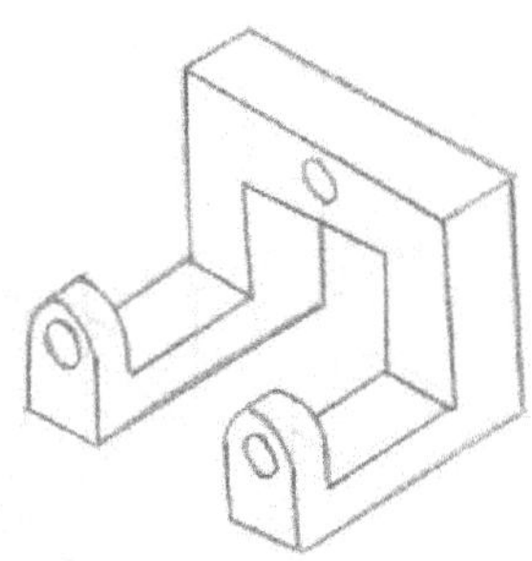

Darken the lines and clean up edges

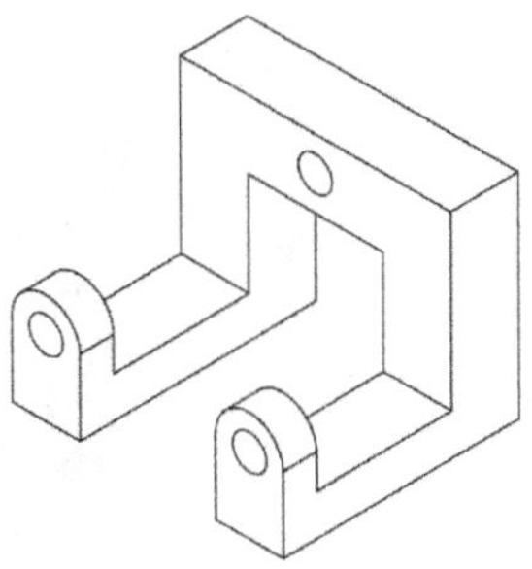

Final result drawn with CAD program

Figure 2.9 Sketching process

Figure 2.10 shows a sketch of a truck. Notice that some shading has been added, but was not required to reflect the object's fundamental geometry. A sketch and a photo of a disassembled pen is shown in Figure 2.11. See Figure 2.12 for other sketching examples.

Figure 2.10 Sketch of truck with some shading added

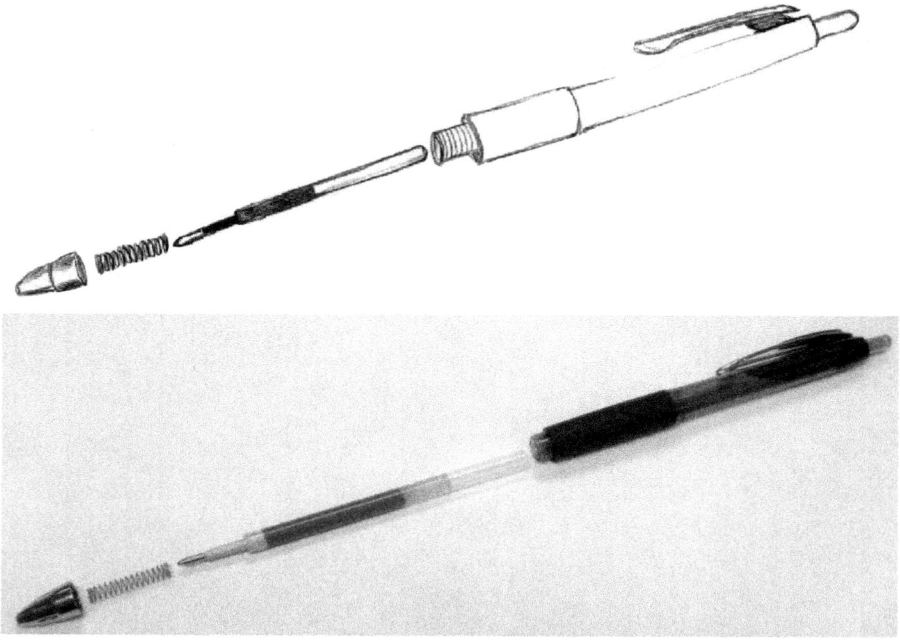

Figure 2.11 Sketch and photograph of disassembled pen

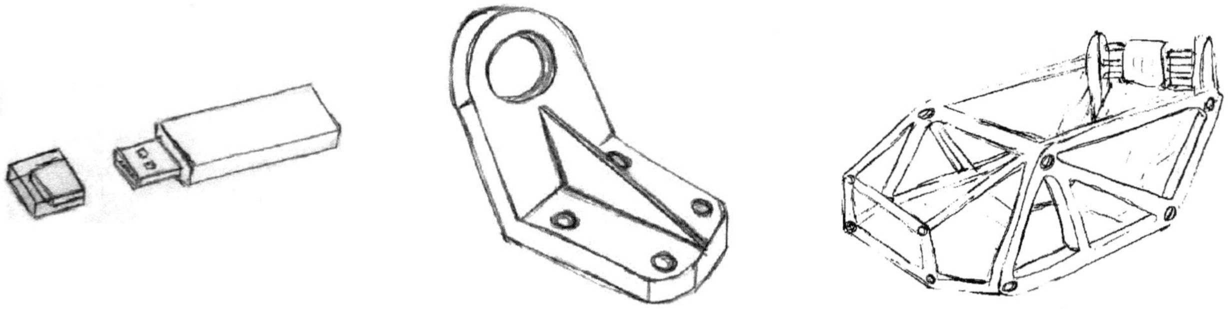

USB drive Mounting bracket Protective frame for race car

Figure 2.12 Sketching examples

Practice Test

1. What is the standard engineering lettering style?

A) Sans Serif
B) Single Stroke Gothic
C) Times New Roman
D) A and C
E) None of the above

2. The darkness of a specific pencil lead is dependent on ______

A) How hard you press down on the pencil.
B) The amount of clay added to the graphite.
C) The dullness of the pencil lead.
D) All of the above.
E) None of the above.

3. Engineering sketching is ______

A) A form of communication.
B) Done without the aid of drawing instruments.
C) A legal document.
D) All of the above.
E) None of the above.

4. Engineering sketching is important because ______

A) It shows ownership of ideas.
B) It conveys information to another who will construct the object.
C) Helps the designer think through possible problems.
D) All of the above.
E) Only A and B.

5. Engineering sketching ______

A) Consists of simple, geometric shapes.
B) Is not needed if the designer has access to a CAD workstation.
C) Is artistic in scope.
D) All of the above.
E) None of the above.

Problems

For the following objects, use engineering paper (or grid paper) to sketch accurate representations. Try the black line drawings first, then move to the photographs of various simple objects

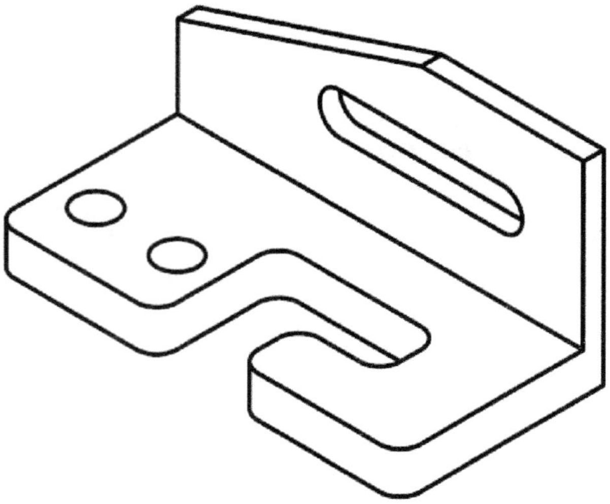

1. Bracket

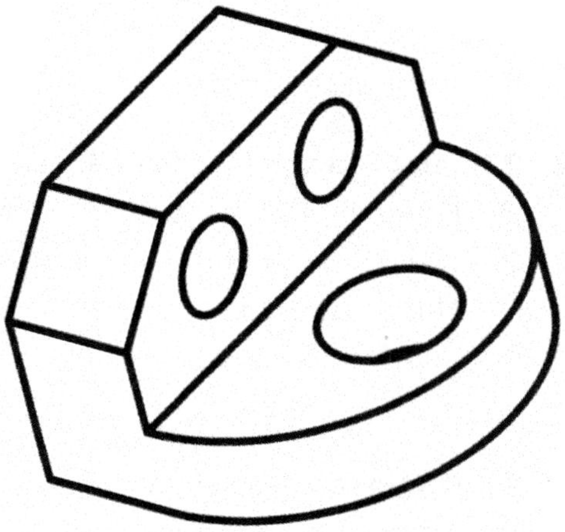

2. Widget

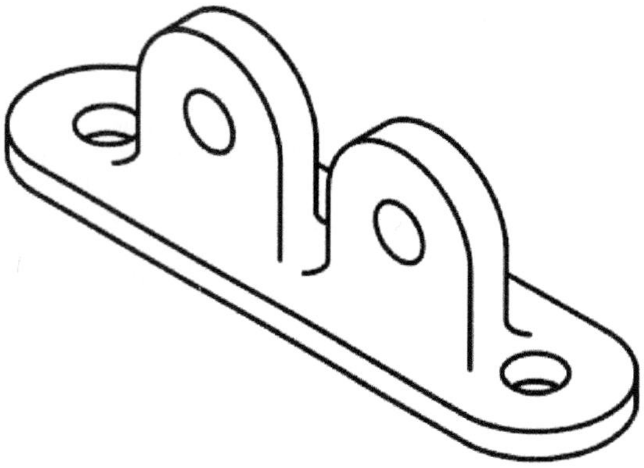

3. Cleat

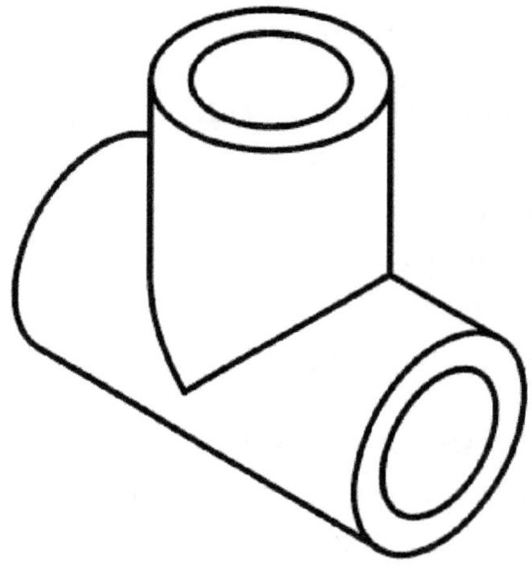

4. Tee connection

5. Stapler

6. Two hole punch

7. Vice

8. Toy motorcycle

9. Toy tractor

10. Imprinting device

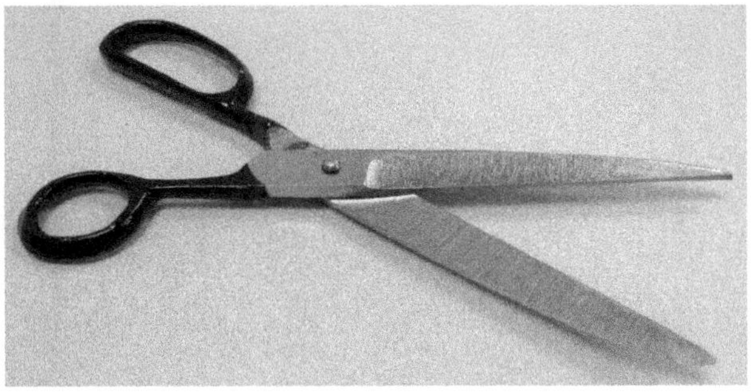

11. Scissors

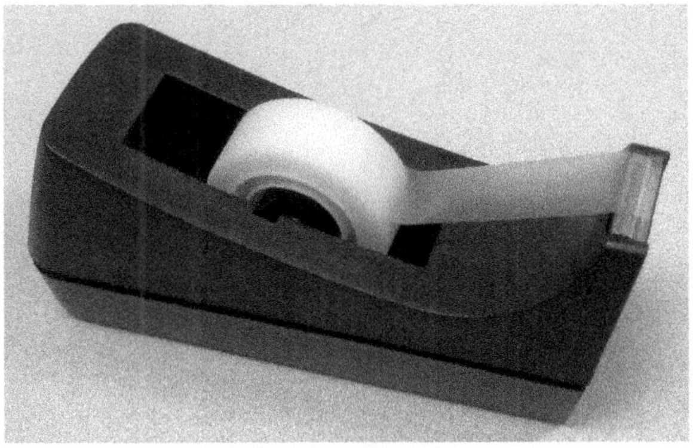

12. Tape dispenser

13. Coffee Mug

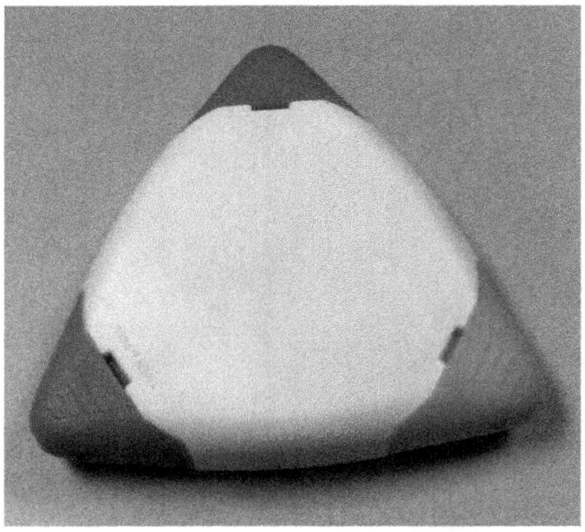

14. Highlighter

15. Sharpener

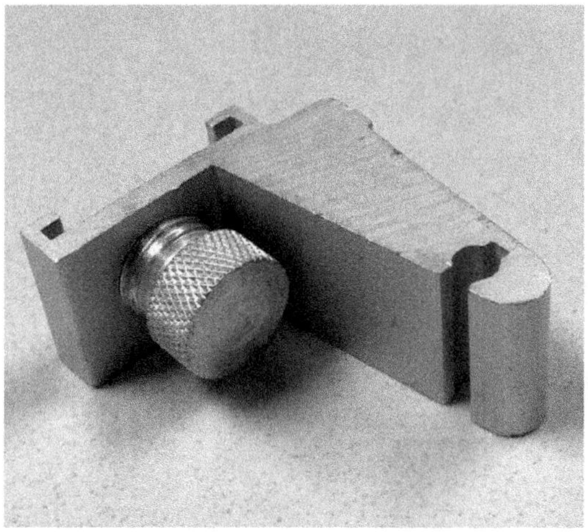

16. Bracket

USING THE EXAMPLE LETTERS AND NUMERALS AS A GUIDE, REPEAT THE LETTERS AND SENTENCES BELOW.

A B C D E F G H I J K L M N O P Q R S T U V W X Y Z

1 2 3 4 5 6 7 8 9 0

A	B	C	D	E	F	G
H	I	J	K	L	M	N
O	P	Q	R	S	T	U
V	W	X	Y	Z		
0	1	2	3	4	5	6
7	8	9				
$\frac{1}{2}$	$\frac{3}{4}$	$\frac{5}{6}$	$\frac{7}{8}$	$\frac{9}{16}$	$\frac{17}{32}$	$\frac{59}{64}$

THE QUICK BROWN FOX JUMPS OVER THE LAZY DOG.

ENSURE THAT YOU PROVIDE PROPER SPACING BETWEEN LETTERS.

SINGLE STROKE GOTHIC (SANS SERIF) MUST BE USED IN ENGINEERING DRAWINGS.

ALL LETTERS IN ENGINEERING DRAWINGS MUST BE UPPER CASE.

SKETCHING IS A POWERFUL COMMUNICATION TOOL.

AUTHOR:		SCALE:
TEAM:	SECTION:	DATE:

Orthographic Projections

3
CHAPTER

1. Introduction

Throughout the ages, humans have struggled to represent the world by recording images on various media. Whether animals on cave walls, maps of the Earth, or mechanisms drawn on paper, the overwhelming obstacle has always been how to represent three-dimensional objects accurately on two-dimensional media.

The representation of objects on paper is based on projection methods. Different projection methods exist, and will be discussed in detail later, but the orthographic system is most useful in engineering. Albrecht Dürer, a German artist from the Renaissance, was credited with the first orthographic maps of the Earth and a book introducing drawing human proportions from various viewpoints. Gaspard Monge, a French drafter from the 18th century and father of descriptive geometry, formalized the orthographic system while working at a military school.

Orthographic Projection is a system where an object is represented from different directions parallel to the faces of the object. Orthographic Projection is useful in engineering, especially at manufacturing stages, because it provides accurate detail and dimensions of the object without distortion. The ability to mentally visualize and manipulate objects in three dimensions is essential to understand and produce accurate orthographic drawings.

2. Orthographic Theory

Orthographic projections create a number of two-dimensional views of the object using three orthogonal planes. These principal planes; horizontal, frontal, and profile, intersect each other forming 90° angles (see Figure 3.1).

Orthographic projection can be understood by imagining the object to be drawn placed inside a glass box (see Figure 3.2). The edges of the object are projected perpendicularly onto the outer panes of the box producing six two-dimensional images (see Figure 3.3). By unfolding the glass box, six orthographic views of the object are created (see Figure 3.4). Notice that the glass box must be unfolded in a standard manner to obtain the standard arrangement of orthographic views, where the top view is directly above the front view, the right-side view is directly to the right of

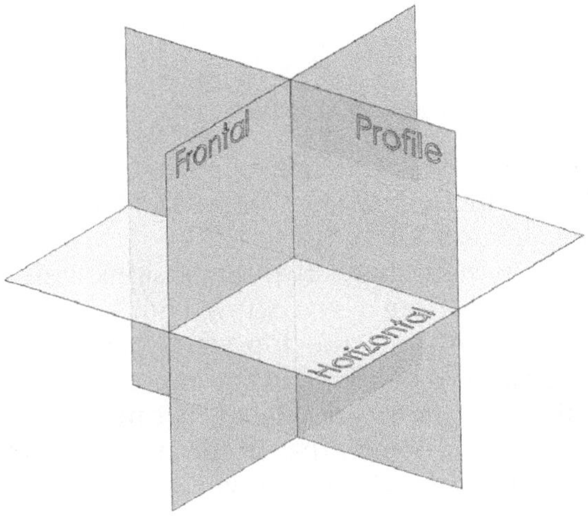

Figure 3.1 Principal planes

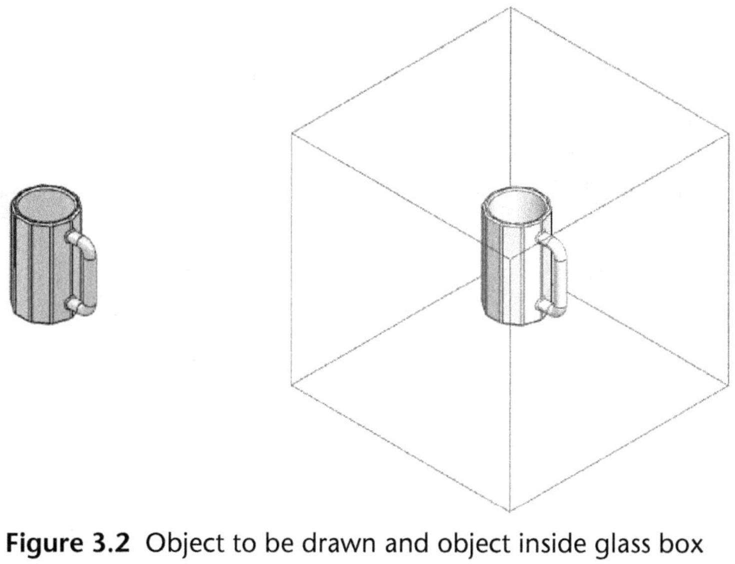

Figure 3.2 Object to be drawn and object inside glass box

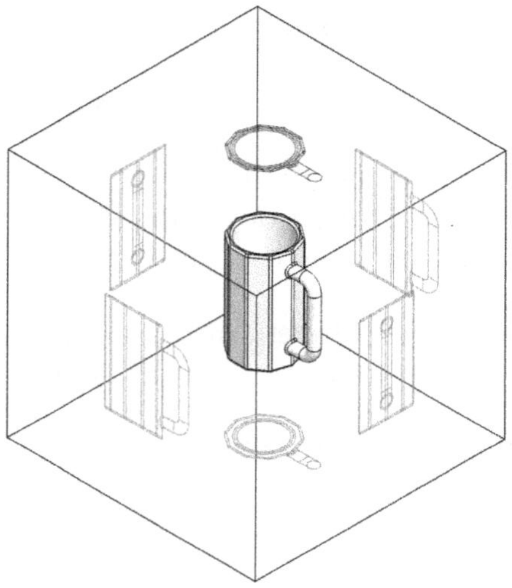

Figure 3.3 Projections of the mug onto the glass box

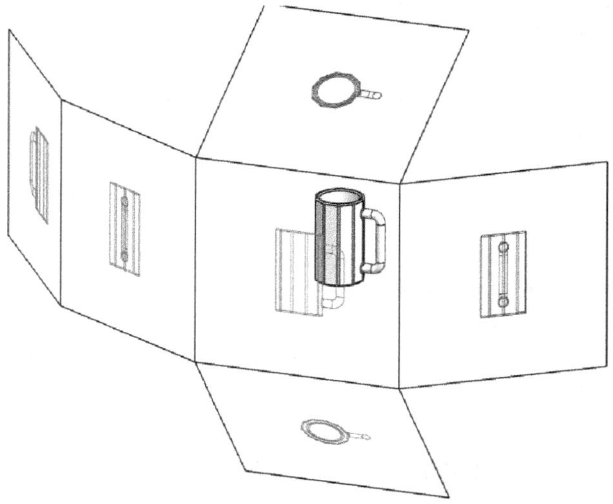

Figure 3.4 Glass box unfolding

the front view, the left side view is directly to the left of the front view, the bottom view is directly below the front view, and the back view is placed directly to the left of the left side view (see Figure 3.5).

Each view shares two dimensions with another (front and back views show width and height, top and bottom views show width and depth, and left and right side views show height and depth). The front and top view share the dimension of width, the front and side view share the dimension of height, and the top and side view share the dimension of depth. If the views and the projection planes are related, it can be seen that the horizontal plane shows the top and bottom views of the object, the profile plane shows right and left side views, and the frontal plane shows front and back views. The views associated with each plane are similar, but mirror images of each other, with certain features being visible on one side, but obscured in the other.

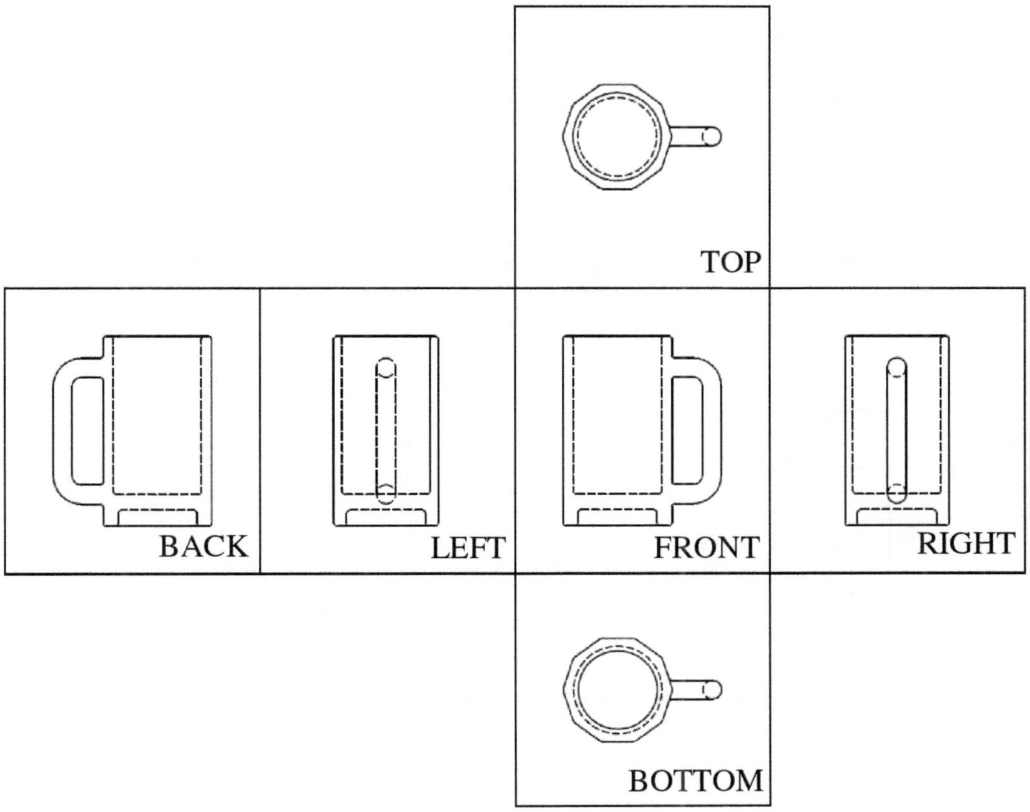

Figure 3.5 Orthographic views of a glass mug

2.1 Foreshortening

Foreshortening is the phenomenon that exists when an object extends away from the viewer, making a feature appear smaller than it actually is. This concept is often seen in orthographic projection, since each view only illustrates two dimensions. The example shown in Figure 3.6 illustrates that in the front view, the slanted plane that extends toward the back appears shorter than it is, while in the top view, the true distance is represented, but only the edge is shown. This concept is discussed in greater detail when auxiliary views are explained.

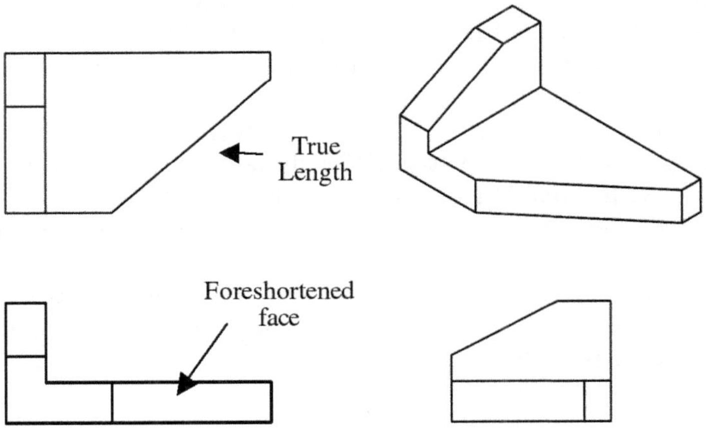

Figure 3.6 Foreshortening example

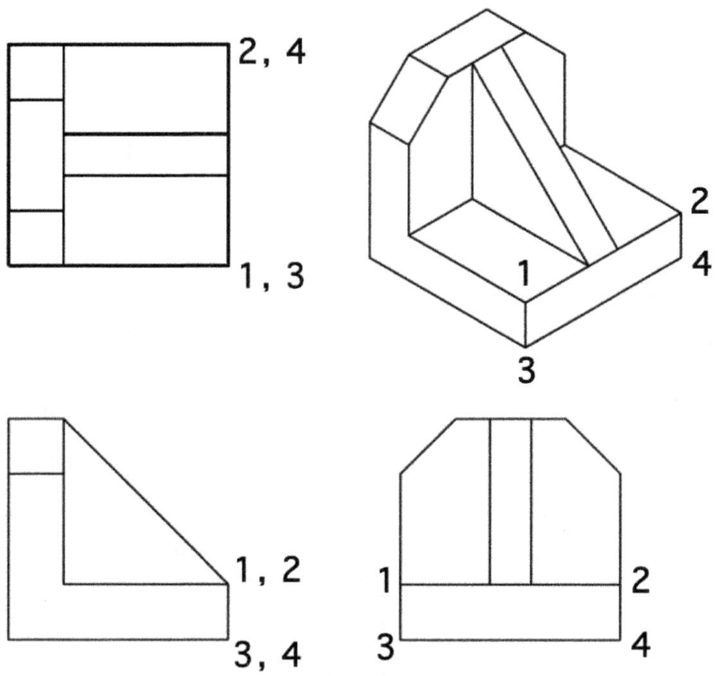

Figure 3.7 Representation of faces by edges and points

Depending on which orthographic view is observed, different features are illustrated with faces, lines, or points. The object shown in Figure 3.7 has each corner of one of its faces numbered for illustrative purposes. The surface 1-2-3-4 is represented as a surface in the right side view, but is shown as an edge (or line) in the top and front view. As a reminder, no single orthographic view contains sufficient geometric information to adequately define most objects. In general, when verifying that the orthographic views drawn are correct, an important clue is that a corner in one view always corresponds to an edge view representing the same feature in another view. Some exceptions may exist, such as the top point of a pyramid, in which a point does not have a corresponding edge view.

2.2 Third Angle vs. First Angle Projection

There are two ways of projecting an object onto the panes of an enclosing glass box; First Angle and Third Angle projections. In the United States Third Angle Projection is preferred, which can be visualized by placing the observer outside the glass box, looking at the two dimensional image projected on each side of the box. In Third Angle Projection, the top view is directly above the front view with the right side view to the right of the front view.

Other countries use First Angle Projection, in which the viewer can be visualized as being inside the box with the object, looking at each two dimensional image being projected on the inside of each pane of the glass box. In First Angle Projection, the right side is to the left of the front view and the top view is placed beneath the front view. See Figures 3.8 and 3.9 to see First and Third Angle Projections illustrated.

On all manufacturing drawings, a projection symbol is placed to indicate the projection method used. These can be seen in Figure 3.10. The arrangement of this symbol illustrates the differences between projection methods and is important to ensure that an object is constructed properly.

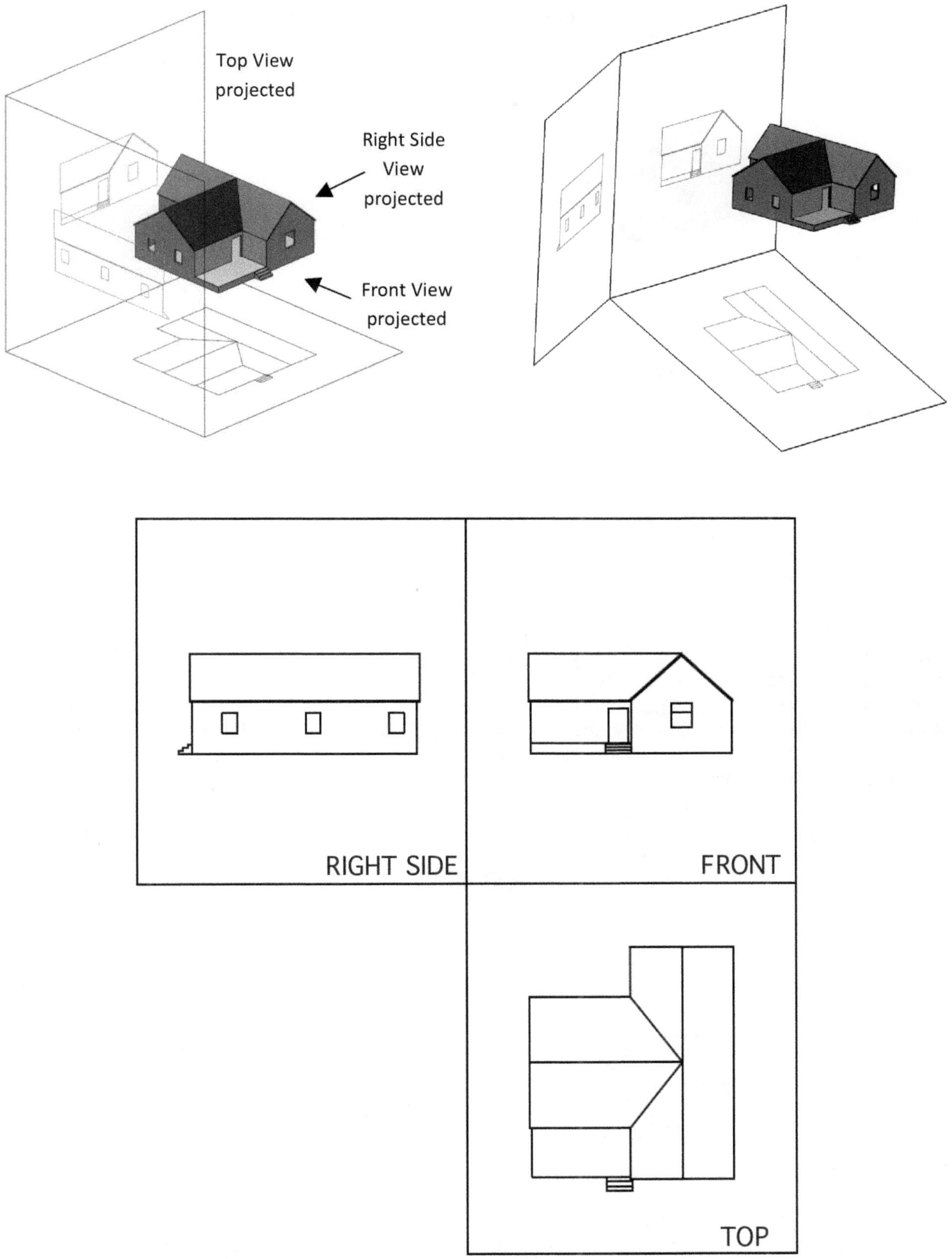

Figure 3.8 First Angle Projection

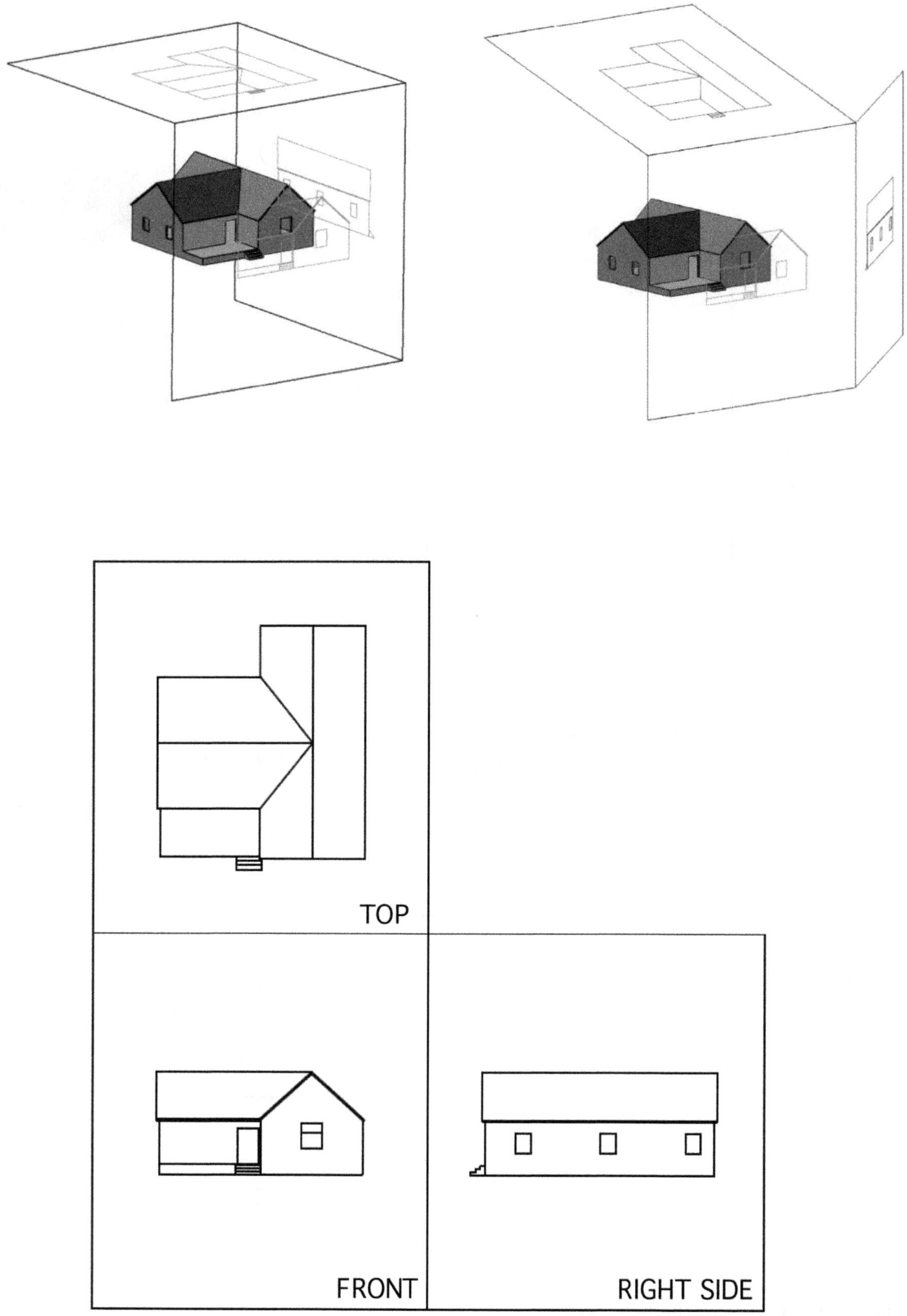

Figure 3.9 Third Angle Projection

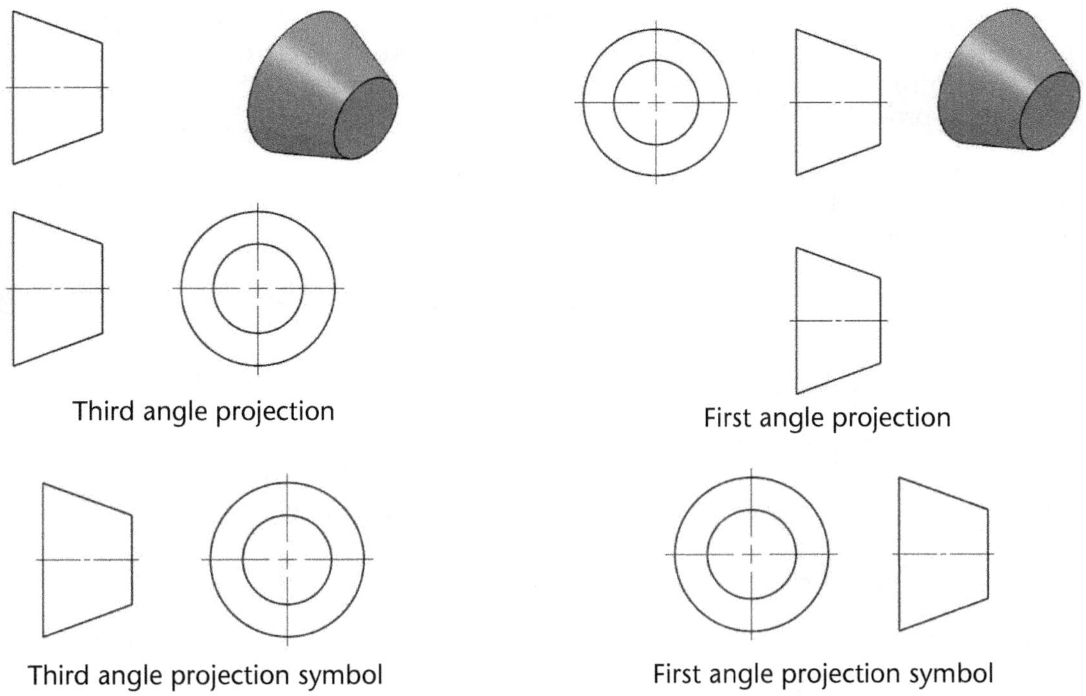

Figure 3.10 Symbols used to indicate the type of projection

3. Number of Views

Since each orthographic view shows only two dimensions of an object, it is easy to understand that more than one orthographic view is required to fully describe a three-dimensional object. A multi-view-drawing is essentially a group of different orthographic views arranged in a standard manner that can be used to represent and visualize all three dimensions of an object. Although unfolding the glass box produces six orthographic views of the object inside, some information is redundant so it is not necessary to draw all six views. The question now is how many views are required to fully describe any object?

In general, only the minimum views necessary to adequately portray the object are drawn. The amount of views necessary is dependent upon the complexity of the object. Occasionally, for very simple objects, one view plus note can be used (see Figure 3.11). More often, also for simple objects, two views are used (see Figure 3.12). In most cases, however, three orthographic views are

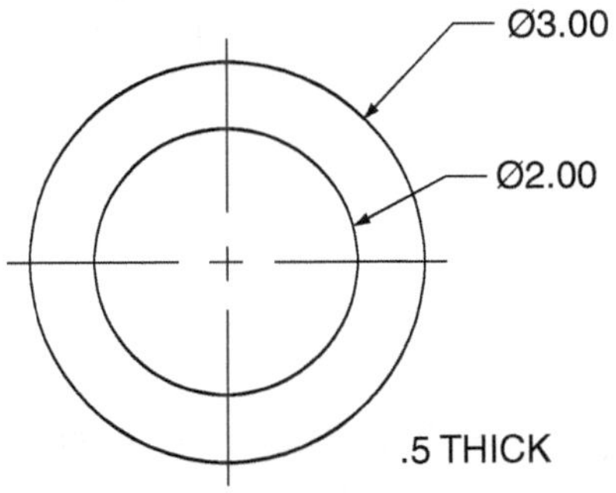

Figure 3.11 One view and a note for very simple objects

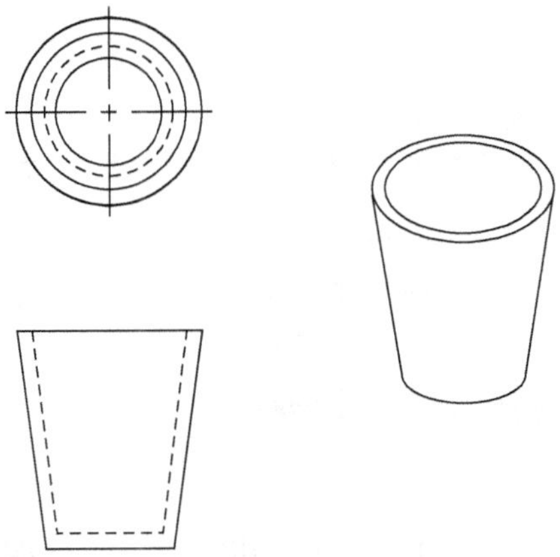

Figure 3.12 Two views for simple objects

required. The example shown in Figure 3.13 illustrates a situation where multiple right side views are possible given incomplete geometric information provided by the front and top views. For complex objects with intricate internal detail, other views can be provided to simplify and facilitate the visualization process.

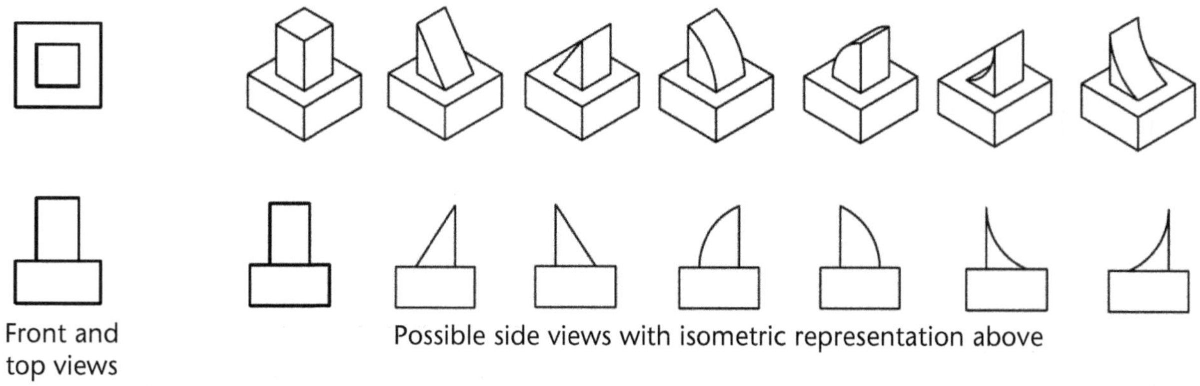

Figure 3.13 Possible right side views

The question now is which three views should be drawn? With Third Angle projection, front, top, and side views are commonly selected. It is imperative that all views maintain proper orthographic alignment. The top view must be placed above the front view and must be the same width. The right side view must be to the right of the front view and must have the same height. Also, the top view cannot have greater depth than the right side view. When constructing orthographic views, depth is projected at a 45° angle to make placement of features in the top view easier. After construction of the top view, the views can be spaced appropriately to ensure room for dimensions. The drawing shown in Figure 3.14 illustrates this concept.

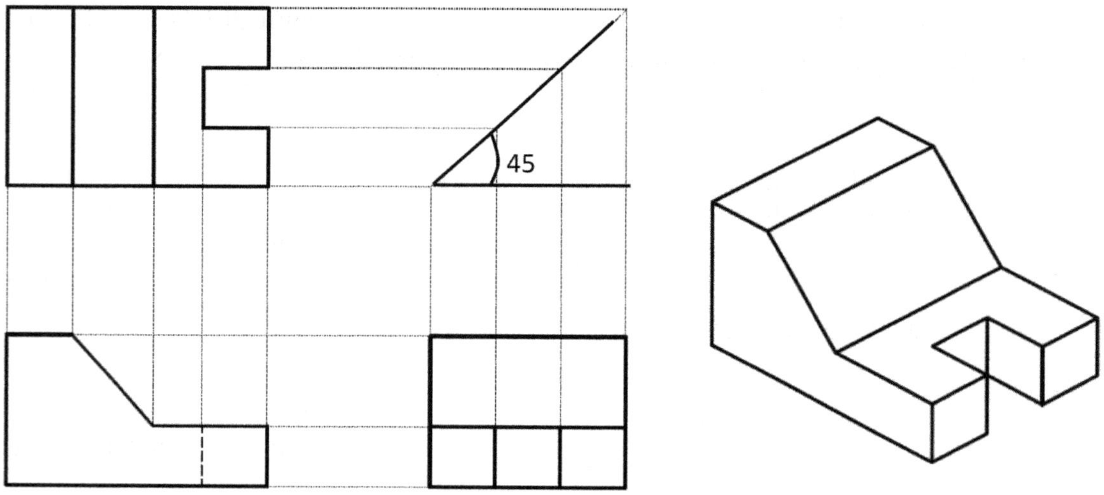

Figure 3.14 Depth projected at a 45° angle

4. Selection of Views

When observing a 3D object, which factors determine which view is which? There are several criteria that have to be met when making this decision.

1- The most descriptive view (MDV) needs to be the front view. The MDV is the one that reveals most of the object's unique features and general shape, which is hinted at by arced or slanted lines (Figure 3.15).

2- If the object has an obvious top, then that should be the orthographic top view. In the mug example, someone obviously drinks from the top of the mug (Figure 3.16).
3- The depth dimension should be minimized in order to conserve space (Figure 3.17).
4- The side view chosen should have fewer hidden lines than its corresponding mirror image (Figure 3.18).

A dilemma often occurs when all the above listed criteria cannot be met. As an example, oftentimes the top of an object is also the most descriptive view. Or the most descriptive view creates a situation where the top view has depth maximized. In such cases, experience should be called upon to try to satisfy the most criteria, realizing that others will be violated.

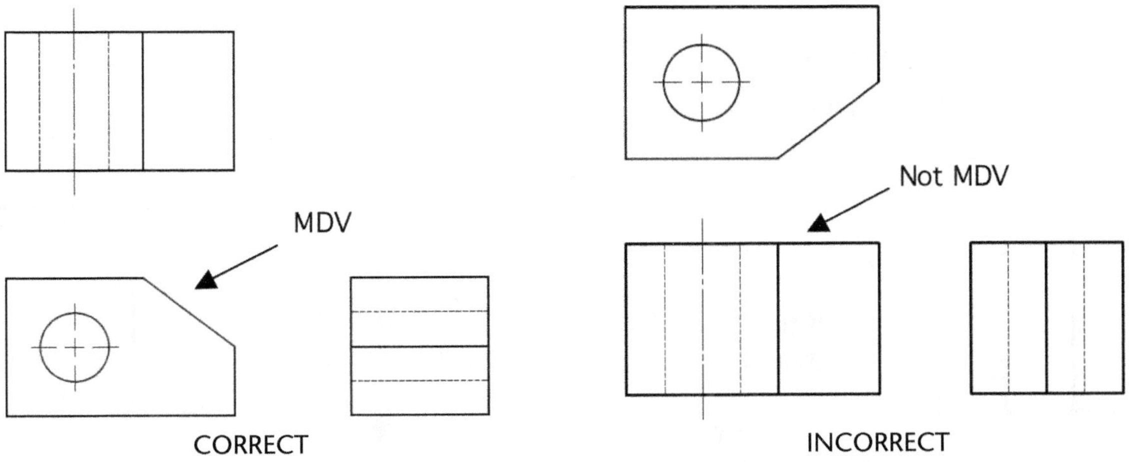

Figure 3.15 Selection of the most descriptive front view (MDV)

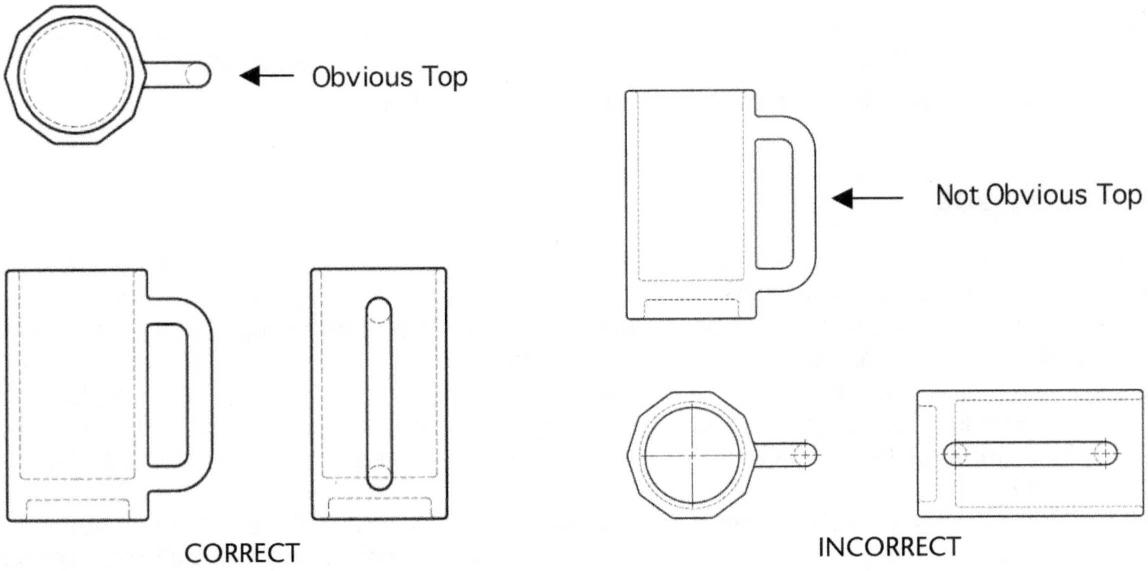

Figure 3.16 Obvious top view

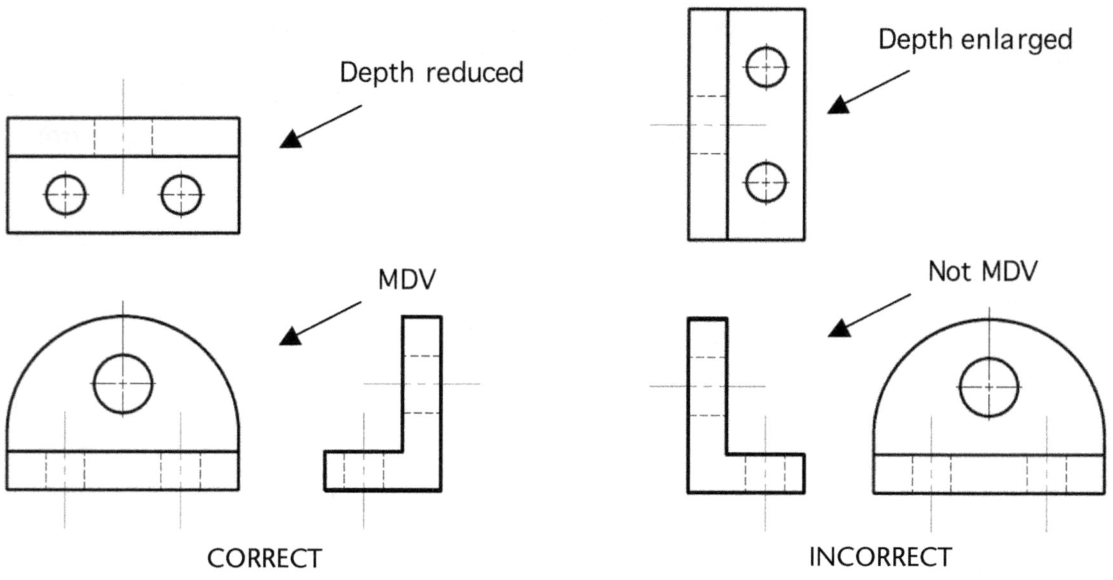

Figure 3.17 MDV and depth

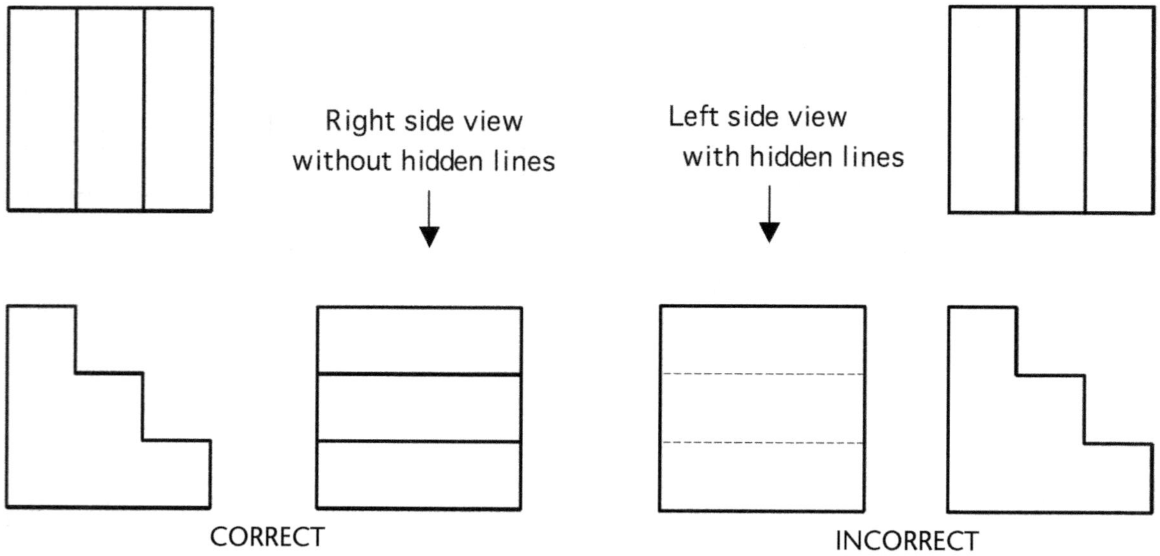

Figure 3.18 Side view that minimizes the number of hidden lines

5. Line Types

In orthographic drawings, different line types are used to describe various physical features. Visible lines used to define edges of objects are shown as continuous, thick, straight lines (ex. 0.5 mm thick). Hidden lines, used to define features that are obscured behind visible surfaces, are shown as dashed lines, usually half as thick as visible lines. Center lines, used to define centers of circular features such as holes, cylinders, or arcs are shown as a long line, followed by a gap, a shorter line, another gap, and another long line. Center lines are also thin lines (see Figure 3.19). Construction or Guide lines are light lines, barely visible, used to create geometry used to assist in making drawing the object easier.

Center lines cross each other when marking the center of circular objects, such as holes. Center lines extend approximately 1/8″ past the visible line boundary, commonly referred to as center line overshoot. Center lines representing circular objects in a rectangular view, are drawn halfway between each boundary line (see Figure 3.20). Center lines are also drawn to show concentricity with a semicircular arc, but with shortened ends (see Figure 3.21).

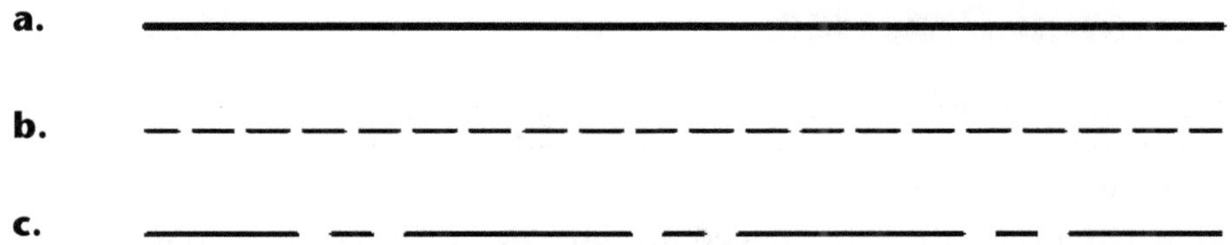

Figure 3.19 Alphabet of lines. a) Visible Line, b) Hidden Line, c) Center Line.

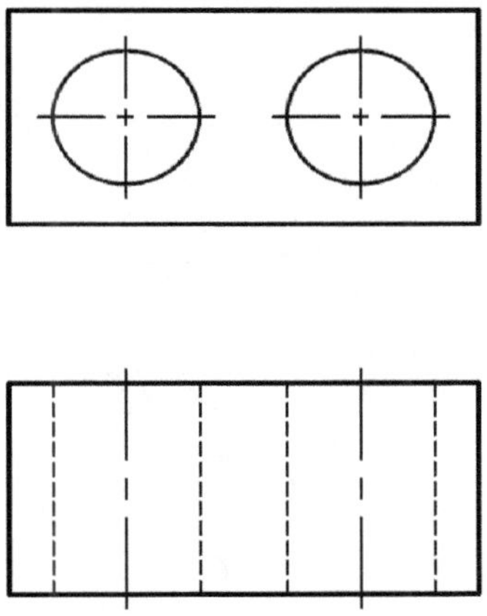

Figure 3.20 Center lines

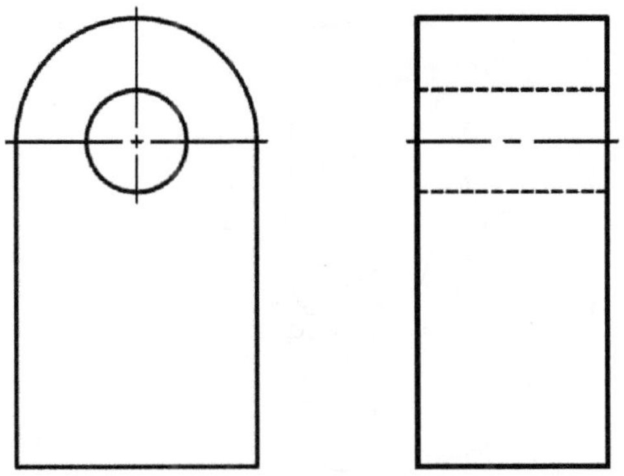

Figure 3.21 Concentric center lines

5.1 Precedence of Lines

There are occasions when lines representing various features coincide, and a decision has to be made as to which line to draw. In general, visible lines take precedence over hidden lines. Hidden lines take precedence over center lines (see Figure 3.22). Oftentimes, visible lines are drawn to show the beginning or ending of specific features, such as fillets. If a feature transitions smoothly, such as a semi-circular arc, the tangent edges are omitted for clarity. Many parametric modeling software programs add these lines, commonly referred to as tangent edges, and these can be removed easily. This topic will be discussed in greater detail later.

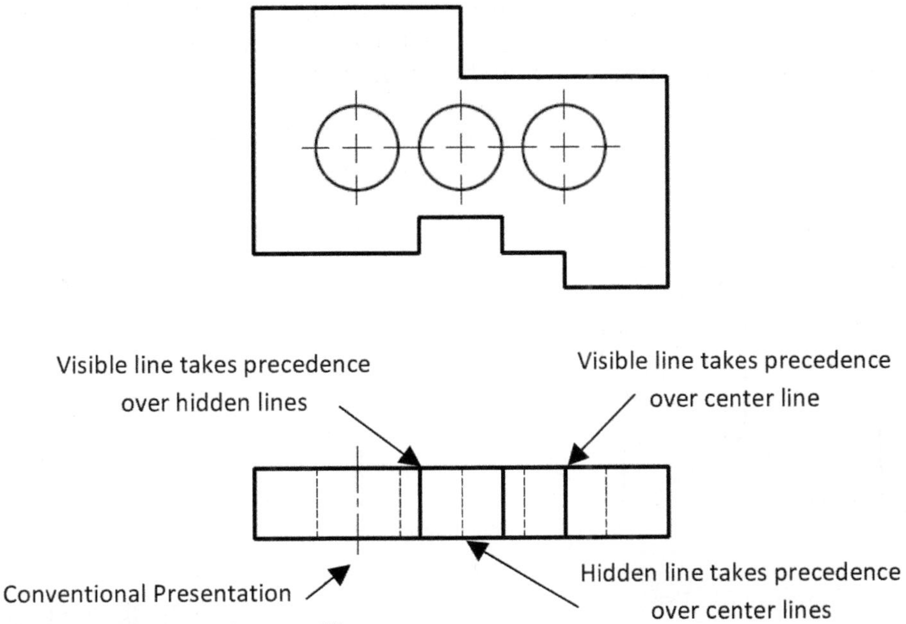

Figure 3.22 Precedence of lines

6. Conventional Practices

Oftentimes, it is useful to violate the rules of orthographic projection in order to provide a more aesthetically pleasing drawing that also provides adequate information for manufacture. In a case where a plate with equally spaced holes and ribs exists (spaced at 120°), it is useful to show in the front view the holes and ribs spaced at 180°, even though it is inaccurate, so a confusing front view is not drawn. The example in Figure 3.23 illustrates this concept. Remember, no one view provides all information about an object, so correct rib and hole placement have to obtained in the top view.

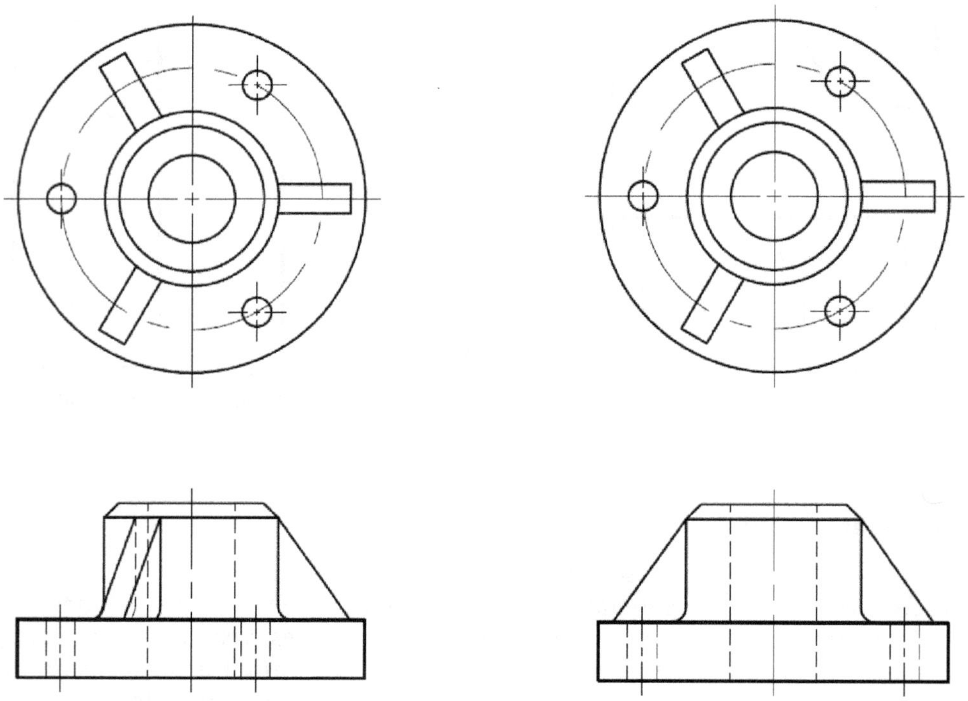

True Orthographic Projection

Orthographic Projection after applying Conventional Rotations

Figure 3.23 Conventional rib placement

Runouts are occurrences where a cylindrical object merges with a straight object and the transition zone needs to be illustrated. Runouts are shown by a visible line that stops somewhere in the middle of the transition zone. This concept is shown in Figure 3.24.

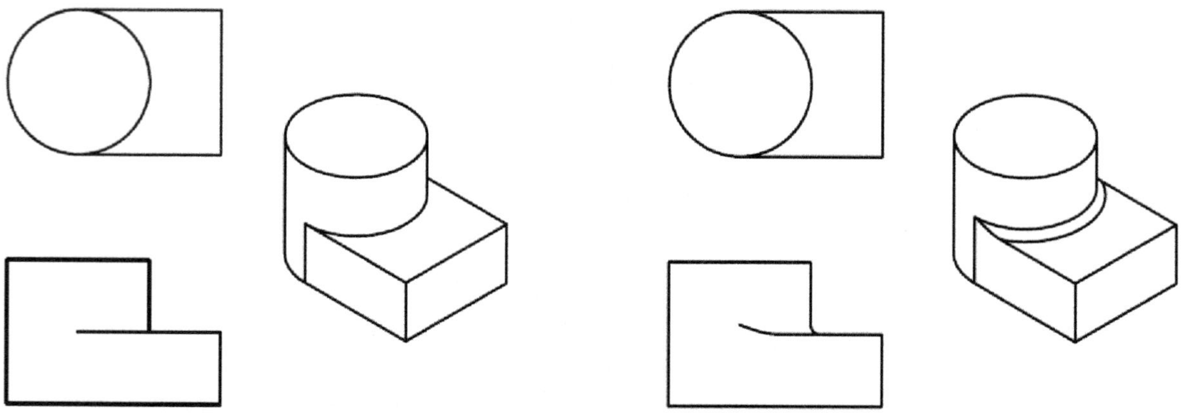

Runout without fillet

Runout with fillet

Figure 3.24 Runouts

Practice Test

1. **How many orthographic views are required to illustrate an object?**
   A) Six
   B) Depends on the object
   C) The minimum necessary to adequately show the object's geometry
   D) B and C
   E) None of the above

2. **The thickest line drawn in an orthographic drawing is ______**
   A) Center line B) Hidden line C) Visible line D) Dimension line E) None of the above

3. **If a center line and a visible line coincide, the ____________ line is drawn.**
   A) Center
   B) Visible
   C) Hidden
   D) Both lines are drawn on top of each other
   E) None of the above

4. **The ________________ should be the most descriptive view.**
   A) Front view B) Right side view C) Top view D) Auxiliary view E) None of the above

5. **Which dimension should be minimized in orthographic views?**
   A) Height B) Width C) Depth D) Length E) Does not matter

6. **________________ are drawing methods used to provide a clearer understanding of an object, even if orthographic rules are violated.**
   A) Conventional Practices
   B) Orthographic Violations
   C) Projection Practices
   D) Orthographic alignment must always be enforced.
   E) None of the above.

Problems

For the following objects, use your engineering paper to construct appropriate orthographic views. Each grid square on your engineering pad is 0.10". Ensure that you choose the correct front view.

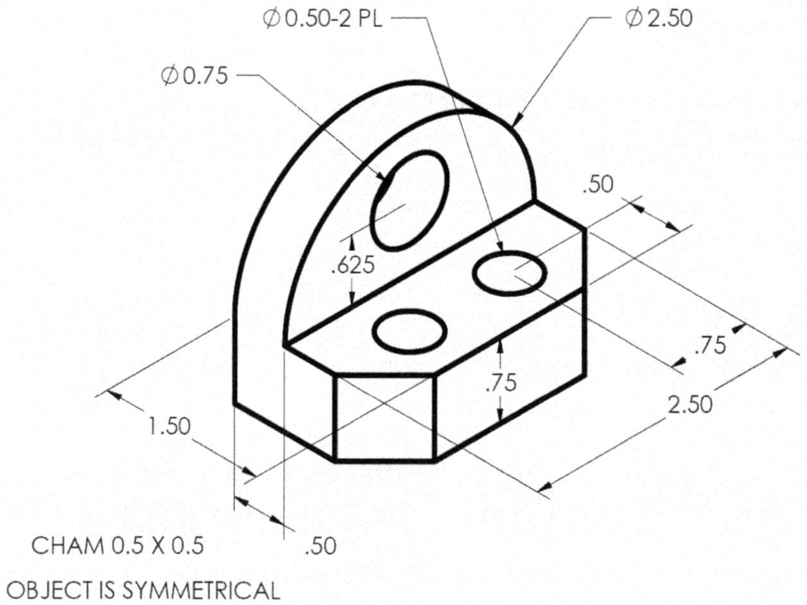

Problem 1—Widget

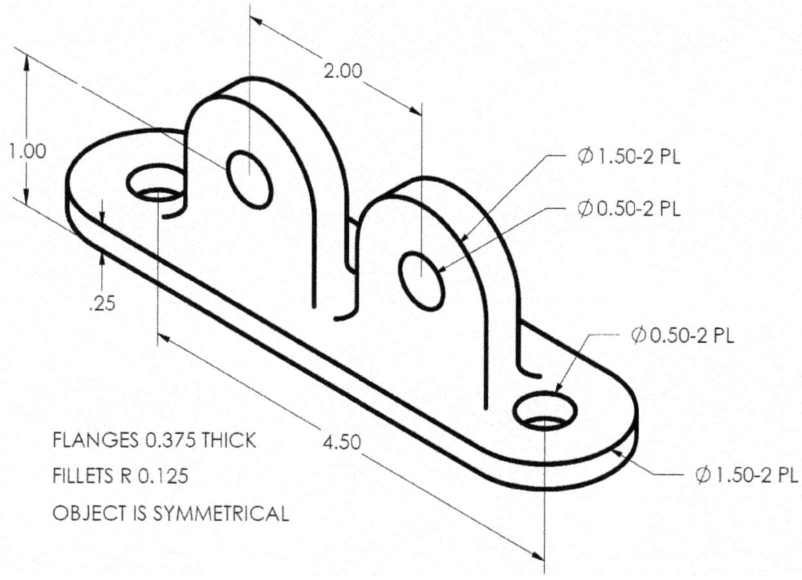

Problem 2—Cleat

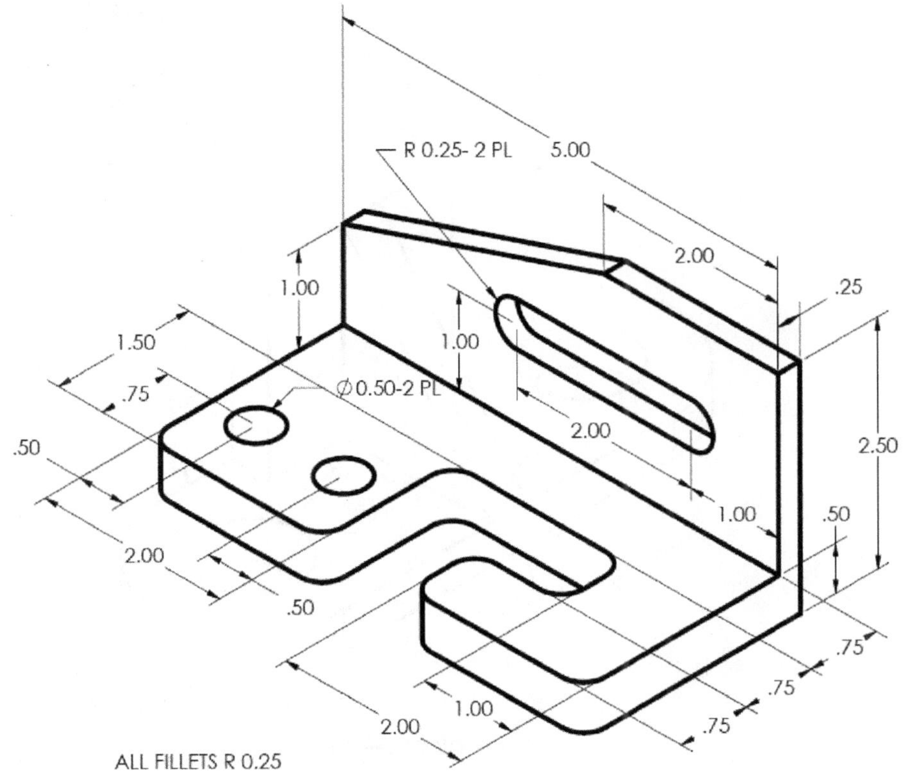

Problem 3—Bracket

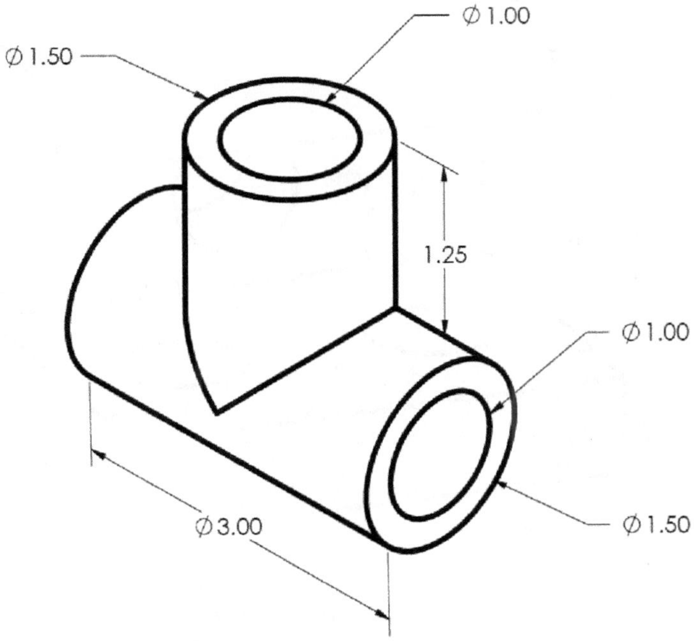

Problem 4—Tee Connection

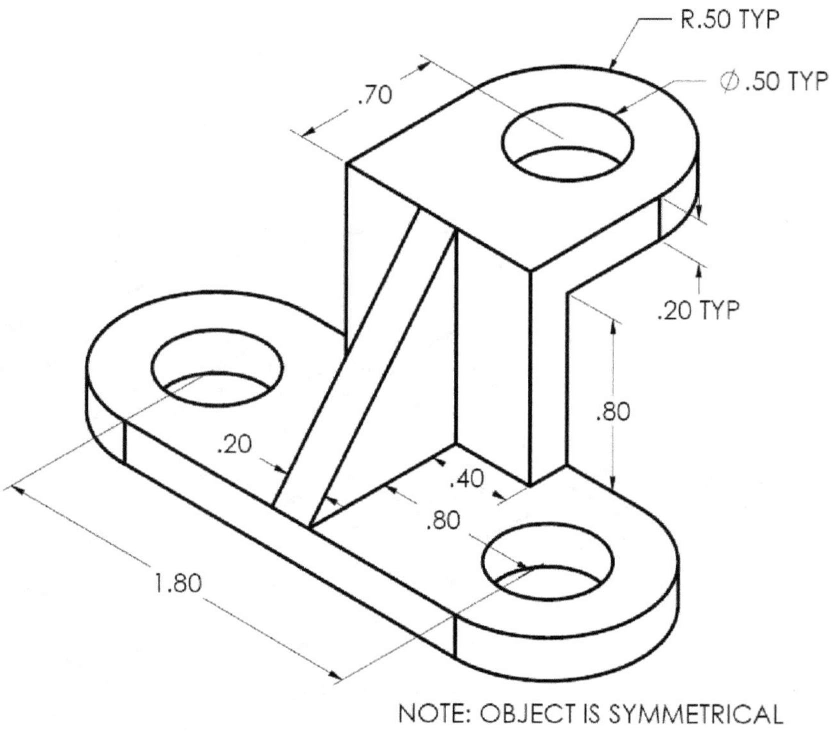

Problem 5—Bracket

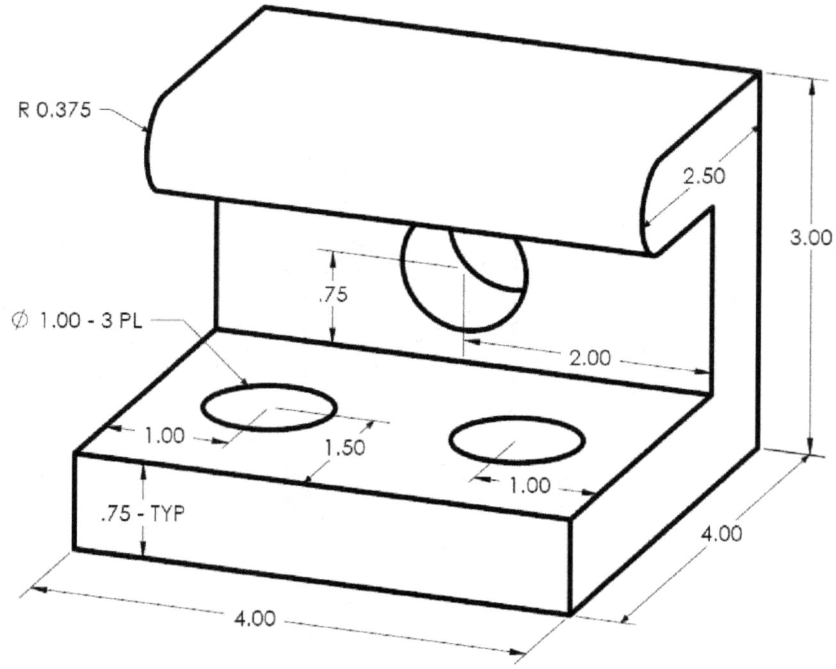

Problem 6—Stabilizer

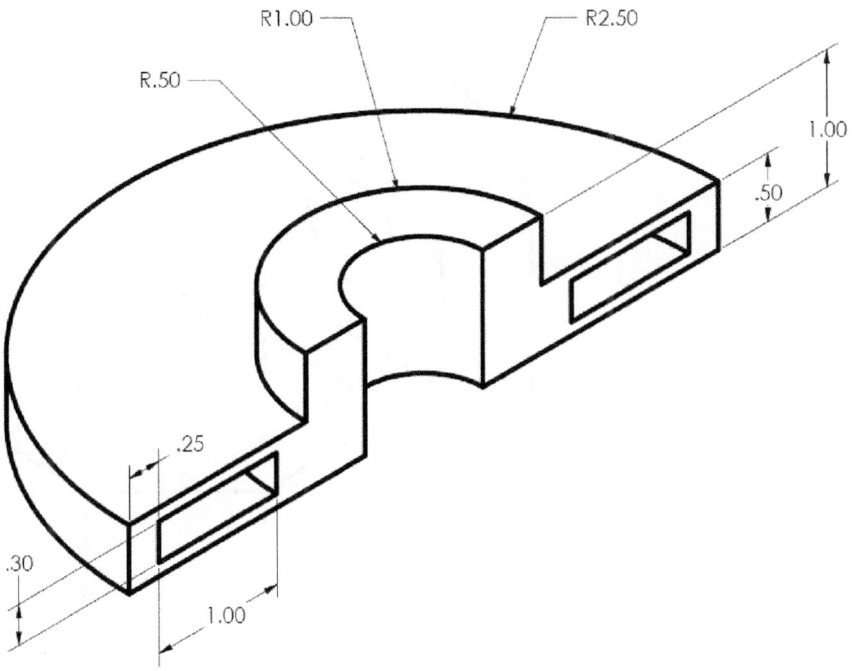

Problem 7—Disc

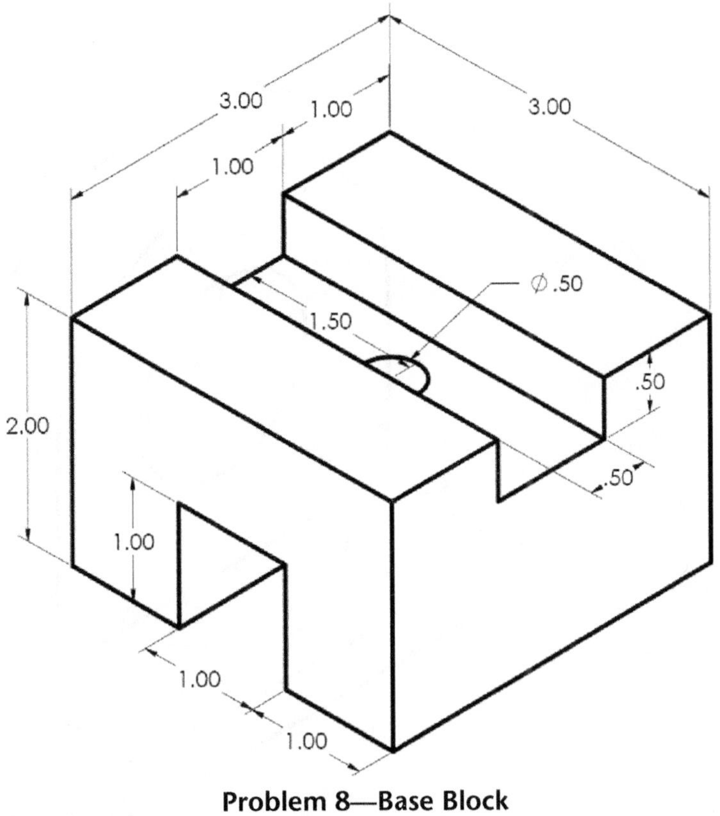

Problem 8—Base Block

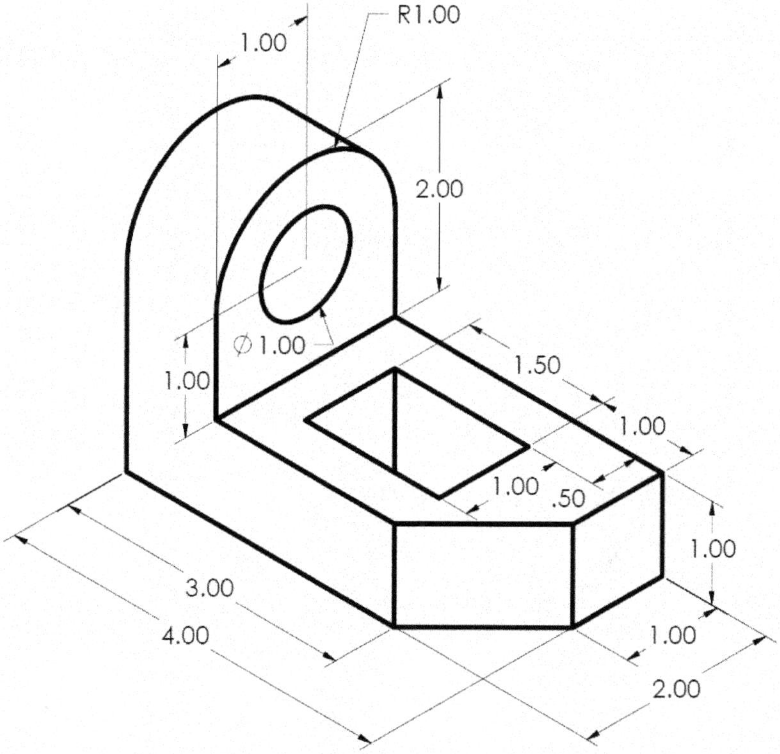

Problem 9—Corner Bracket

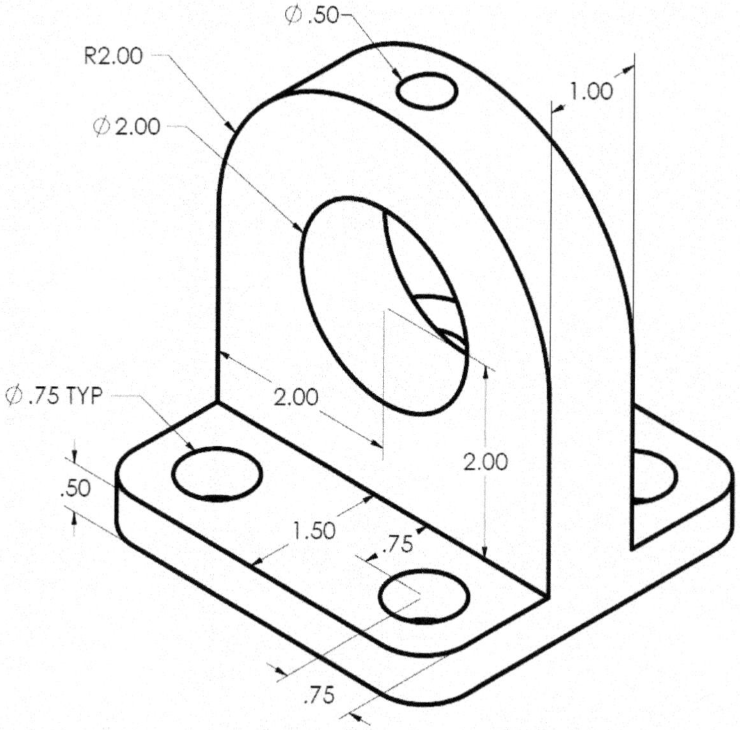

Problem 10—Cable Eyelet

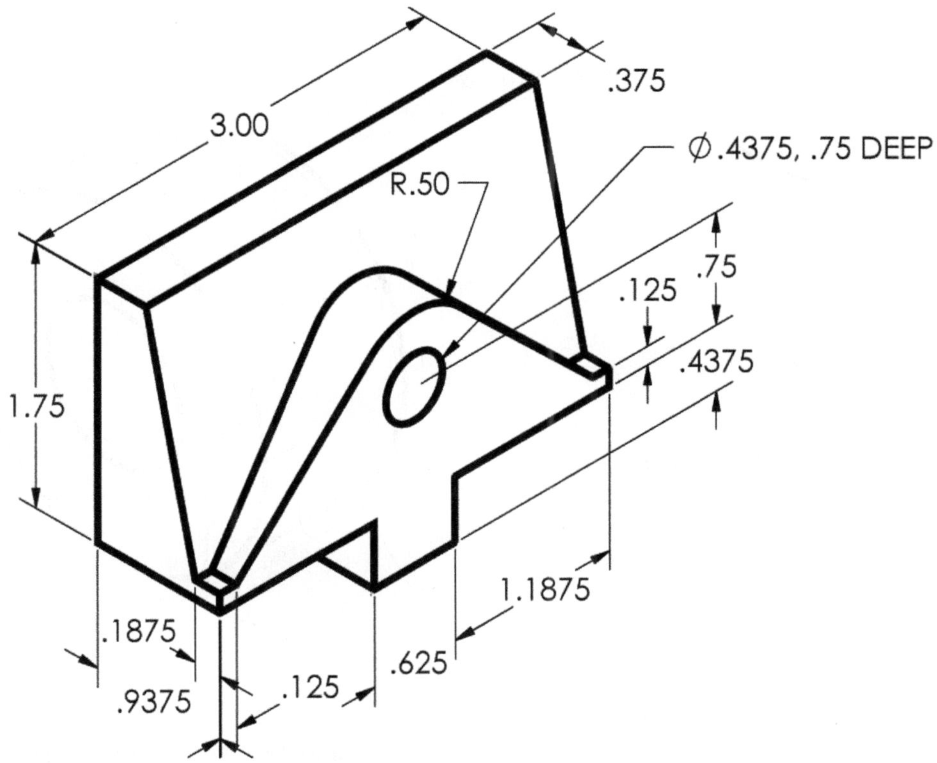

Problem 11—Sliding Jaw

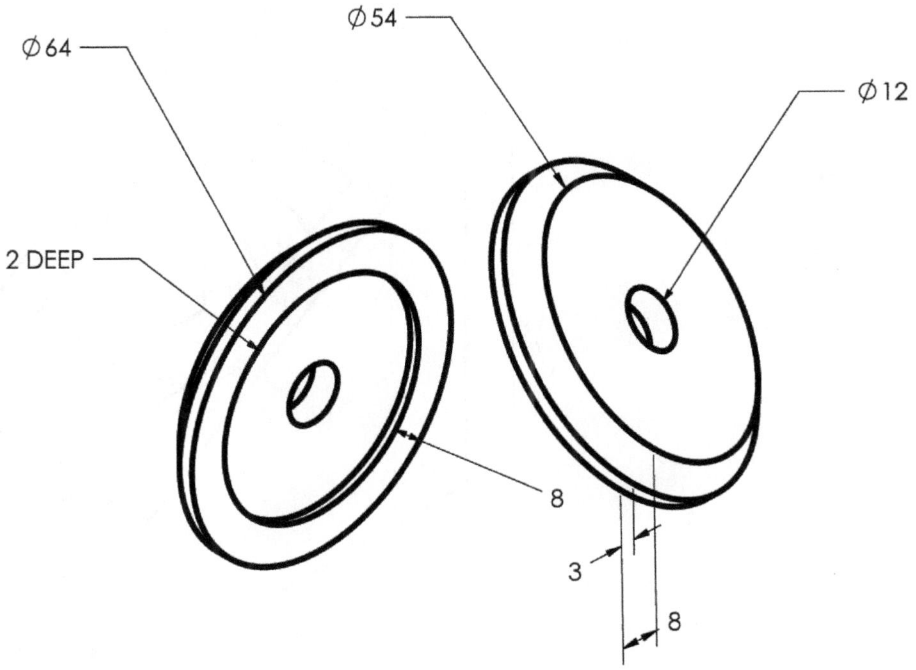

Problem 12—Arbor Mount

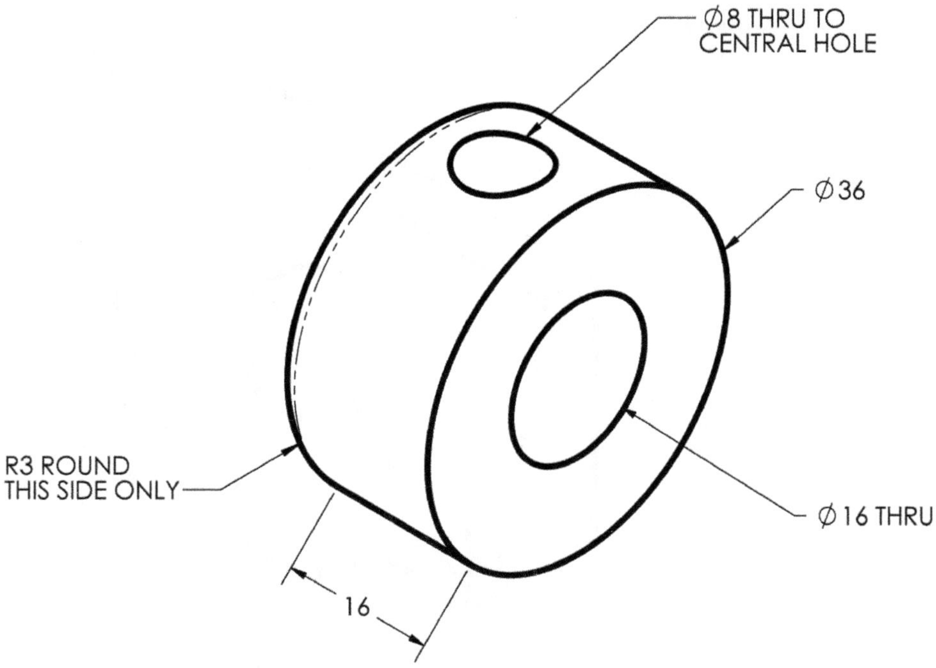

Problem 13—Retaining Collar

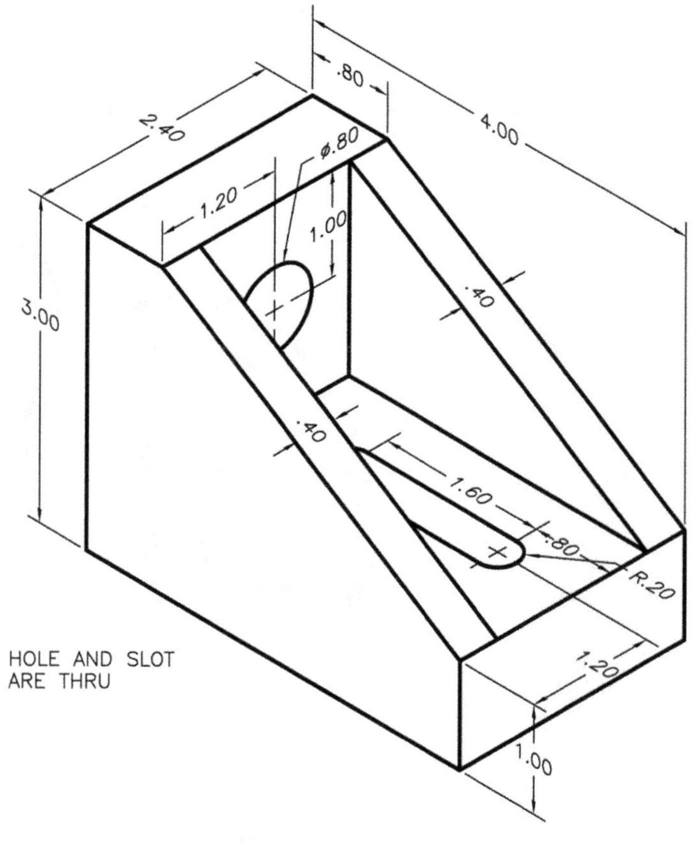

Problem 14—End Support

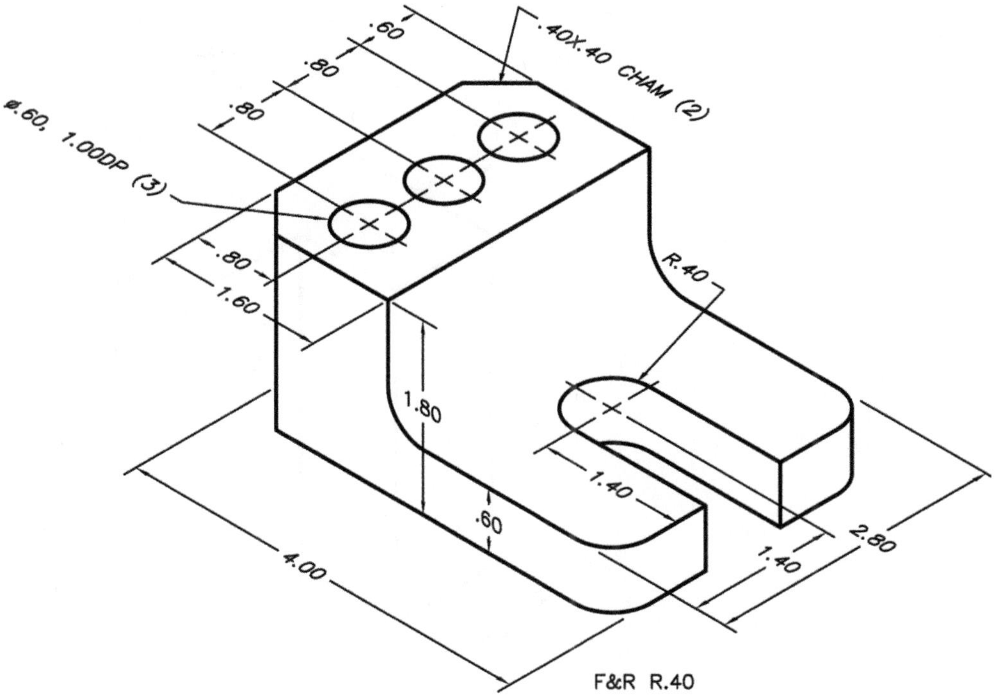

Problem 15—Alignment Mount

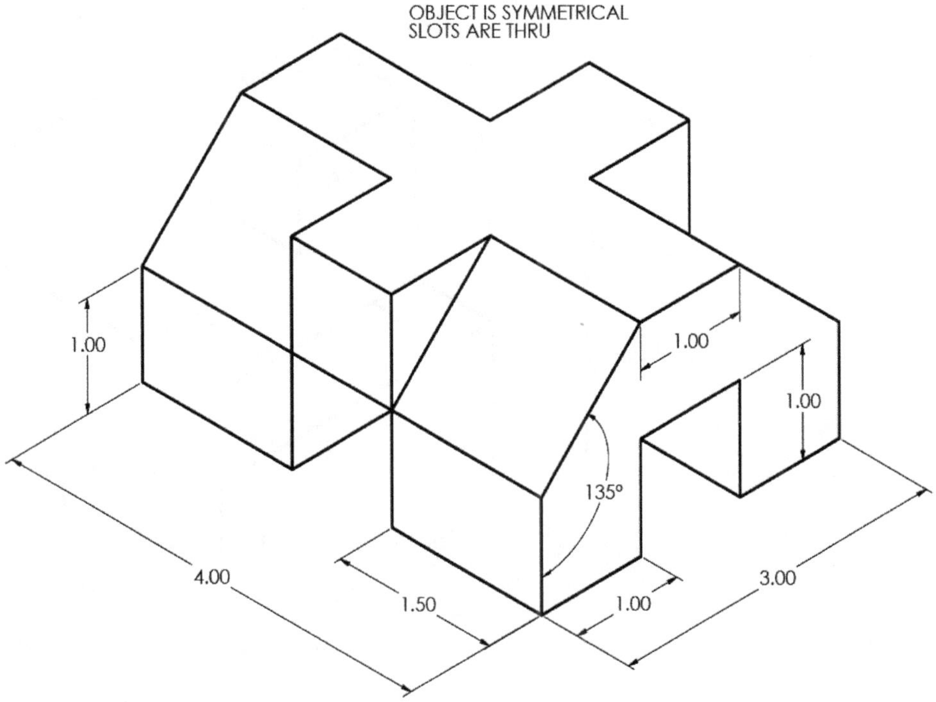

Problem 16—Shelving Spacer

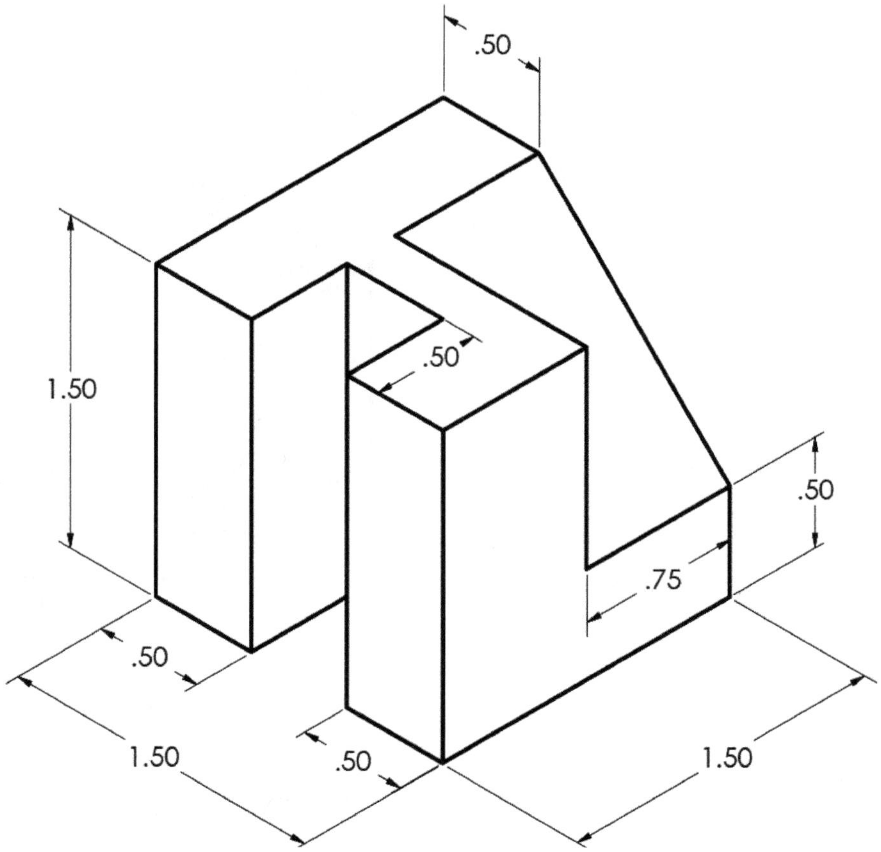

Problem 17—Shim Guide

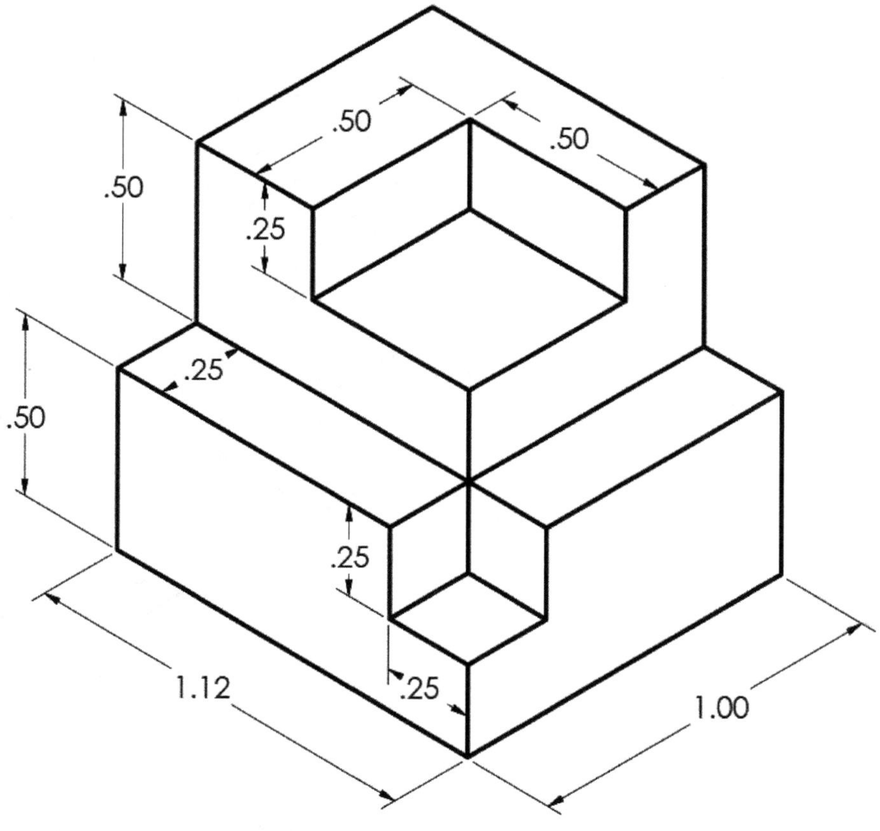

Problem 18—Offset Block

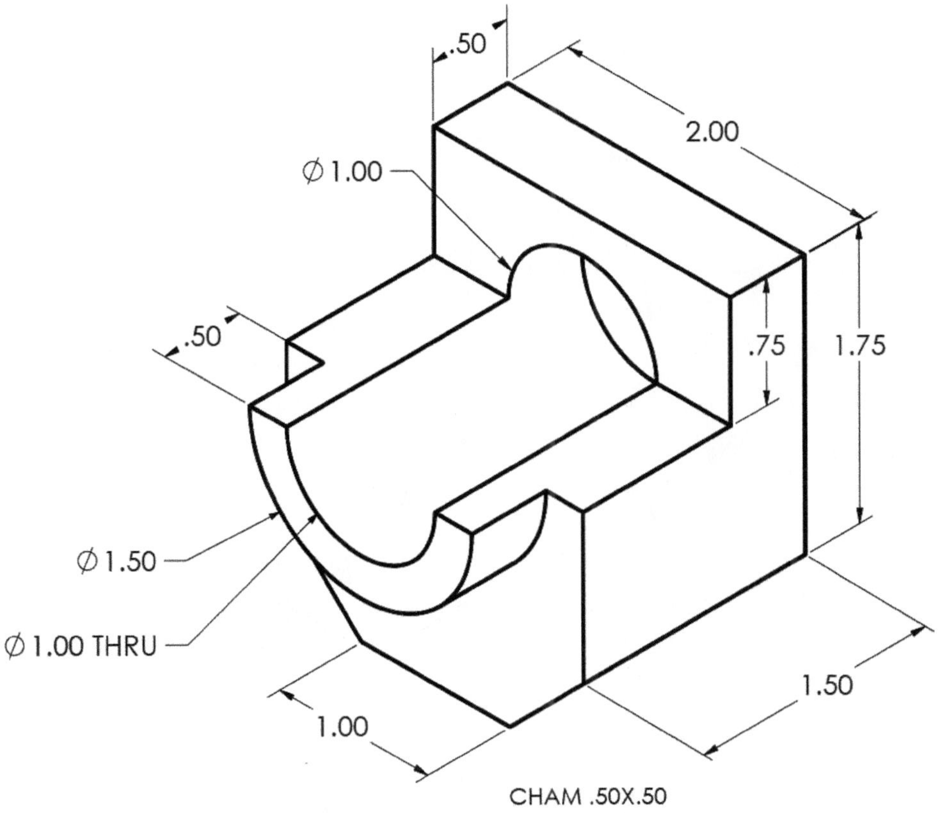

Problem 19—Shaft Support

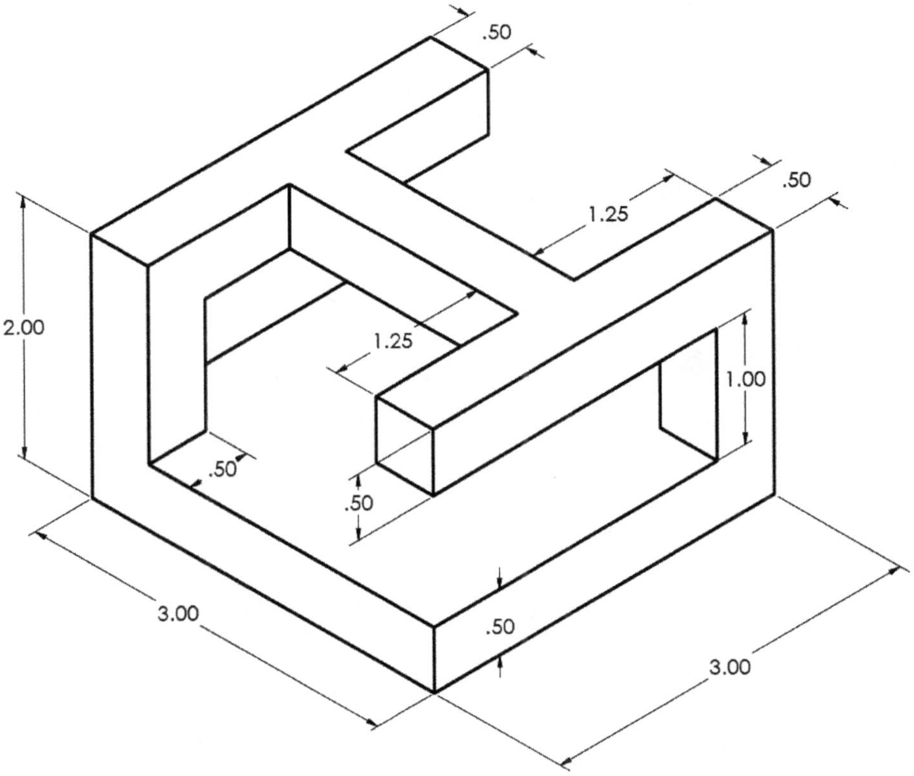

Problem 20—Frame

Pictorials

4
CHAPTER

1. Introduction

Technical illustrations provide detailed information about the shape, size, and material of a product or design and they need to be accurate, clear, and unambiguous. Although objects are three-dimensional, representing 3D objects accurately on 2D media, such as paper, is not trivial and requires special care.

The precise representation of an object on a plane (called picture plane or projection plane) from a particular point of view is called true projection. Think of the picture plane as a paper located between the viewer and the object. A drawing is created by connecting the points where the lines of sight from the viewer's position, traveling to the different points on the object, intersect the projection plane (see Figure 4.1). Drawing true projections of objects is a laborious process, so alternative projection methods have been developed to simplify this task. This simplification is accomplished by taking angles and measurements to produce a projection that resembles the true projection of the object.

One of these projection methods has already been discussed in the previous chapter; orthographic drawings use three principal orthographic planes (horizontal, frontal, and profile) as picture planes to produce several views of an object. For each view, the observer is located at infinity in front of the projection plane, hence the lines of sight travel parallel to each other in the direction of the object and hit the picture plane perpendicularly. Each orthographic projection generates a drawing that shows only two dimensions. Therefore, more than one drawing is required to fully describe any object.

Pictorials are a type of technical drawing that shows several faces of an object at once, much as they would be captured by a camera. Pictorial drawings are often more clearly understood than orthographic drawings, especially for non-technical audiences. As opposed to multi-view drawings, where the viewer

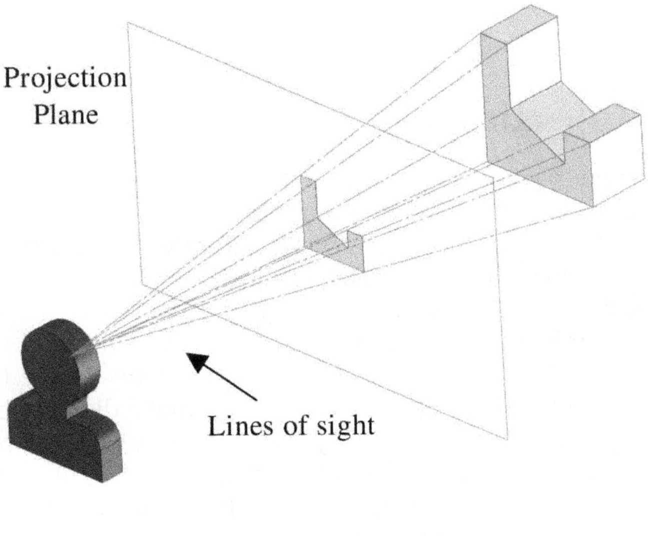

Figure 4.1 True projection of an object

is required to mentally visualize and combine the individual views into one 3D object, pictorials show all three dimensions of an object in one single drawing. Pictorial drawings are commonly used in conjunction with orthographic drawings. They are very useful in the early stages of the design process, as they mimic the way we represent objects in our mind and how we visualize spatial relationships.

2. Types of Projections

There are four major projection techniques used in engineering: orthographic, axonometric, oblique, and perspective, as shown in Figure 4.2. Axonometric, Oblique, and Perspective projections are pictorials, showing all three dimensions of an object in a single drawing. Orthographic projections require multiple 2D drawings to describe any object. Orthographic, Axonometric, and Oblique projections are parallel projections, which means that lines that are parallel on the real object will remain parallel on the drawing as well. This simplification is a result of theoretically locating the viewer infinitely far away from the picture plane and assuming that the lines of sight travel parallel to each other and perpendicular to the projection plane.

Perspective projections are non-parallel projections: lines that are parallel in the real object are not necessarily shown as parallel in the perspective drawing. Perspective projections reproduce the effect that distant objects appear smaller than objects that are closer to the observer and represent how an eye perceives the world. Lines that are parallel in nature appear to converge towards a single point, depending on the position of the viewer.

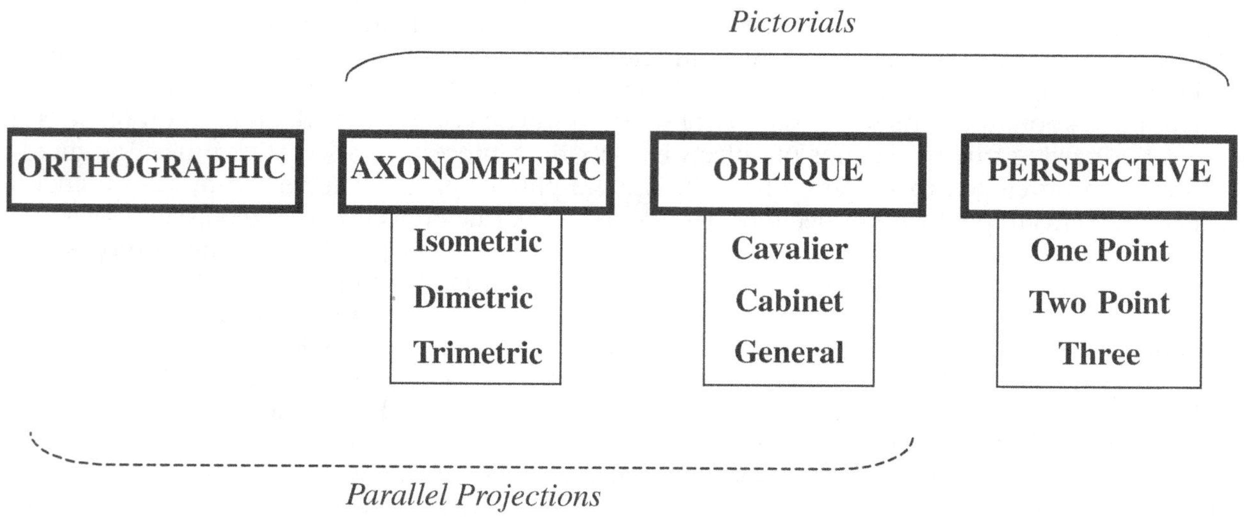

Figure 4.2 Classification of projection techniques

3. Axonometric Projections

Axonometric projection is a type of parallel projection where the observer is theoretically located at infinity and the object is rotated and tilted around an axis (called an axonometric axis) relative to a projection plane in order to display all three dimensions (see Figure 4.3). There are infinite numbers of angles and rotations that can be used to draw objects using this type of projection, but only a few of these angles are useful in practice. Such angles are optimized and standardized so the process of creating axonometric drawings is simple and fast and the results are accurate and reasonably realistic. In this case, the term "reasonably realistic" refers to visual realism, or how close the drawing resembles what our eyes see. Unlike perspectives, axonometric projections make distant objects appear the same size as nearby objects. This simplification is not representative of how humans perceive the world, but by sacrificing "visual realism" and making the drawing appear distorted, an advantage is gained: the actual dimensions of the real object can be measured directly from the drawing if the scale is known.

Axonometric projections are classified by the orientation of the primary axes relative to the projected face of the object. This orientation determines the amount of distortion for each dimension. When an axis is oriented parallel to the projected face, dimensions along that axis are shown true size. Otherwise, dimensions appear shorter than their true length. The closer an axis comes to being perpendicular to the projected face, the shorter it becomes. This orientation can be observed through the three angles formed by the primary axes on the projected object. If all three angles are equal (120°) an Isometric projection is created. If two of the angles are equal, the projection is Dimetric. If all three angles are different, the projection is Trimetric. See Figure 4.4 for an example of axonometric projections using a cube, and then using a more complex object. Notice in the isometric drawing of the object how a basic cube can still be used to identify the primary axes.

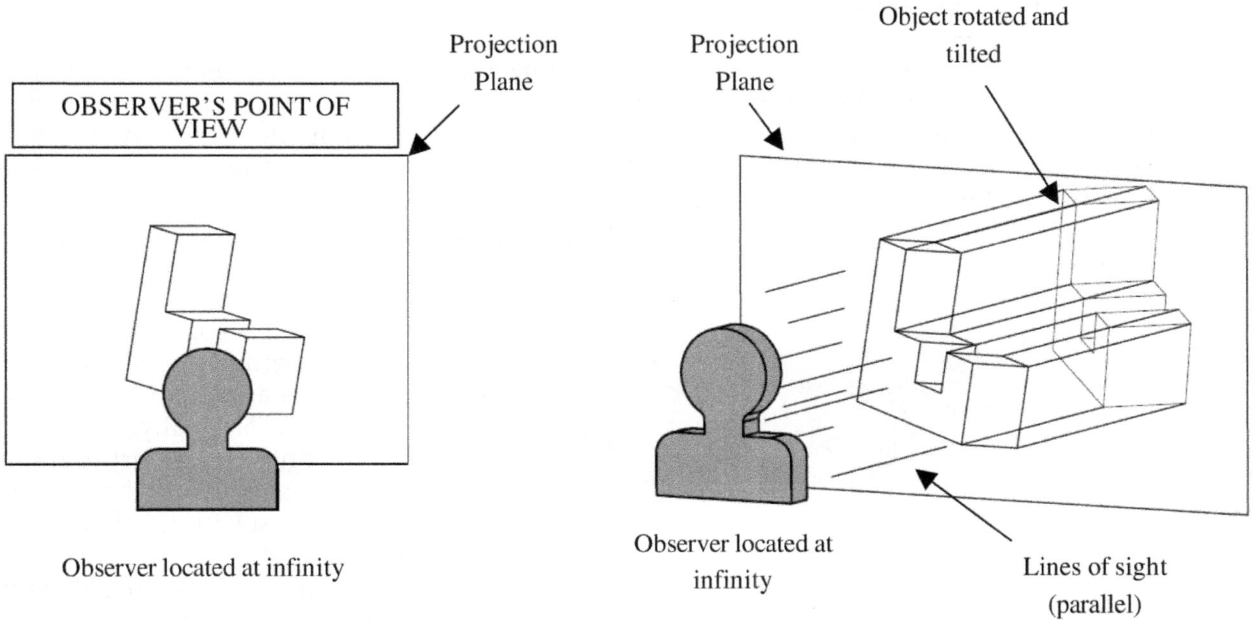

Figure 4.3 Axonometric projection

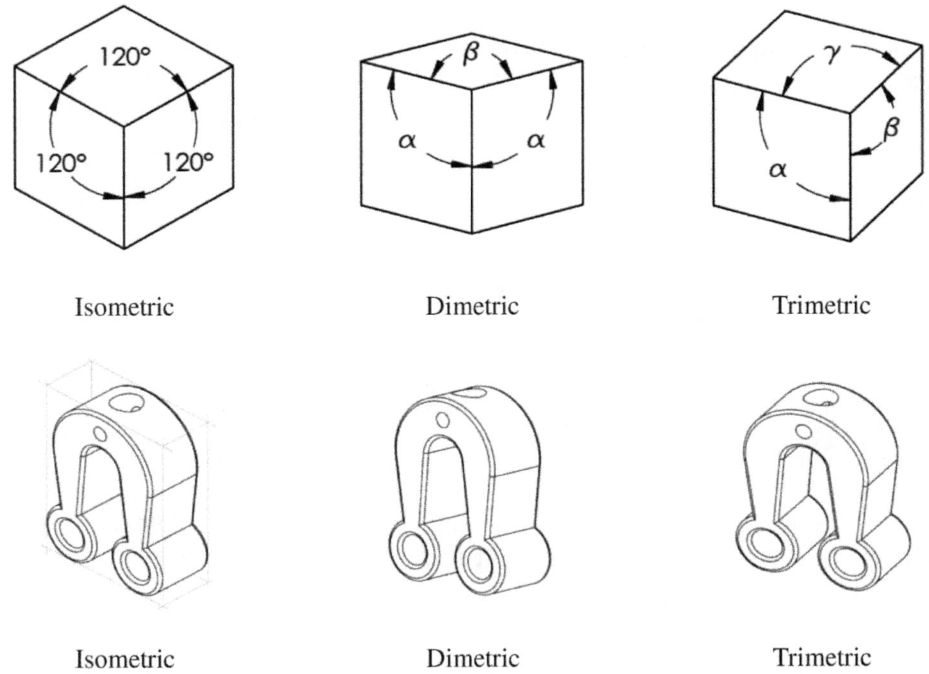

Figure 4.4 Three types of axonometric projections

3.1 Standard Practices for Axonometric Drawings

Orientation There are many different ways axonometric axes can be oriented relative to the projection plane and still maintain the relationships and angles. In general, axonometric projections are drawn with the main axes above the horizontal. The same rules used in orthographic projections to select the most descriptive orientation apply to axonometric drawings. Examples of different orientations of the same object using an isometric projection are shown in Figure 4.5.

Lines In axonometric drawings, hidden lines often create more confusion than they provide information, especially for complex models (see Figure 4.6 for an example). Hidden lines are normally omitted in axonometric drawings unless the features they represent are absolutely required to fully illustrate the object. As it happens with orthographic drawings, the use of hidden lines can generally be minimized by selecting the correct orientation for the object. Center lines are also omitted, unless additional clarity is needed to show concentricity. As a rule, only axonometric drawings with dimensions and/or tolerances for manufacturing purposes will include center lines. Tangent edges are normally omitted on orthographic drawings, because they can mislead the viewer by giving the false impression of a visible line. However, tangent edges are typically shown in axonometric projections. Omitting tangent edges in axonometric drawings often has severe effects on the quality of the drawing and how the information is perceived (see Figure 4.6 for an example). Most 3D CAD applications will show tangent edges by default when axonometric pictorials are created.

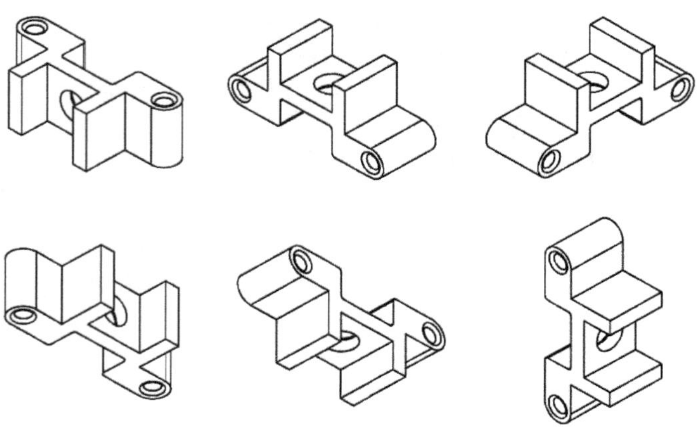

Figure 4.5 Different orientations of the same object using isometric projections

Dimensions It is a common practice in production drawings to include orthographic drawings and pictorials of all the parts of a design. Normally, dimensions are included only in the orthographic views and pictorials are used as an illustration tool to assist with the visualization of the object.

Occasionally, axonometric drawings need dimensioning and general dimension standards apply. All elements of a dimension must be positioned on the same plane (normally axonometric planes) as the line that is being dimensioned. Dimensioning practices will be discussed later. See Figure 4.7 for an example of a dimensioned isometric drawing.

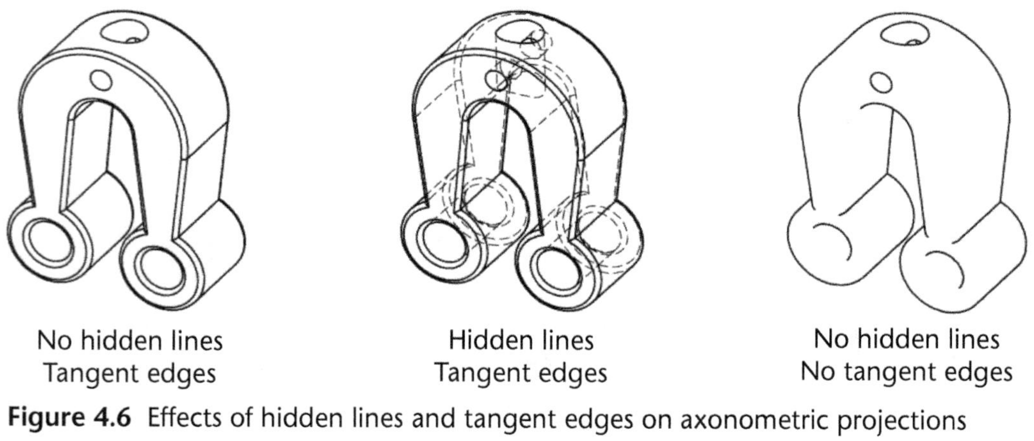

Figure 4.6 Effects of hidden lines and tangent edges on axonometric projections

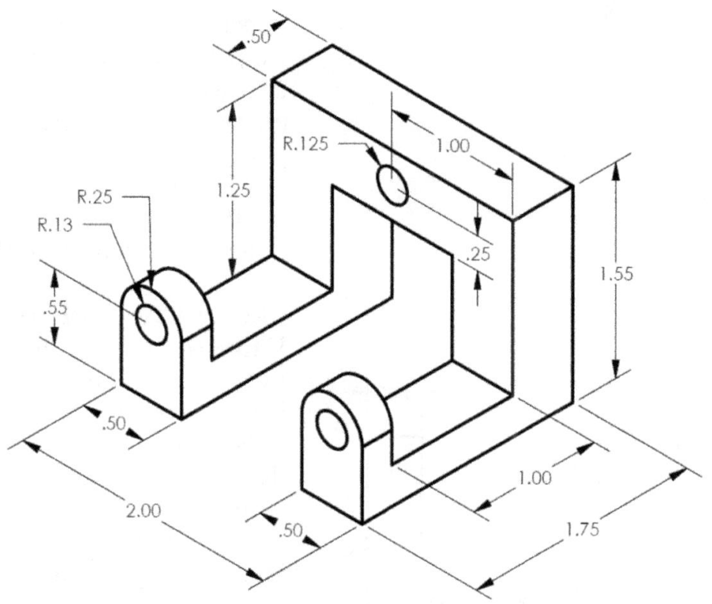

Figure 4.7 Dimensioned isometric drawing

3.2 Isometric Projections

Isometric projection is the most popular type of axonometric projections used in engineering. Because of its popularity, many tools are readily available to assist in the sketching and construction of isometric drawings. Even most computer aided design (CAD) packages nowadays incorporate isometric grids, snaps, viewpoints, cameras, and templates optimized for this type of projection.

Isometric projections are built by locating the observer at infinity looking perpendicular to the picture plane. The object to be drawn is first rotated 45° about a vertical axis and then tilted forward 35.27° as shown in Figure 4.8. These values cause the projected object to be oriented in such a way that the three primary axes (in this case called isometric axes) form equal 120° angles among themselves and 30° angles with the horizontal (see Figure 4.9).

The orientation of the object in isometric projection also causes an overall scale reduction of approximately 80% in the projected drawing. Although this is not a major concern, for practical purposes, a full scale isometric is always preferred, since immediate information about the true dimensions of the object is given. Because of the specific orientation of the axes, isometric projections keep drawings proportionate in size, so drawings do not appear distorted. In addition, lines and edges drawn parallel to the isometric axes will always appear true size.

There are three major disadvantages when using isometric drawings. The first is that all lines and edges that are not parallel to the isometric axes will not appear true size and therefore will not be as easy to draw as isometric lines. The second drawback is that circular features on the real object will

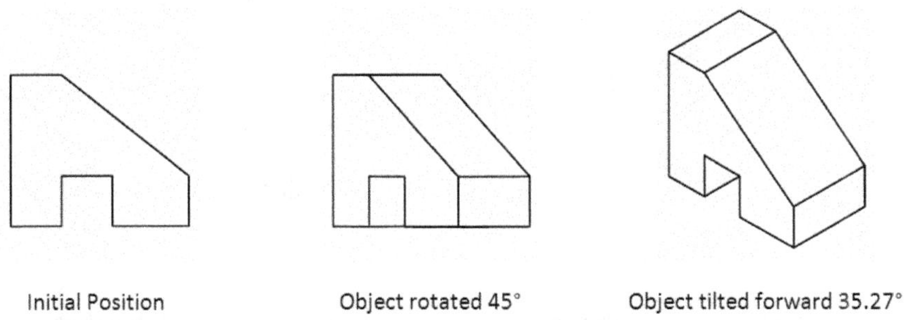

Figure 4.8 Creation of an isometric projection

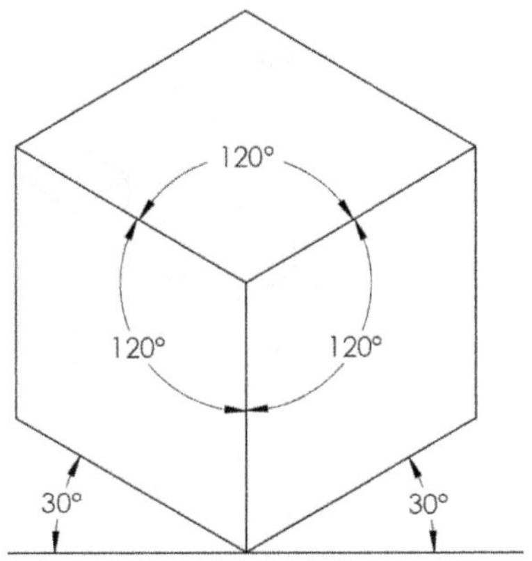

Figure 4.9 Isometric cube

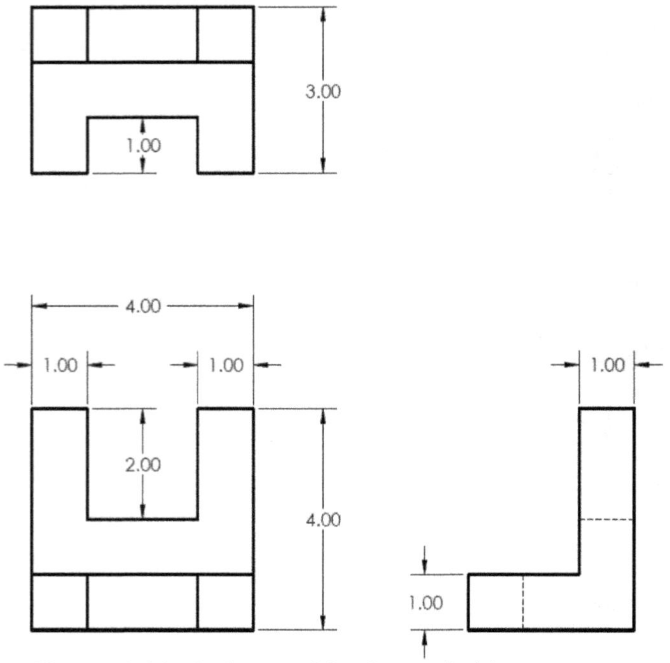

Figure 4.10 Orthographic views of object

appear as ellipses on all faces in the isometric drawing. Although there are templates and tools to assist with the construction of ellipses, it is difficult to draw and orient them correctly in an isometric drawing. Finally, curved surfaces and irregular features on the object must be located with coordinates to properly plot them on the drawing.

Sketching Techniques for Isometric Drawings The most common technique needed to sketch isometric drawings uses basic cubes or construction boxes as visual aids. Isometric sketching begins with the definition of the isometric axes. Although isometric grids and drafting triangles are commonly used to facilitate the task, several freehand techniques can be employed. Examine the object shown in Figure 4.10 to gain an understanding of this process. Assume an isometric grid where lines are spaced 1 unit from each other.

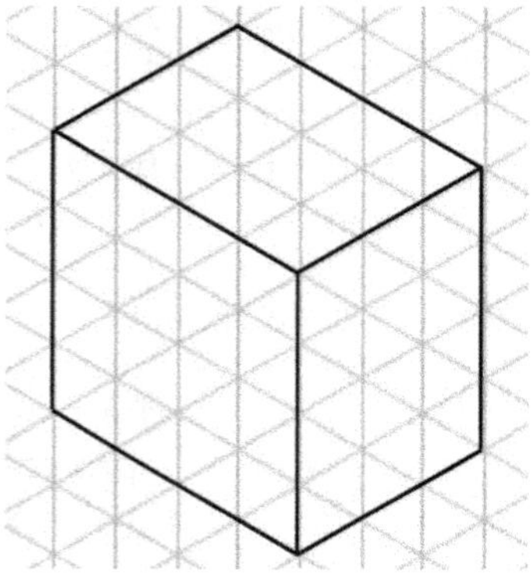

1. Build a box with the overall size of the object to be drawn (in this case 4×4×3).

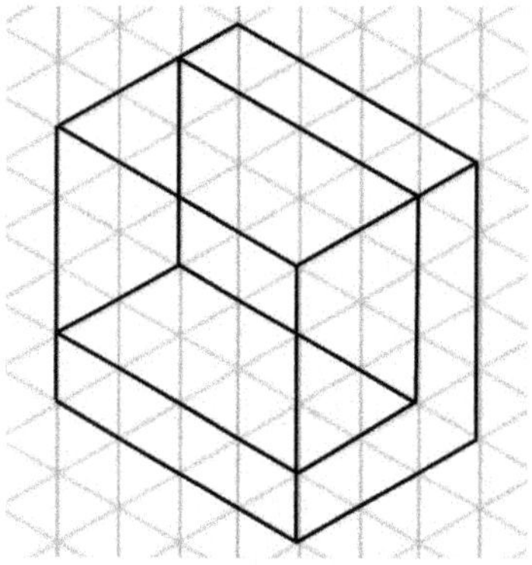

2. Sketch the overall shape or outline using the orthographic views. In this case, the right side view offers great detail.

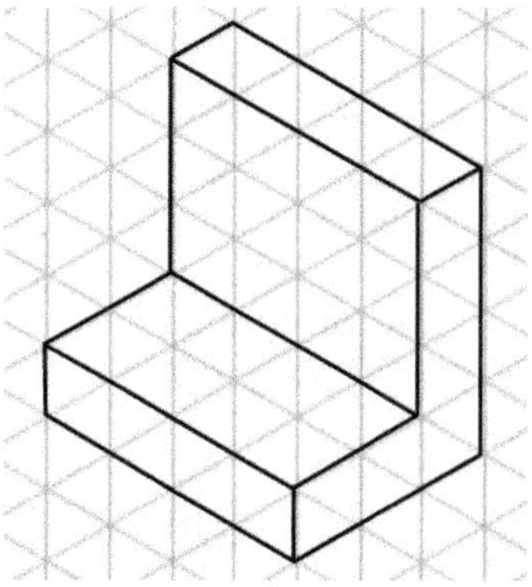

3. Delete the lines that are no longer needed.

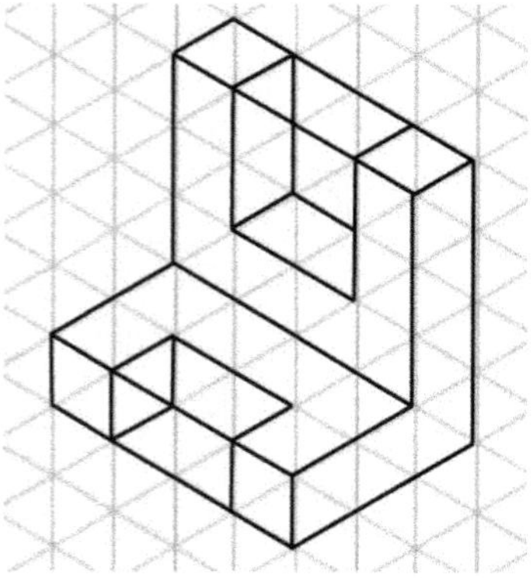

4. Add details to the overall shape using the other orthographic views. In this case, front and top views.

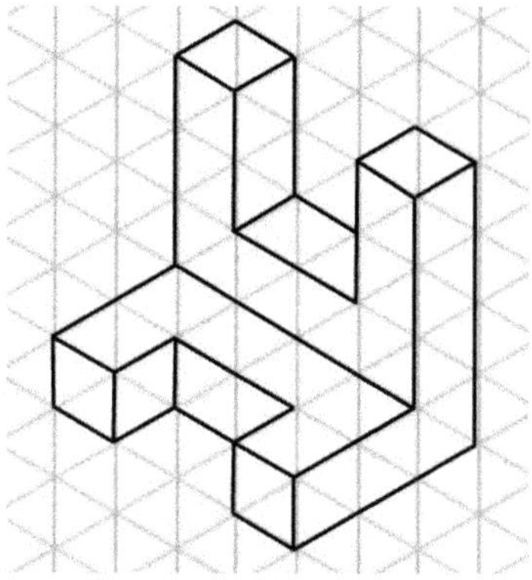

5. Delete the lines that are no longer needed.

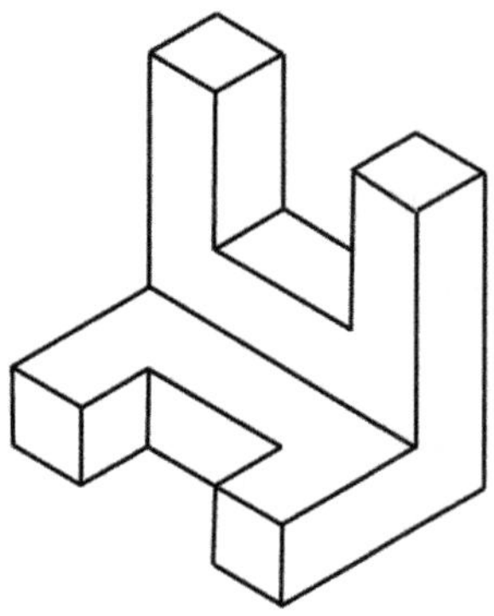

6. Completed drawing.

Angles and Non-Isometric Lines A non-isometric line is a line that is not parallel to any isometric axis. An observation can be made that any line in an orthographic drawing that is not parallel or perpendicular to an orthographic plane will be a non-isometric line in the isometric drawing. Any angle other than 0°, 90°, 180°, and 270° creates a non-isometric line. In summary, angles always appear true size in multi-view drawings but never in isometric drawings.

A common technique used to represent angles and non-isometric lines in isometric drawings is to start the drawing by sketching all the isometric lines (normal lines in the multi-view drawing, and thus appearing true size in the pictorial). Non-isometric lines can be added by connecting the endpoints of the existing isometric lines with straight lines. This process is illustrated in Figure 4.11.

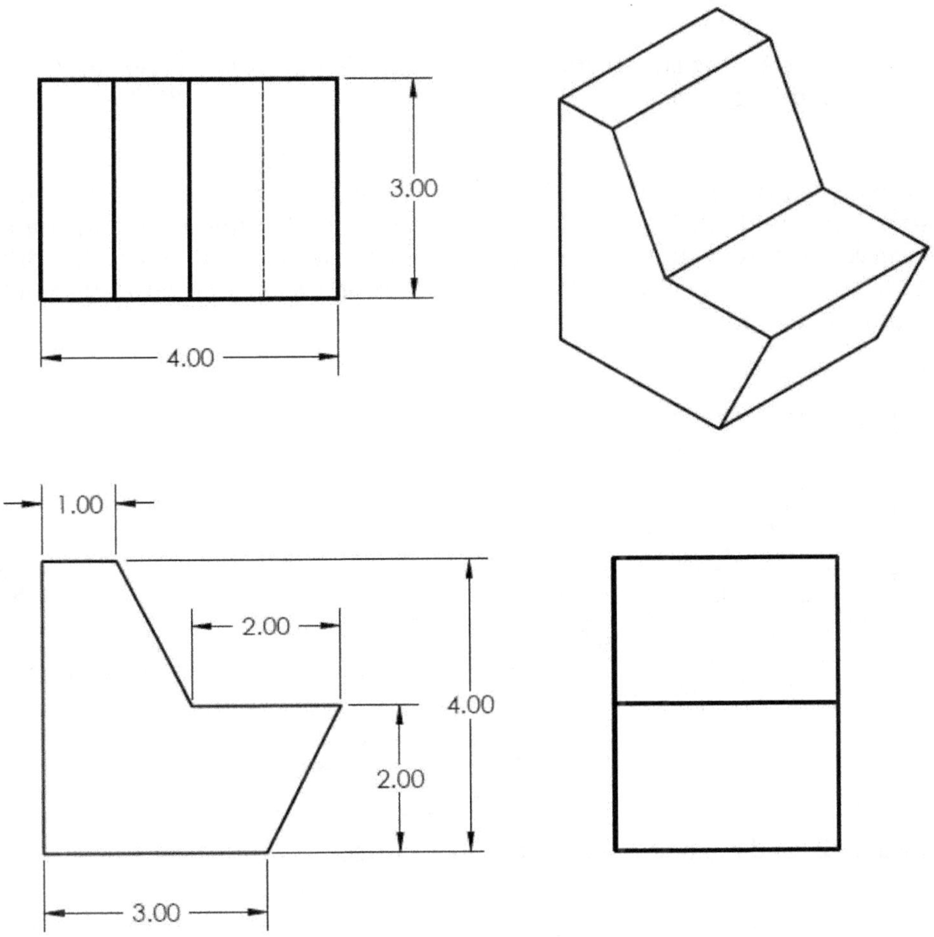

Figure 4.11 Orthographic views of object with angled features

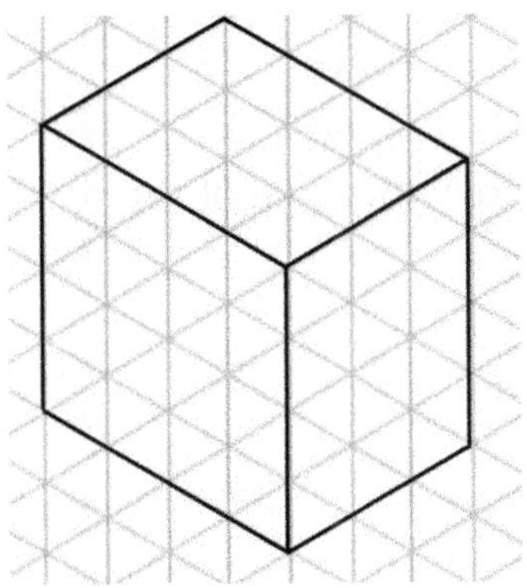

1. Build a box with the overall size of the object to be drawn (in this case 4×4×3).

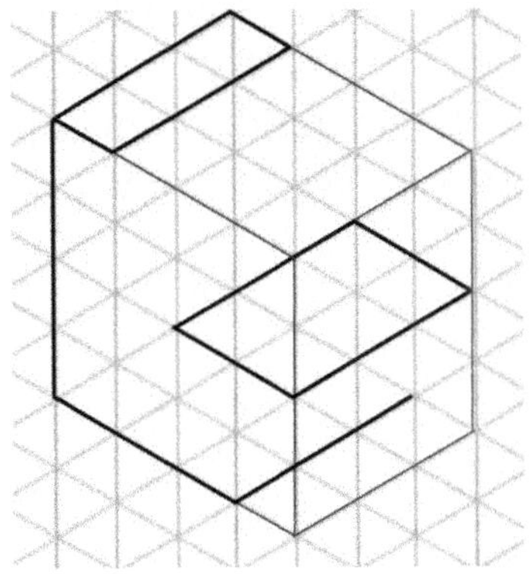

2. Sketch the overall shape or outline (isometric lines only) using the orthographic views. In this case, the front view offers great detail.

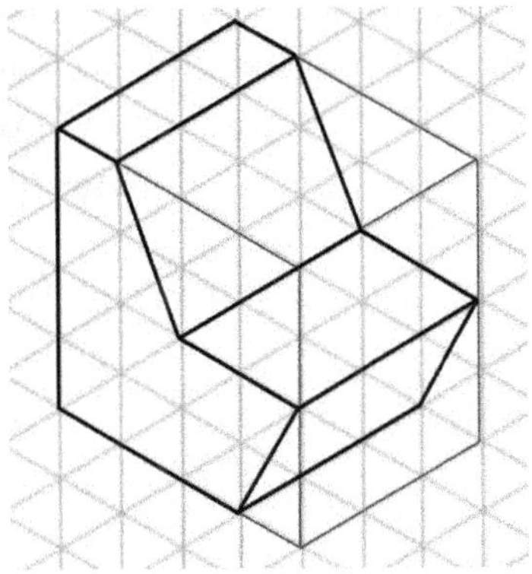

3. Connect endpoints of existing lines to create non-isometric lines.

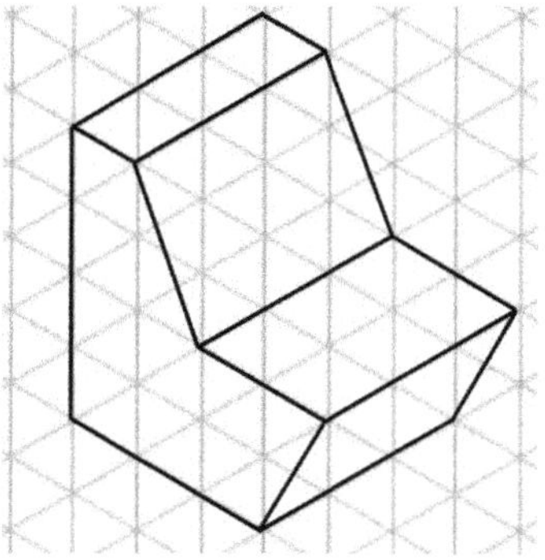

4. Delete the lines that are no longer needed.

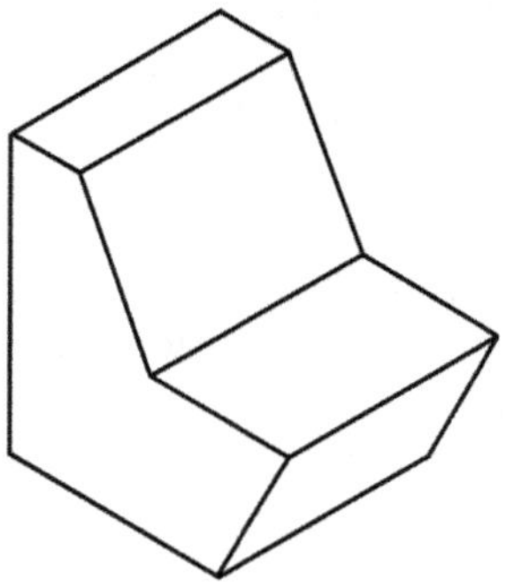

5. Completed drawing.

Circles and Curved Features The general method for creating irregular curved features in isometric drawings is the grid method. This method consists of constructing a grid on the orthographic view where the curved feature is shown. The intersections of the curved line with the divisions in the grid give a sequence of control points that can be used to plot the curve on the isometric drawing. See Figure 4.12 for an example.

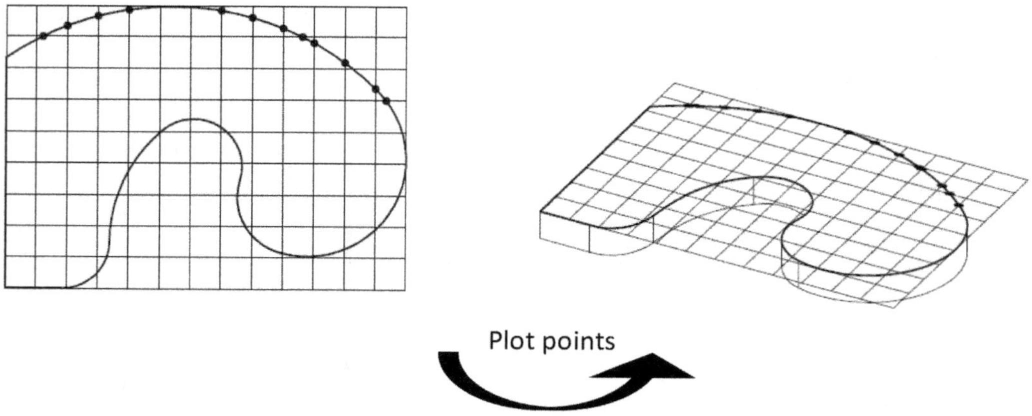

Figure 4.12 Grid method to sketch curves and irregular features

The same technique can be used to sketch full circles in isometric drawings. However, since circles will resemble ellipses (see Figure 4.13) an easier method can be used. An isometric square (construction box) can be created such that each side equals the diameter of the desired circle. The method is based on how isometric ellipses are connected to their construction boxes:

- Ellipses always touch the midpoint of each side of the isometric construction box.
- The major diameter of the ellipse is always aligned with the longest body diagonal of the construction box.

See Figure 4.14 for a representation of an ellipse inside its construction box.

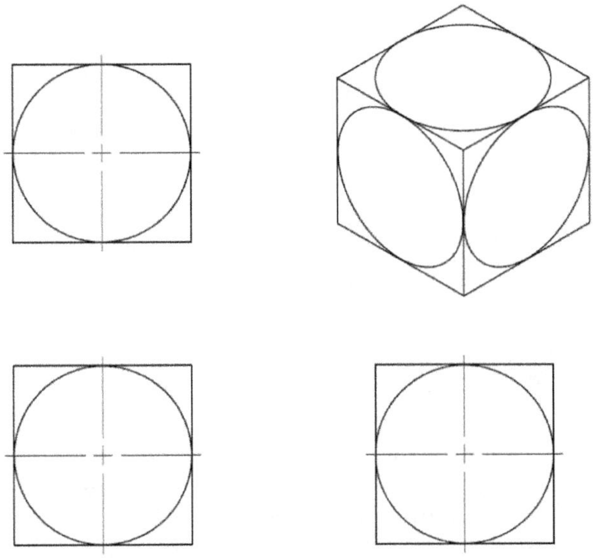

Figure 4.13 Circles appear as ellipses in isometric drawings.

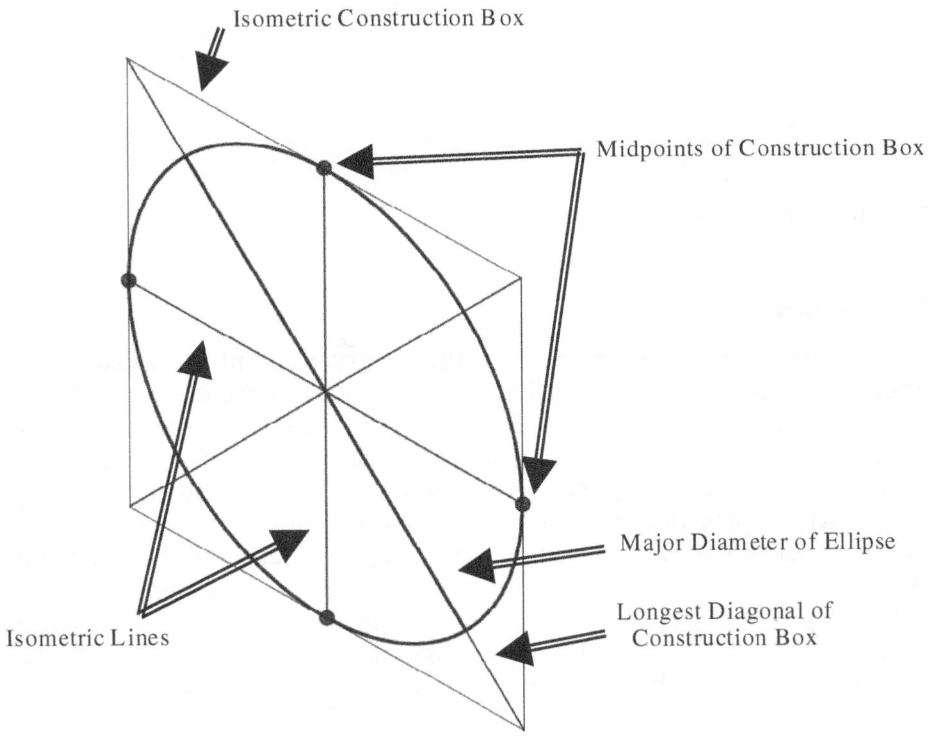

Figure 4.14 Isometric ellipse inside its construction box

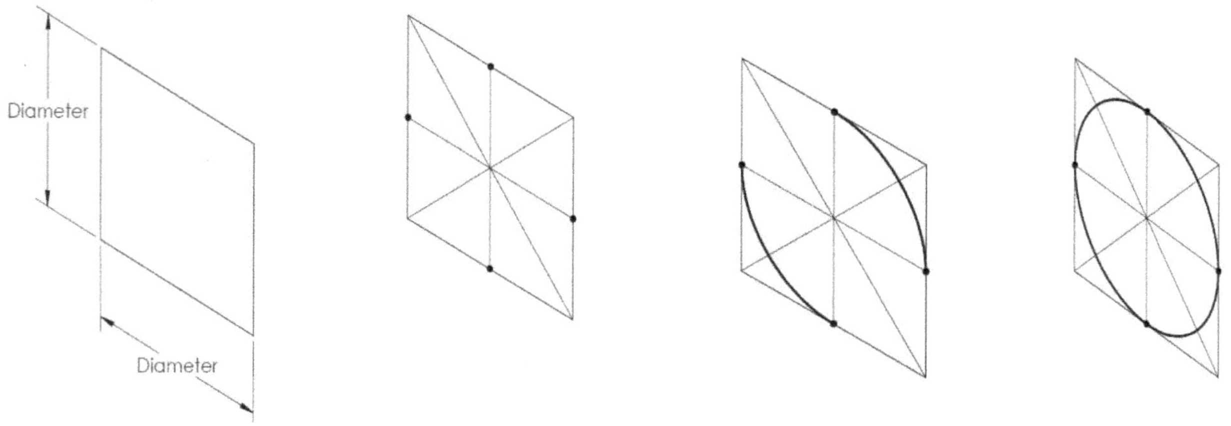

1. Sketch the construction box using the diameter of the circle.

2. Trace the diagonals of the box and find the midpoint of each side.

3. Sketch arcs to build the ellipse.

4. Finish drawing.

Parts with holes and/or cylindrical features are common examples where isometric circles must be sketched. Isometric cylinders are commonly shown standing upright. To sketch these features, begin with a construction box and sketch the ellipses on the top and bottom of the cylinder or hole. Finally, connect the ellipses with tangent lines and delete the construction lines when the drawing is finished (See Figure 4.15).

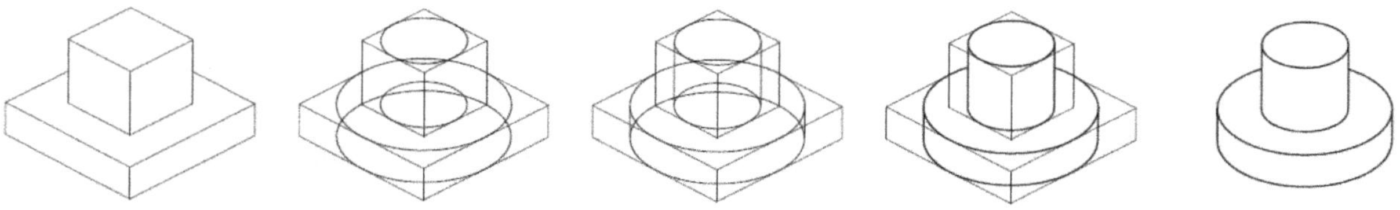

Figure 4.15 Cylindrical features in isometric drawings

3.3 Dimetric Projections

Dimetric projections are constructed by locating the observer at infinity looking perpendicular to the picture plane. The object to be drawn is first rotated about a vertical axis and then tilted forward so that only two of the three primary axes form equal angles with the projection plane. The first rotation of the object before it is tilted occurs at 45°, so that one of the diametric axes is vertical. This way, the two equal angles will appear on either side of the vertical line creating a symmetric dimetric projection, which provides for easier visualization. If one of the equal angles and the third angle are placed on either side of the vertical, an asymmetric dimetric projection can be produced.

Infinite tilt angles can be used to create dimetric projections. In general, any angle other than 35.27°, which would produce an isometric, is acceptable. The most common configuration uses 105° for the equal angles (thus, 150° for the third angle) and 15° for the receding angles (the angles formed by the dimetric axes and the horizontal). See Figure 4.16 for an example of different dimetric projections.

3.4 Trimetric Projections

Despite the fact that trimetric projections are considered the most visually pleasant type of axonometric projections, they are not frequently used because of the difficulty needed to draw them. Determination of dimensions in trimetric drawings requires three scales because all three axes are oriented at different angles.

Trimetric projections are constructed by locating the observer at infinity looking perpendicular to the picture plane. The object to be drawn is first rotated about a vertical axis and then tilted

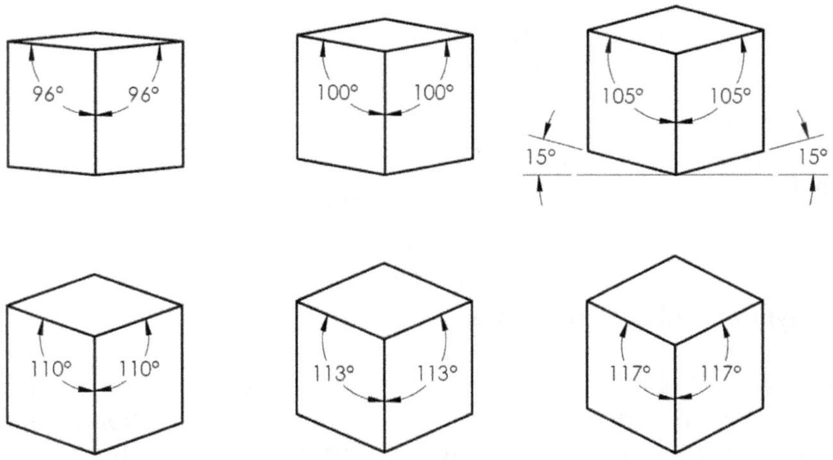

Figure 4.16 Different angles for dimetric projections

forward so that the angles formed by the three primary axes and the projection plane are different. The first rotation of the object before it is tilted can occur at any angle other than 45°, which would produce a dimetric or isometric projection, depending on the tilt angle.

Similar to dimetric projections, infinite tilt angles can be used to generate trimetric projections (see Figure 4.17 for some examples). The most popular orientation uses 105°, 135°, and 120° angles for the axes and 15° and 45° for the receding angles.

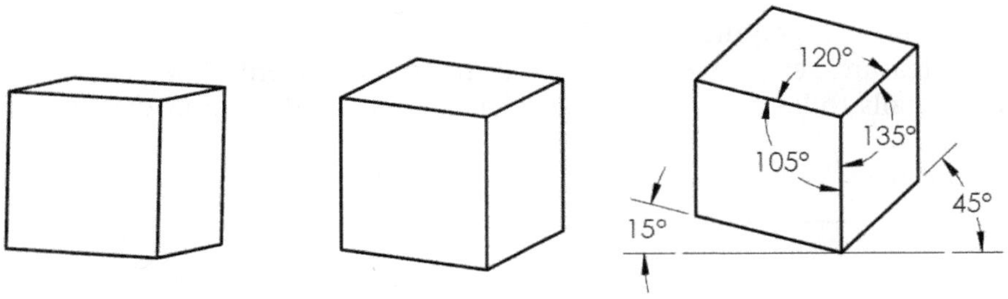

Figure 4.17 Different rotation and tilt angles for trimetric projections. The dimensioned drawing represents the most common trimetric orientation.

4. Oblique Projections

Oblique drawings are a distinctive type of parallel projections where the most descriptive face of the object is projected parallel to the projection plane, thus appearing true size, and depth is represented using angled, parallel lines drawn to one side of the front face.

Oblique pictorials are simple to create, but their use is not as widely extended in production environments as isometric drawings due the amount of distortion that occurs. The use of oblique drawings is normally limited to simple parts and objects, or complex objects with simple top and side surfaces. However, because of their simplicity, oblique pictorials are a great tool to quickly communicate ideas in early stages of the design process.

All types of oblique projections represent the most descriptive view true size and the depth using any receding angle between 0° and 90°. Similar to dimetric and trimetric drawings, an infinite number of angles can be used to produce oblique drawings. Regardless, angles that are less than 30° or greater than 60° create excessive distortion. Traditionally, 30°, 45°, and 60° angles have been used because of the availability of standard drafting triangles and other drawing instruments that simplified the calculation of angles.

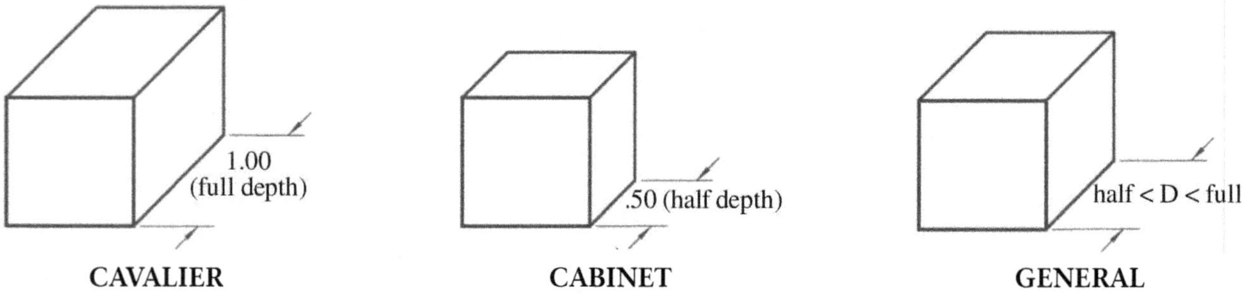

Figure 4.18 Three types of oblique projections

There are three types of oblique pictorials based on how the depth of the object is represented on the drawing (see Figure 4.18). If depth is represented true size, the projection is called cavalier oblique. The advantage of this type of projection is that true sizes can be obtained directly from the drawing. The drawback is that it creates a great amount of distortion due to the absence of foreshortening.

When depth is scaled to half size, a cabinet oblique projection is created. This type of projection reduces the distortion created by the cavalier oblique so the drawing appears more realistic. If depth is shown using any value between half and true size, a general oblique is produced. General oblique pictorials are normally used with specific objects where a cavalier oblique makes the object's depth appear too long and a cabinet oblique makes the object's depth too shallow.

4.1 Standard Practices for Oblique Drawings

Although oblique drawings are easy to draw, general guidelines must be followed to optimize quality and reduce the level of distortion.

Front View As with orthographic projections, the view that gives the most information about the object (shape, size, features, etc) must be selected as the oblique front view. This guideline allows complex features and details to be drawn more easily, accurately, and without distortion. (See Figure 4.19 for an example.)

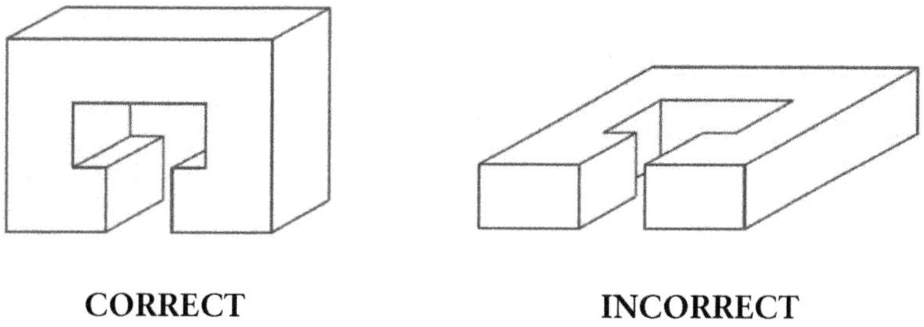

Figure 4.19 Selection of oblique front view

Receding Angle As mentioned earlier, receding angles less than 30° and greater than 60° create severe distortion, especially cavalier projections. Therefore, angles between 30° and 60° should be chosen. (See Figure 4.20 for an example.)

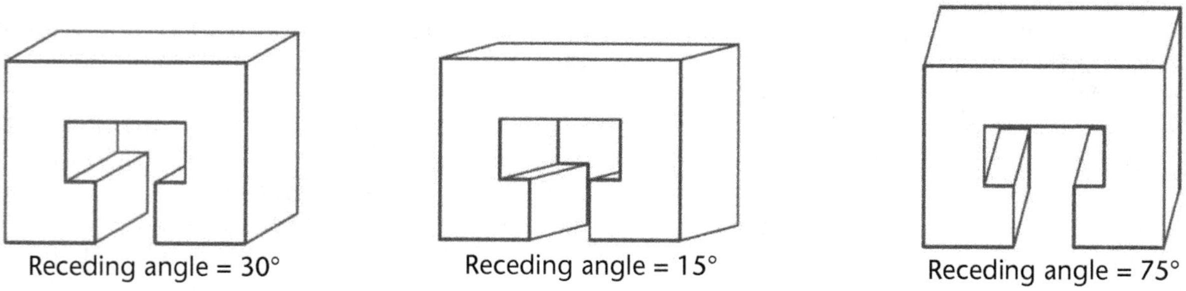

Figure 4.20 Effects of the receding angle in oblique pictorials

Orientation In oblique drawings, irregular and circular features cause more overall distortion than regular faces and shapes. For this reason, such features should be placed parallel to the projection plane whenever possible. (See Figure 4.21 for an example.)

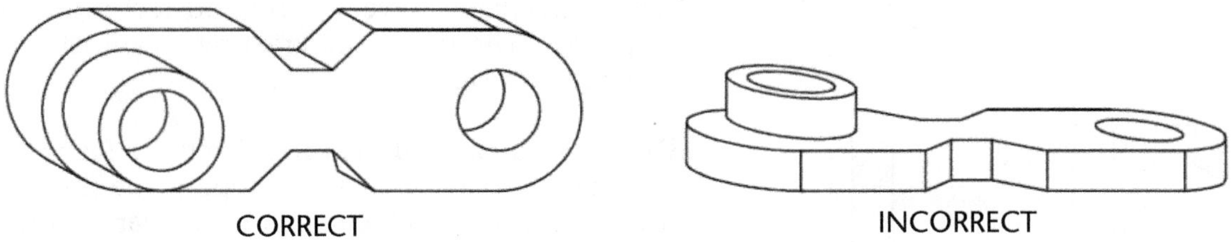

Figure 4.21 Orientation of oblique drawings

Lines The same rules used in axonometric drawings apply to obliques: hidden lines are normally omitted unless the features they represent are required to fully illustrate the object (see Figure 4.22 for an example). Center lines are also omitted unless needed to show concentricity. Tangent edges are typically shown; although in some cases the effect of omitting tangent edges in oblique drawings is not as severe as it is in axonometric drawings.

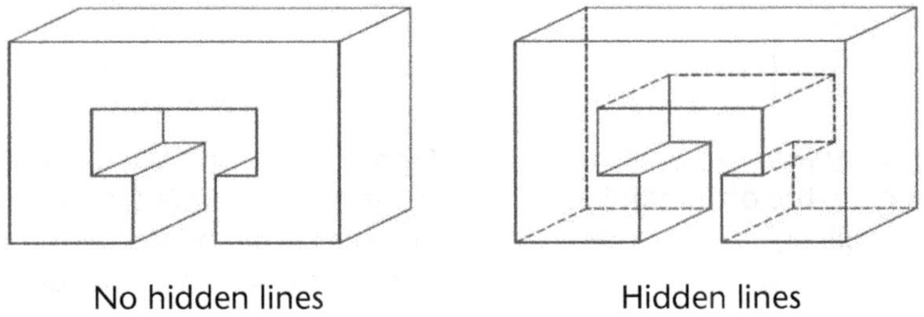

Figure 4.22 Effects of hidden lines in oblique pictorials

Dimensions Since most descriptive features will be placed on the front face of the oblique drawing, most dimensions will also be located on this plane. For those dimensions not in the front plane, the same dimensioning practices used in axonometric drawings apply. (See Figure 4.23.)

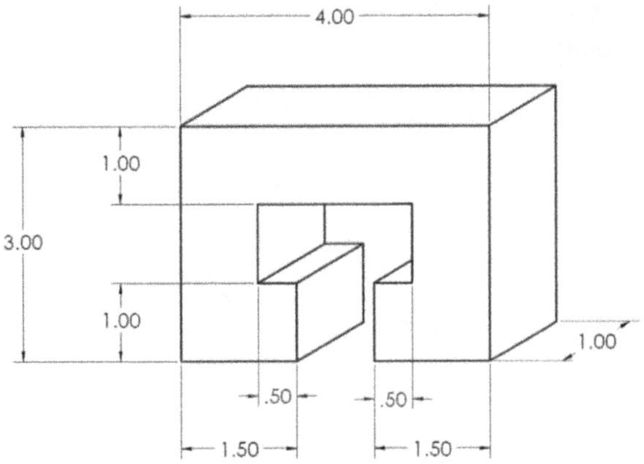

Figure 4.23 Dimensioned oblique pictorial

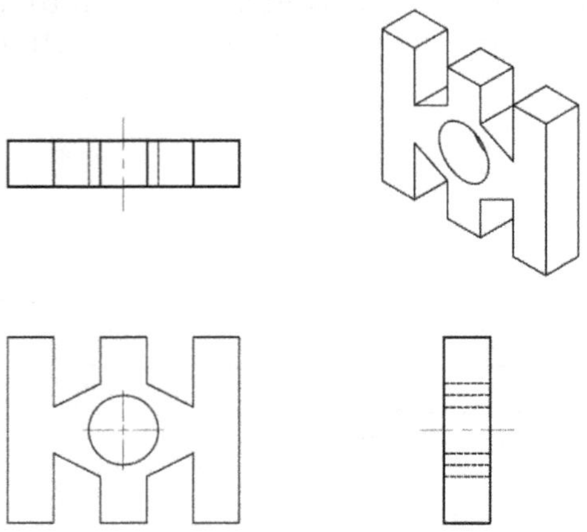

Figure 4.24 Object for oblique pictorial

4.2 Sketching Techniques for Oblique Drawings

Oblique sketching begins with the selection of the front view. Although grids and drafting triangles are readily available, several freehand techniques can be employed. The process of creating a cabinet oblique is illustrated in Figure 4.24.

Circles and Curved Features Irregular features in the oblique front view will always appear true size. However, it is difficult to sketch curves and circular features in the receding planes as they appear distorted. The grid method employed to sketch curved features in isometric drawings can also be used in obliques. To avoid distortion, cylinders drawn as obliques are traditionally shown lying on their side. (See Figure 4.25 for an example.)

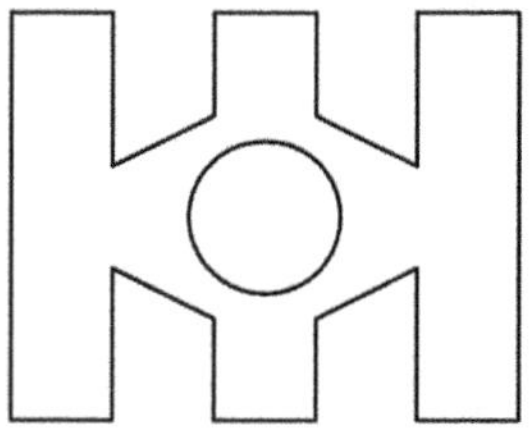

1. Sketch the oblique front view true size using the orthographic views. In this case, the orthographic front view offers the most detail.

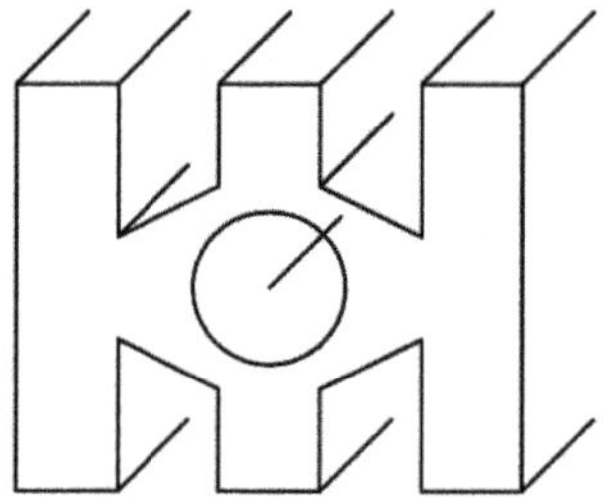

2. From every point, sketch lines at the oblique angle selected (in this case, 45°). In this drawing, the lines are half the size of the lines in the multi-view drawing (cabinet oblique).

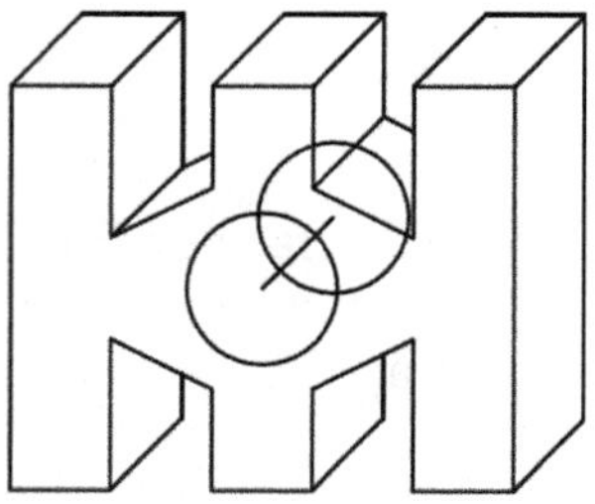

3. Connect the endpoints of the sketched lines to complete the back side of the object. Also, sketch another circle to determine the visibility of the back of the hole.

4. Delete the lines that are no longer needed.

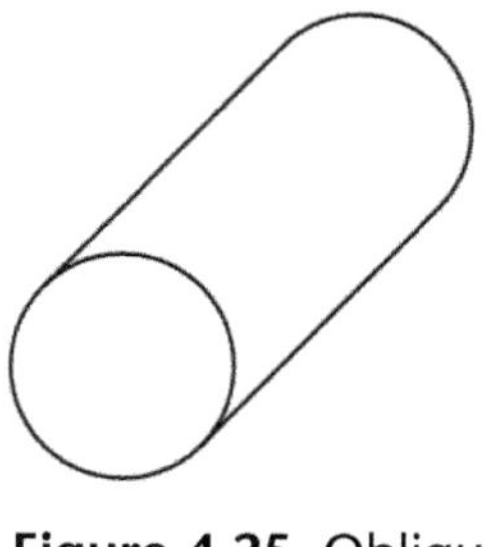

Figure 4.25 Oblique cylinder

5. Isometric vs. Oblique

When representing simple objects with straight edges and flat surfaces, isometric and oblique pictorials look equally good assuming all the general sketching guidelines discussed above are followed. However, oblique pictorials create a great amount of distortion when representing objects with curved and circular features. See Figure 4.26 for an example. In this case, isometric pictorials are preferred.

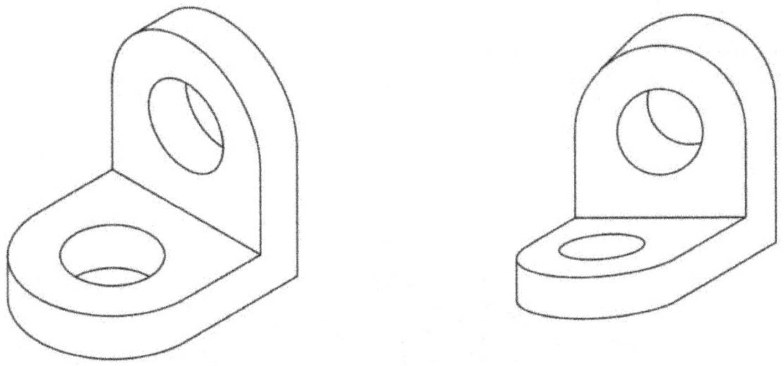

Figure 4.26 Isometric pictorial vs. oblique pictorial

6. Perspective Projections

A perspective projection results when the observer is not located at infinity, but at a finite distance from the projection plane. This projection causes several visual effects on the drawing projected on the picture plane. First, parallel lines in the object will not remain parallel on the drawing. In fact, all lines that are not in a plane parallel to the picture plane will converge toward a single point (called vanishing point). Secondly, objects that are closer to the viewer will appear larger than those further away.

Perspective projections are the most realistic projections. The way objects are represented in perspective projections is the way eyes perceive the world. However, perspective drawings contain the most visual errors of any standard projection method; lines or angles with the same dimension on the object will have different dimensions on the drawing. Perspectives are difficult to construct, especially circular features, and plotting distances and calculating true dimensions in perspective are time consuming tasks. The use of perspectives in engineering is commonly limited to drawings with illustrative and presentation purposes only, where accuracy is not required, and approximations are frequently used in order to simplify the process.

There are different factors involved in producing a perspective drawing. These factors take into account the position of the observer (called station point) relative to the object to be drawn, the position of the ground relative to the horizon, and the position of the object relative to the projection plane (See Figure 4.27). These factors, along with the orientation of the object, will determine how many vanishing points will be used in the drawing. In fact, perspective projections are classified based on the number of vanishing points.

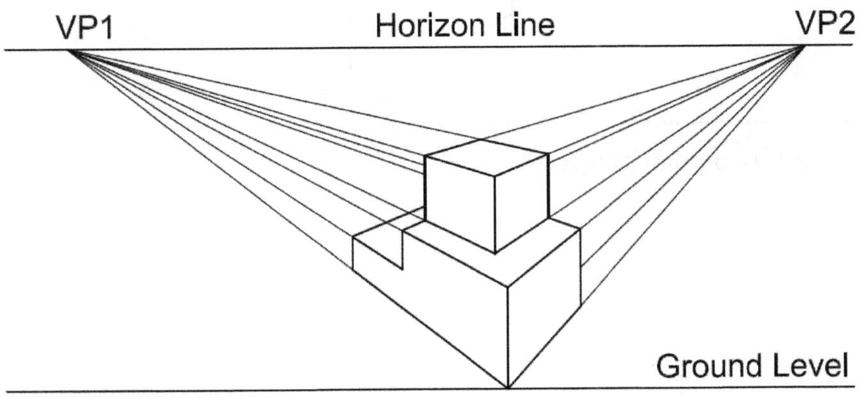

Figure 4.27 Example of perspective projection

6.1 One-Point Perspective

One-Point perspectives show one face of the object in true size (actually, all faces that are parallel to the projection plane). Lines representing depth converge toward one single vanishing point. See Figure 4.28 for an example.

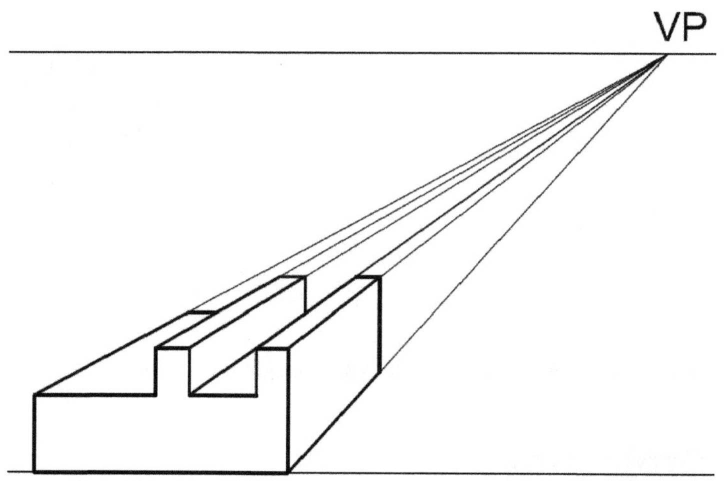

Figure 4.28 One-point perspective

6.2 Two-Point Perspective

Two-Point perspectives show the object positioned at a certain angle relative to the projection plane. Lines representing depth will converge to one of the vanishing points located on the horizon line. In addition, all vertical edges of the object are parallel to the projection plane. See Figure 4.29 for an example.

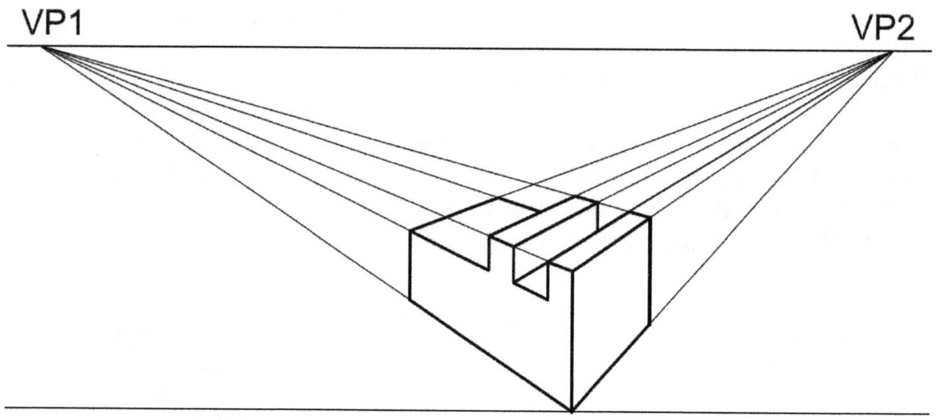

Figure 4.29 Two-point perspective

6.3 Three-Point Perspective

Three-Point perspectives show the object positioned at a certain angle relative to the projection plane, making lines that show depth converge to one of the vanishing points, as with two-point perspectives. In this case, all vertical edges of the object will converge to a third vanishing point located above or below the object. See Figure 4.30 for an example.

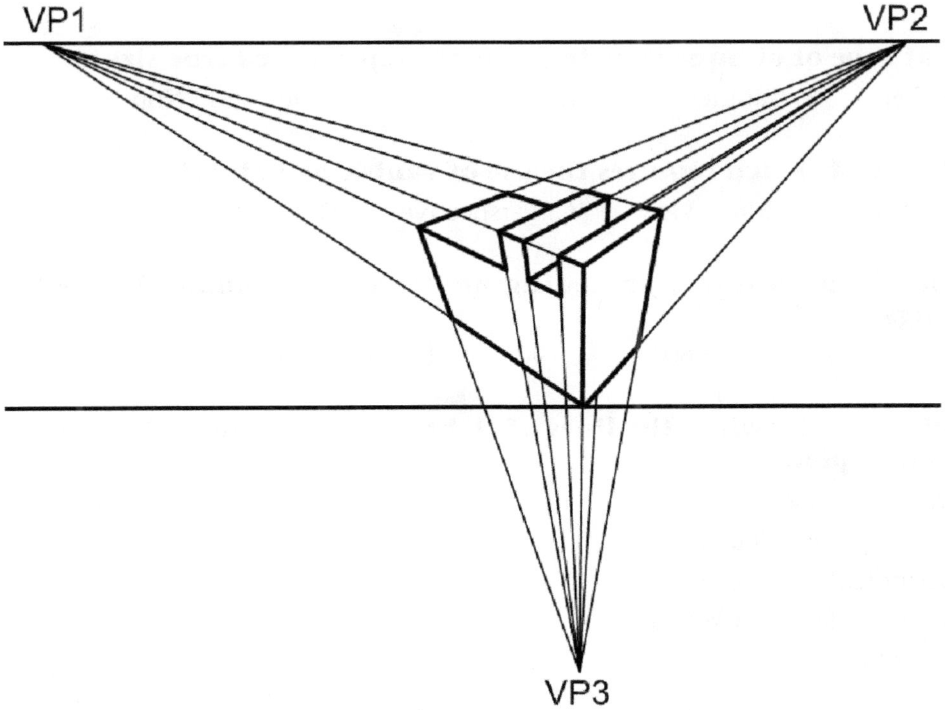

Figure 4.30 Three-point perspective

Practice Test

1. **Which type of sketch is not a pictorial?**
   A) Trimetric B) Cavalier oblique C) Multi-view D) One-point perspective E) Axonometric

2. **Which of the following is not a parallel projection?**
   A) Cabinet oblique B) Multi-view C) Dimetric D) Isometric E) Two-point perspective

3. **In what type of oblique drawing is depth represented true size?**
   A) Cavalier B) Cabinet C) General D) Trimetric E) None of the above

4. **What type of sketch requires the use of vanishing points?**
   A) Isometric B) Multi-view C) Perspective D) Oblique E) None of the above

5. **Which of the following receding angles is not commonly used in oblique drawings?**
   A) 30° B) 45° C) 60° D) 90° E) None of the above

6. **In a dimetric drawing, the lines of sight travel _____ to each other and _____ to the picture plane.**
   A) Parallel, parallel
   B) Parallel, perpendicular
   C) Perpendicular, parallel
   D) Perpendicular, perpendicular
   E) None of the above

7. **Circles in a real object will appear as _______ in an isometric drawing.**
   A) Circles B) Ellipses C) Depends on the orientation D) Arcs E) None of the above

8. **What is the value of the isometric axis angle?**
   A) 60° B) 90° C) 120° D) 180° E) None of the above

9. **In isometric drawings, lines that are not parallel to any of the isometric axes will appear _____.**
   A) Foreshortened B) True size C) Hidden D) Stretched E) A and D are correct

10. **Which is the most realistic type of pictorial?**
    A) Isometric B) Perspective C) Oblique D) Multi-view E) None of the above

Problems

For the following objects, use your engineering paper to construct appropriate pictorial views as indicated by your instructor. Ensure that you choose the correct front view.

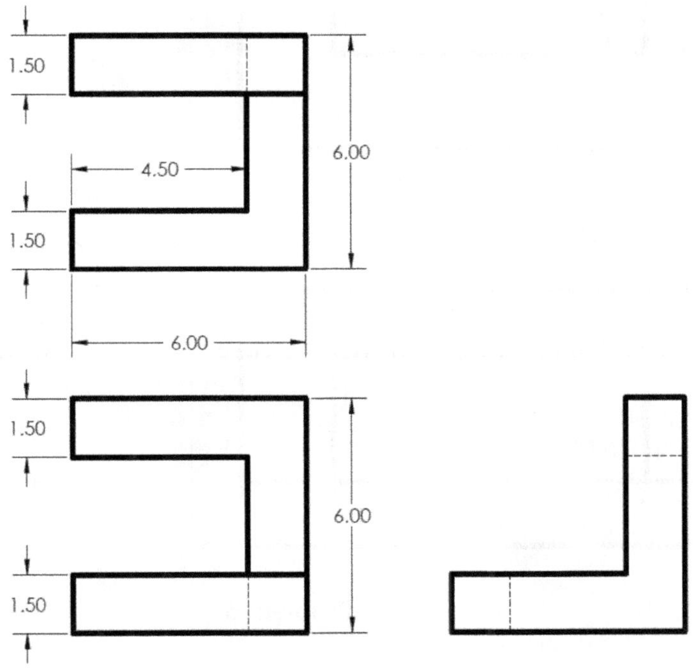

Problem 1

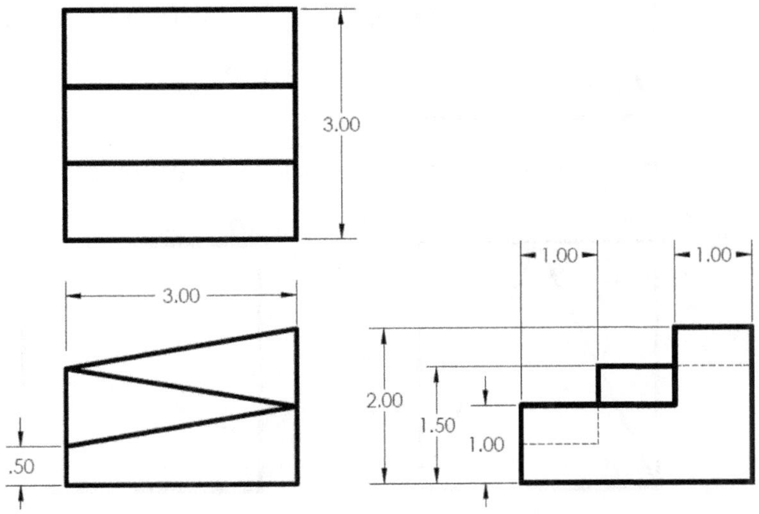

Problem 2

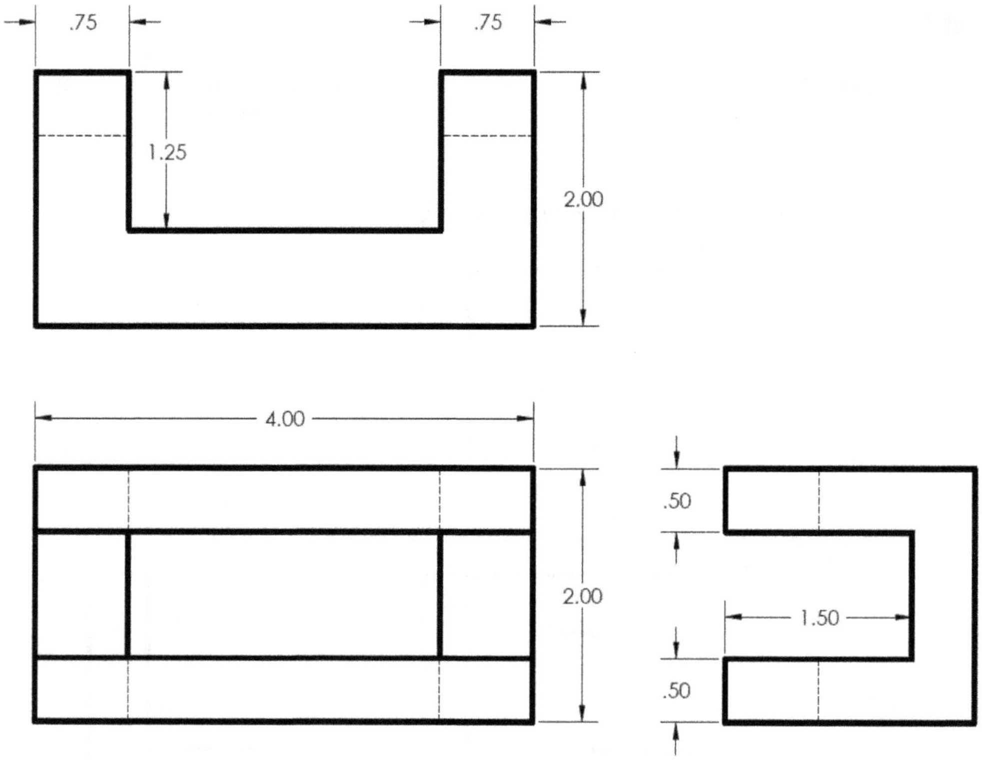

Problem 3

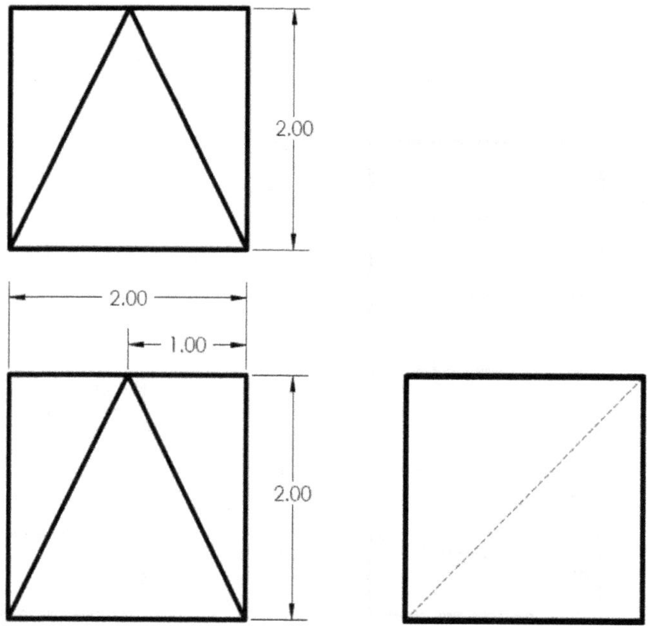

Problem 4

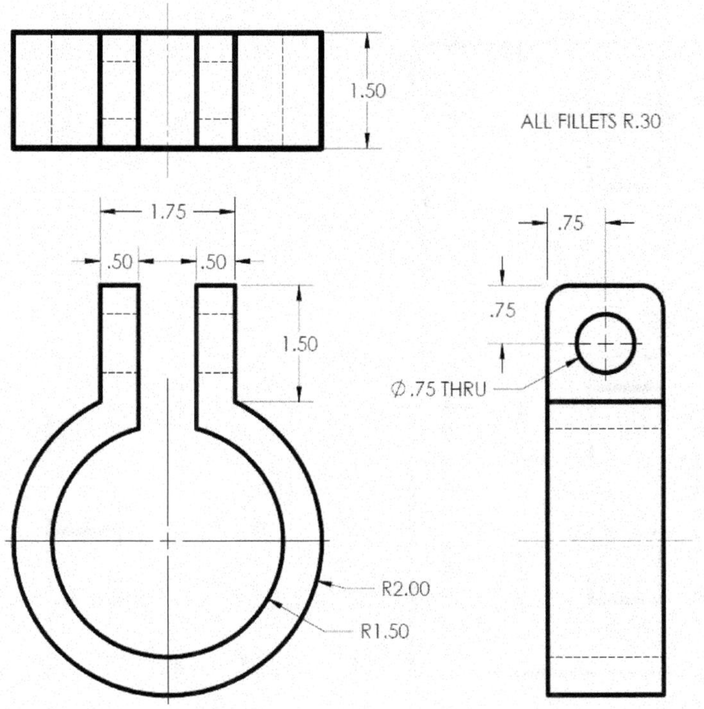

Problem 5

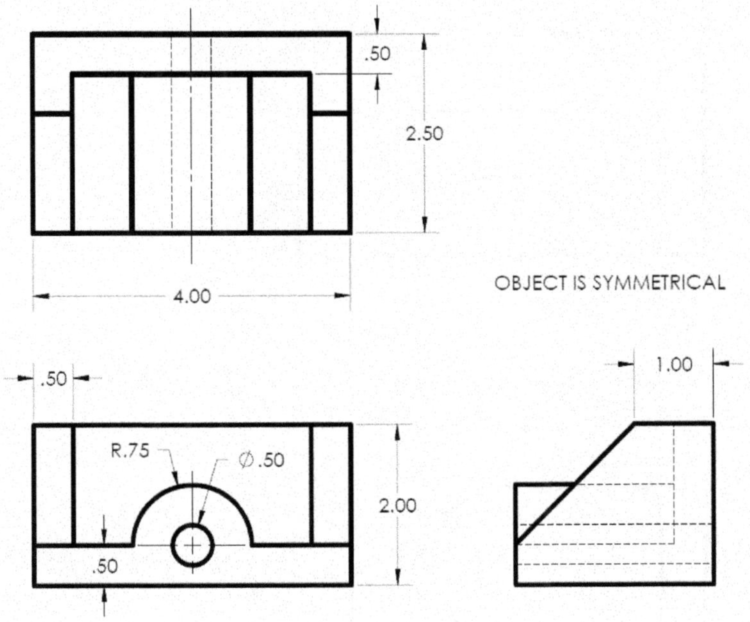

Problem 6

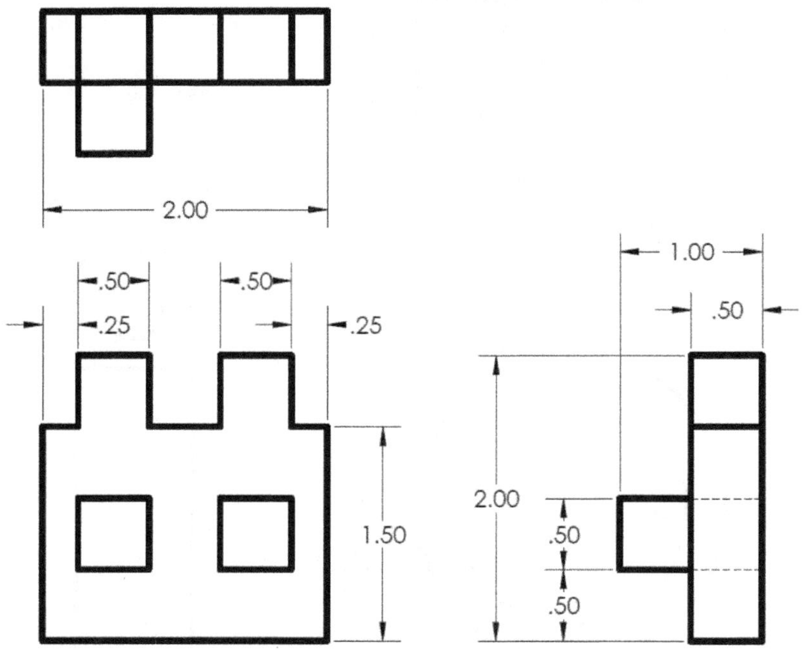

Problem 7

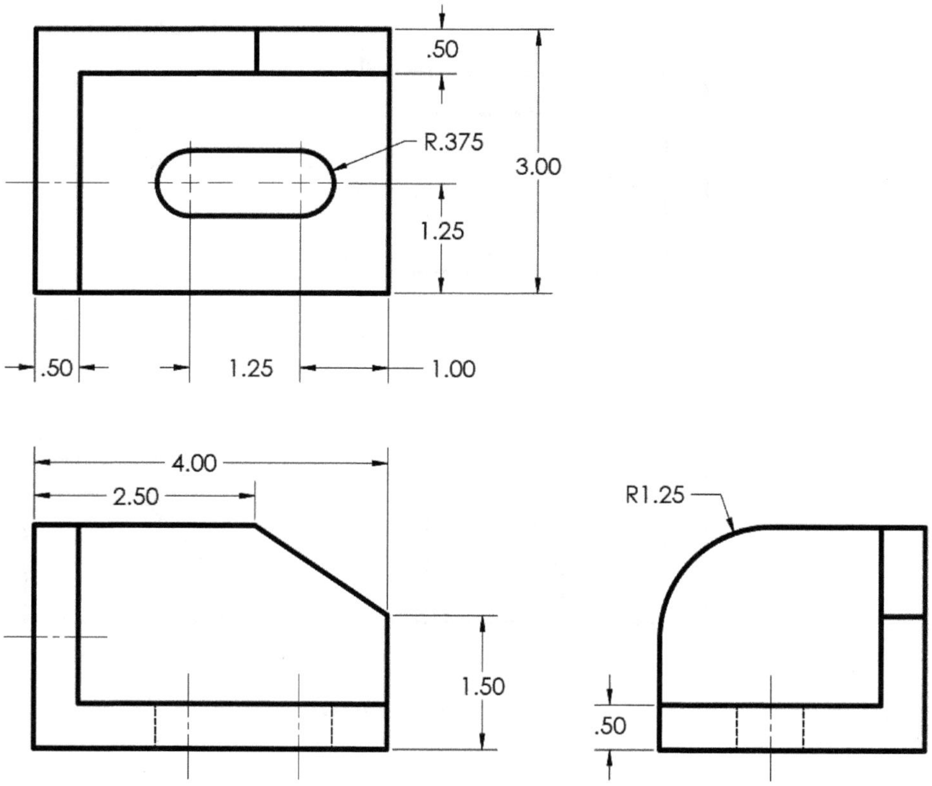

Problem 8—Corner Support

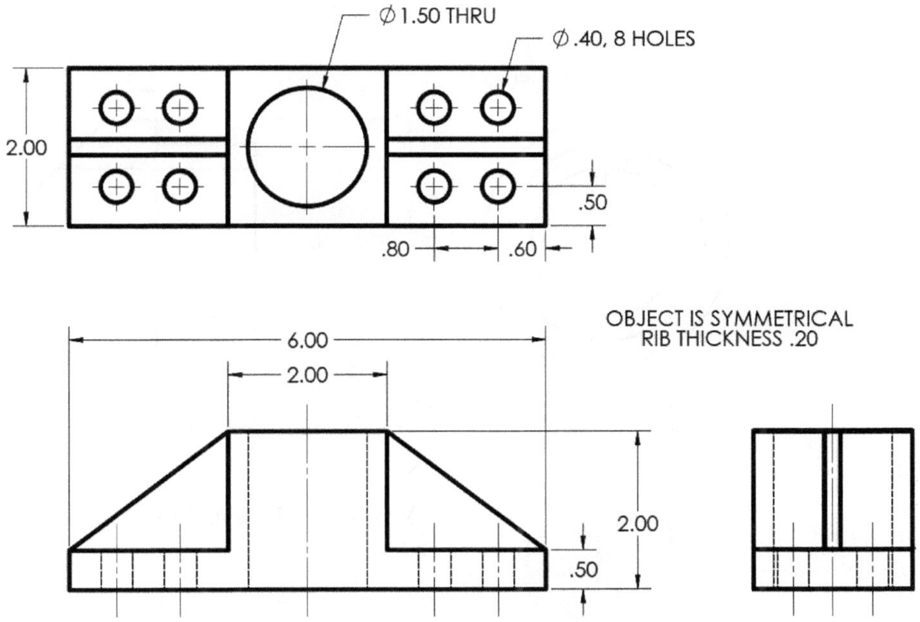

Problem 9—Pole Support

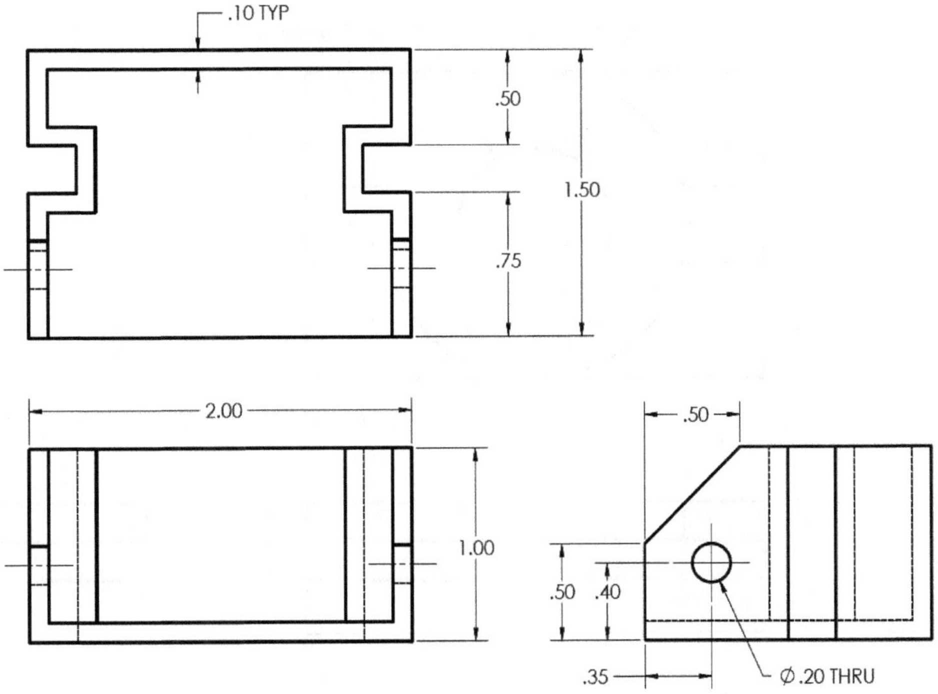

Problem 10—Platform

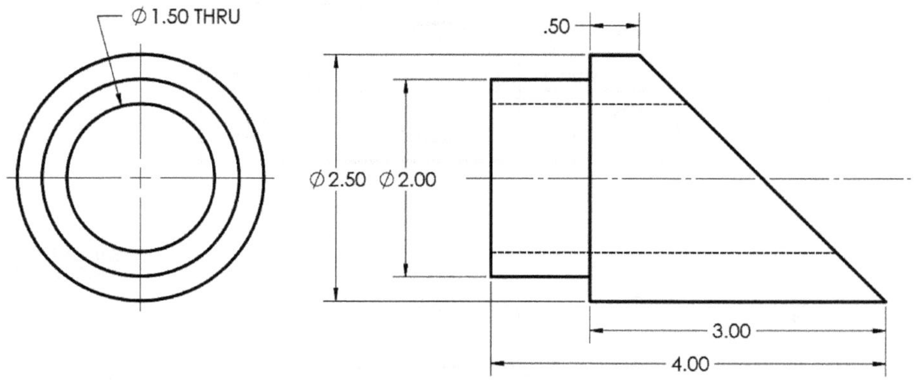

Problem 11—Wedge

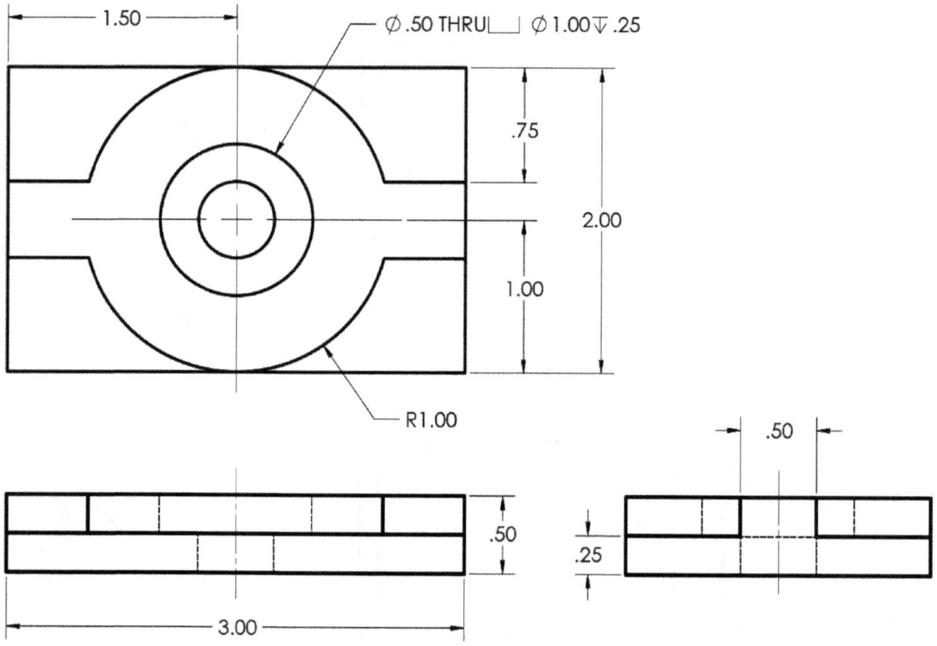

Problem 12—Support

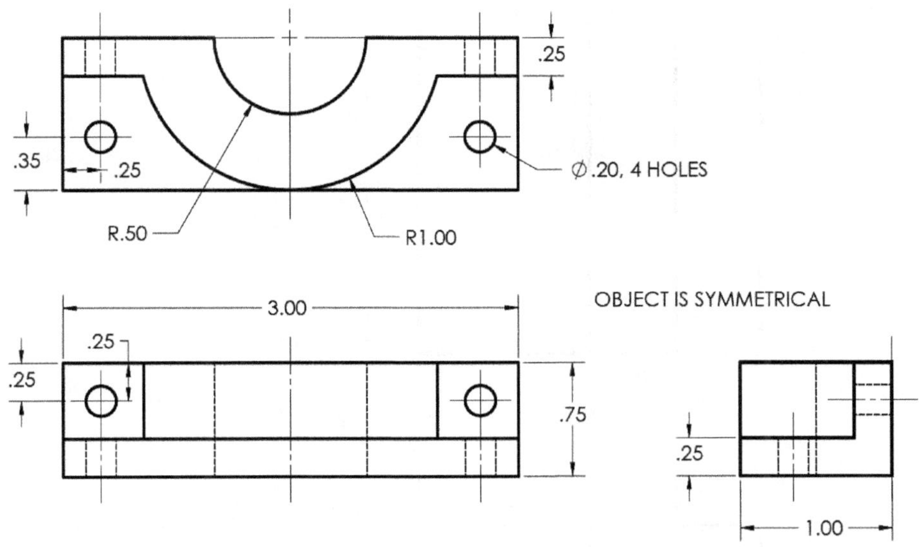

Problem 13—Bracket

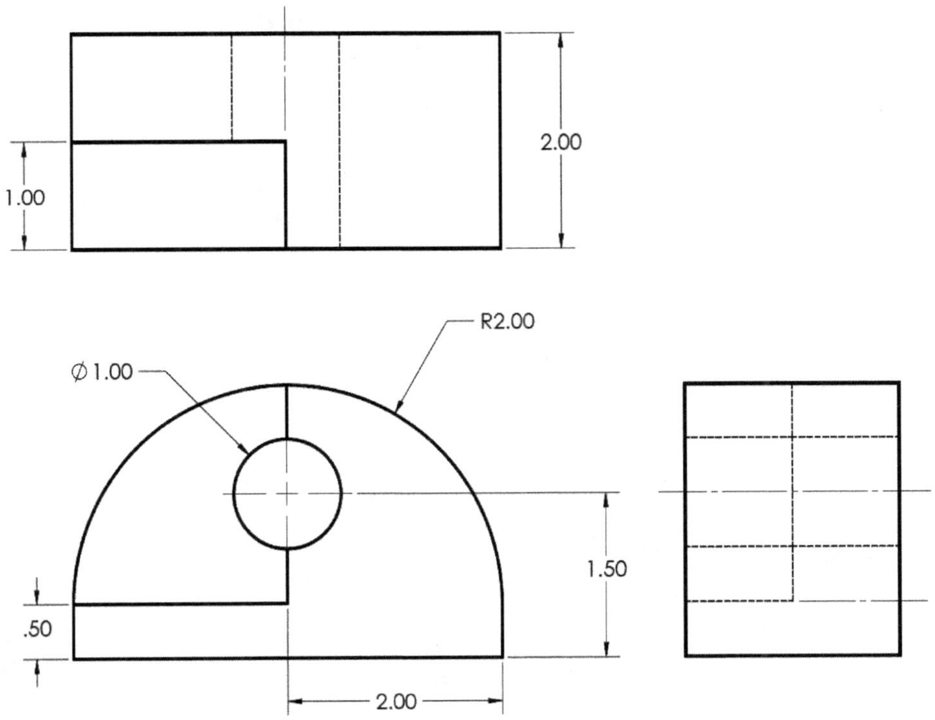

Problem 14

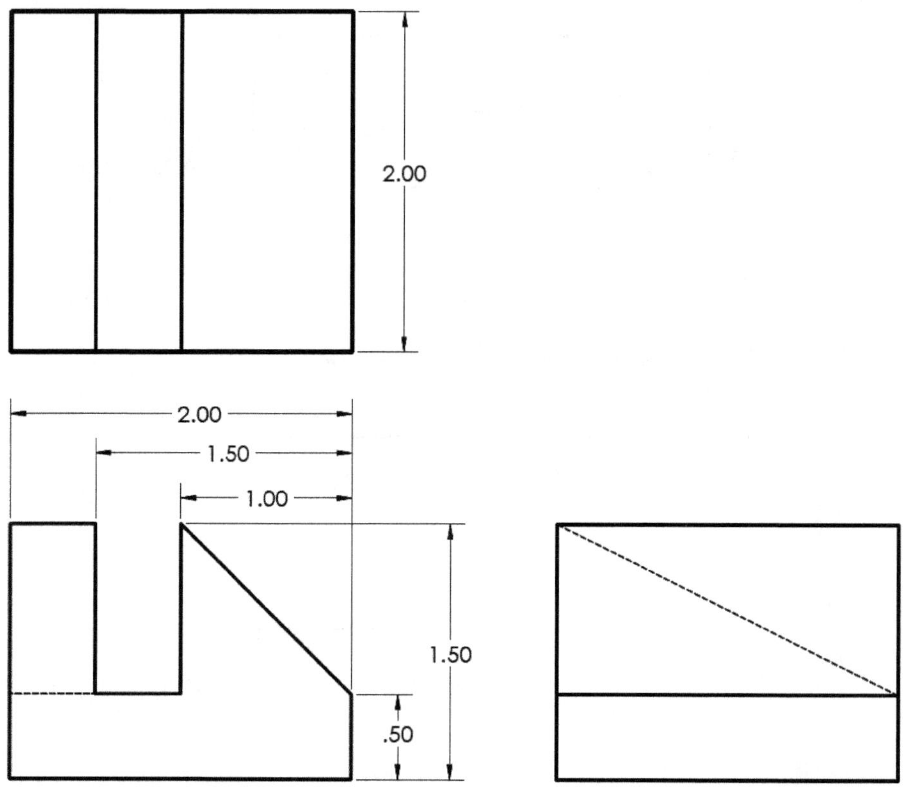

Problem 15

5
CHAPTER

Special Views

1. Introduction

Orthographic projection provides an accurate way to represent 3D objects on paper. The main advantage of using orthographic projection is that objects are represented true size and without distortion. Although the six standard orthographic views provide geometric information necessary to represent most 3D objects, there are situations where this configuration alone may not be the most appropriate one. Examples of these situations include objects with slanted surfaces or parts with intricate internal detail. In order to represent such features in a simple and efficient manner, and to facilitate the visualization process, special views are constructed. In some cases, these special views replace certain conventional orthographic views, and in other cases they are provided as additional views to the standard projections. Special views are classified as auxiliary and sectional views.

2. Auxiliary Views

Auxiliary views are used to represent the true size and shape of objects with sloping surfaces. As discussed earlier, any surface that is not parallel to any of the three principal orthographic planes (horizontal, frontal, and profile) will suffer the effects of foreshortening, and therefore will not be shown true size using the conventional views. The object shown in Figure 5.1 illustrates a situation where a hole is bored through a sloping surface. In all views, the sloping surface appears foreshortened and the hole appears elliptical even though it is circular. In this case, an auxiliary view showing the true shape and size of the sloping surface and the hole would be more practical.

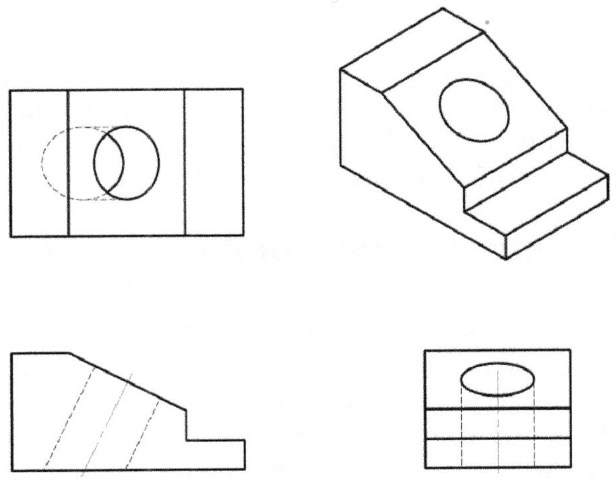

Figure 5.1 Conventional orthographic views without auxiliary view

An auxiliary view is essentially any view of an object other than one of the primary orthographic views or a pictorial that provides geometric information about the true size and shape of a slanted surface of an object. This information is particularly useful for manufacturing and production purposes.

Auxiliary views are drawn normal to a sloping surface, so that the feature can be seen true size and shape. Using the "glass box" concept from previous chapters, an auxiliary view is

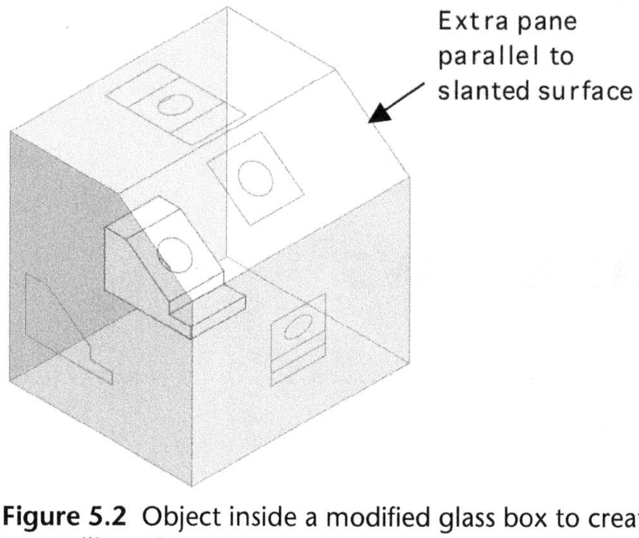

Figure 5.2 Object inside a modified glass box to create an auxiliary view

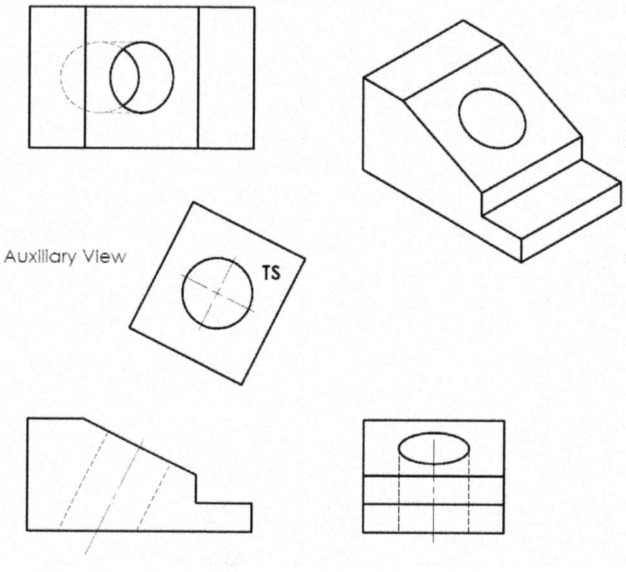

Figure 5.3 Standard orthographic views with auxiliary view showing true size of hole

created by adding an extra pane to the glass box so that the new pane is parallel to the slanted surface of the object. (See Figure 5.2.) The resultant multi-view drawing with the auxiliary view after unfolding the modified glass box is shown in Figure 5.3. Notice how the auxiliary view is created by looking perpendicular to the slanted surface.

2.1 Considerations for Auxiliary Views

There are two types of primary auxiliary views: full and partial. Full auxiliary views show all surfaces of the object projected onto the auxiliary plane (see Figure 5.4). Partial auxiliary views show only the visible lines that correspond to the inclined surface. All other visible lines associated with the view are not drawn.

Full auxiliary views are uncommon. All surfaces other than the inclined one do not provide any useful information in the auxiliary view. What's more, they will not appear true size in the auxiliary view and can create confusion in the viewer. For this reason, partial auxiliary views are usually preferred.

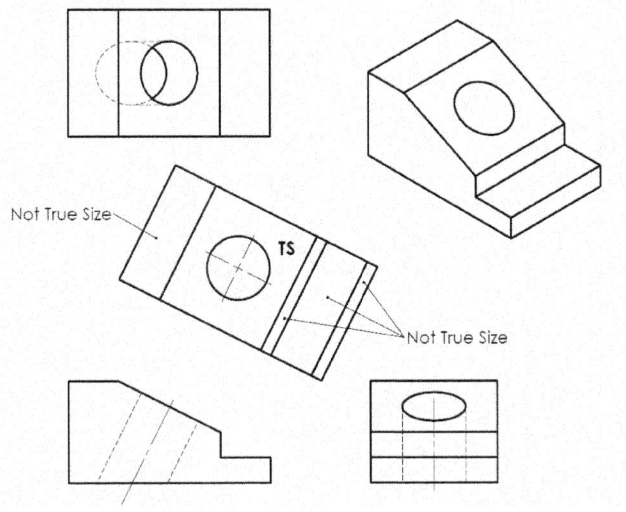

Figure 5.4 Example of a full auxiliary view

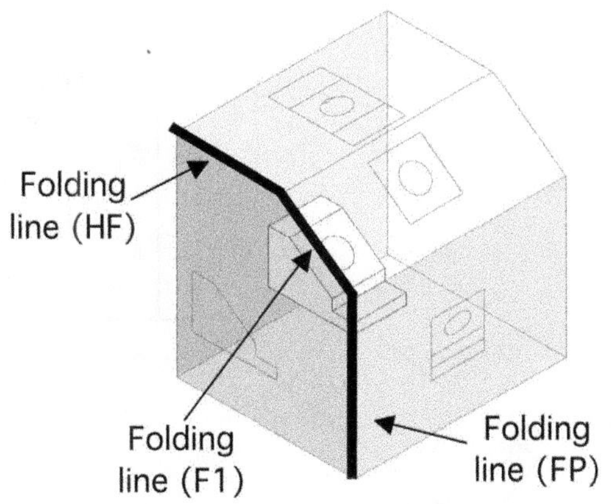

Figure 5.5 Glass box showing folding lines

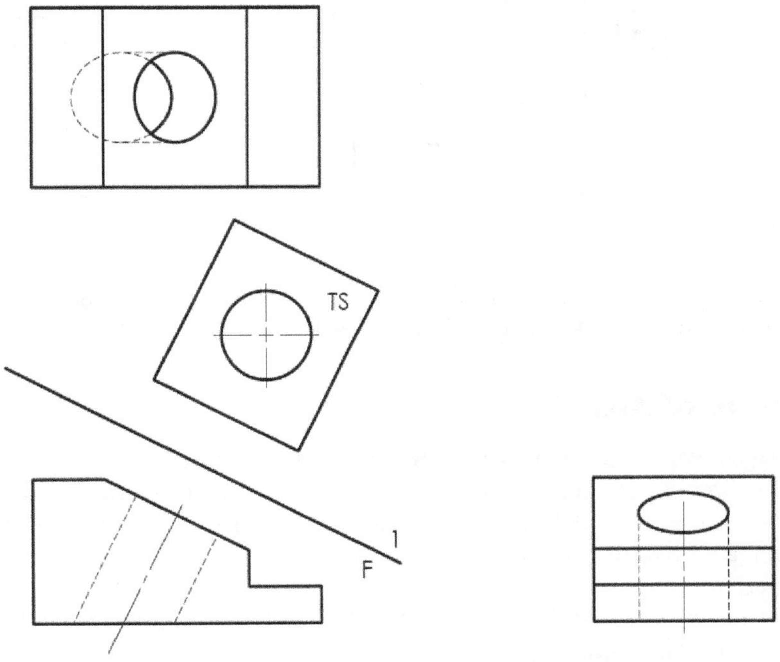

Figure 5.6 Orthographic views showing folding line for auxiliary view

For drawings with auxiliary views, it is common practice to include and label the folding lines used to create the views. A folding line is an edge of the glass box along which a pane is unfolded. The folding line utilized to unfold the auxiliary pane enables one to transfer measurements when constructing the auxiliary view. Folding lines for standard orthographic planes are conventionally labeled with the initials of the two planes they connect. For example, in Figure 5.5, Line HF represents the folding line between the horizontal and the frontal planes (Front and Top views). Folding line FP represents the folding line between the frontal and profile planes (Front and Side views). Folding lines for auxiliary views are represented with the capital letter H, F, or P (Horizontal, Frontal, or Profile) followed by a number representing in sequence the auxiliary view constructed. This concept is illustrated in Figures 5.5 and 5.6. Labeling these lines correctly is important, especially for drawings

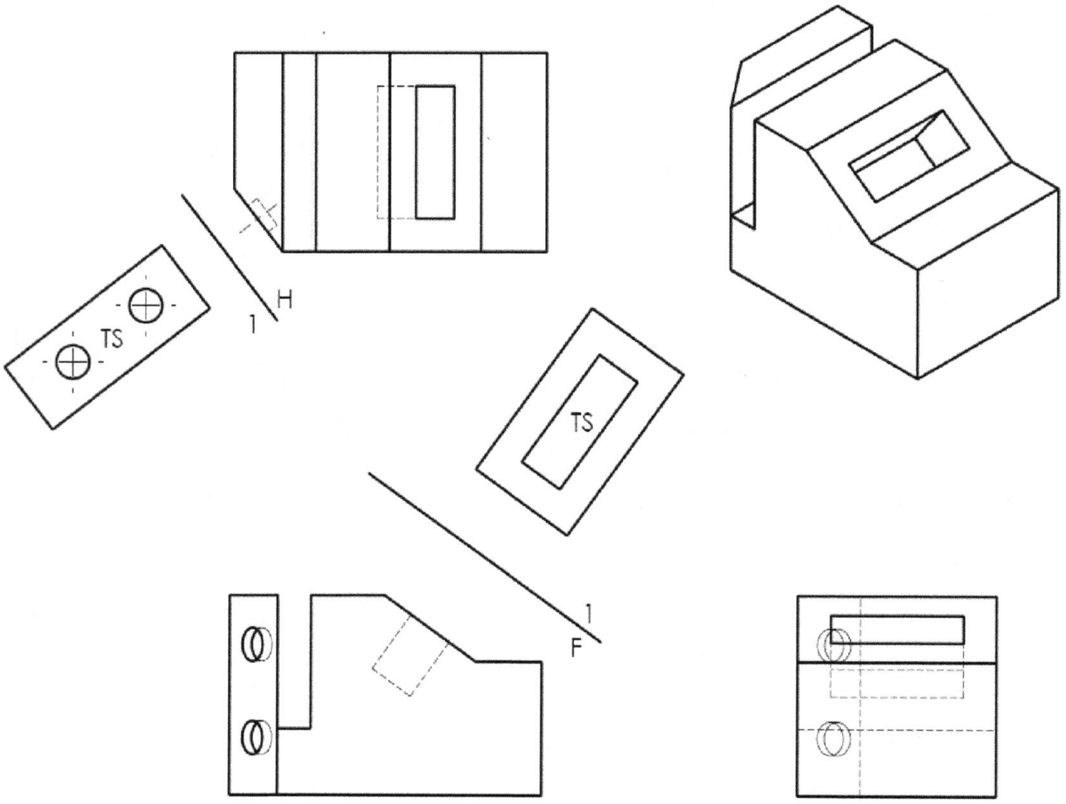

Figure 5.7 Drawing with multiple auxiliary views

with multiple auxiliary views (see Figure 5.7). Finally, when the drawing is finished, the letters "TS" must be placed on the auxiliary view to indicate that the view is true size.

2.2 Construction of Auxiliary Views

Creating auxiliary views by hand is a somewhat time-consuming task. In modern 3D CAD packages, however, the process of creating auxiliary views has been drastically simplified. In fact, when an entire 3D model is provided to the manufacturer, auxiliary views may not even be necessary. The manufacturer can rotate and manipulate the 3D object at will to obtain the true sizes and dimensions needed.

The construction process (if done by hand) is illustrated using the orthographic view of the model represented in Figure 5.8.

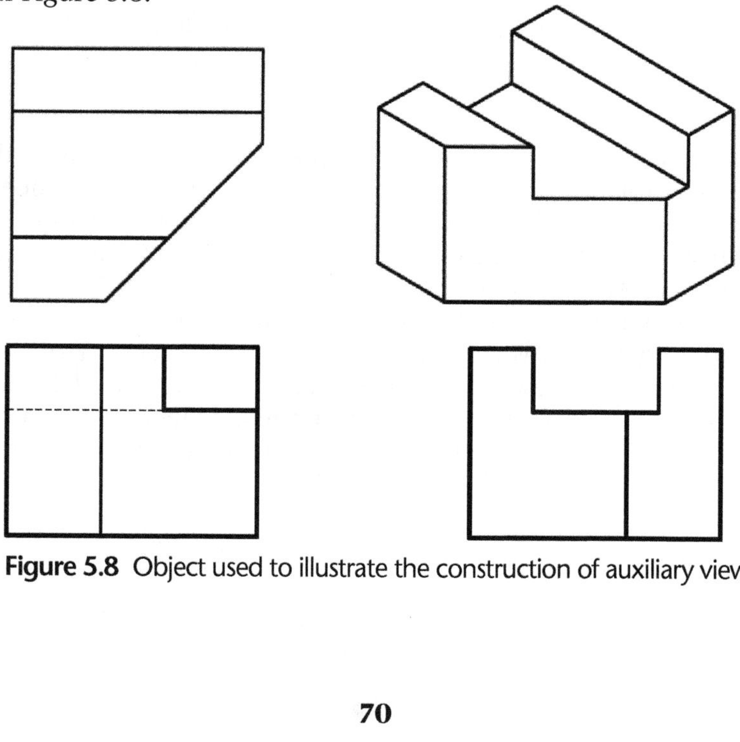

Figure 5.8 Object used to illustrate the construction of auxiliary views

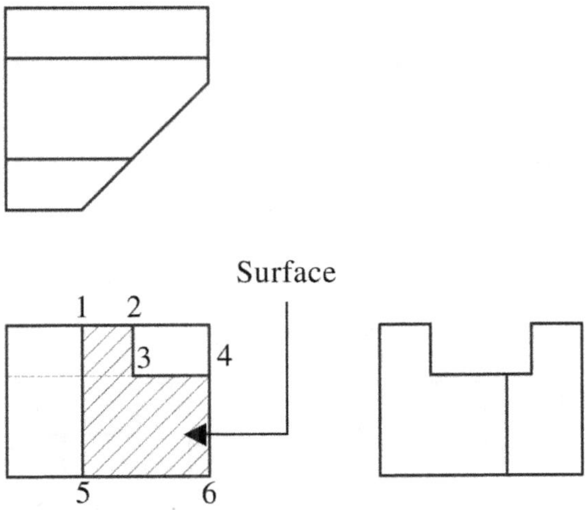

1. Select the view where the misshapen surface appears as a surface and not as an edge. Number the corners of this surface. In this case, either the front or right side view can be used.

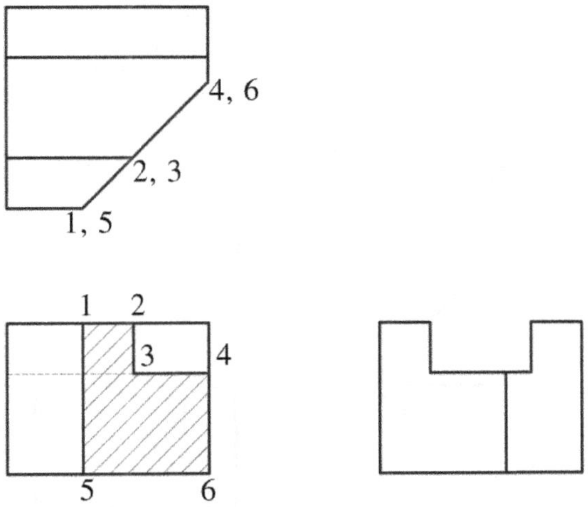

2. Transfer the numbers to the view where the surface appears as an edge.

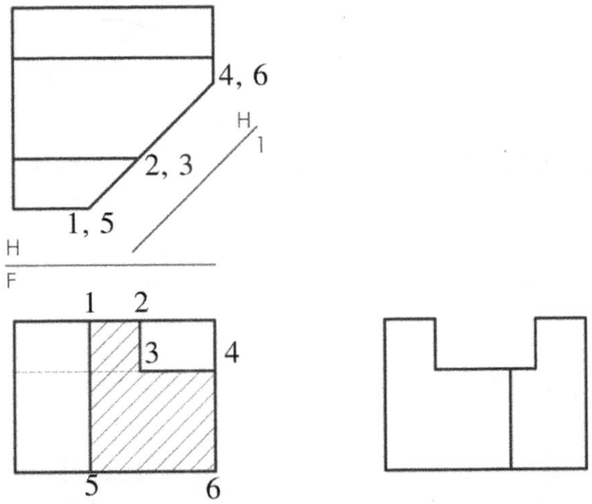

3. Define folding lines.

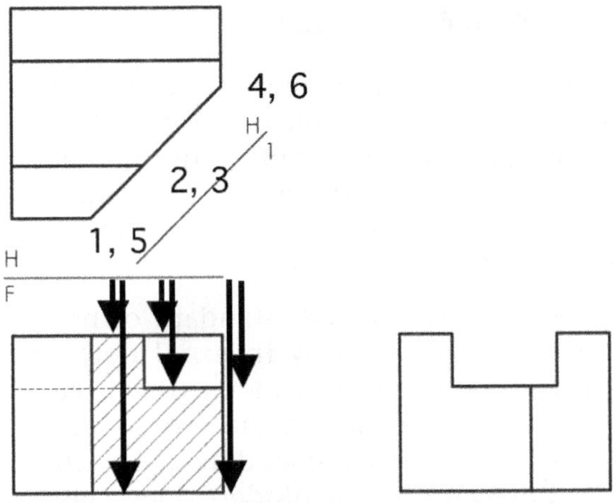

4. Measure the distance between the orthographic folding line (HF in this case) and each numbered corner of the surface.

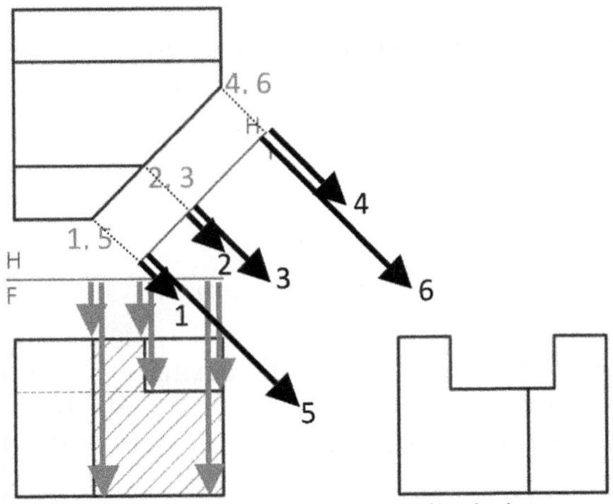

5. Transfer the distances perpendicularly to the auxiliary folding line.

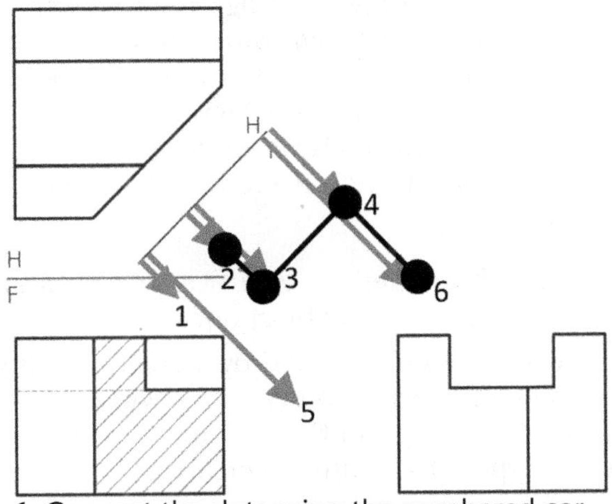

6. Connect the dots using the numbered corners from step 1 as a reference (for example, corner 4 is connected to corners 3 and 6, corner 3 is connected to corners 4 and 2, etc).

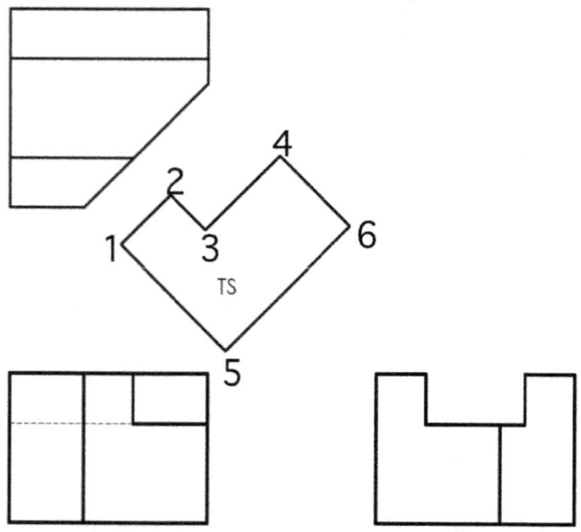

7. Finish connecting the dots and add "TS" for true size on the true surface.

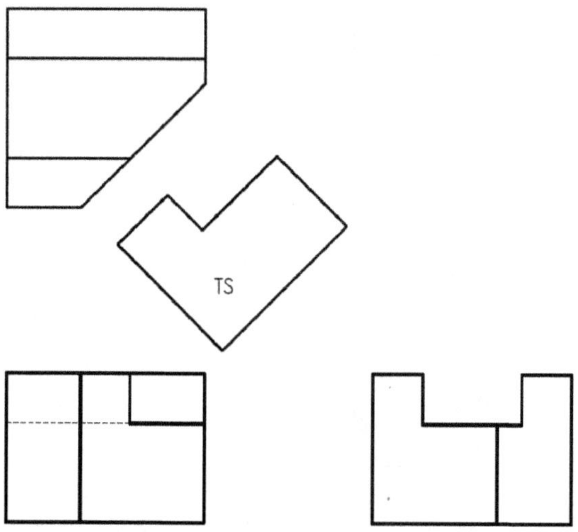

8. Complete drawing.

3. Sectional Views

Sectioned orthographic views are a special type of view used to facilitate the visualization of complex parts with sufficient internal detail where standard orthographic views are confusing or not clear. Section views serve three purposes:

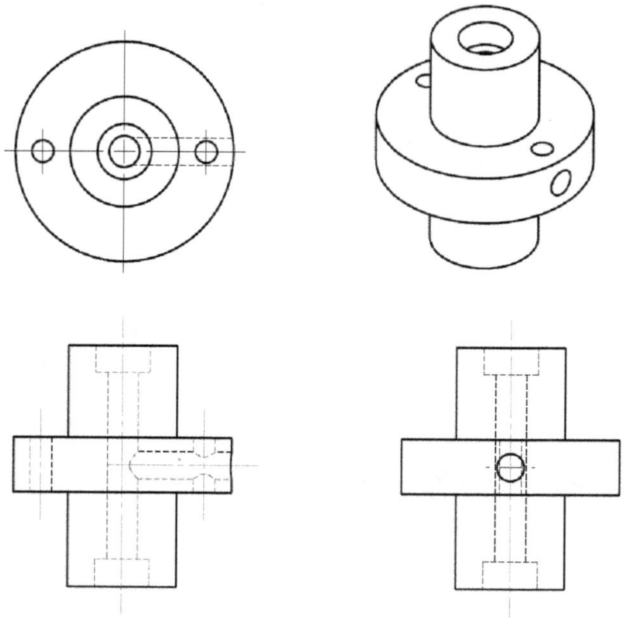

Figure 5.9 Use of hidden lines to show internal detail in standard orthographic views

- Show internal detail. Standard orthographic views show internal detail through the use of hidden lines (see Figure 5.9). Although the use of hidden lines is acceptable, drawings with a large number of hidden lines may become very confusing. Section views are used to minimize this problem and better represent internal features of objects.
- Replace complex orthographic views. In general, orthographic views with large numbers of hidden lines can be replaced with section views. In some cases, section views may also be a series of additional views depending on the type.
- Describe materials. By using standard and conventional practices, section views can show what material an object or an assembly is made of.

Section views are created by visually cutting away part of an object so the desired internal feature can be more easily observed. Obviously, there are certain rules that must be followed when performing this process so the drawing is clear and unambiguous. Depending on how the cut is done, different types of sections are created.

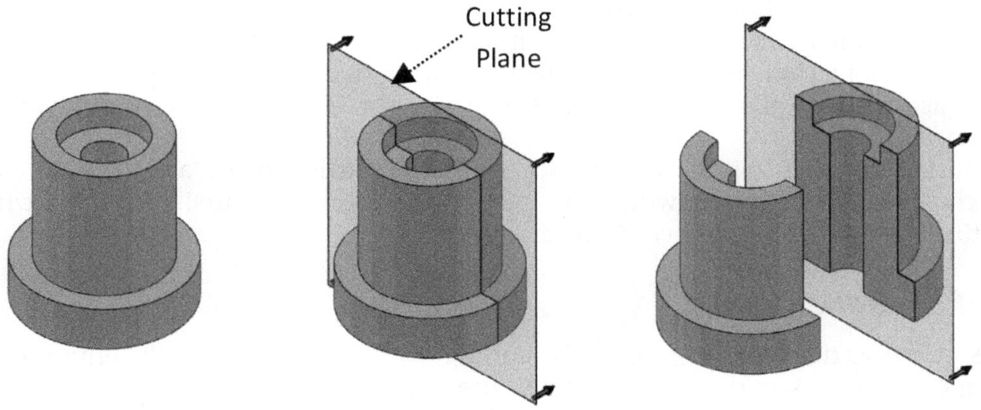

Figure 5.10 Cutting plane to create a sectional view

3.1 Cutting Plane

The first concept to understand when studying sections is the cutting plane. The cutting plane is essentially a flat surface used to perform a cut on the object. Think of the cutting plane as an imaginary blade utilized to slice an object (see Figure 5.10). The cutting plane defines what part of the object is shown and drawn, and what part is omitted. Obviously, the cutting plane can be placed in an infinite number of positions, thus creating different types of cuts, but only a few of those positions are useful in engineering. Defining the cutting plane should be the first step when creating sectional views.

The cutting plane is normally parallel to one of the principal projection planes (horizontal, frontal, and profile) and the line of sight is represented by arrowheads perpendicular to the cutting plane. In multi-view drawings, the cutting plane is shown in a view where it appears as an edge (see Figure 5.11). The cutting plane line is similar in appearance to a center line, except with two short

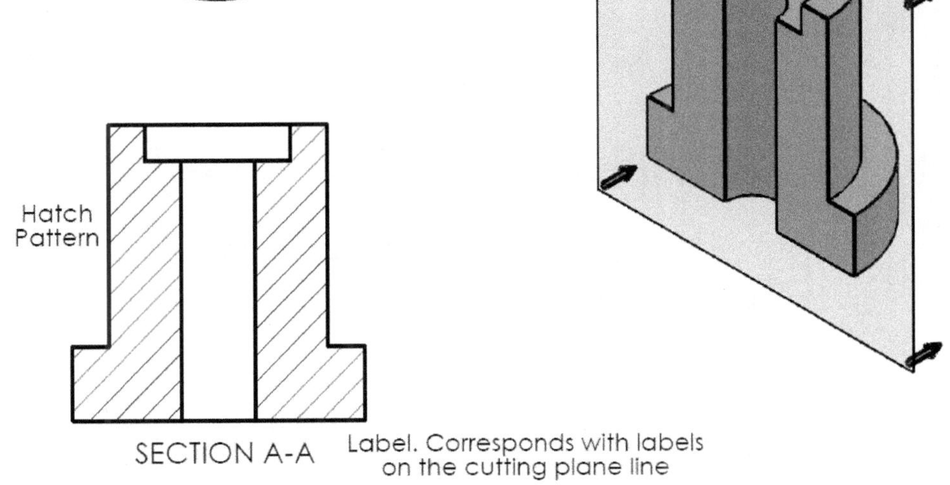

Figure 5.11 Top view with cutting plane line and sectional view replacing standard front view

dashes instead of one. The thickness of a cutting plane line is generally the same as or slightly greater than the visible object line. If the cutting plane line coincides with another line in the drawing, the cutting plane line takes precedence. It is important to label cutting plane lines, especially if more than one sectional view is used in the drawing. Cutting plane lines are typically labeled using two letters, one at each end of the line. The sectional view is then labeled accordingly. See Figure 5.11 for an example.

The solid surfaces of the object that intersect with the cutting plane are represented with a shaded or hatched pattern in the drawing. This pattern indicates the actual surface that has been cut by the cutting plane. Hatch patterns will be discussed later.

3.2 Types of Sectional Views

Full Sections A full section is created when the cutting plane passes through the entire object. It is the most common type of section view. Basically, the object is cut in two pieces, with the piece "behind" the cutting plane shown, and the piece "in front" of the plane omitted. See Figures 5.12 and 5.13 for an example. When a full section is created, hidden lines in the object that become visible in the remaining part after the cut is performed are shown. Lines that are still hidden in the remaining part after the cut is performed are omitted. This concept is illustrated in Figure 5.14.

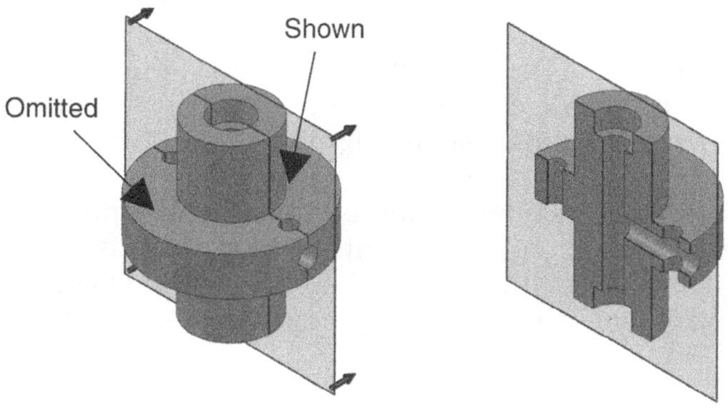

Figure 5.12 Construction of a full section

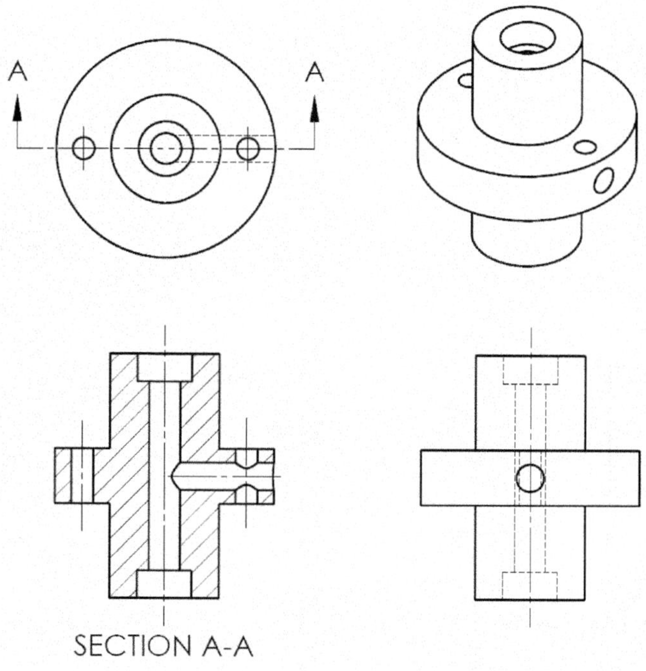

Figure 5.13 Orthographic views with full section

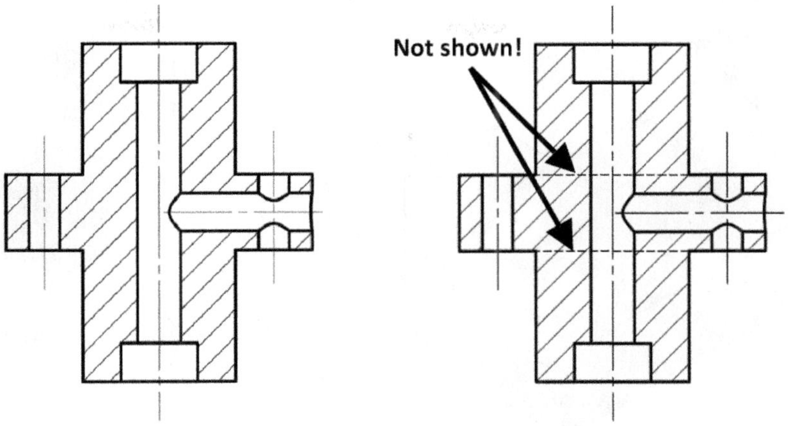

CORRECT
Hidden lines are not shown in sectional views.

INCORRECT
According to orthographic standards this sectional view would be the correct one, hidden lines are conventionally omitted in full sections for clarity.

Figure 5.14 Hidden lines in full sections

Offset Sections An offset section is a variation of a full section in which the cutting plane is "bent" 90° to make it pass through certain features of the object. This type of section is useful for asymmetric objects. In these cases, full sections would not reveal all the internal detail of the object and many features would still be obscured (see Figure 5.15). By "bending" the cutting plane, a more efficient cut can be made that reveals all features of the object (see Figures 5.16 and 5.17). Lines that represent edges along which the cutting plane is bent are not shown in the sectional view (see Figure 5.18).

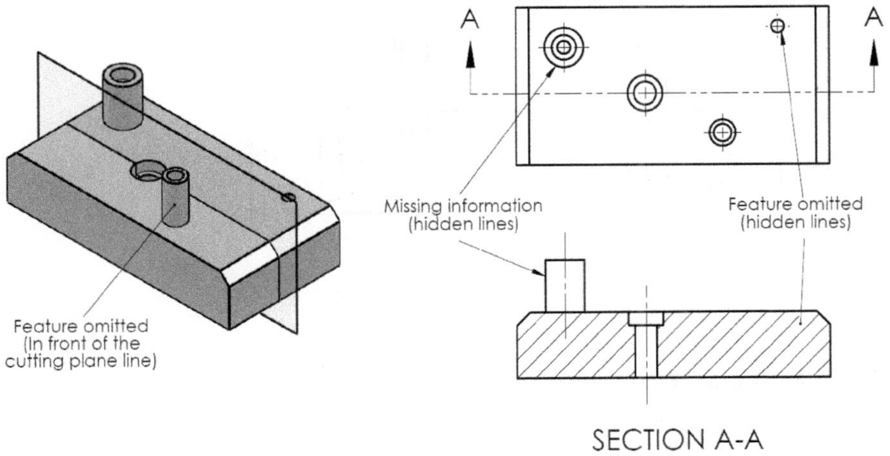

Figure 5.15 Situation where a full section is not appropriate

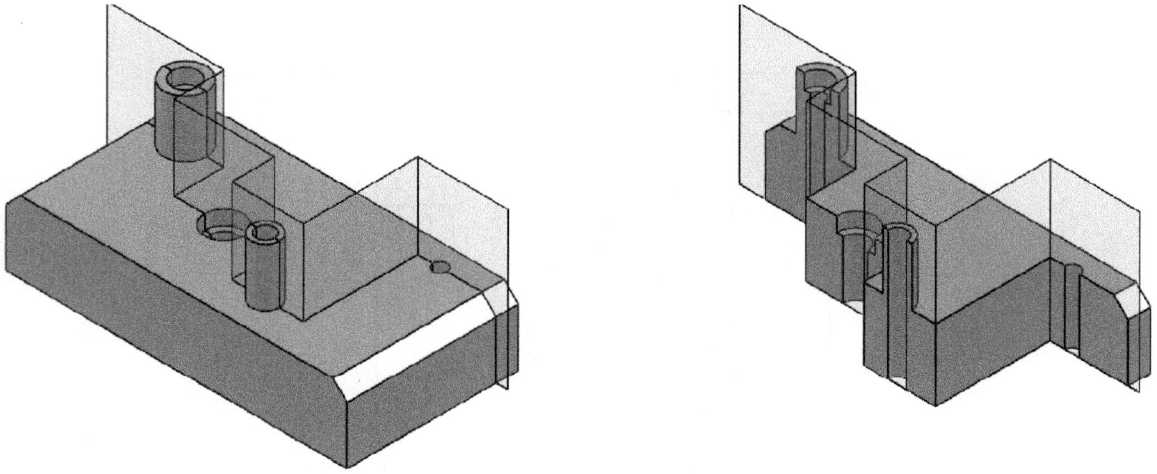

Figure 5.16 Modified cutting plane to create offset section

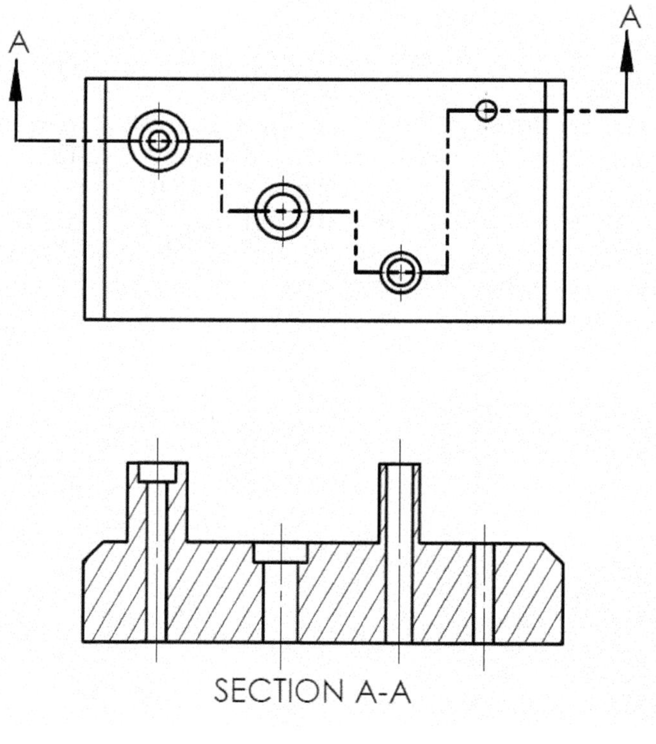

Figure 5.17 Offset section

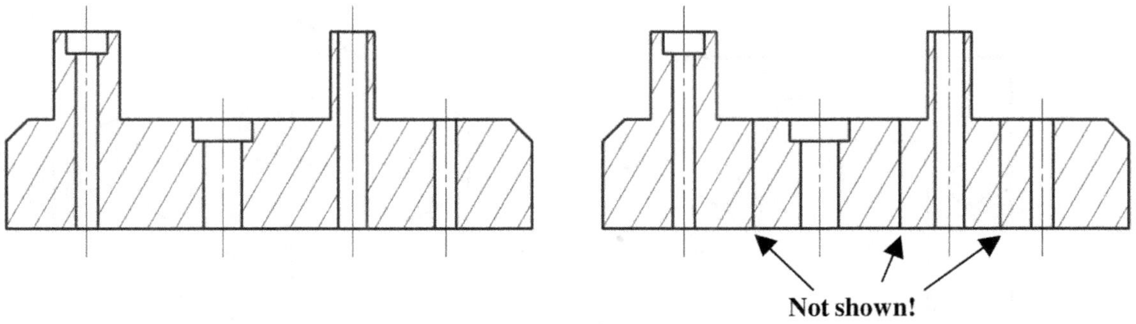

Note: Top view and cutting plane line are shown in the previous figure

CORRECT	INCORRECT
Offset edges along which the cutting plane is bent are not shown in the sectional view.	Although according to orthographic standards this sectional view would be the correct one, offset edges are conventionally omitted in the sectional view.

Figure 5.18 Offset edges of the plane are not shown in the section view

Half Sections Half sections are created by removing a quarter of the object, which reveals only half of its internal detail. The cutting plane is bent at a 90° angle such as the first part is perpendicular to the viewing plane and the second part is parallel (See Figures 5.19 and 5.20). Half sections are often used to represent objects that are symmetrical.

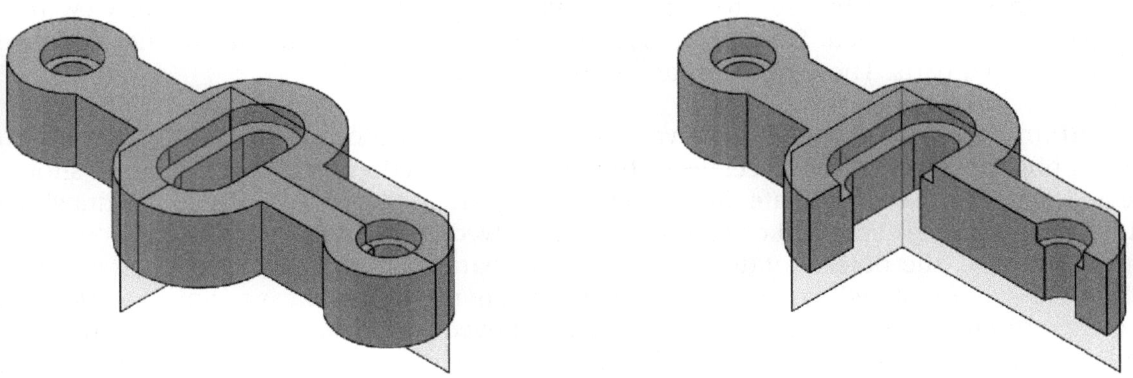

Figure 5.19 Half section

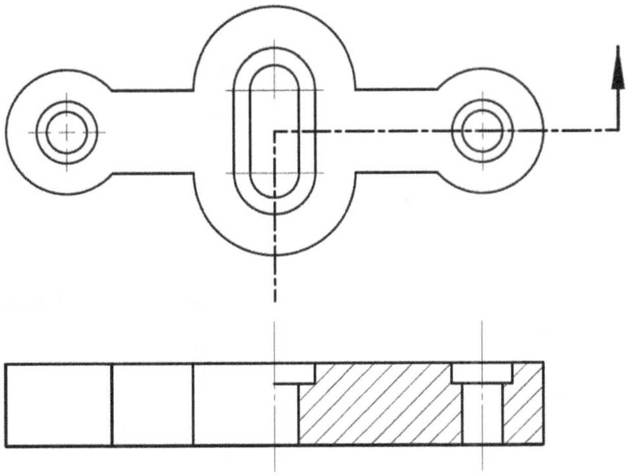

Figure 5.20 Orthographic views with half sections

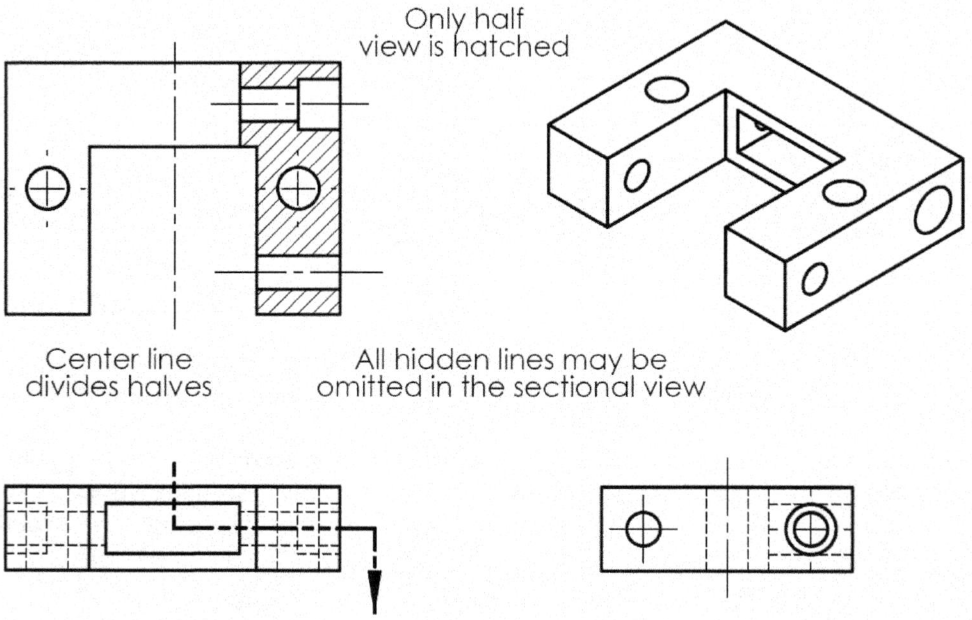

Figure 5.21 Rules to create half sections

Several rules must be followed when creating half sections. First, it is not necessary to label the cutting plane line. Second, a center line is used to divide the sectioned part of the view from the non-sectioned part, even if the center of the part does not include a circular feature. As a general rule, centerlines can also be used to represent symmetry. Third, only half of the view (the sectioned part) is cross hatched. Finally, all hidden lines must be omitted, both in the sectioned and the non-sectioned parts. These concepts are illustrated in Figure 5.21.

Revolved and Removed Sections Revolved and removed sections are cross sectional views of an elongated object rotated in the direction of the plane of projection with the purpose of showing its shape and material and to facilitate dimensioning. They are used to add clarity to the drawing without adding the complexity of full or half sections. Revolved and removed sections can be pictured as very thin slices of the object rotated to make them parallel to one of the orthographic planes. Revolved sections are placed over the object after they are rotated, whereas removed sections are offset a certain distance from the orthographic view. Removed sections are often placed on centerlines extended from the section cuts. See Figure 5.22 for examples of revolved and removed sections.

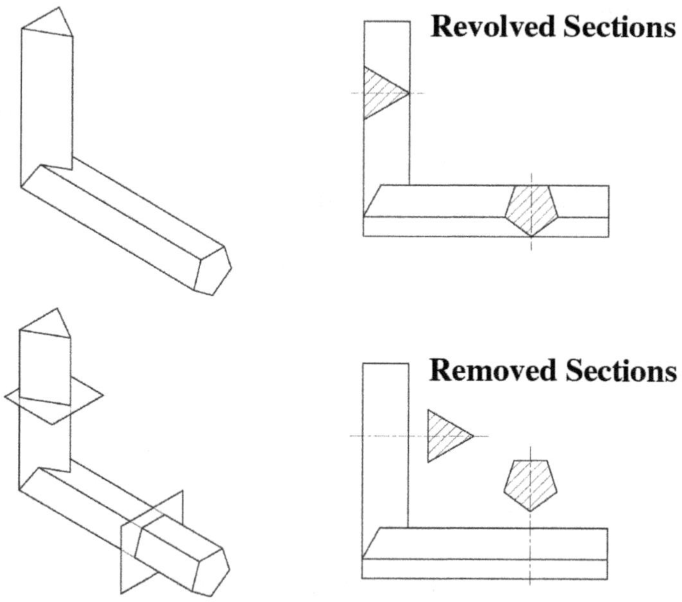

Figure 5.22 Revolved and removed sections

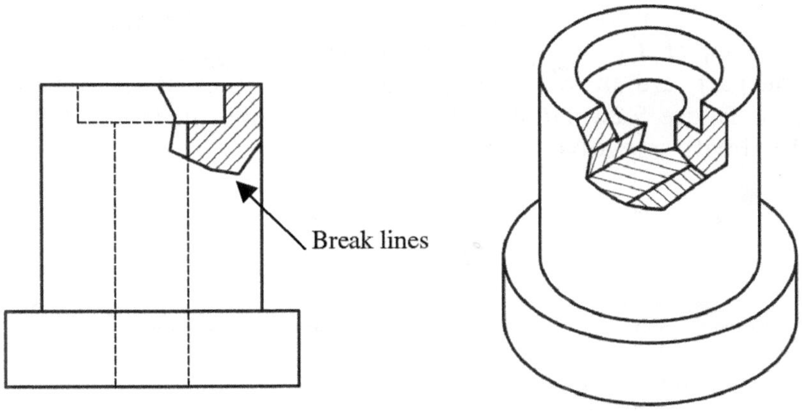

Figure 5.23 Broken-out section

Broken-Out Sections A broken-out section is a type of section view utilized to show only a small portion of the interior detail of an object when full or half sections are not necessary. Broken-out sections present an object with a portion broken away as if the outside surface was hit with a hammer. The depth of the cut depends on the internal feature of interest. No cutting plane is required. Instead, arbitrary break lines are used. See Figure 5.23 for an example.

3.2.1 Other Types of Section Views

Broken Views Broken views are used to show extremely large or long objects at reasonable sizes in the drawing and to facilitate dimensioning. Broken views are created by using two break lines located close to the ends of the object and removing the long portion of the part, which contains no important information, located in between the break lines. In this way, the drawing can be shown at a larger scale with a better view of the details providing more space for dimensions. See Figure 5.24 for an example.

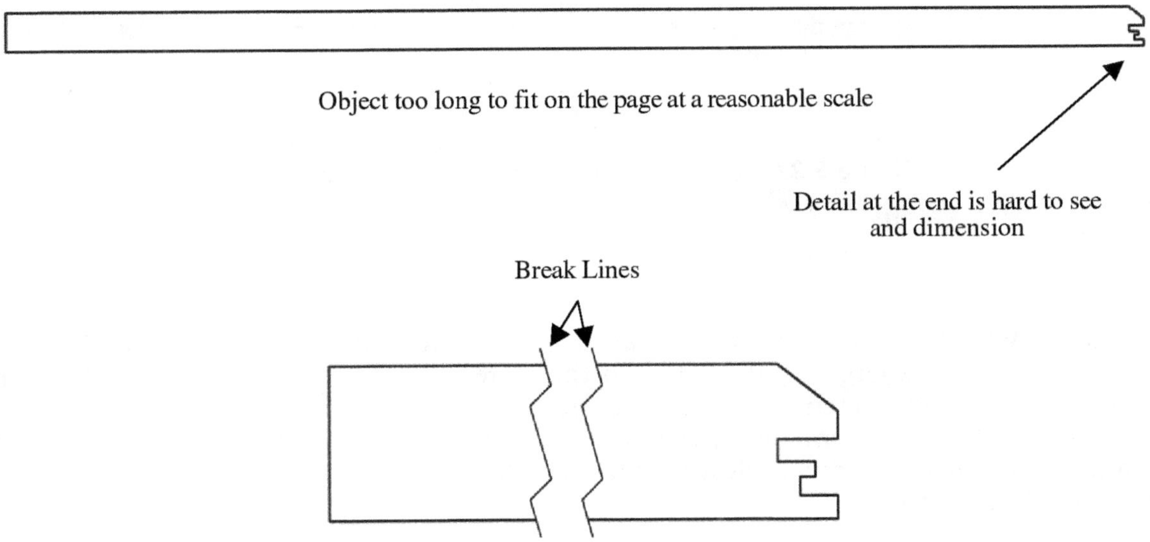

Figure 5.24 Broken view

Different conventional symbols can be used for break lines, depending on the type of break or the material of the object. Common long break symbols, used when the break is performed longitudinally to the part, are shown in Figure 5.25. Examples of short break lines based on material are shown in Figure 5.26. Finally, the example shown in Figure 5.27 illustrates how a break can be combined with a revolved section.

Figure 5.25 Examples of conventional breaks

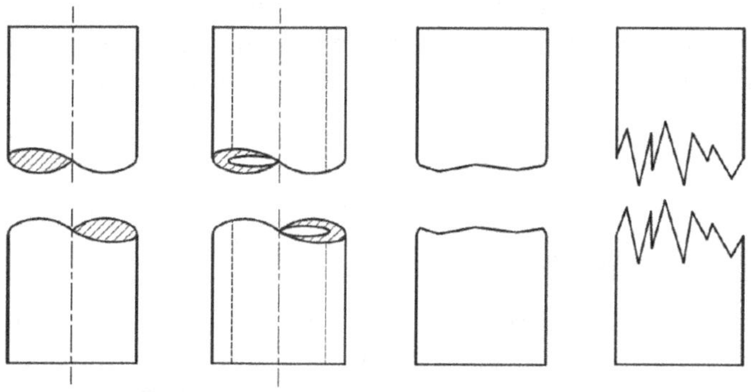

Figure 5.26 Short conventional breaks based on material

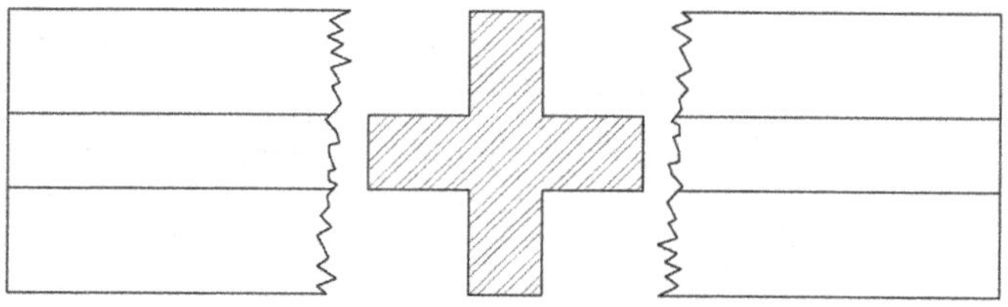

Figure 5.27 Revolved section with short break lines

Detailed Views Detailed views are similar to broken views in that they are used to show details of an object at a larger scale. Although there are different styles of detailed views, they are generally created by sketching a circle around the desired detail on a regular orthographic view and showing the content inside the circle on a separate view at a larger scale. Detailed views must be labeled properly, indicating the scale, to avoid confusion. See Figure 5.28 for an example of a detailed view.

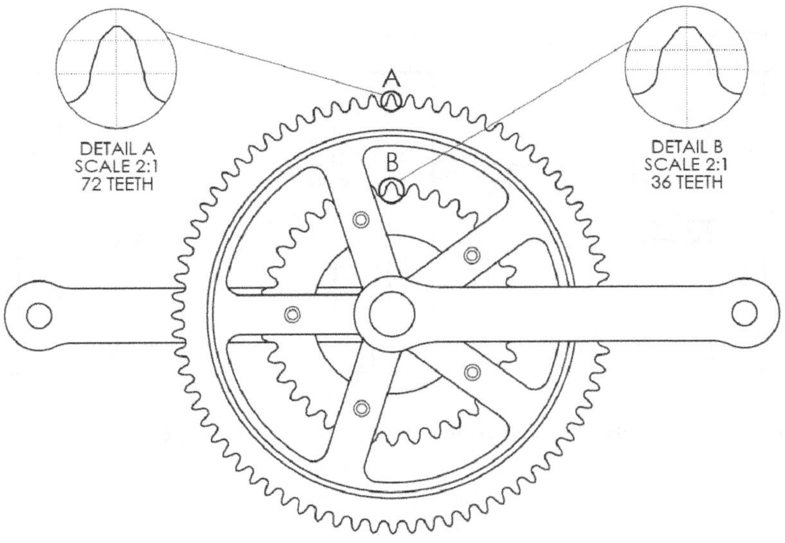

Figure 5.28 Examples of detailed views

Sectioned Pictorials Occasionally, sectioned pictorials are provided for clarity. Sectioned pictorials are particularly common in assembly drawings, where orthographic views of multiple sectioned parts may be confusing. In sectioned pictorials, the cutting plane is not shown. Hatch patterns are shown and the general practices used in regular section views also apply to sectioned pictorials. See Figure 5.29 for examples of sectioned pictorials.

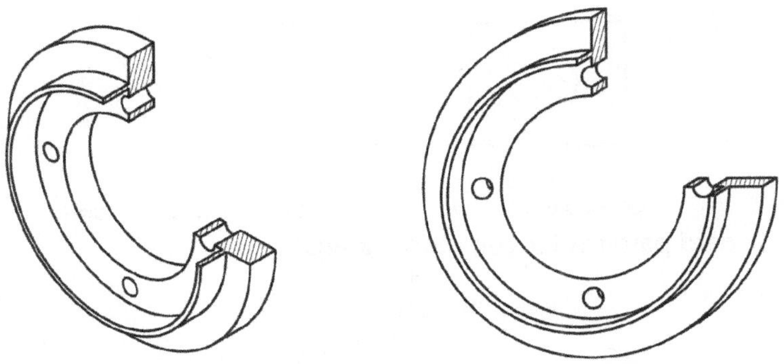

Figure 5.29 Sectioned isometric and sectioned oblique of the same object

3.3 Using Hatch Patterns to Describe Materials

As mentioned earlier, one purpose of sectional views is to describe the material that an object or assembly is made of (see Figure 5.30). This description is accomplished through the use of hatch patterns. Hatch patterns represent the solid area of the object where the cutting plane slices the object.

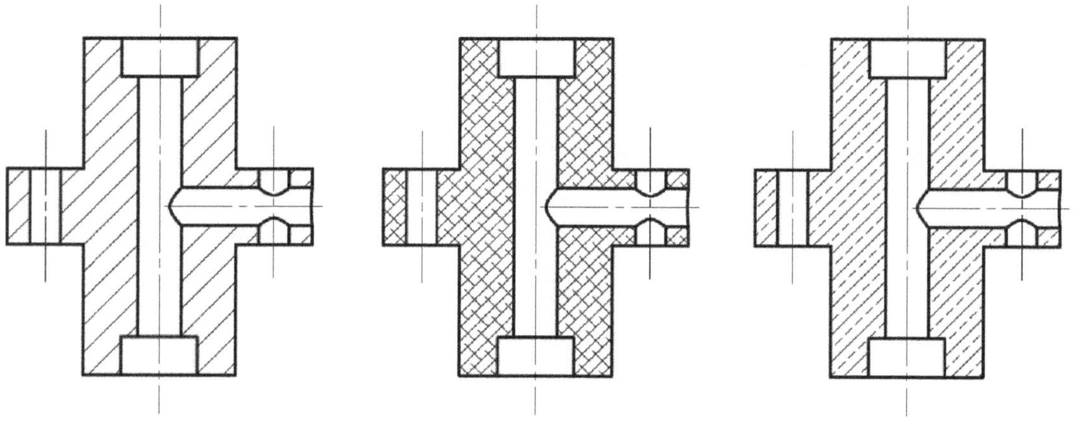

Figure 5.30 Different hatch patterns represent different materials

Patterns for the most common materials are dictated by ANSI standards (see Figure 5.31) and they are used in many engineering sectors. Patterns for specific materials, however, can be driven by industry standards or even company practices.

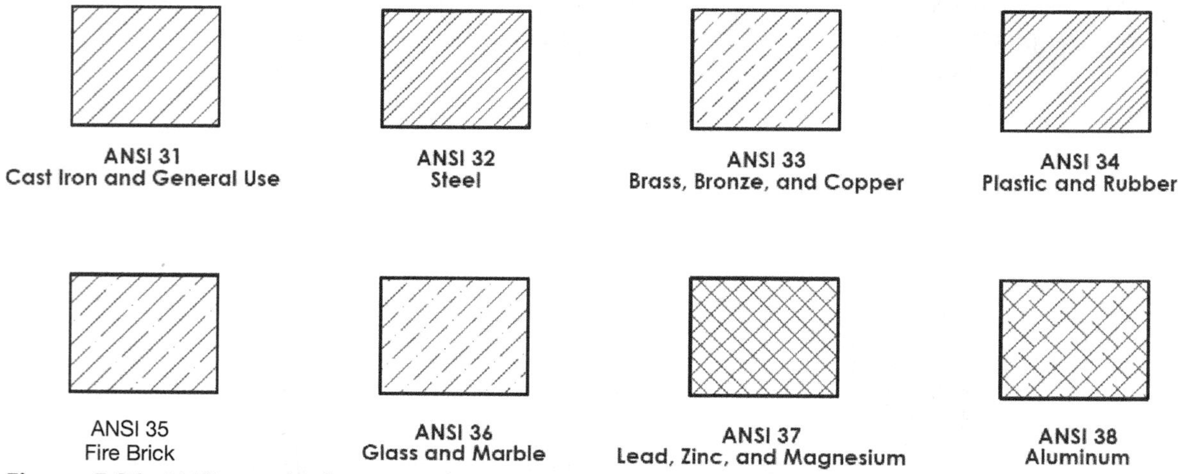

Figure 5.31 ANSI standard patterns for common materials

In addition to using the correct pattern for the material that is being represented, there are several rules that must be followed when hatching sectional views. These rules take into account factors like denseness, orientation, and scaling of patterns to avoid confusing or ambiguous drawings.

First, hatch pattern lines within a sectioned area must be parallel and thinner than visible lines. The thickness of the hatch lines depends on the drawing and the spacing between lines, but it should be no thicker than visible lines. Second, hatch spacing should be small enough so that no lines of the pattern are misread as edges and big enough so the material represented by the pattern can be recognized easily. Hatch pattern spacing normally ranges from 1/16″ to 1/8″. Solid patterns should be avoided. Third, hatch lines should not be parallel or perpendicular to any lines on the drawing that correspond to a major feature (see Figure 5.32).

For sectioned assemblies, different parts should be hatched at different angles, otherwise, there may be confusion distinguishing different components of the assembly (see Figure 5.33).

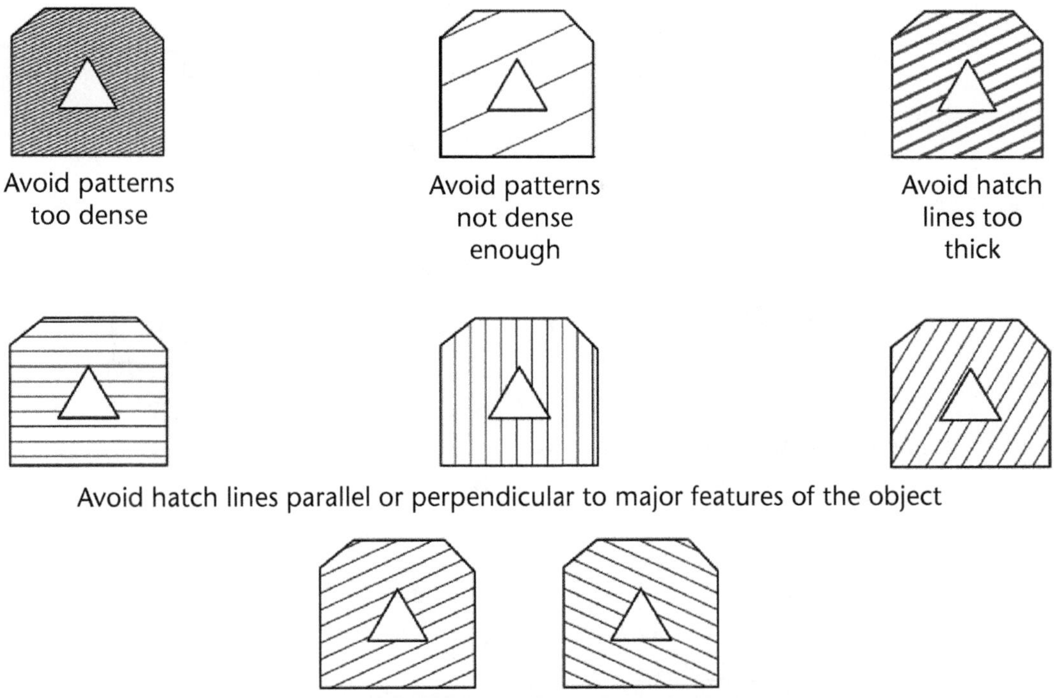

Figure 5.32 Good and bad practices for hatch patterns

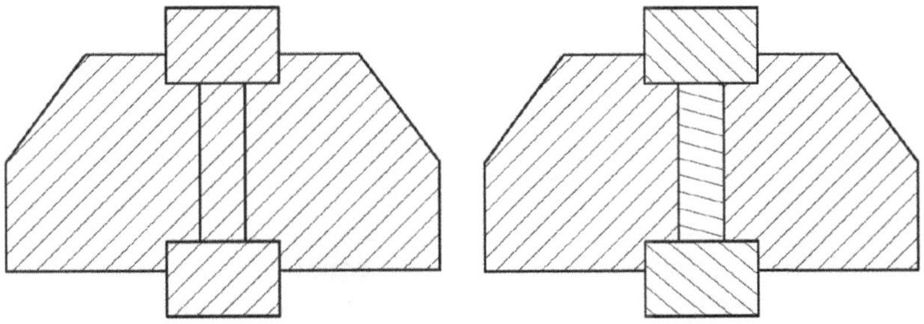

Figure 5.33 Different components of an assembly should be hatched at different angles.

4. Conventional Practices for Sectional Views

Conventional practices are drawing methods that are used, even when they violate standard rules of orthographic projection. These practices can considerably simplify drawings and some of them have been studied in previous chapters (Orthographics).

Conventional Rotations The same rules used for conventional rotations in orthographic projections apply to sectional views. Because of the geometry of certain objects, true orthographic views can be confusing to the viewer. In such cases, it is acceptable to rotate certain features of the object in the sectional view so that the drawing is more easily understood. When conventional rotations are applied, the sectional view is often referred to as an aligned section. The drawing shown in Figure 5.34 illustrates the case of an object with two equal arms; however, the true sectional view would give a mistaken impression that the arms of the object are not equally

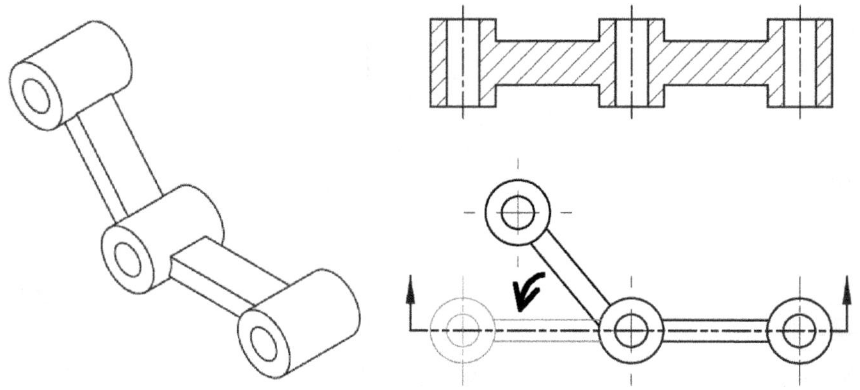

Figure 5.34 Aligned sectional view

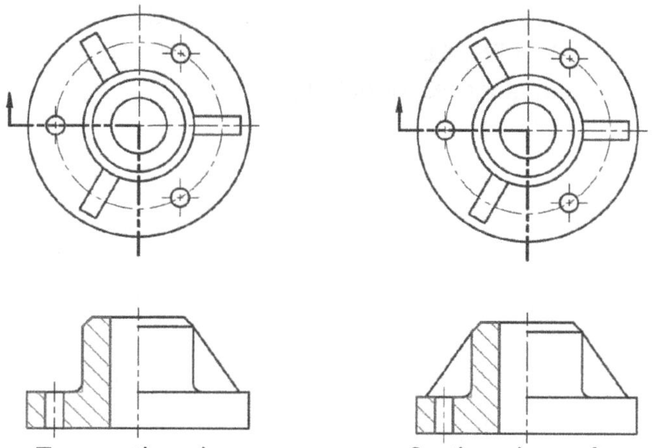

True section view gives the impression that the ribs are not equally spaced

Section views after applying conventional rotations

Figure 5.35 Conventional rotations

spaced. In such cases, an aligned section is more appropriate. The example shown in Figure 5.35 illustrates the case of a plate with holes and ribs equally spaced at 120°. It is useful to show in the section view the holes and ribs spaced at 180°, even though it is inaccurate, so a confusing sectional view is not drawn.

Features Not Hatched There are certain features that are not hatched even though the cutting plane passes through them. Some of these features include gaskets, shafts, pins, roller bearings, and most vendor items such as bolts, washers, shafts, nuts, and screws. See Figure 5.36 for an example.

Features like ribs, webs, lugs, and spokes are usually not hatched. Ribs and webs are thin, flat features that provide structural support to an object. Spokes are the rods or braces joining the hub to the

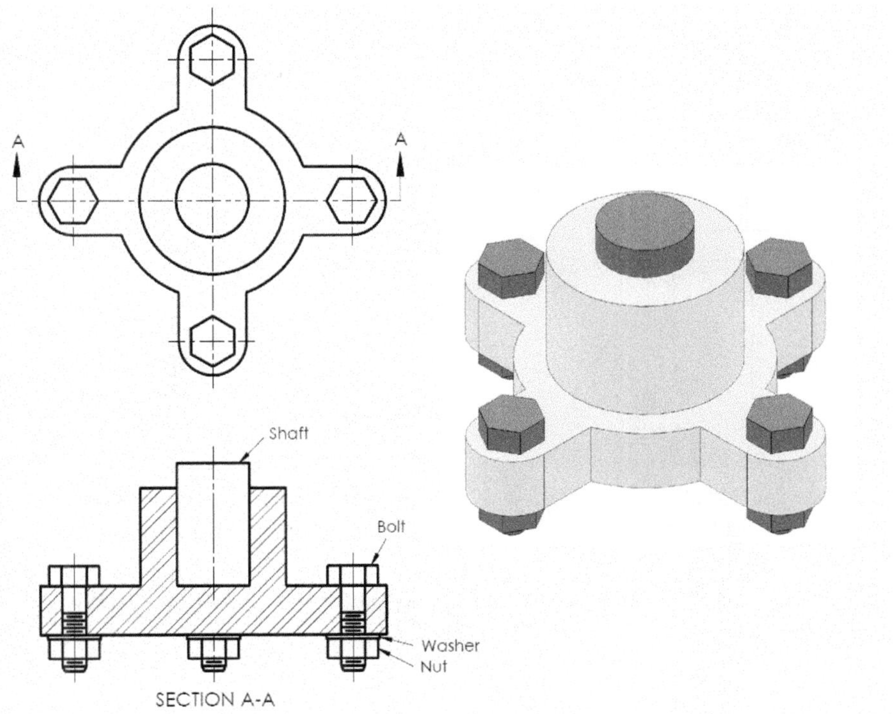

Figure 5.36 Parts that are not hatched

rim of a wheel. A lug is a flange added to an object for attachment to another part. See Figure 5.37 for examples of these features. Ribs and webs are occasionally hatched if the cutting plane cuts crosswise through them (see Figures 5.38 and 5.39).

Figure 5.37 Example of ribs, webs, spokes, and lugs

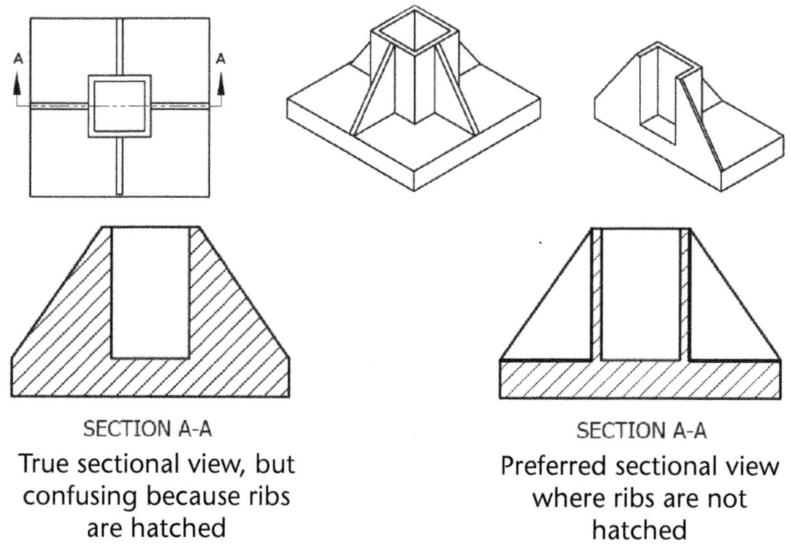

Figure 5.38 Full section where cutting plane passes longitudinally to the ribs

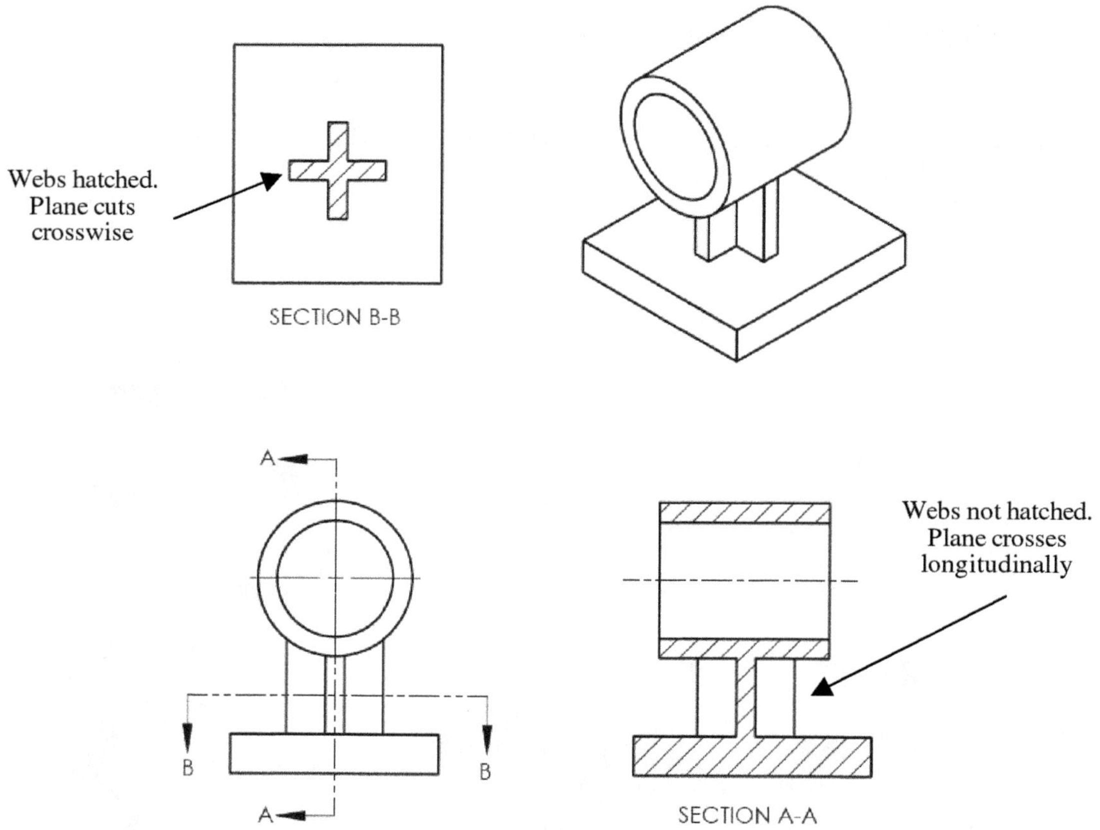

Figure 5.39 Webs are hatched only when the cutting plane crosses perpendicularly.

Multiple Section Views It is possible to have multiple section views of an object in the same drawing. In this case, it is essential to label all section views and cutting planes correctly so the viewer can identify what features are being examined. Additionally, multiple section views do not need to align with the adjacent orthogonal views. In general, the first section remains aligned with the orthogonal view in which it was created and the rest of the sections are shown next to the first sectional view. See Figure 5.40 for an illustration of this concept.

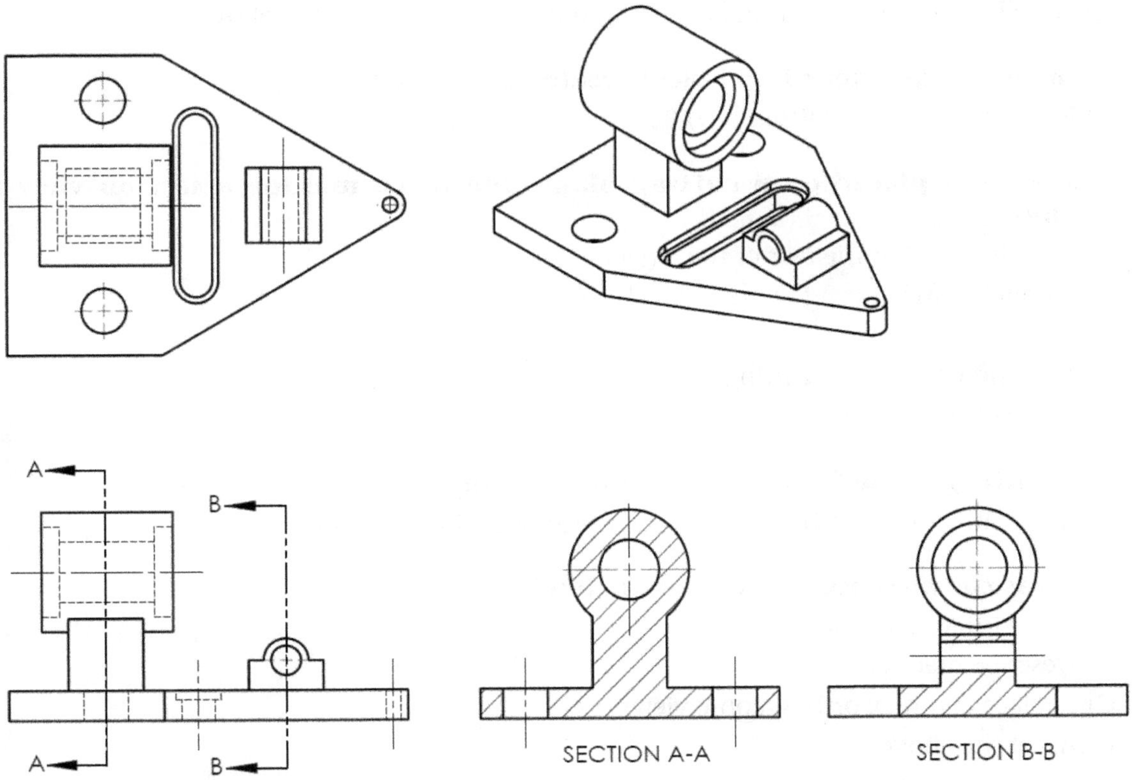

Figure 5.40 Drawing with multiple section views

Practice Test

1. **What types of views show true size projections of surfaces that do not appear true size in any standard orthographic view?**
   A) Sectional B) Auxiliary C) Profile D) Inclined E) None of the above

2. **What type of auxiliary views show only the visible lines that correspond to the inclined surface and omit all other line associated with the view?**
   A) Partial B) Full C) Primary D) Secondary E) None of the above

3. **Removing a quarter of an object creates a _____ section**
   A) Full B) Half C) Offset D) Removed E) None of the above

4. **The arrows placed on a cutting plane line when making a section view are always _____**
   A) Pointing toward the rear of the object
   B) Parallel to the cutting plane
   C) On one end only
   D) Perpendicular to the cutting plane
   E) None of the above

5. **The cutting plane for a section view must appear in the _____ view.**
   A) Side B) Front C) Top D) Any of the above E) None of the above

6. **What is the purpose of a sectional view?**
   A) Show internal detail
   B) Describe materials
   C) Replace complex orthographic views
   D) All of the above
   E) None of the above

7. **Which of the following parts are not cross-hatched in a sectioned assembly?**
   A) External parts
   B) Internal parts
   C) Threaded fasteners
   D) All parts in a sectioned assembly are cross-hatched
   E) None of the above

8. **The hatching symbol used for ____ is also used to represent a generic material.**
   A) Cast iron B) Basic C) Steel D) Concrete E) Copper

9. **If an object is symmetrical, and to save space or time, what is the most appropriate sectional view that can be drawn?**
   A) Full B) Half C) Offset D) Partial E) None of the above

10. **Sectioned lines (crosshatch) are usually spaced ____**
    A) 1/32″ to 1/16″ B) 1/16″ to 1/8″ C) 1/8″ to 1/4″ D) 1/4″ to 1/2″ E) None of the above

Problems

For the following objects, use your engineering paper to construct appropriate orthographic and auxiliary views. Each grid square on your engineering pad is 0.10″. Ensure that you choose the correct front view.

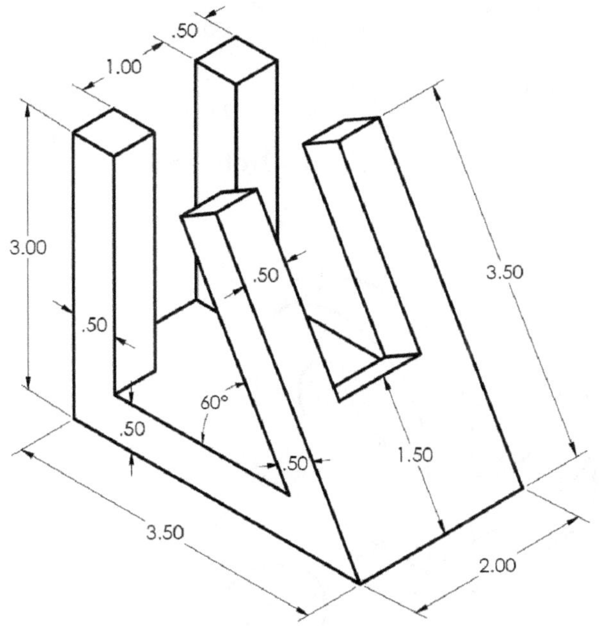

Auxiliary views—Problem 1

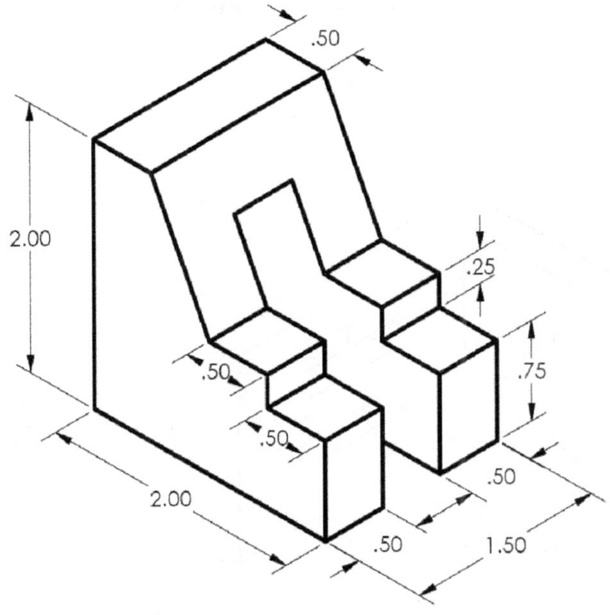

Auxiliary views—Problem 2

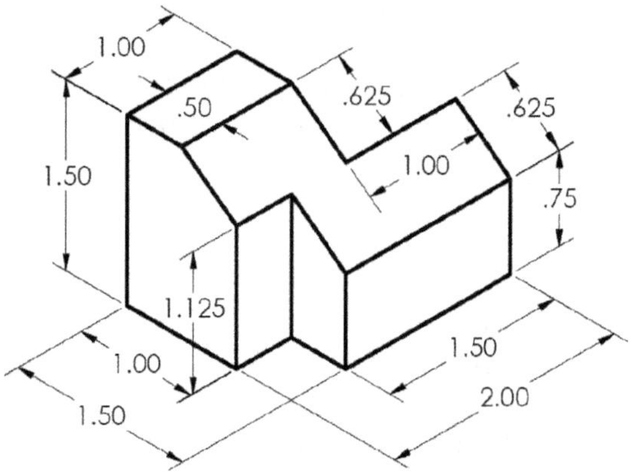

Auxiliary views—Problem 3

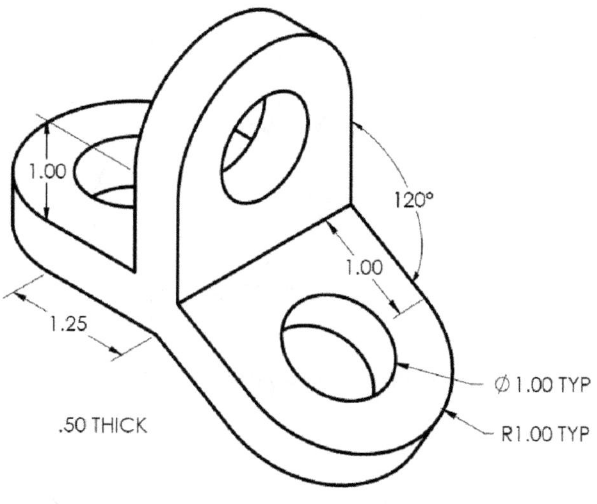

Auxiliary views—Problem 4

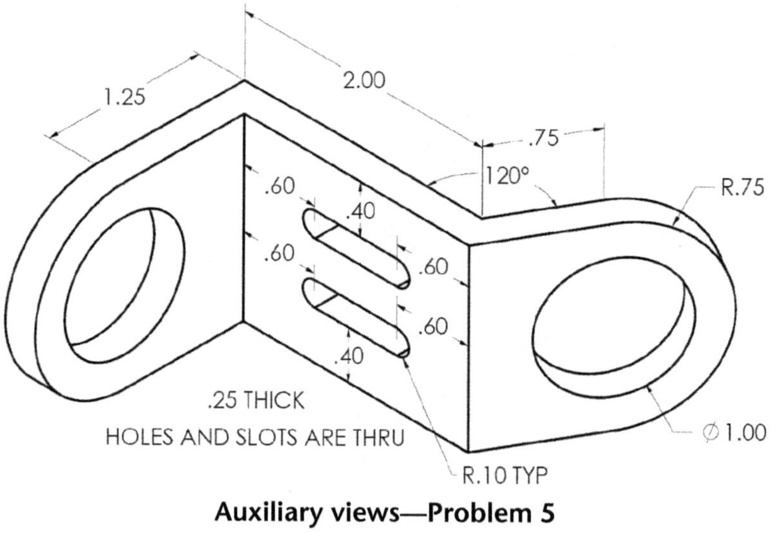

Auxiliary views—Problem 5

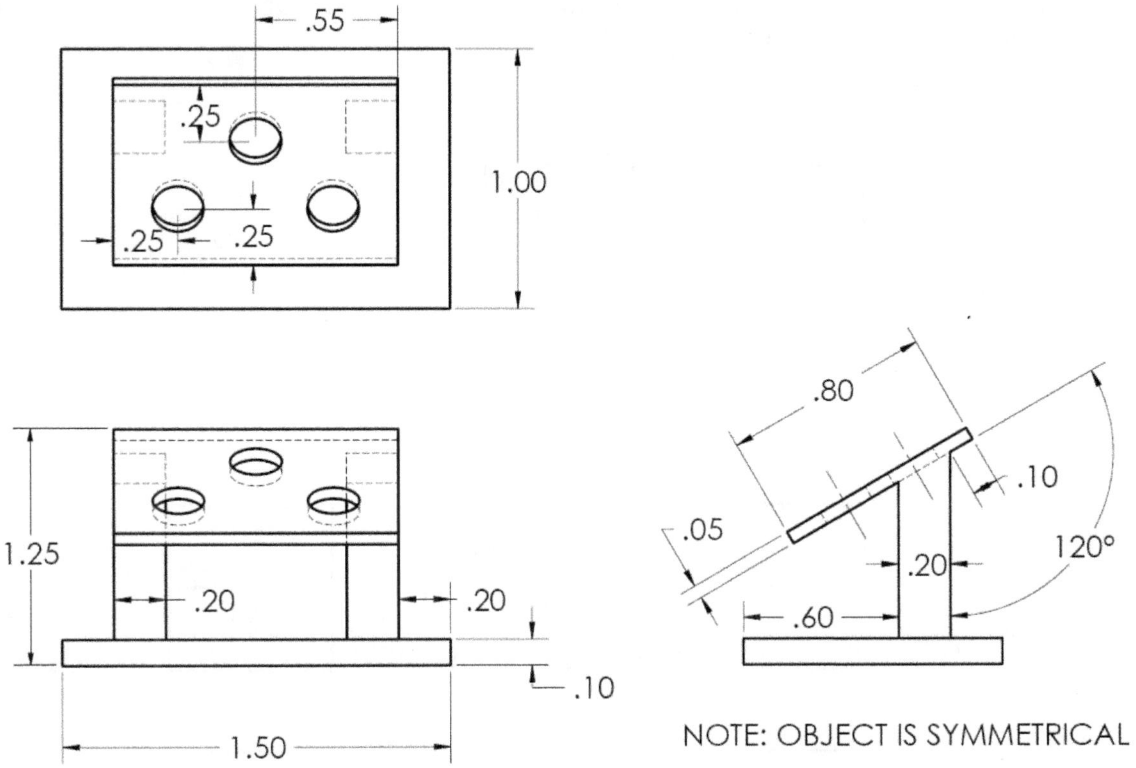

Auxiliary views—Problem 6

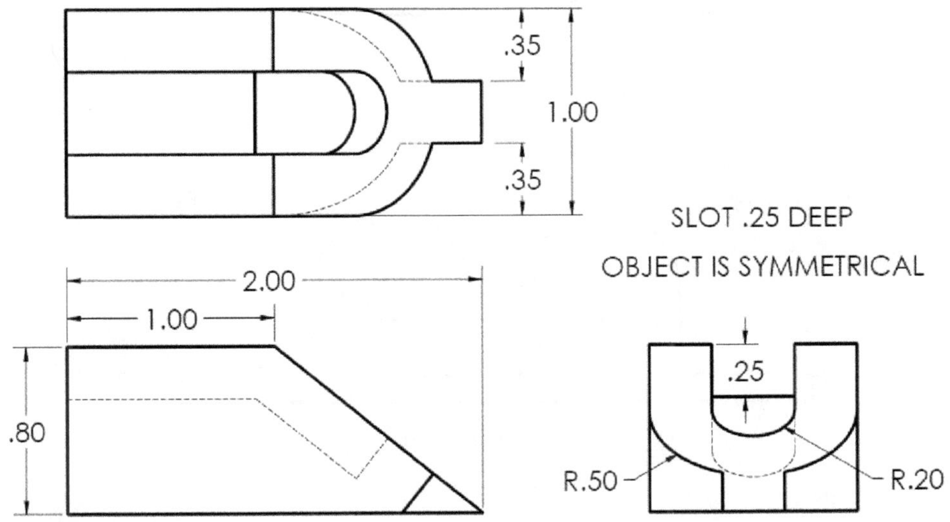

Auxiliary views—Problem 7

For the following objects, use your engineering paper to construct appropriate sectional views. Each grid square on your engineering pad is 0.10″.

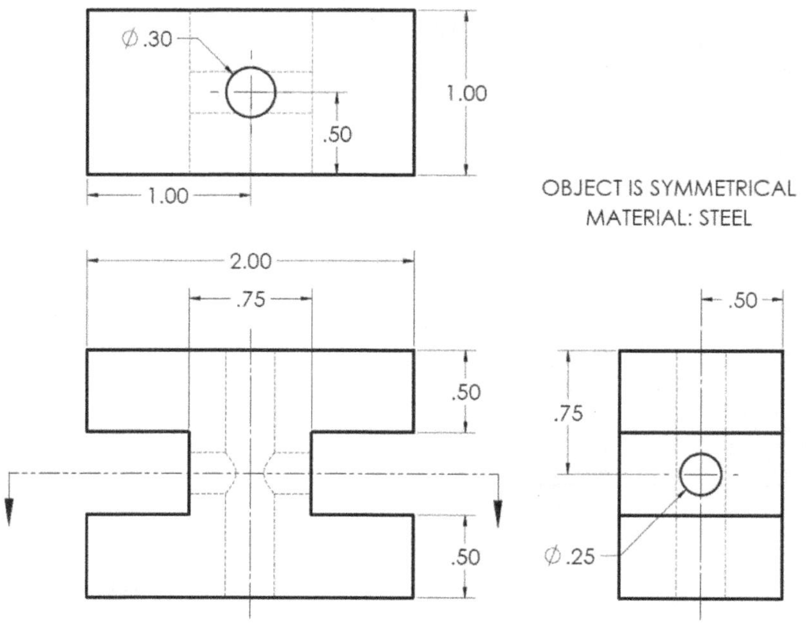

Section views—Problem 1

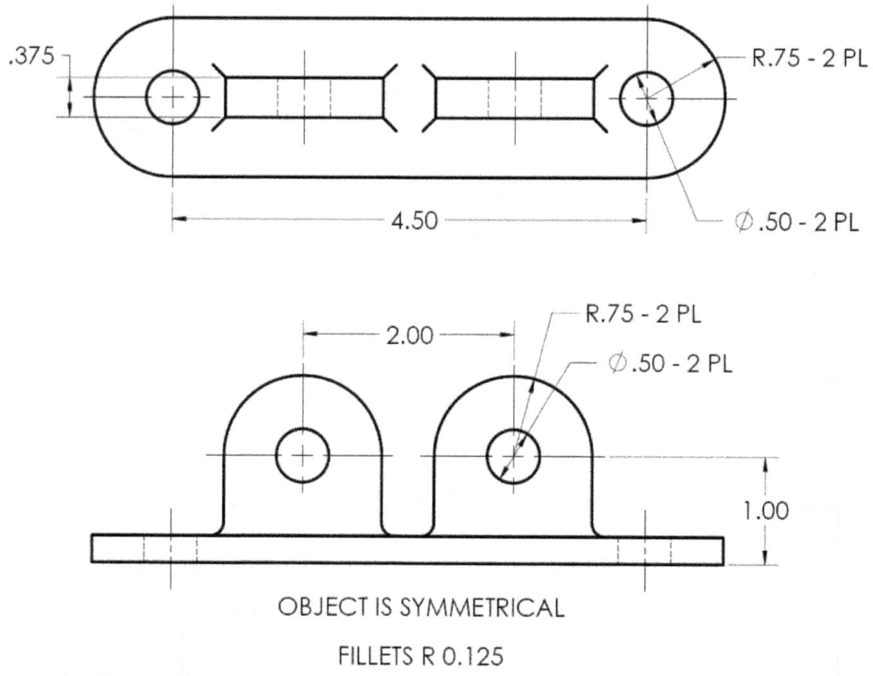

Section views—Problem 2

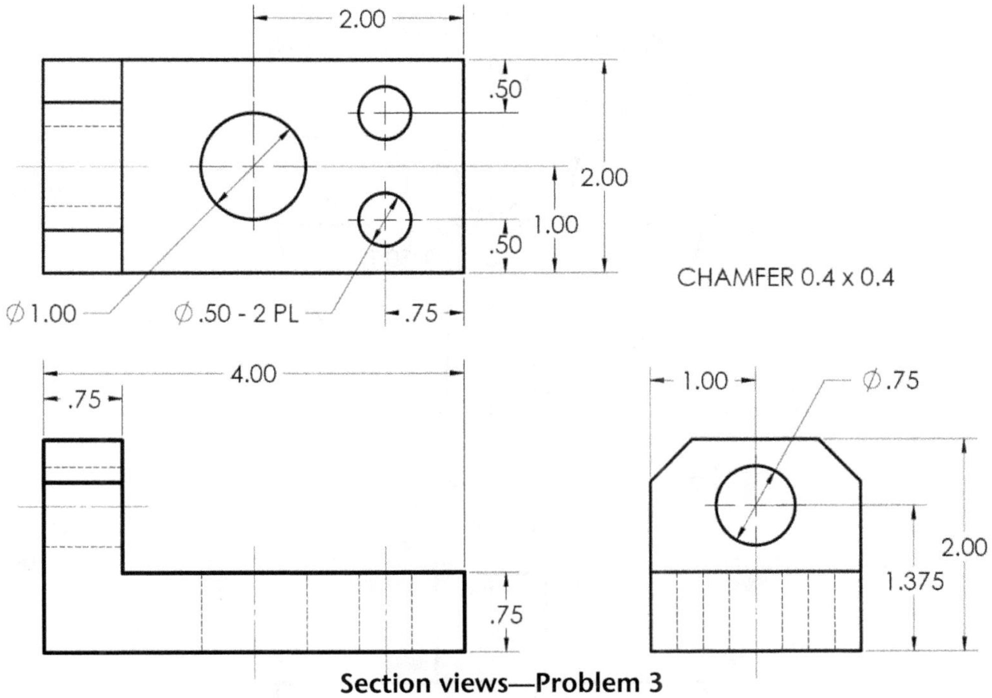

Section views—Problem 3

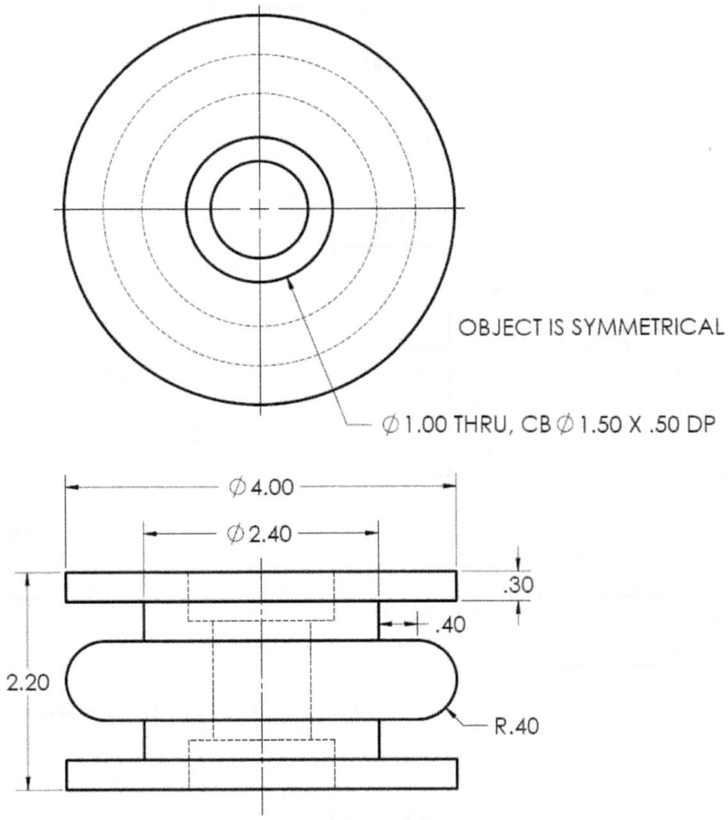

Section views—Problem 4

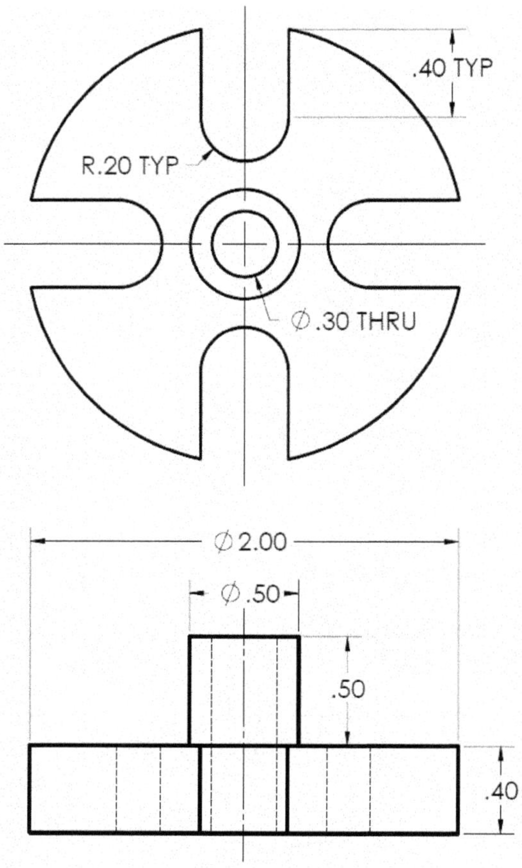

Section views—Problem 5

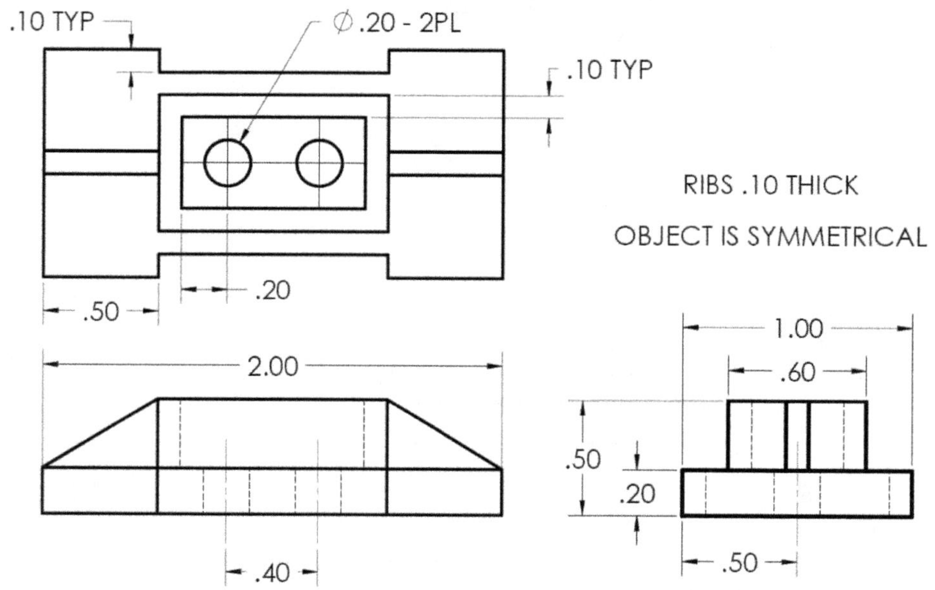

Section views—Problem 6

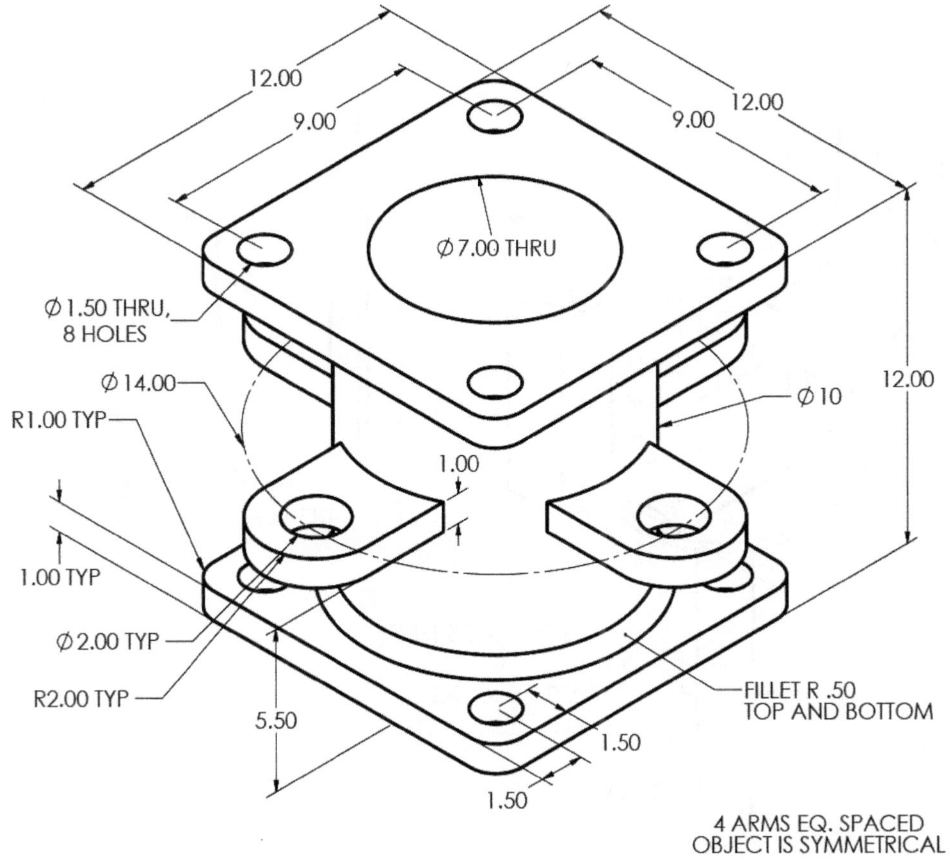

Section views—Problem 7

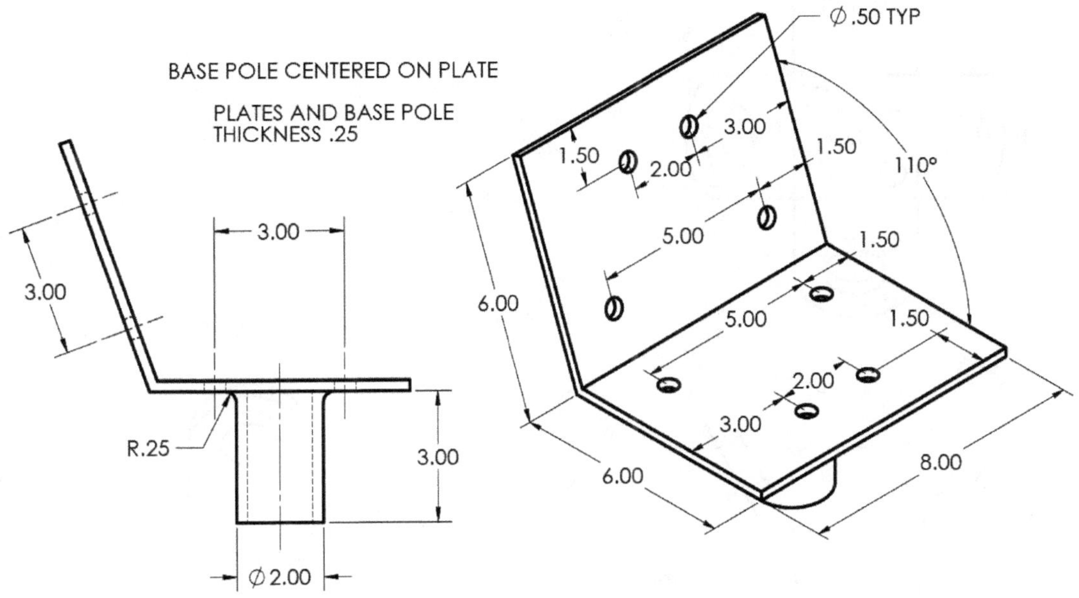

Auxilary views—Problem 8

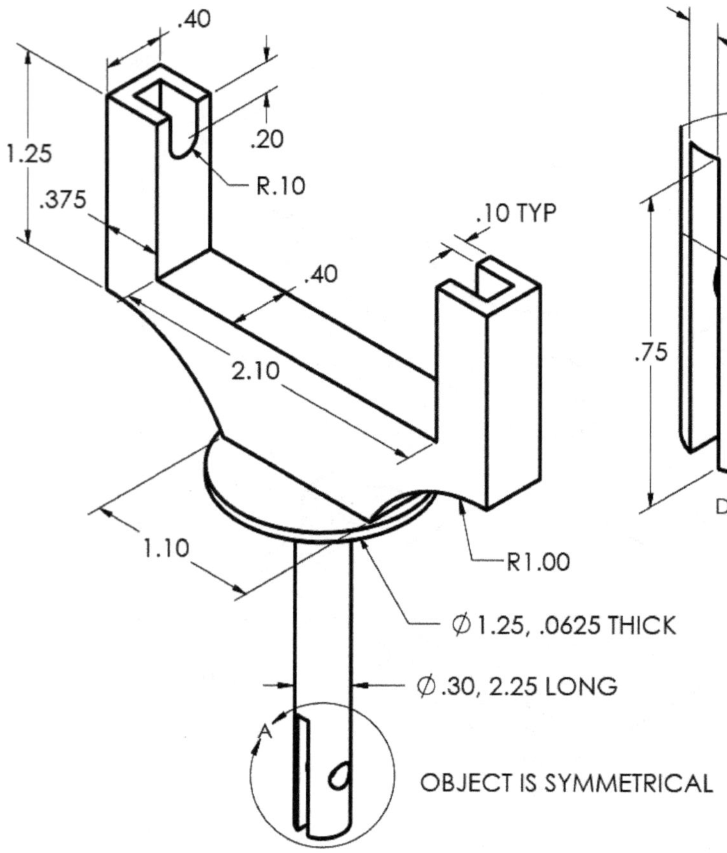

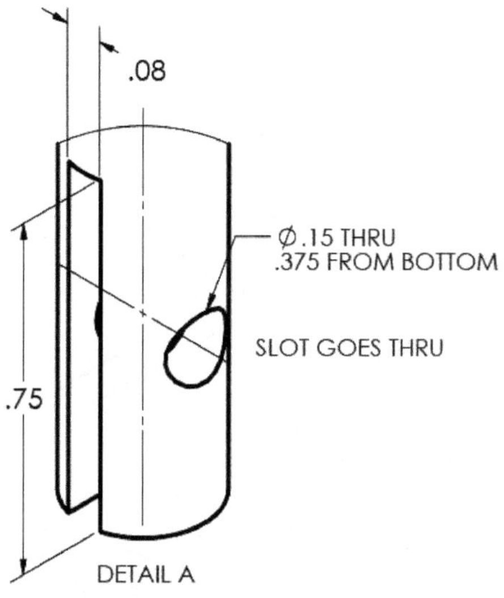

Section views—Problem 8

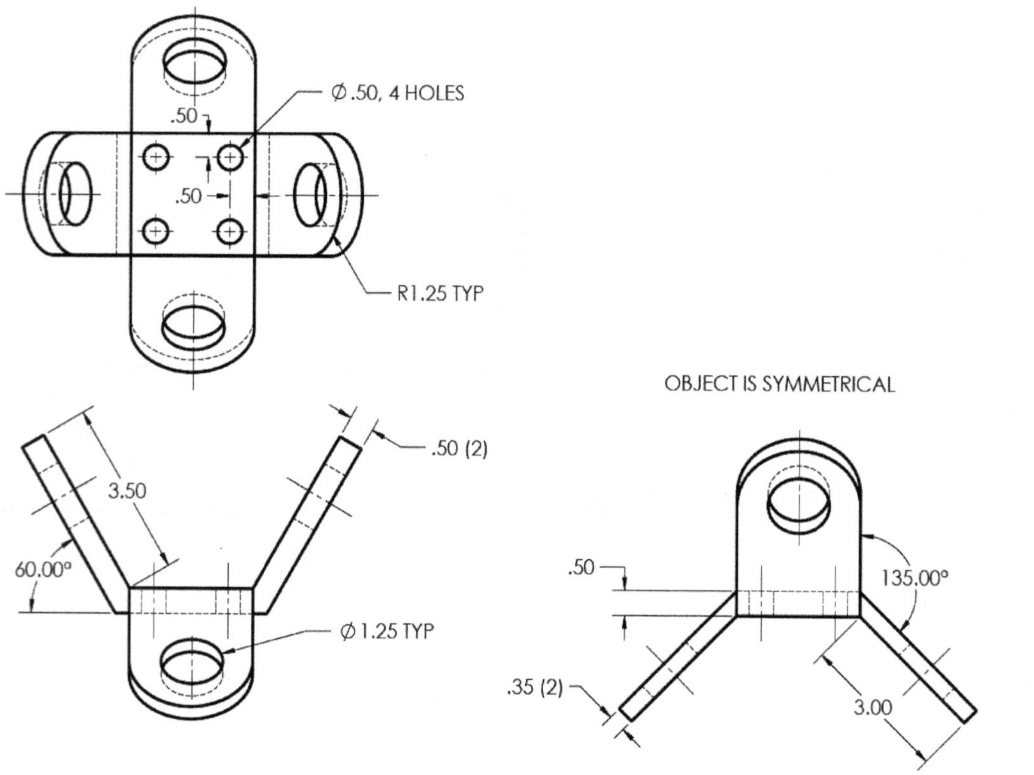

Auxilary views—Problem 9

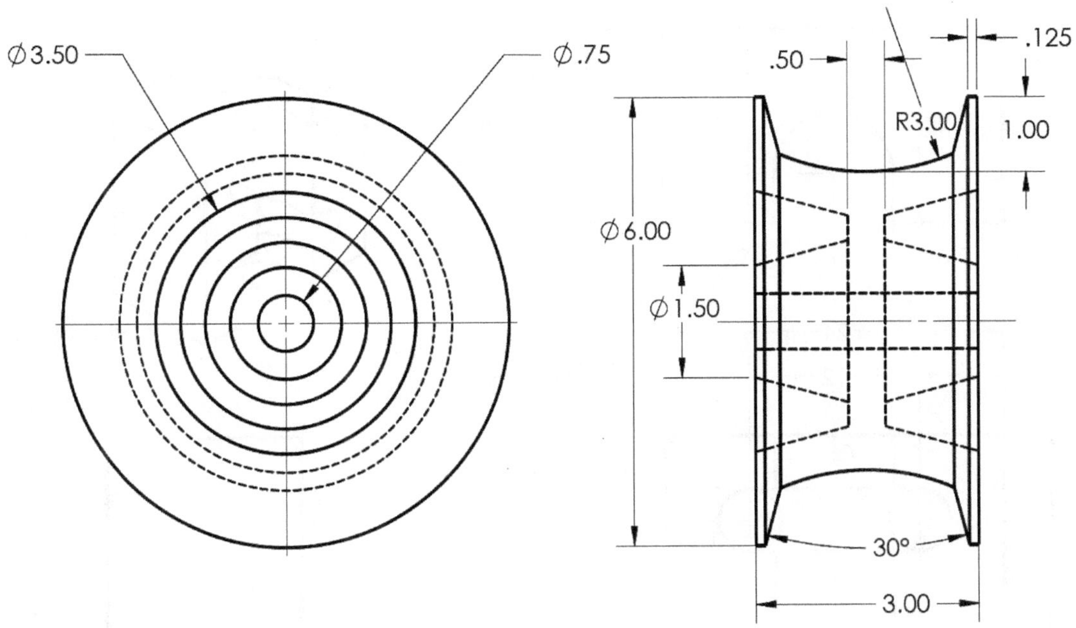

Section views—Problem 9

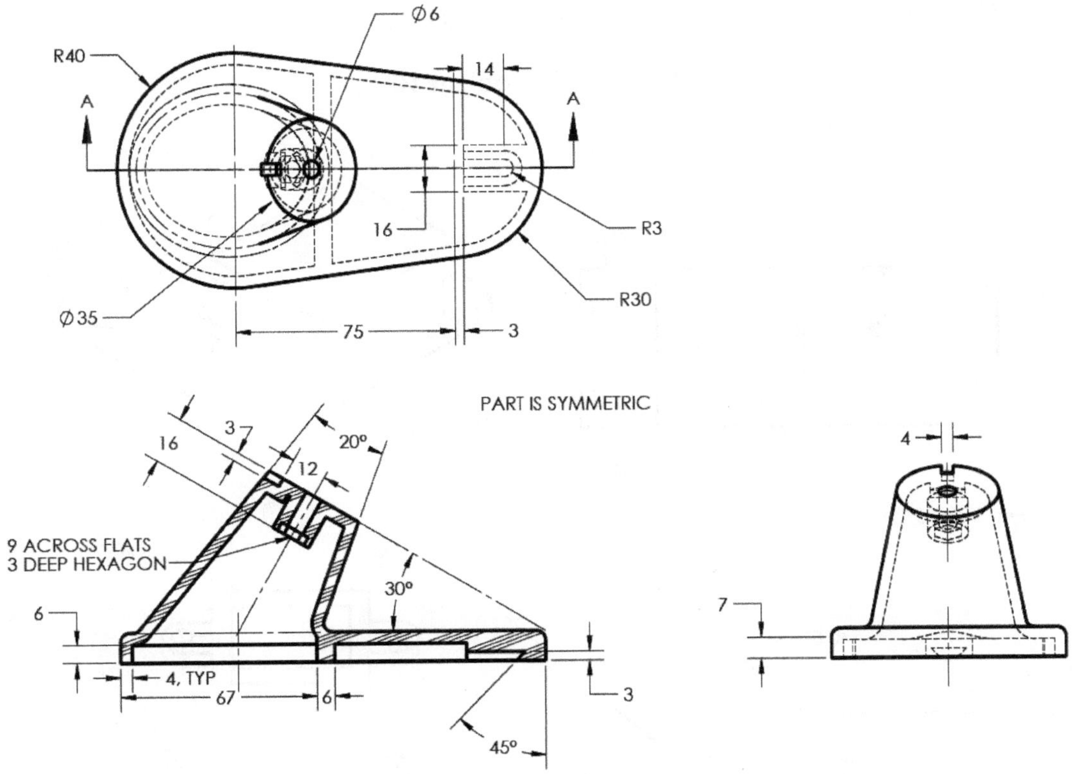

Auxilary views—Problem 10

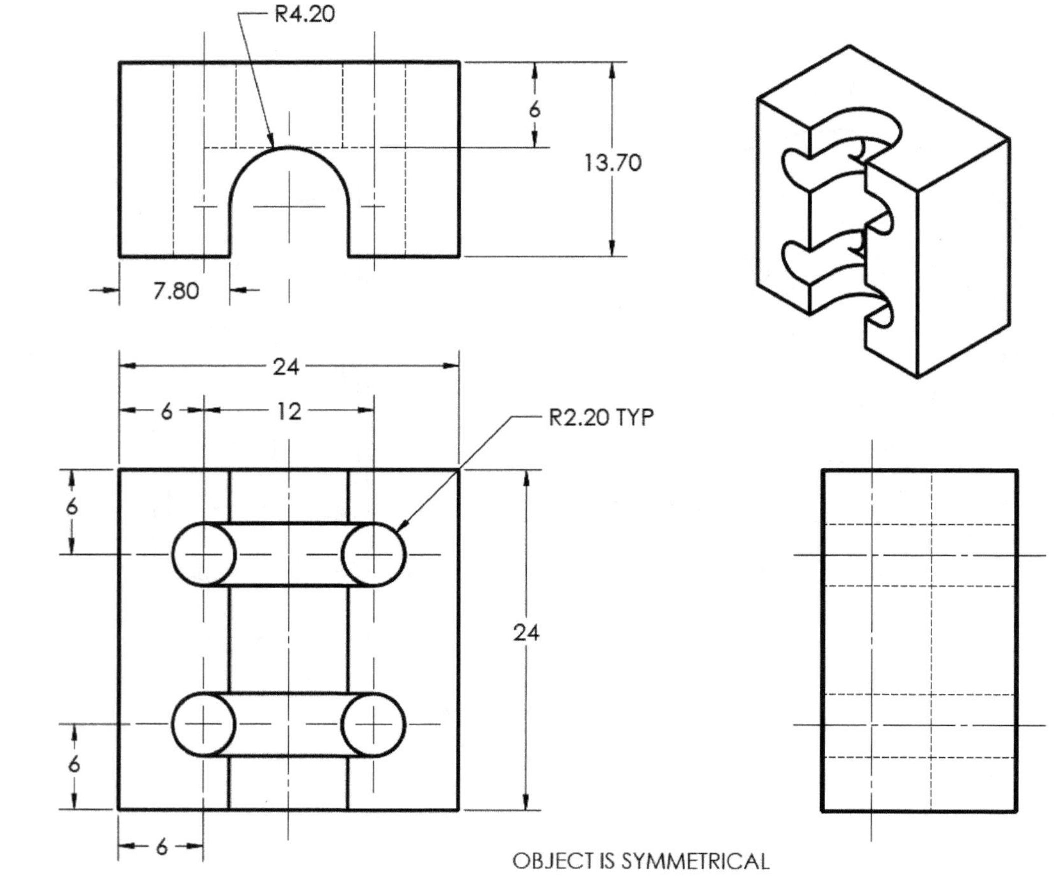

Section views—Problem 10

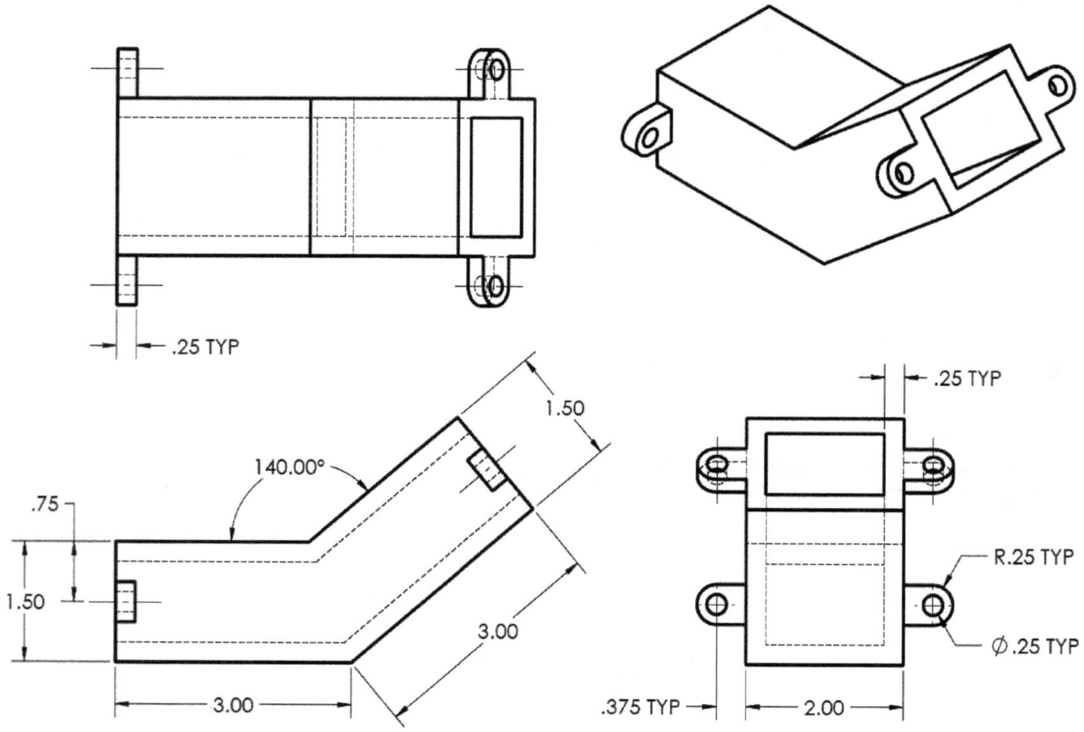

Auxilary views—Problem 11

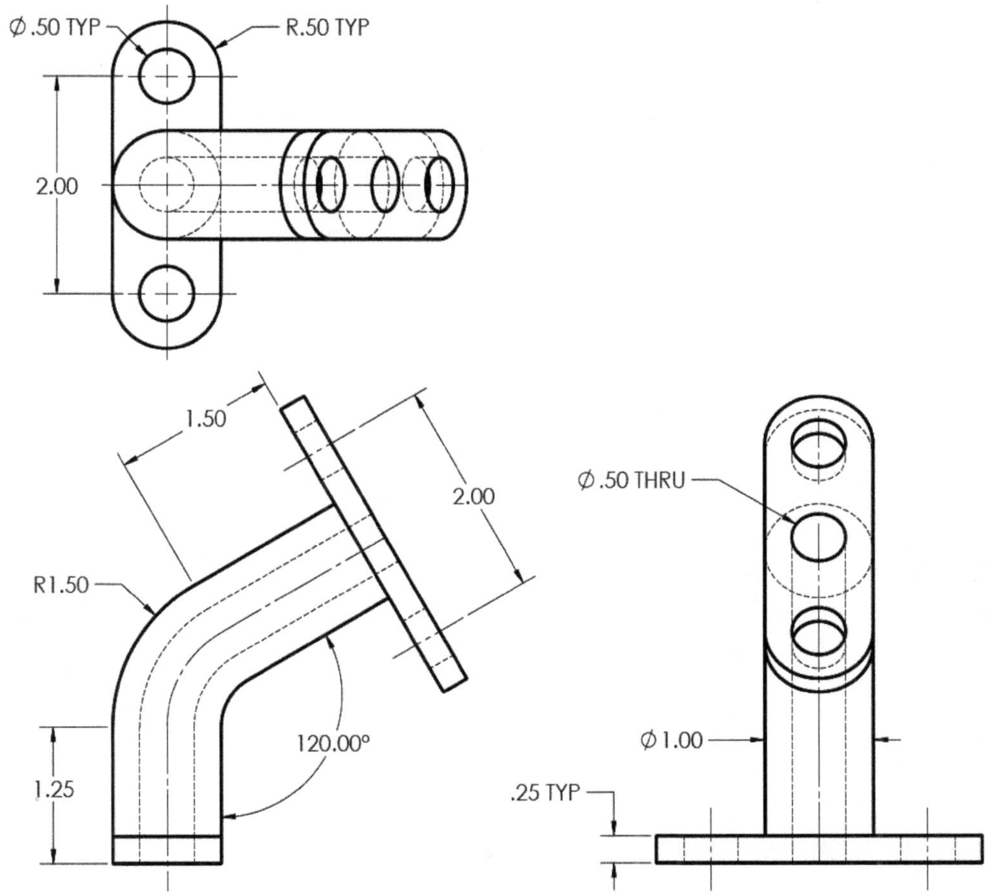

Auxilary views—Problem 12

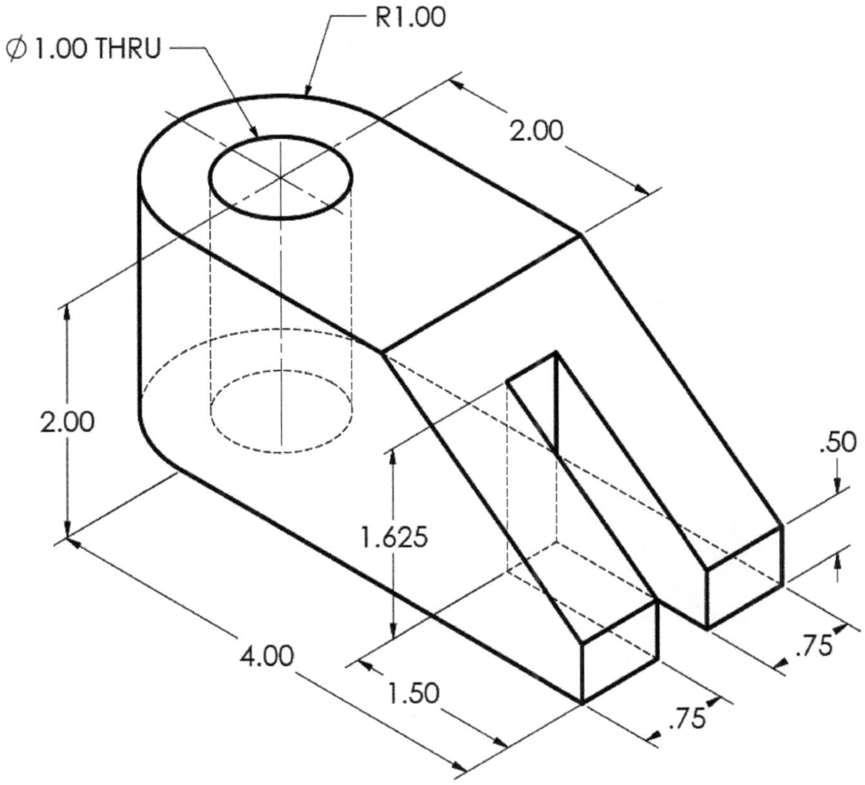

Auxilary views—Problem 13

6

CHAPTER

Scaling

1. Introduction

Most of the time designers need to draw items which are either too large to fit on standard sheets of paper, such as a city map or the layout plan of a building, or too small to be able to represent every detail precisely, such the mechanism of a wrist watch. In such cases, a method called scaling provides a system to accurately resize real world objects and represent them on paper. The following chapter discusses the process for successful scaling and the correct use of scaling tools.

For an object to be accurately scaled, the dimensions need to be reduced by the same proportion in all directions (x, y, and z-axes). As an example, if a map is scaled properly, it will be resized in the same proportion in the north-south and east-west directions, otherwise distortion would occur.

A scale is defined as the ratio between the dimensions of a model and the dimensions of the actual object being represented. It is indicated by two numbers. The first of them refers to the units on the drawing and the second refers to the actual dimensions of the object. For example, a scale defined as 1 inch to 10 feet specifies that for every inch that is measured on the drawing, 10 feet in the actual object are represented.

In theory, any scale is acceptable in a drawing and selecting a particular scale depends on the designer. However, there exist certain scales that are considered standard and should always be used if possible. If a non-standard scale is chosen for a drawing (ex. 1 = 3.1415), then the distance measured on the drawing has to be multiplied against the scale to achieve the proper measurement, which can be a laborious process and prone to errors. When an object is scaled to standard proportions, a set of instruments, also called scales, can be used in the shop or at the worksite to take dimensions directly from the drawing. In order to reduce confusion in this chapter, the word "scale" will be lower case if referring to the proportion of an object and the first letter will be capitalized ("Scale") if referring to the actual measuring instrument.

Scales are six sided measuring instruments, usually made of plastic or wood, where each side provides a set of measuring ticks based on a particular standard scale. Three Scales are used in the engineering profession: Engineer's, Metric, and Architect's scales (see Figure 6.1).

2. Engineer's Scale

The engineer's Scale is nothing more than a Scale which subdivides an inch into specific segments (10, 20, 30, 40, 50, and 60, depending on the specific scale being used). As an example, the engineer's 10 Scale subdivides each inch into ten segments; the engineer's 20 Scale subdivides each inch into twenty segments, and so forth. Although in the engineer's Scale, the inch is the primary unit, it can also be used for feet and even miles. While a little confusing at first, this system is designed to simplify taking and transferring measurements.

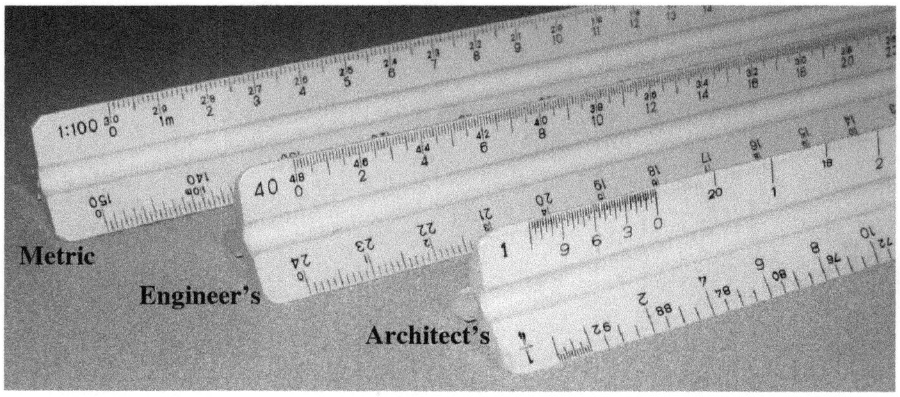

Figure 6.1 Metric, Engineer's, and Architect's Scales

The engineer's Scale appears in the following format: 1 = X, where "X" is a standard scale.

For example, in a scale 1 = 20, the "1" is read and understood to be one inch equals 20 inches (if no units are specified, else it could be 20 feet, 20 miles, etc.). When taking a measurement from the Scale, each segment translates into one inch. If the scale was 1 = 20 feet, then each segment would translate into 1 foot.

Scaling instruments can be used even with drawings that use scales that are not directly available on the standard instruments. As long as the scale of the drawing is based on multiples of 10, then the correct measurement can be read from the Scale without any additional mathematics, by just shifting the decimal place, either to the left or right. If the scale was 1 = 2 feet, then each segment would translate into 0.1 feet. If the scale was 1 = 200 miles, each segment would represent ten miles.

Example 1:
In the following figure (Figure 6.2), assume the measured scale is 1 = 10. Therefore, one inch = ten inches. The line extends 43 marks, each at 1/10 inch, so the line is 43 inches. Each tenth mark is numbered. At a scale of 1 = 10 feet, the line would be 43 feet. If the scale is 1 = 1, the line would be 4.3 inches.

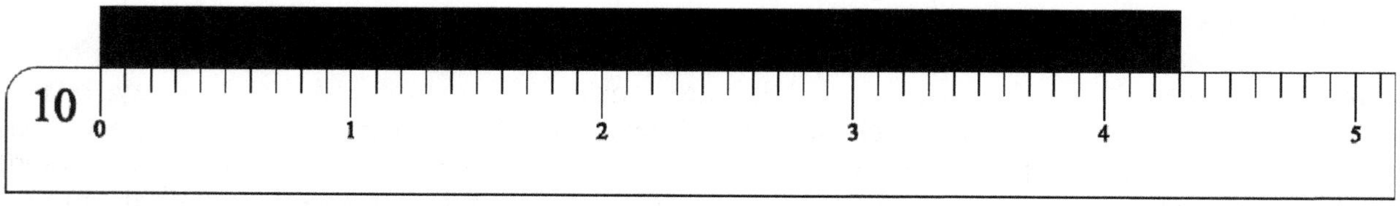

Figure 6.2 Example using the engineer's 10 Scale

Example 2:
In the following figure (Figure 6.3), assume the measured scale is 1 = 500. Therefore, one inch = 500 inches. Since a 1 = 500 Scale is not provided, the 1 = 50 Scale is used, with the measured distance altered to reflect the larger scale. The line extends 88 marks, each at 1/50 inch (at a scale of 1 = 50), but since the required scale is 1 = 500 (ten times larger), each mark will represent a distance of 1/5 inch, so the line would be 880 inches. Another method would be to measure the line as if the scale were indeed 1 = 50 (distance = 88 inches), then move the decimal point one place to the right (distance = 880 inches).

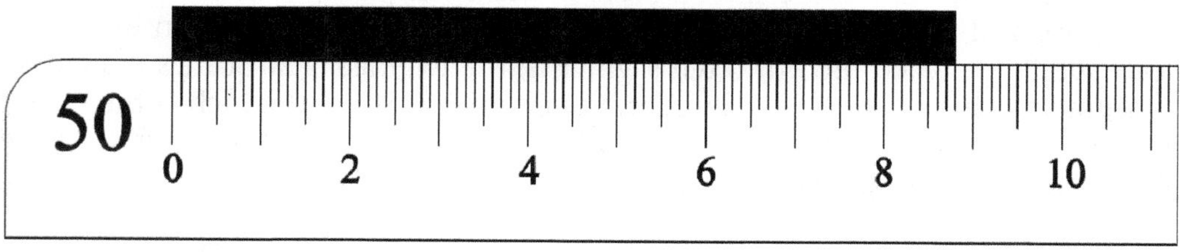

Figure 6.3 Example using the engineer's 50 Scale

3. Metric Scale

The metric Scale is a Scale which reduces by a specific proportion a metric tape measure. The metric Scale is an exact ratio, not a segmentation of distances as with the engineer's Scale. Common metric Scales exist for 1:100, 1:200, 1:250, 1:300, 1:400, and 1:500.

The metric Scale appears in the following format: 1:X, where "X" is a standard scale.

For example, in a scale 1:100, the scale is read as "one to one hundred" and means that one drawn unit equals 100 real world units. Note that all measurements read from the metric Scale are in meters, while the distance is commonly reported in millimeters. As with the engineer's Scale, each scale can be used for scales of multiples of 10, as long as the decimal point is shifted. For example, the 1:100 Scale can be used for 1:1, 1:10, or 1:0.001.

Example 3:
In the following figure (Figure 6.4), assume the measured scale is 1:100. The line extends 6.3 m, which is reported as 6,300 mm. If the required scale was 1:10, the distance would be 630 mm. If the required scale was 1:1000, the distance would be 63,000 mm.

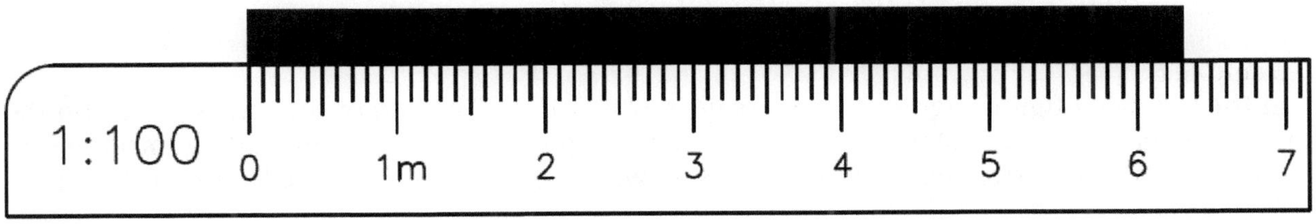

Figure 6.4 Example using the metric's 100 Scale

Example 4:
In the following figure (Figure 6.5), assume the measured scale is 1:5. A 1:5 Scale is not provided, so the 1:500 Scale is used, with decimal place adjustments. The line extends 18.75 m, at a scale of 1:500, which would be reported as 18,750 mm. The required scale is 100 times smaller, so the decimal place is moved two places to the left (or the previous value is divided by 100) to reflect the true distance, which is 187.5 mm.

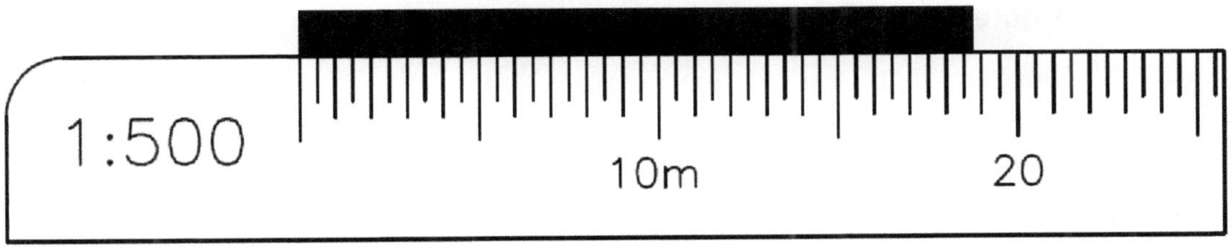

Figure 6.5 Example using the metric's 500 Scale

4. Architect's Scale

The architect's Scales differs from the other Scales discussed in that the number on the left of the decimal or equal sign (1:100 or 1 = 100) is specified as some fraction of an inch and the number to the right of the equal sign is always 1'-0". The architect's Scale exists for the following:

3/32" = 1'-0"	1/8" = 1'-0"	3/16" = 1'-0"	1/4" = 1'-0"	3/8" = 1'-0"
1/2" = 1'-0"	3/4" = 1'-0"	1" = 1'-0"	1-1/2" = 1'-0"	3" = 1'-0"

As an example, the 1/8" = 1'-0" scale is numerically 1 = 96, because there are 96 1/8" segments in 1'-0". When measuring distances with the architect's Scale, instead of placing the Scale at the left end of the line to be measured (0 on the x-axis) and measuring in the positive direction (along a number line), the end point of the line should be placed on the exact foot marker, with the measured line overlap extending into the more subdivided zone to ascertain the fraction of a foot. With the architect's Scale, the given scale cannot be altered easily by moving a decimal point.

Example 5:
In the following figure (Figure 6.6), the scale is 1/2" = 1'-0". The end of the line is placed on the foot marker (4' in this instance), with the remaining portion of the line extending into the subdivided zone. The extended section matches up with the 3" marker, so the length of this line is 4'-3".

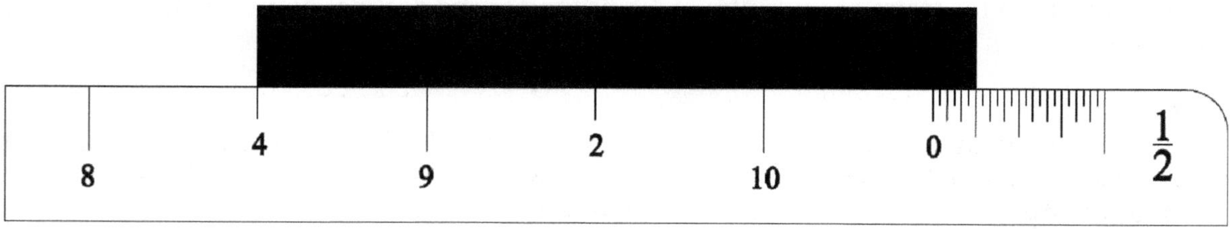

Figure 6.6 Example using the architect's 1/2" = 1'0" Scale

Example 6:
In the following figure (Figure 6.7), the scale is 1/8" = 1'-0". The end of the line is placed on the foot marker (7' in this instance), with the remaining portion of the line extending into the subdivided zone. The extended section matches up with the 10" marker, so the length of this line is 7'-10".

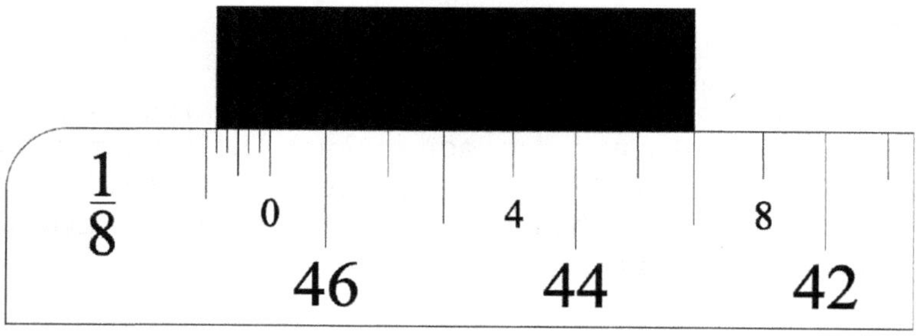

Figure 6.7 Example using the architect's 1/8" = 1'0"

Practice Test

1. **A line needs to be measured at a scale of 1/8″ = 1′-0″. Which Scale should be used?**
   A) Metric Scale
   B) Engineer's Scale
   C) English Scale
   D) Architect's Scale
   E) None of the above, because 1:350 is not a standard scale.

2. **A line needs to be measured at a scale of 1:300. Which Scale should be used?**
   A) Metric Scale
   B) Engineer's Scale
   C) English Scale
   D) Architect's Scale
   E) None of the above, because 1:300 is not a standard scale.

3. **A line needs to be measured at a scale of 1 = 2000 ft. Which Scale should be used?**
   A) Metric Scale
   B) Engineer's Scale
   C) English Scale
   D) Architect's Scale
   E) None of the above, because 1:2000 is not a standard scale.

4. **A line needs to be measured at a scale of 1:350. Which Scale should be used?**
   A) Metric Scale
   B) Engineer's Scale
   C) English Scale
   D) Architect's Scale
   E) None of the above, because 1:350 is not a standard scale.

5. **It is acceptable if the scale is different for each direction.**
   A. True B. False

6. **The engineer's Scale is based on pure ratios.**
   A. True B. False

7. **The metric Scale gives measurements in meters, but it is standard to report in millimeters.**
   A. True B. False

8. **The architect's Scale needs to be used for standard scales only.**
   A. True B. False

INDICATE THE LENGTH OF EACH DIMENSION ON THE RUGBY FIELD BELOW. THE SCALE IS 1:1000.

A. ______________________

B. ______________________

C. ______________________

D. ______________________

E. ______________________

F. ______________________

G. ______________________

H. ______________________

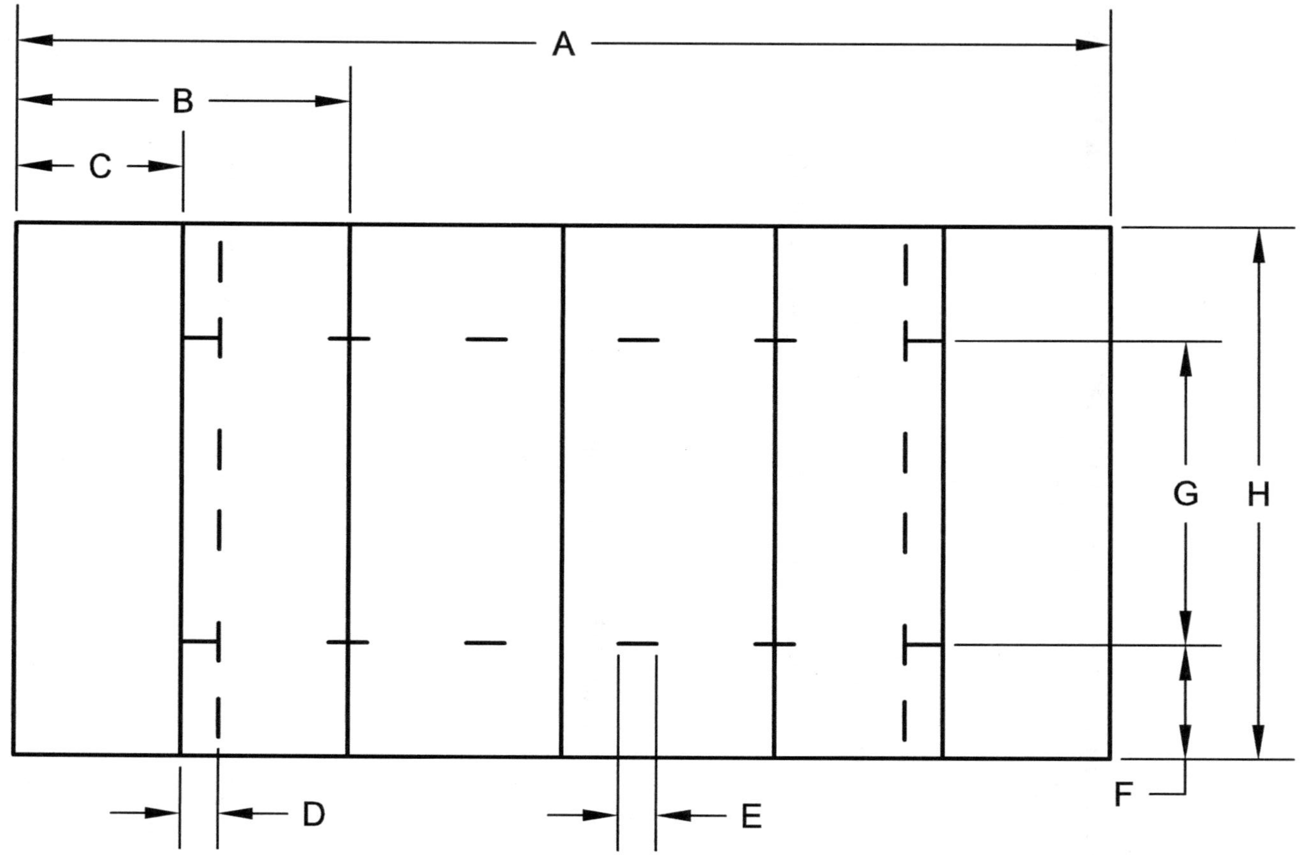

AUTHOR:	SCALE:
TEAM: SECTION:	DATE:

INDICATE THE LENGTH OF EACH DIMENSION ON THE DETAIL OF THE FOOTBALL FIELD BELOW. THE SCALE IS 1/32" = 1' - 0".

A. ______________________

B. ______________________

C. ______________________

D. ______________________

E. ______________________

F. ______________________

G. ______________________

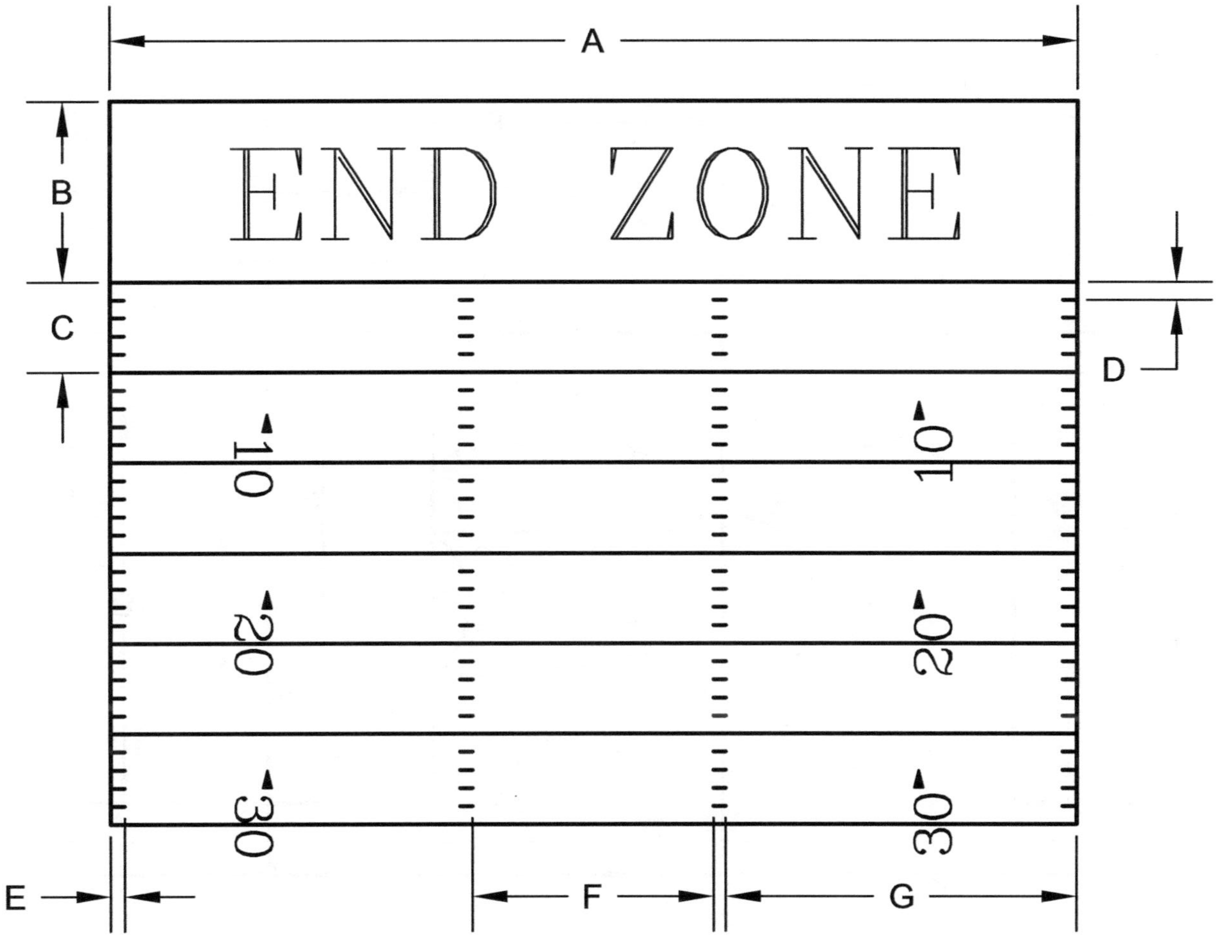

AUTHOR:	SCALE:
TEAM: SECTION:	DATE:

INDICATE THE LENGTH OF EACH DIMENSION ON THE SOCCER FIELD BELOW.
THE SCALE IS 1/64" = 1' - 0".

A. ______________________

B. ______________________

C. ______________________

D. ______________________

E. ______________________

F. ______________________

G. ______________________

H. ______________________

I. ______________________

J. ______________________

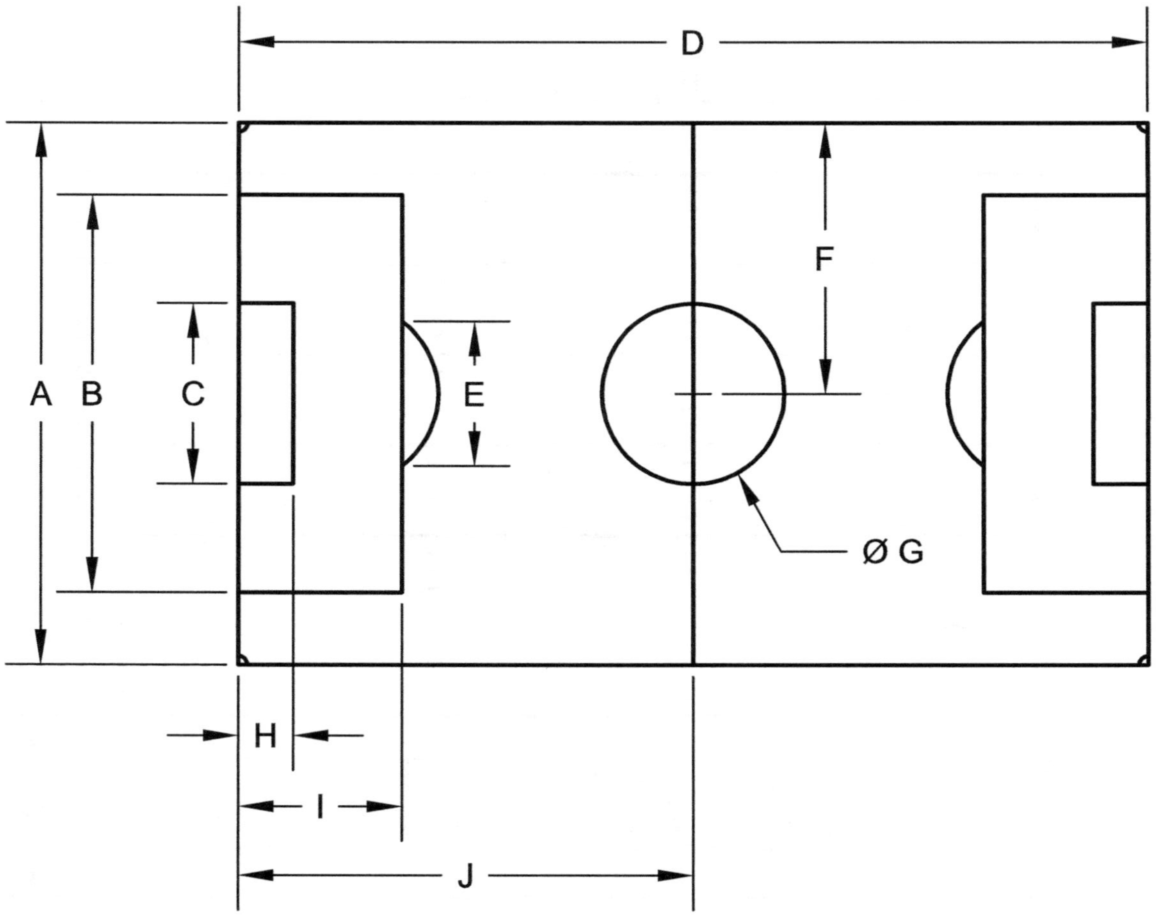

AUTHOR: .... SCALE: ....

TEAM: .... SECTION: .... DATE: ....

INDICATE THE LENGTH OF EACH DIMENSION ON THE TENNIS COURT BELOW.
THE SCALE IS 1=300.

A. ____________________

B. ____________________

C. ____________________

D. ____________________

E. ____________________

F. ____________________

G. ____________________

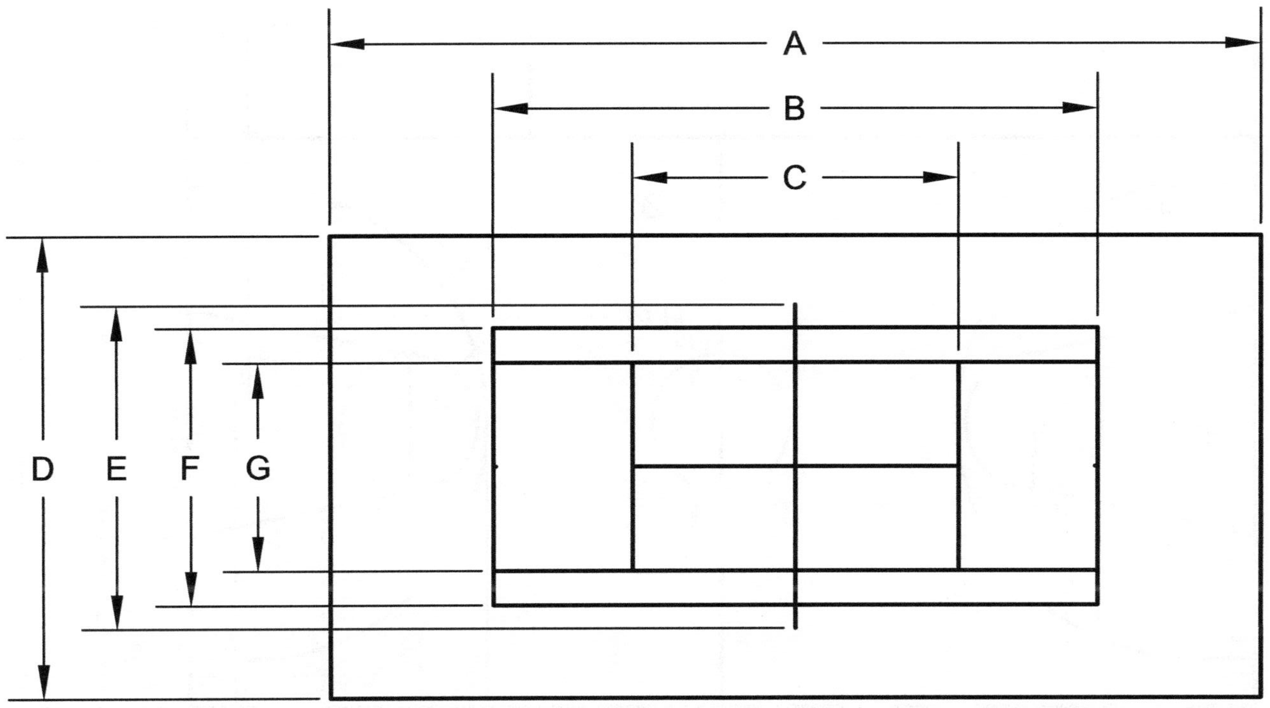

AUTHOR: .... SCALE: ....

TEAM: .... SECTION: .... DATE: ....

INDICATE THE LENGTH OF EACH DIMENSION ON THE INTERNATIONAL BASKETBALL COURT BELOW. THE SCALE IS 1:200.

A. ______________________

B. ______________________

C. ______________________

D. ______________________

E. ______________________

F. ______________________

G. ______________________

H. ______________________

I. ______________________

J. ______________________

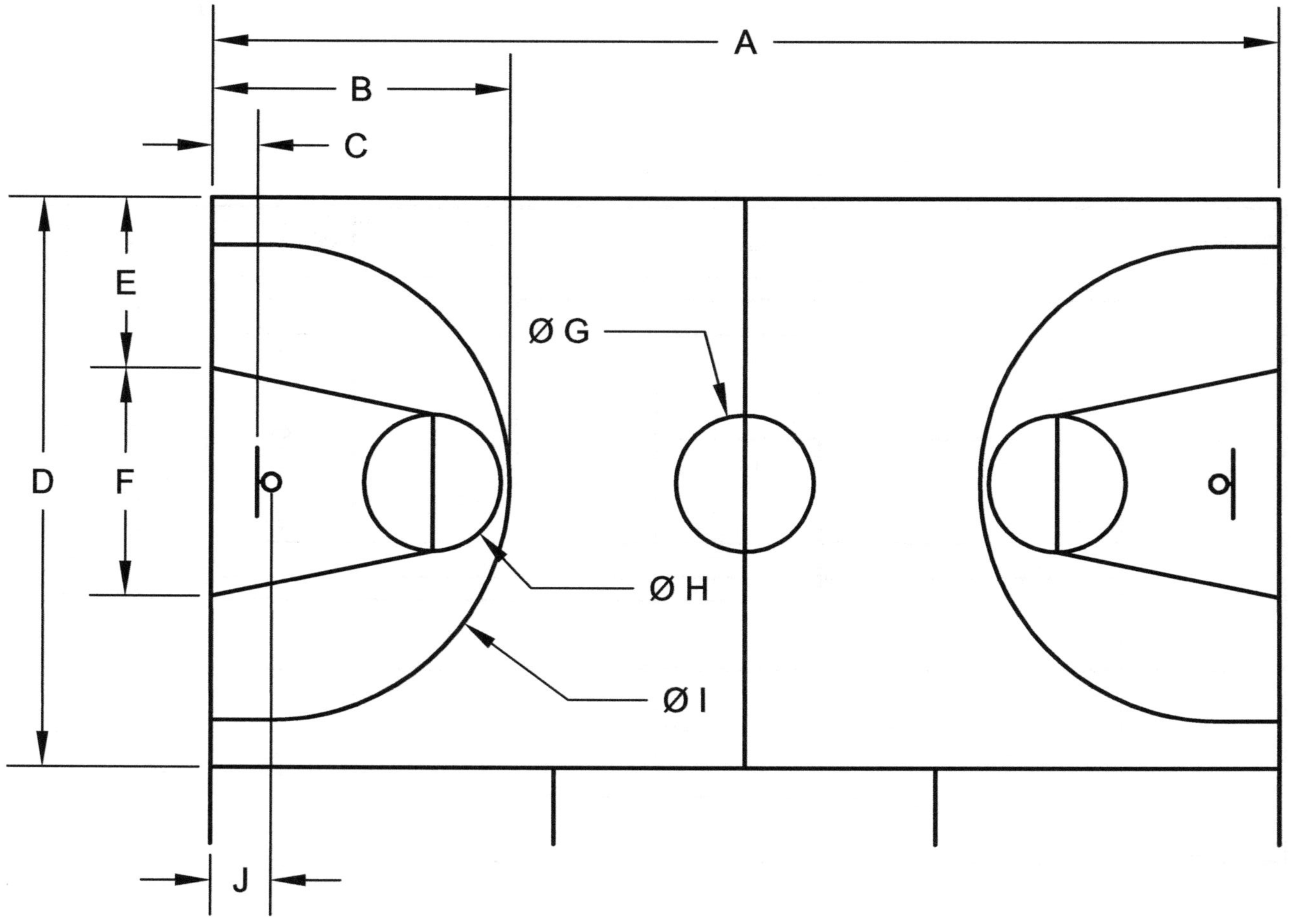

AUTHOR:	SCALE:
TEAM: SECTION:	DATE:

INDICATE THE LENGTH OF EACH DIMENSION ON THE HOCKEY RINK BELOW. THE SCALE IS 1=400.

A. ______________________

B. ______________________

C. ______________________

D. ______________________

E. ______________________

F. ______________________

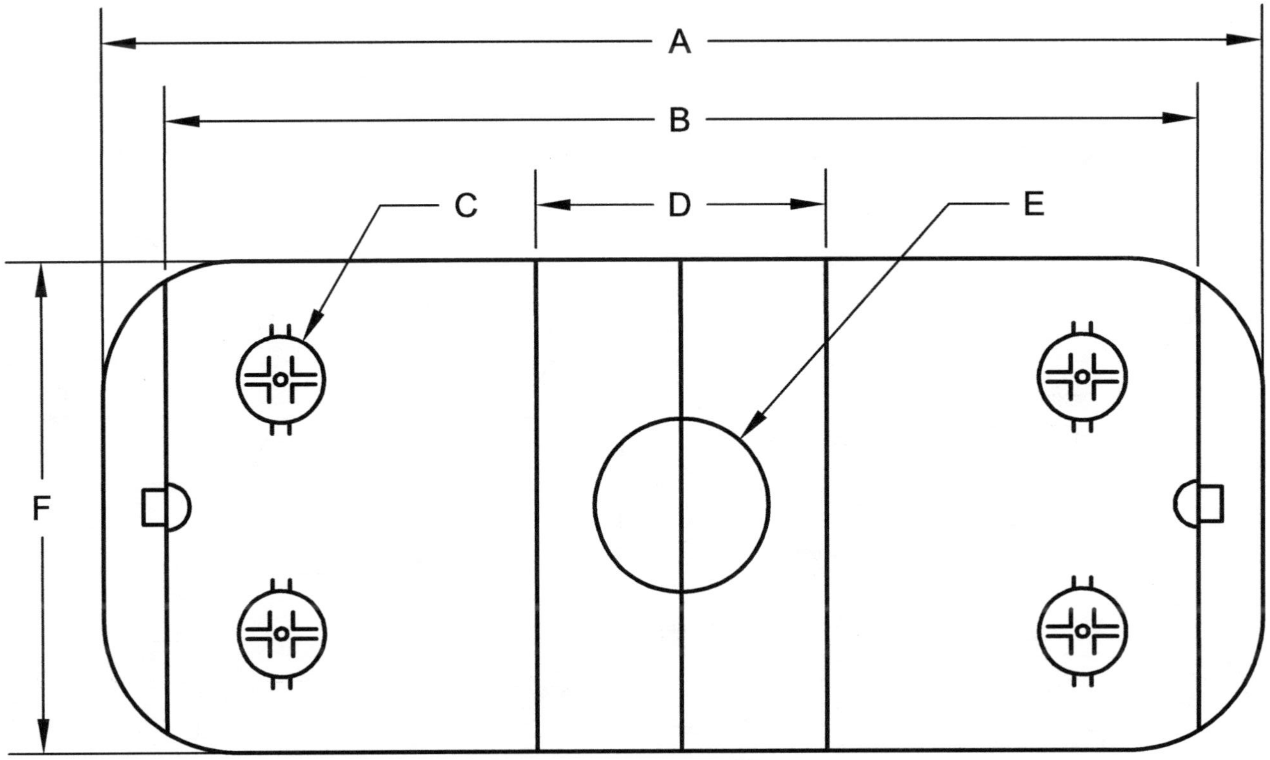

AUTHOR:	SCALE:
TEAM: SECTION:	DATE:

FIND THE DISTANCE BETWEEN THE SCHOOLS AS LISTED. THE MAP SCALE IS1:10,000,000. REPORT DISTANCES IN KILOMETERS.

A. TEX. A&M - AUB.:__________ D. TENN. - ALABAMA:__________ G. ARK. - OLE MISS:__________

B. MISSOURI - FLORIDA:__________ E. MISS. ST. - VAND.:__________

C. LSU - KENTUCKY:__________ F. GEORGIA - S. CAR.:__________

AUTHOR:	SCALE:
TEAM: SECTION:	DATE:

DETERMINE THE DISTANCE BETWEEN THE FOLLOWING CITIES: (A) DALHART TO HARLINGEN, (B) EL PASO TO TEXARKANA, (C) LUBBOCK TO HOUSTON, (D) AUSTIN TO SAN ANGELO, (E) SAN ANTONIO TO DALLAS, (F) DALHART TO EL PASO. ASSUME A MAP SCALE OF1" = 100 MILES:

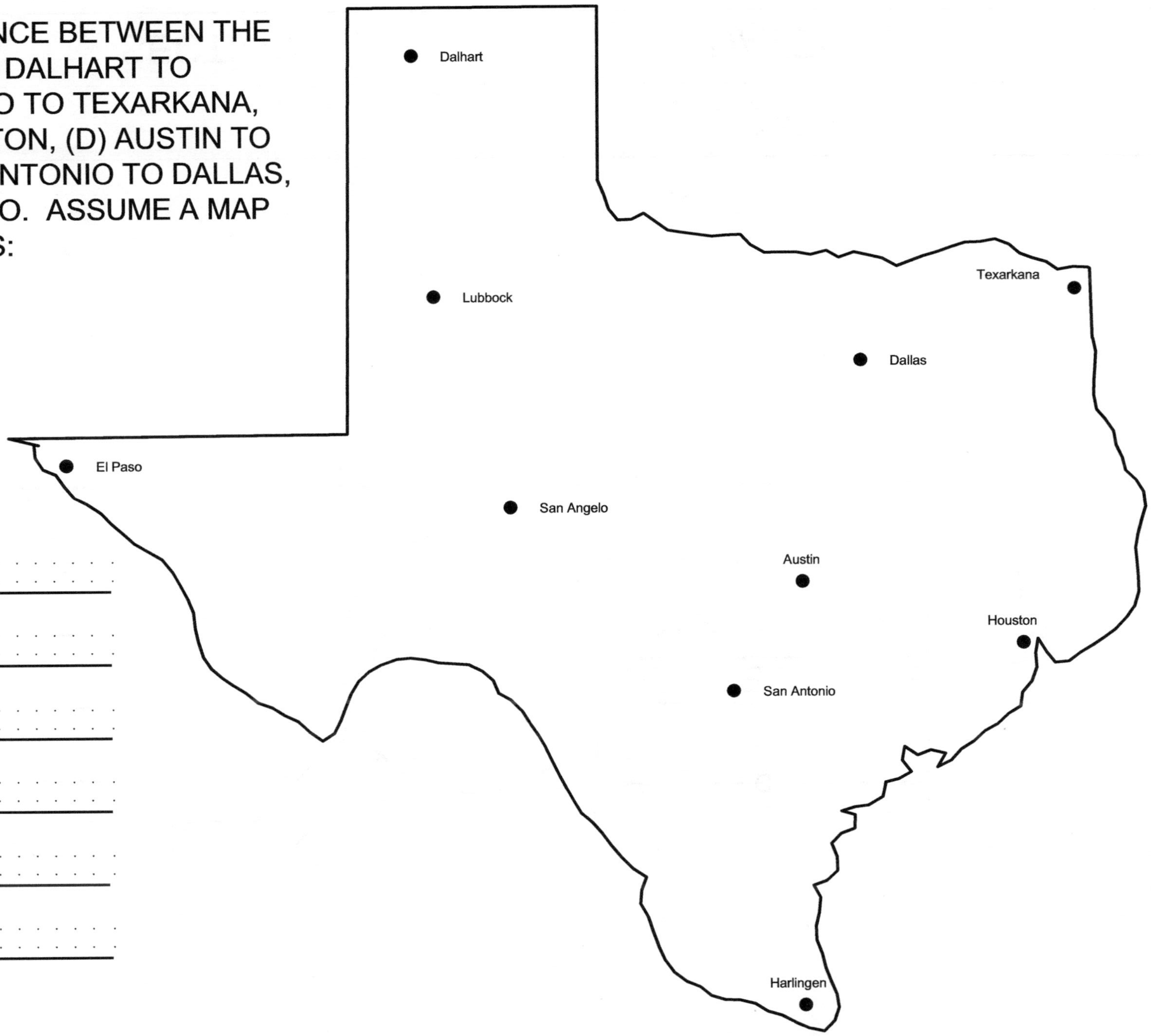

A. ______________________

B. ______________________

C. ______________________

D. ______________________

E. ______________________

F. ______________________

AUTHOR: .... SCALE: ....

TEAM: .... SECTION: .... DATE: ....

INDICATE THE LENGTH OF EACH DIMENSION ON THE HOUSE KEY BELOW. THE KEY IS SHOWN DOUBLE SCALE.

A. ______________________

B. ______________________

C. ______________________

D. ______________________

E. ______________________

F. ______________________

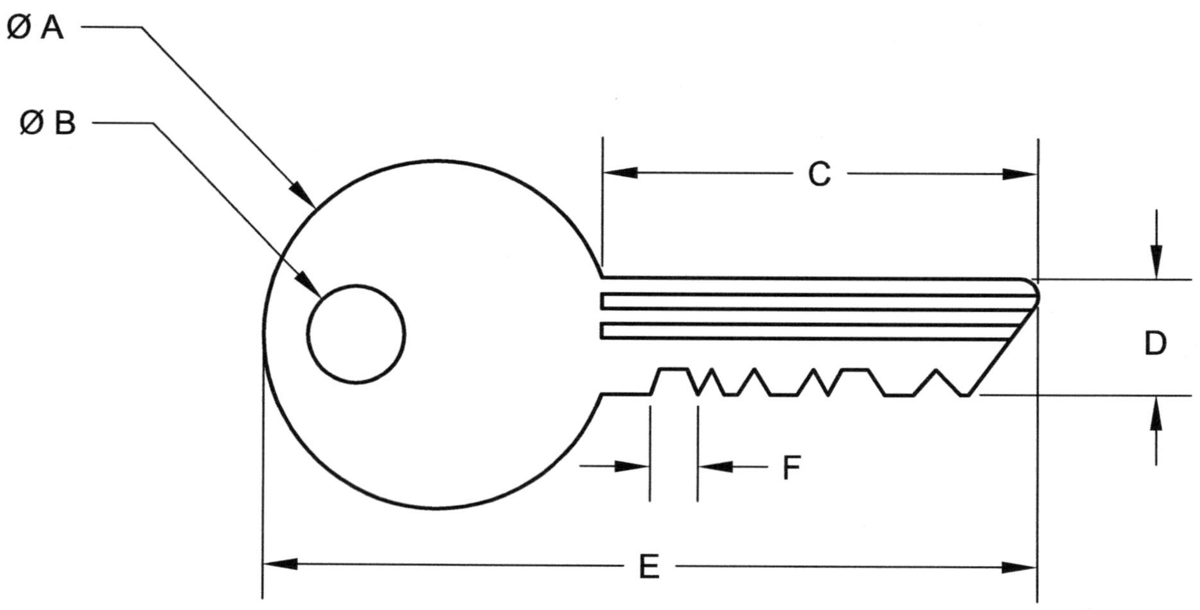

AUTHOR:	SCALE:
TEAM: SECTION:	DATE:

INDICATE THE LENGTH OF EACH DIMENSION ON THE SHIP BELOW. THE SCALE IS 1=1000.

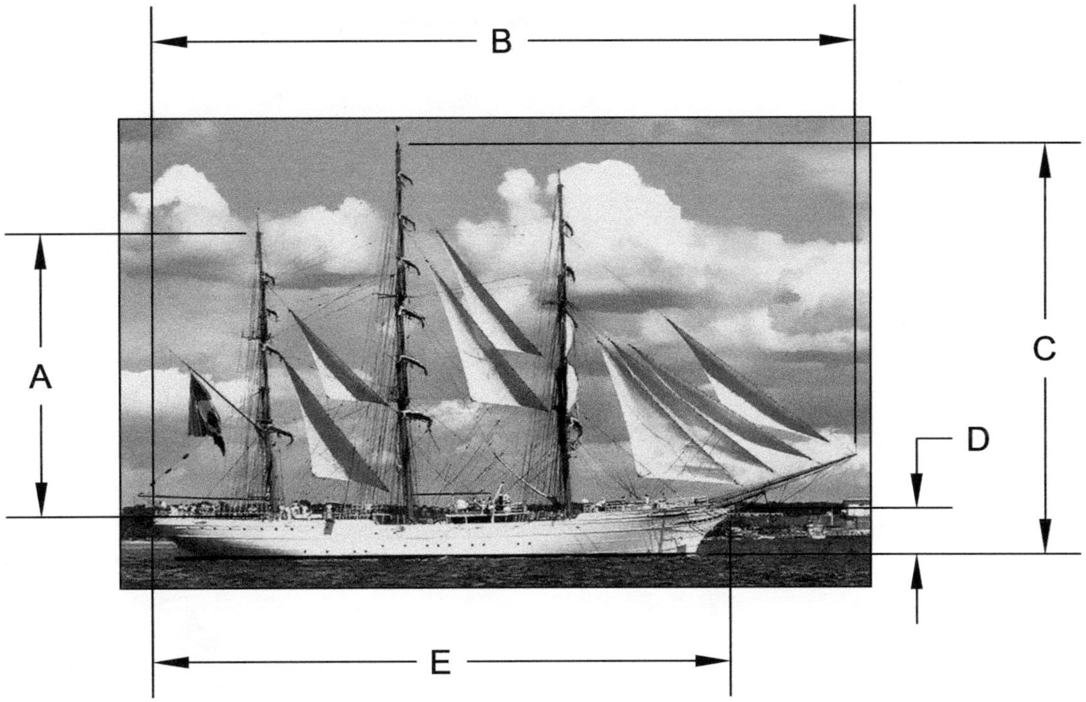

A. ____________________

B. ____________________

C. ____________________

D. ____________________

E. ____________________

AUTHOR:	SCALE:
TEAM: SECTION:	DATE:

CHAPTER 7

Dimensioning

1. Introduction

In order for a design to be manufactured, specific information about its size and shape needs to be conveyed. In general, the exact size and location of all features needs to be explicitly indicated in the drawings. This information could include the overall width of the object, the diameter of a hole, or the radius of a fillet. If this information is incorrect or incomplete, defective parts could be manufactured that could ultimately lead to serious engineering failures.

Most of the information can be expressed with dimensions. Although there are different ways to dimension a drawing, some general guidelines and practices need to be followed in order to avoid possible misunderstandings or misinterpretations. In the United States, these guidelines are defined by the standard ANSI Y14.5M-2009, published by ASME (American Society of Mechanical Engineers).

2. Dimensioning Terminology

Dimensions consist of extension lines, dimension lines, arrowheads, and leaders with elbows. In general, all dimension features are represented by thin lines (0.25 mm). Extension lines extend perpendicularly from the boundaries of the feature being dimensioned, with a small gap (1/16" to 1/8") between the extension line and the visible line so that the extension line is not mistaken for a feature on the object. Dimension lines are located between two extension lines. The dimension value typically appears in the center of the dimension line. Arrowheads are located on both ends of a dimension line to indicate direction. They are usually 1/8" long and 1/24" tall. Leaders are lines with an arrowhead at one end and an elbow at the other used to direct a dimension or a note to a specific feature on the drawing. See Figure 7.1 for an example.

2.1 Dimensioning Styles

Two dimensioning styles can be used: unidirectional or aligned. Unidirectional dimensions always appear in the standard horizontal position, so they are read left to right regardless of the direction of the dimension lines. In aligned dimensions, the text follows the orientation of the feature dimensioned. Aligned dimensions are used almost entirely in architectural drawings. See Figure 7.2.

3. General Rules for Dimensioning

There are numerous rules defined in the standard ANSI Y14.5M-2009 that describe proper ways to dimension a drawing. Although some of these rules have very specific applications, some fundamental practices should always be taken into account.

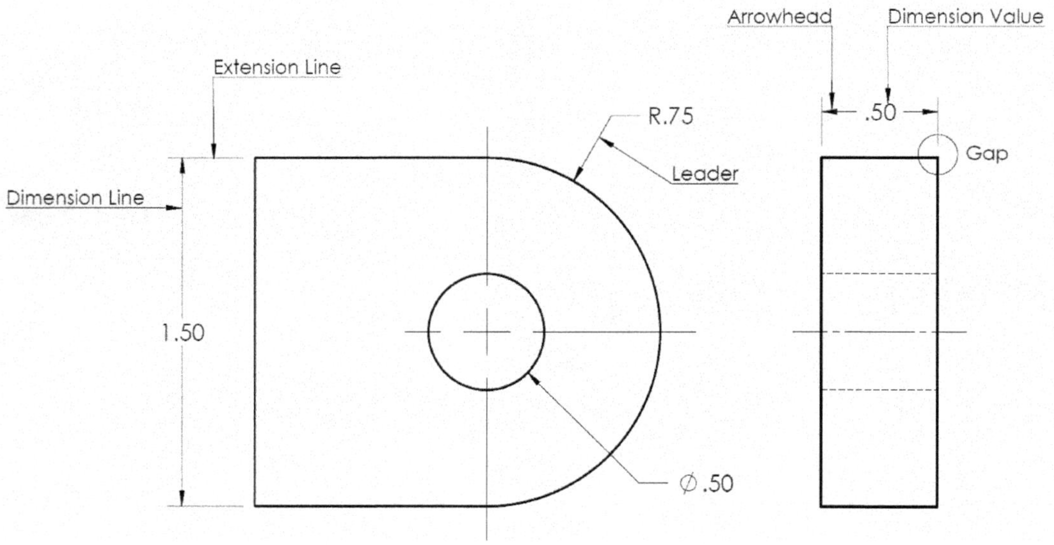

Figure 7.1 Dimension terminology

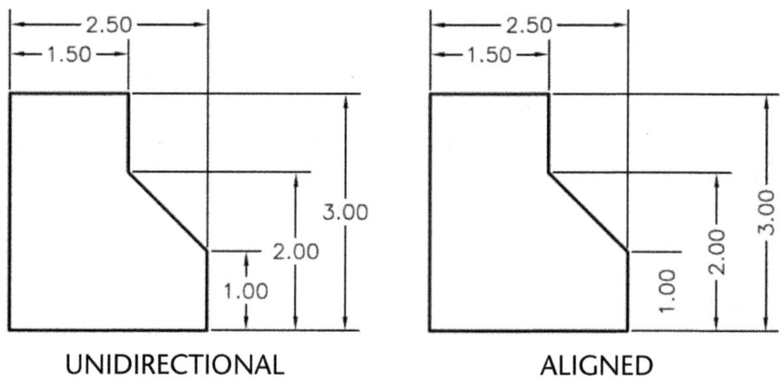

Figure 7.2 Unidirectional and aligned dimensioning styles

- Dimensions are always shown in real world units. Most drawings use either the International System of Units, which is metric (mm is the basic unit), or the US system, which uses English units (inch is the basic unit).
- When dimensions are expressed in inches, do not use the inch mark (") or abbreviation for inches (in). For drawings dimensioned in inches, values less than one inch should not be preceded by a zero. For metric dimensions less than 1 mm, a zero should be placed in front of the decimal point. For metric drawings, the use of the millimeter (mm) notation following the numeral should be omitted, as millimeters are the default units
- Drawings should always be fully dimensioned so the manufacturer has complete information of the design. Incomplete information could result in an inaccurately manufactured design.
- Do not over-dimension your drawings. Avoid giving duplicated or unnecessary dimensions. Only the necessary information required to produce the design should be included.
- Keep the drawing clean and readable. Arrange the dimensions so the drawing is easily understood.

3.1 Dimension Spacing

The first row of dimensions should be placed 3 letter heights away (3/8") from the object, with each successive row being placed 2 letter heights away (1/4"). This spacing allows for clear understanding. See Figure 7.3.

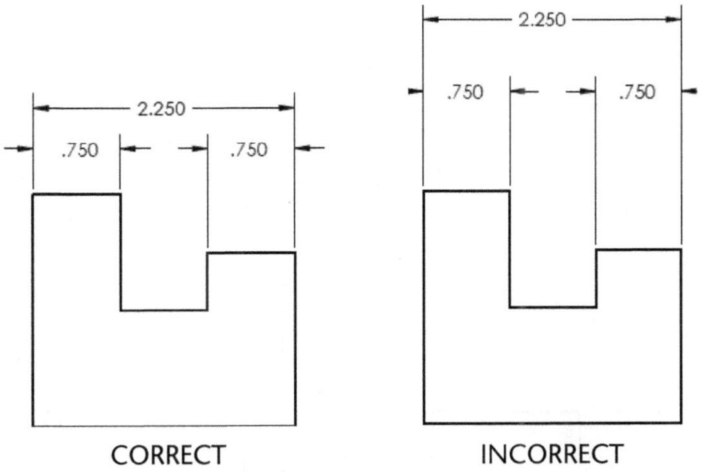

Figure 7.3 Proper dimension spacing

3.2 Do Not Place Dimensions Inside Any View

Always place dimensions away from the view with proper spacing, if possible. In cases where not enough room exists for all required dimensions, space the views further apart to allow for clear understanding of all dimensions. See Figure 7.4.

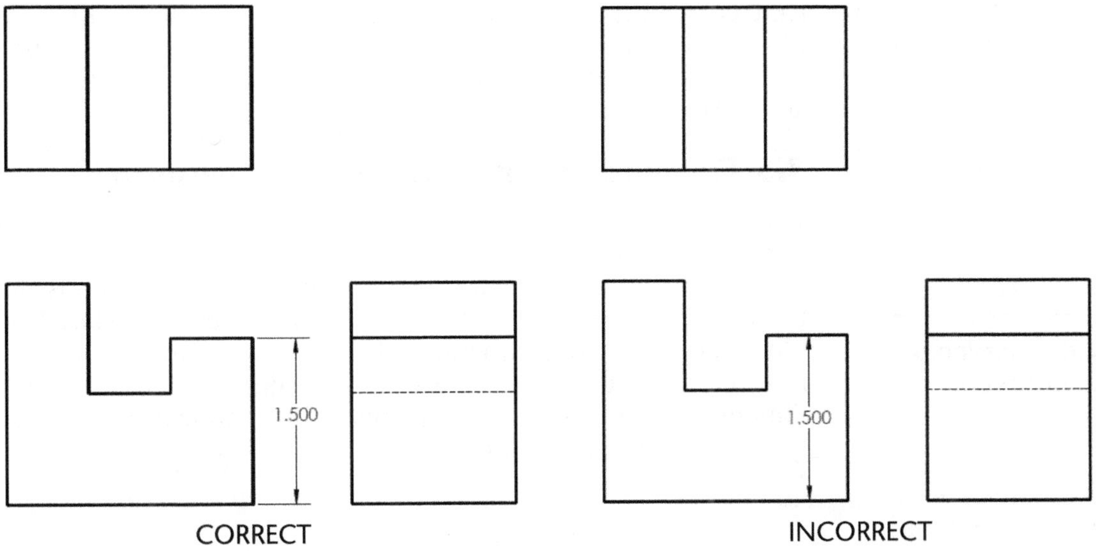

Figure 7.4 Do not place dimensions inside any view.

3.3 Place Dimensions between Views

Place dimensions between orthographic views in order to save space. Ideally, more than one object can be placed on the same page in a manufacturing drawing set. Placing the dimensions between the views can reduce confusion as to which specific dimension belongs to each object. See Figure 7.5.

3.4 Omission of Features

Ideally, a proper dimension should have two extension lines, two dimension lines, two arrowheads, and a numerical dimension centered between the dimension lines. Depending on the size of the feature being dimensioned, there may not be room for all the above dimension features. In

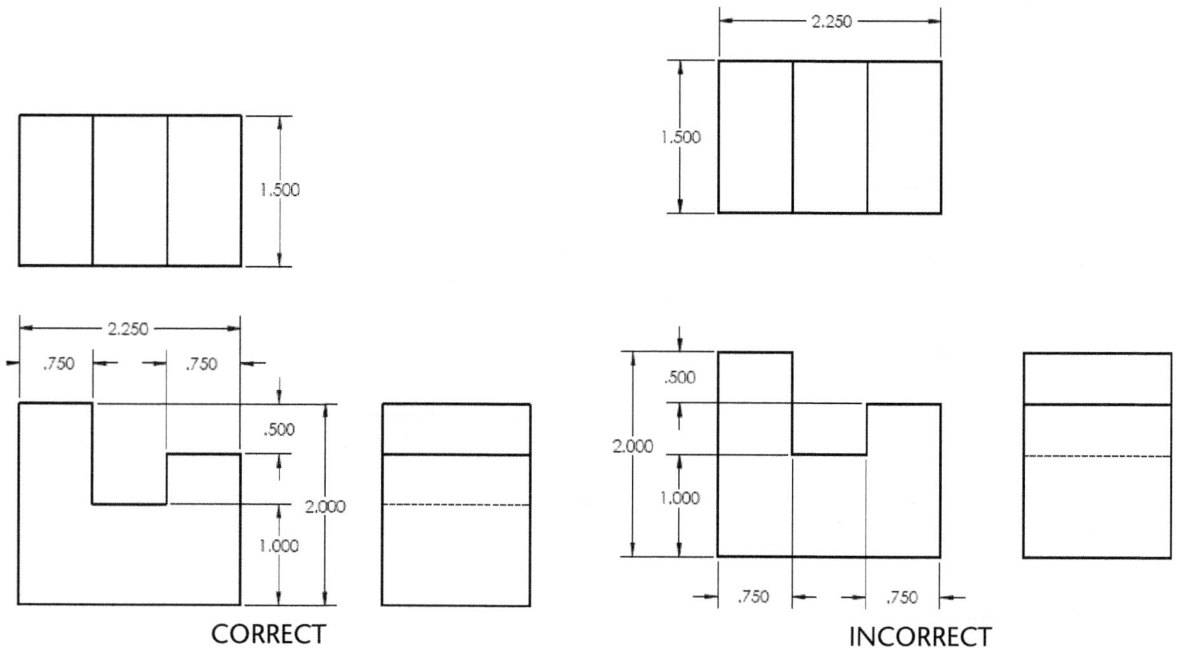

Figure 7.5 Place dimensions between views.

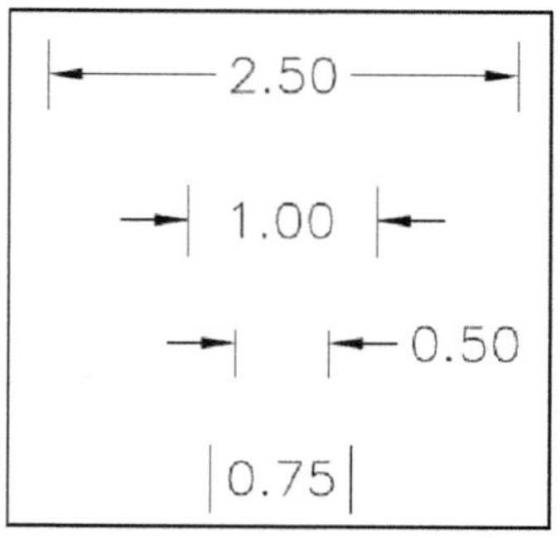

Figure 7.6 Omission of selected dimension features

that case, certain features can be omitted. The number can be placed outside the extension lines with arrowheads pointed inwards, or the dimension can be placed within the extension lines alone. See Figure 7.6.

3.5 Conventional and Baseline Dimensions

Two dimensioning techniques can be used: conventional dimensioning and baseline dimensioning. In conventional dimensioning, an overall dimension is provided with intermediate dimensions closer to the object. In baseline dimensioning, all dimensions start from the same edge of the object, with each feature requiring its own separate row. Baseline dimensions are more accurate than conventional dimensions, since if an error is made in manufacturing, the error is not carried through to the next feature. See Figure 7.7.

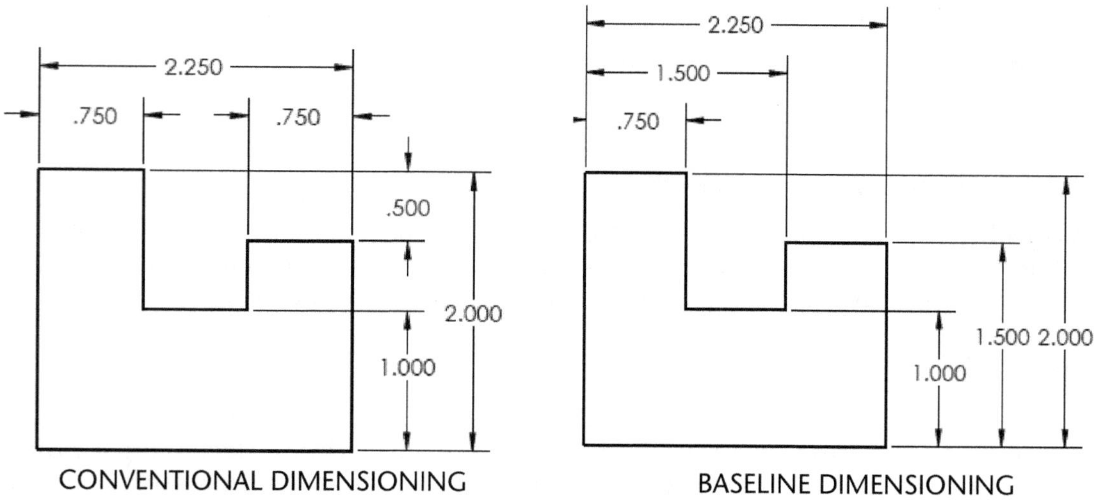

Figure 7.7 Conventional dimensions

If conventional dimensions are used, provide the overall dimension, but leave out one intermediate chain dimension (see Figure 7.8). Theoretically, it does not matter which intermediate dimension is omitted. However, there may be design considerations where certain features require more exact placement. In such cases, omit any other intermediate dimension.

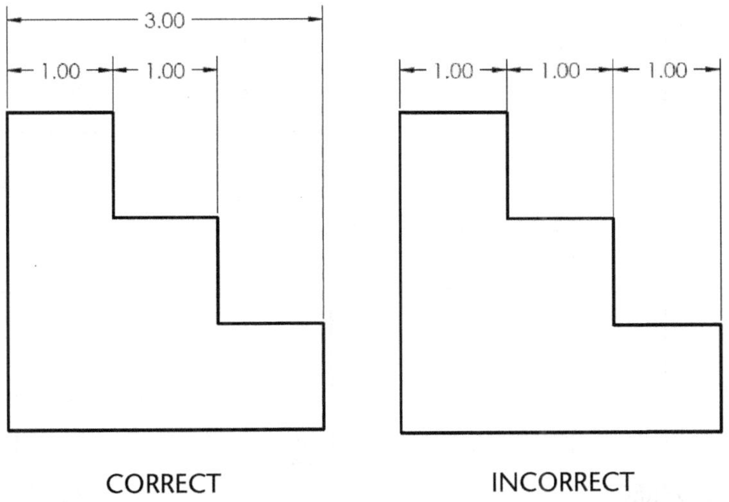

Figure 7.8 Leave out one chain dimension.

3.6 Do Not Cross Dimension Lines

Extension lines can cross visible lines, but not dimension lines. This problem can be avoided by placing the largest dimension furthest away from the object. See Figure 7.9.

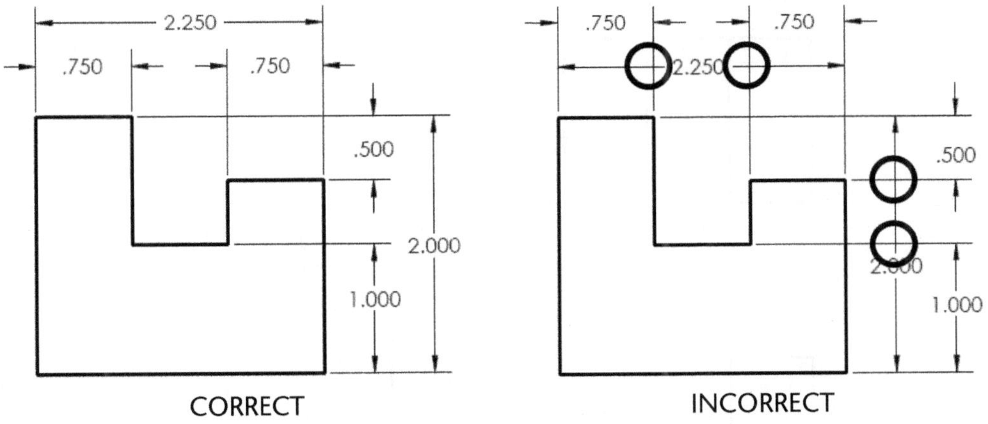

Figure 7.9 Do not cross dimension lines.

3.7 Do Not Over-dimension

Only provide a dimension for each feature once. As an example, if the width of the object is dimensioned in the front view, do not also provide it in the top view. See Figure 7.10.

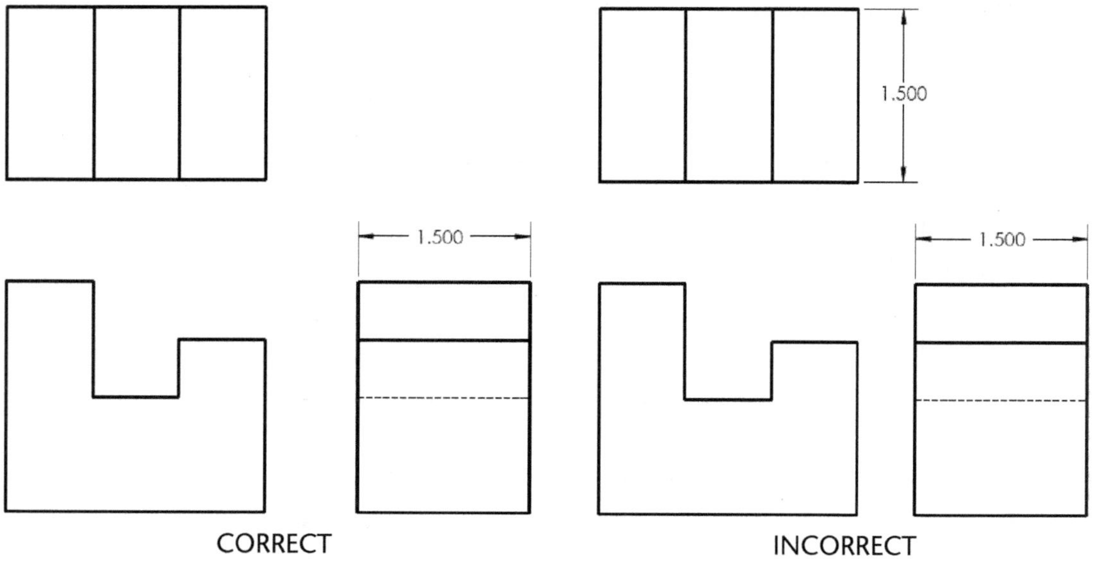

Figure 7.10 Do not provide duplicate dimensions.

3.8 Reference Dimensions

Occasionally, the same dimension can be provided twice in a drawing to eliminate the need for calculations. In these cases, place one of the dimensions in parentheses or provide the text "REF" next to it to specify it as a reference dimension. See Figure 7.11.

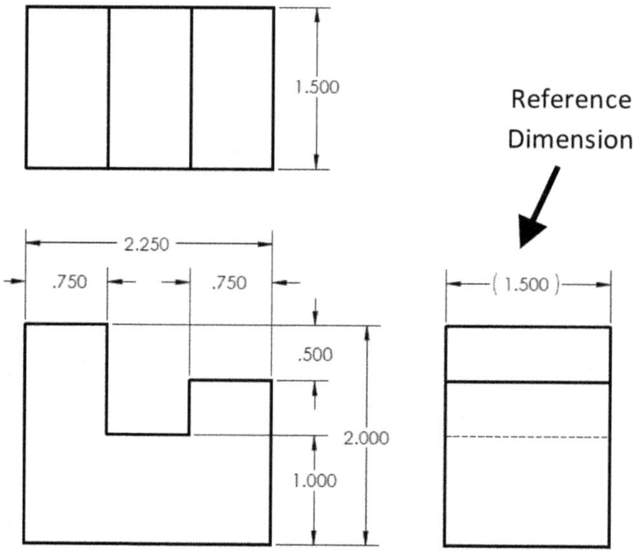

Figure 7.11 Reference dimension

3.9 Do Not Dimension to Hidden Lines

Do not dimension to hidden lines. Always dimension to visible lines. In cases where there are not sufficient visible lines present to adequately dimension an object, because of the object's complexity, create a sectioned orthographic view. Sections will be discussed in a subsequent chapter. See Figure 7.12 for an example.

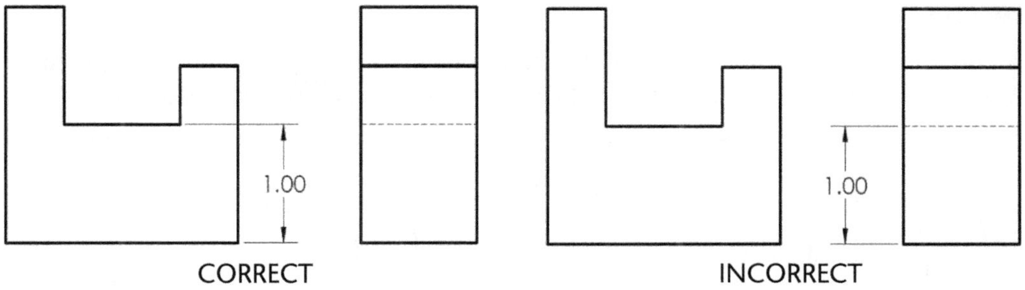

Figure 7.12 Do not dimension to hidden lines.

3.10 Do Not Dimension to T-Joints

Always dimension features in their most descriptive view. An edge view will always appear as a "T," so dimension the feature in the view where the edge appears as a corner. See Figure 7.13.

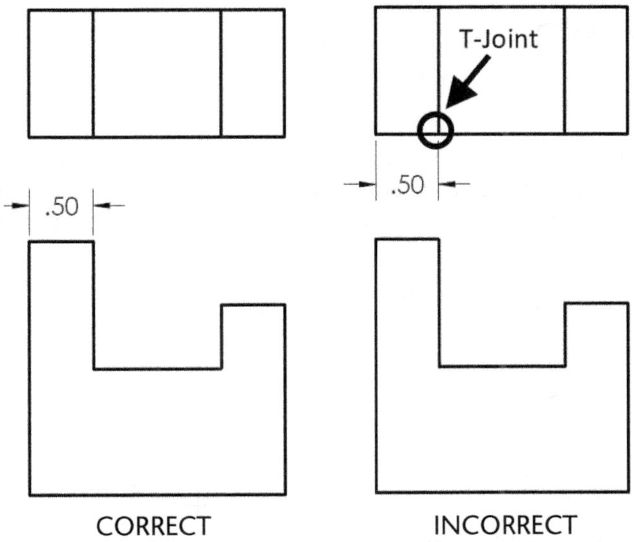

Figure 7.13 Do not dimension to T-Joints.

3.11 Angles

If an object has an angled surface, the angle can be dimensioned. Proper dimensioning requires that the extension lines extend past the visible lines representing the feature. Angled dimensions require plenty of space, so if each vertex of the feature is dimensioned conventionally, angled dimensions are not required. See Figure 7.14.

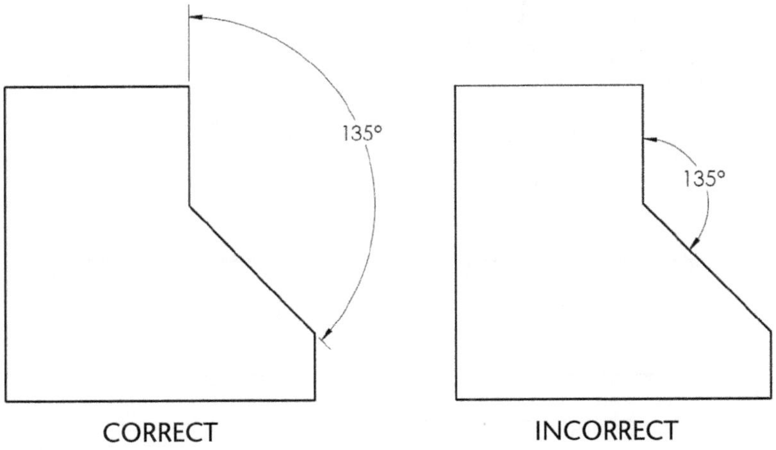

Figure 7.14 Angled dimensions

3.12 Leaders

Leaders are used to direct a dimension or a note to a specific feature on the drawing. Ensure that leaders originate from the end or the beginning of the note, with an elbow provided. See Figure 7.15.

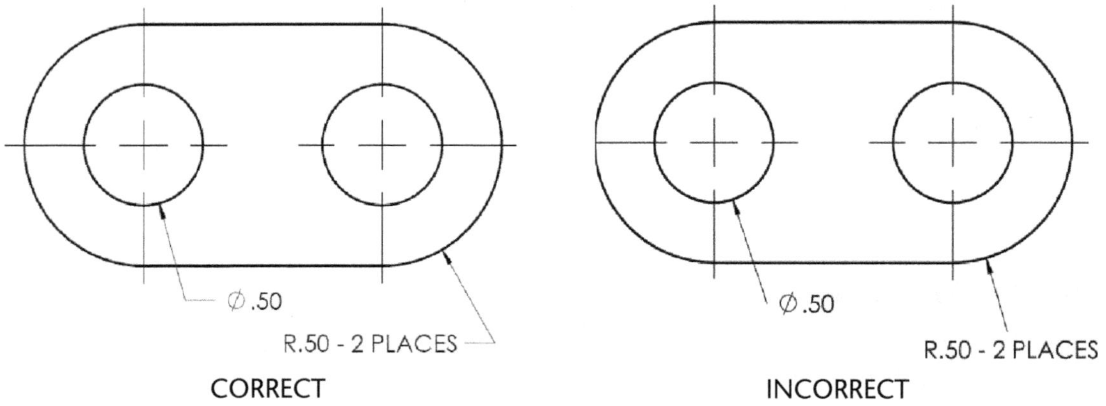

Figure 7.15 Proper use of leaders

3.13 Circles

Circles can be dimensioned with a leader and an arrow pointing toward the center of the circle or by a chord the length of the diameter with arrows pointing outward. Either DIA provided at the end of the dimension or Ø at the beginning of the dimension is acceptable. See Figure 7.16.

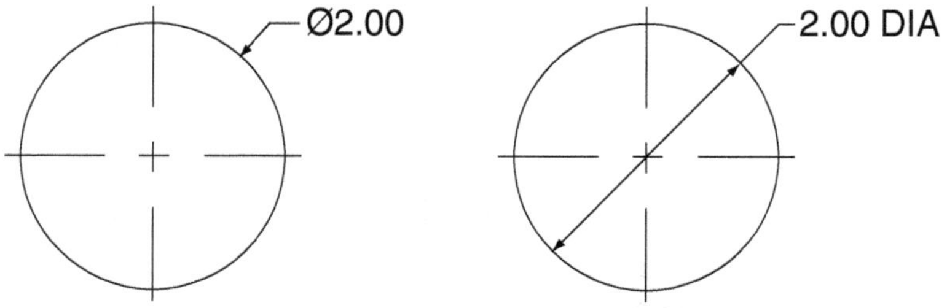

Figure 7.16 Proper circle dimensioning methods

3.14 Holes Are Dimensioned in the Circular View

Holes should be dimensioned in the view where they appear as circles, never where they appear as rectangles; especially if the extents of the hole are represented with hidden lines. See Figure 7.17.

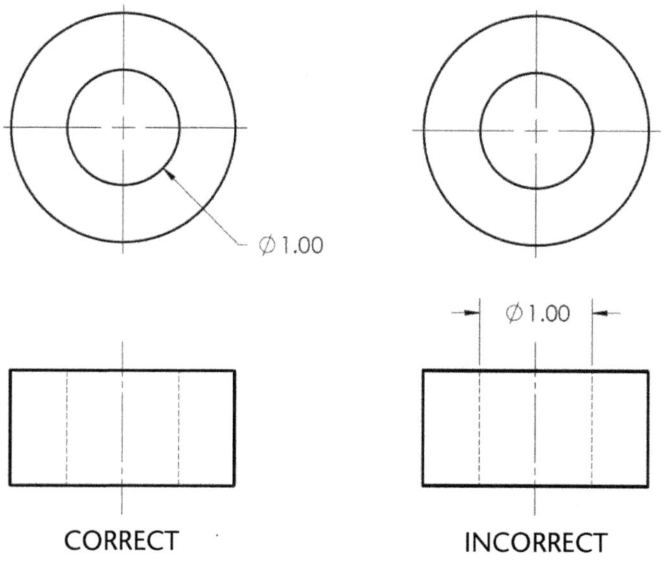

Figure 7.17 Holes should be dimensioned in the circular view.

3.15 Cylinders Are Dimensioned in the Rectangular View

Cylinders should be dimensioned in the view where they appear as rectangles, with a diameter symbol preceding the dimension (either DIA or Ø). The diameter symbol signifies that the linear dimension represents a radial distance. See Figure 7.18.

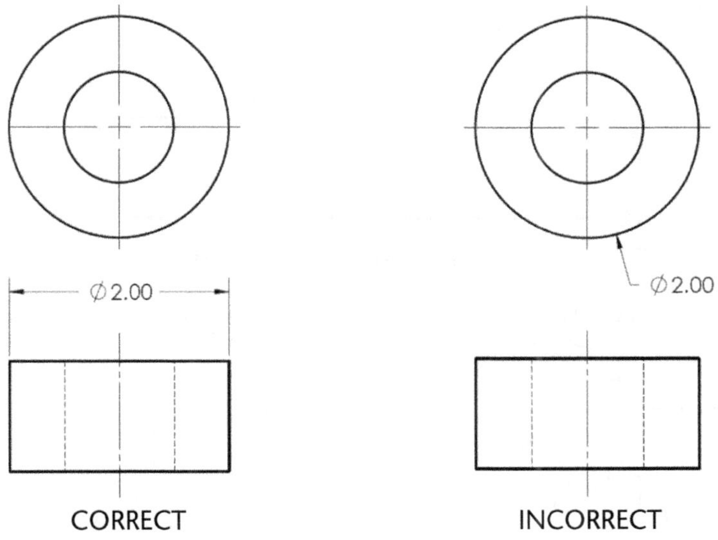

Figure 7.18 Cylinders should be dimensioned in the rectangular view.

3.16 Arcs and Objects with Rounded Ends

Rounded corners (rounds and fillets) and arc features are dimensioned as radii in their rounded views. The abbreviation "R" (for radius) always precedes the dimension (see Figure 7.19). Occasionally, the leader and the radial dimension can be placed inside the views (see Figure 7.20). In general, arcs less than 180° are dimensioned with the radius while arcs greater than 180° are dimensioned with the diameter.

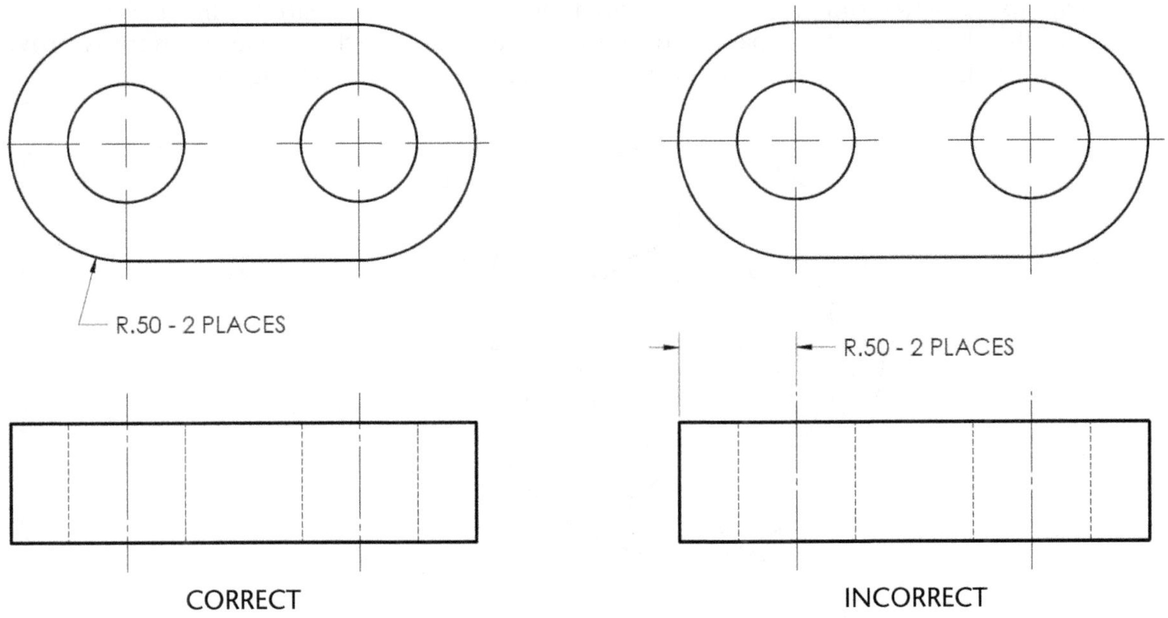

Figure 7.19 Features with rounded ends are dimensioned as radii in their rounded views.

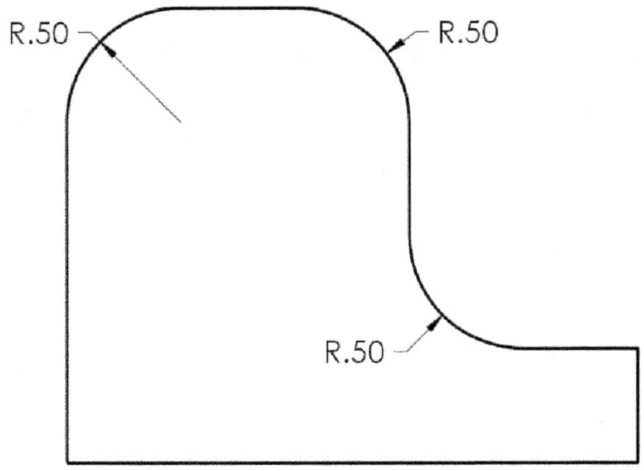

Figure 7.20 Dimensioning rounds and fillets

Always dimension to the center of an arc, rather than the outermost edge. See Figure 7.21.

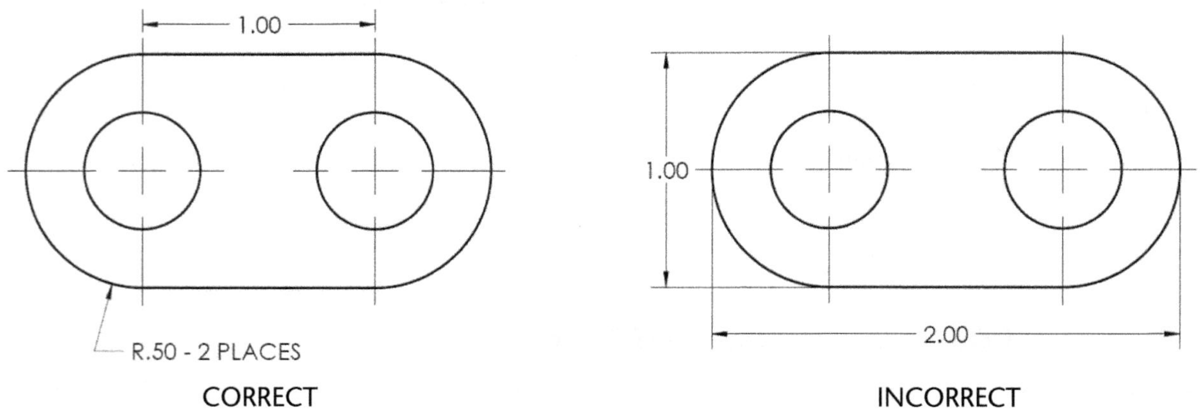

Figure 7.21 Features with rounded ends are dimensioned to their centers.

3.17 Bolt Circles

Bolt circles are useful when dimensioning repetitive holes in a circular pattern, like holes in a wheel hub. A center line is drawn which passes through the centers of each hole and a note is provided specifying that the holes have equal diameters and are equally spaced. See Figure 7.22.

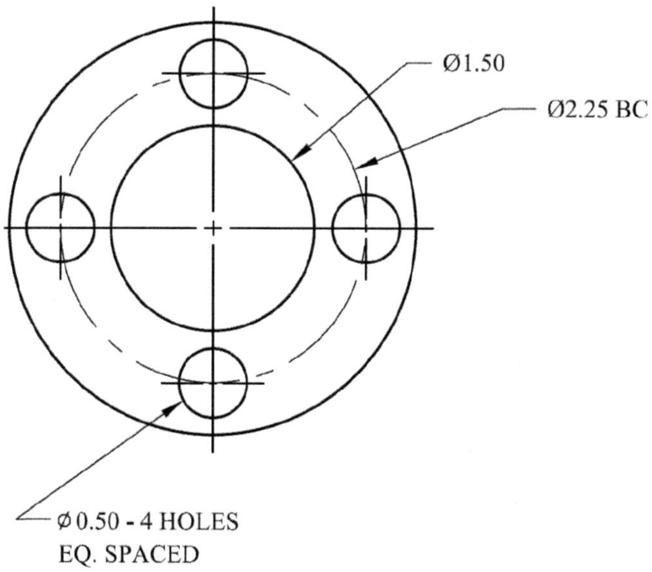

Figure 7.22 Bolt circle

3.18 Repetitive Features

Dimension repetitive features with a note, in order to save space. As an example, if an object has several holes with the same diameter, only dimension one hole's diameter, editing the hole call out to reflect the information about the other holes. See Figure 7.23.

Other standardized notes and abbreviations can also be used for repetitive features. Typical (abbreviated TYP), means that the feature that is being dimensioned is the same as those not dimensioned. Places (abbreviated PL) specifies the number of places that identical features to the one being dimensioned appear in the drawing.

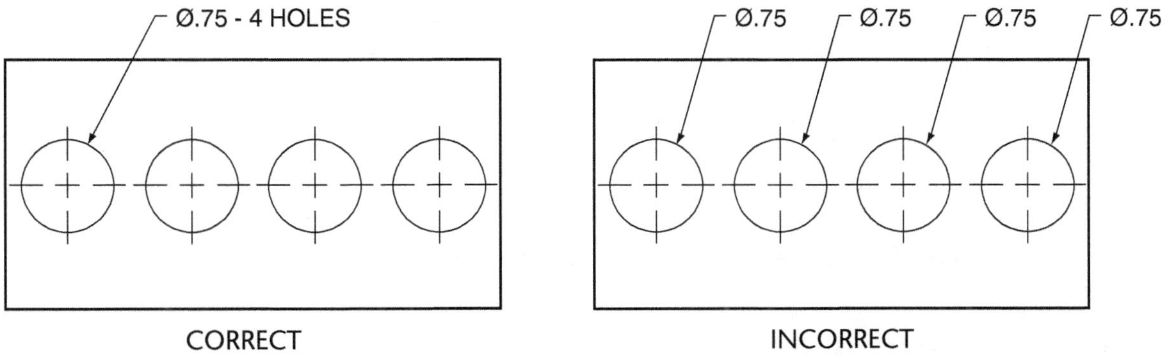

Figure 7.23 Dimension repetitive features with a note

3.19 Chamfers

Chamfers are defined as beveled edges that connect two surfaces. Edges are chamfered in order to remove the sharp edges of machined or wooden parts. These features can be dimensioned by labeling the distances in the vertical and horizontal direction or by listing the horizontal distance with the corresponding angle. See Figure 7.24 for an illustration.

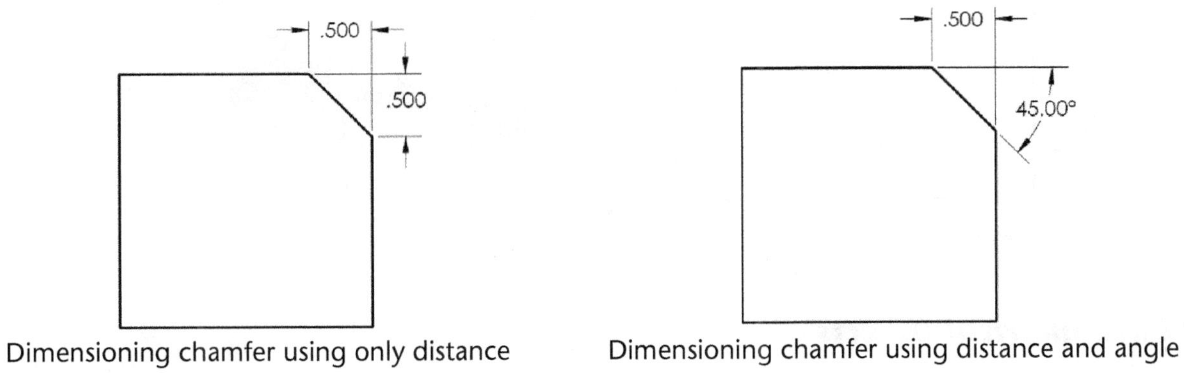

Figure 7.24 Proper dimensioning of chamfers

3.20 Notes

Most drawings require the use of notes, in addition to dimensions, to provide specific information and specifications that would be difficult to represent by a drawing alone or with regular dimensions. Some notes are used to reduce the number of dimensions in objects with repetitive features. Examples of these notes include TYP, 2 PLACES, or EQ. SPACED, mentioned in previous sections.

General notes, usually located at the bottom corner of a drawing are used to provide general information that applies to the whole drawing. Examples of general notes include: ALL FILLETS R.1.00, OBJECT IS SYMMETRICAL, or MATERIAL: ABS PLASTIC.

Chamfers and fillets are typical examples of features that can be dimensioned with notes, especially if all chamfers and fillets on the drawing are the same (see Figure 7.25).

Standard elements, such as machined holes and thread fasteners require specific standard notes that will be discussed in a later chapter.

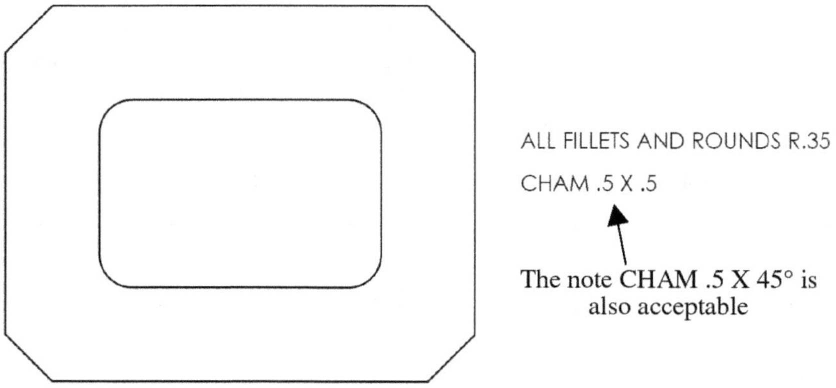

Figure 7.25 Dimensioning chamfers and fillets using notes only

3.21 Finished Surfaces

Parts that are cast, such as engine blocks, have rough surfaces. These surfaces need to be smoothed occasionally so that better contact can be made between parts. In such cases, specific areas (or whole sides) can be ground smooth. This smoothing process is referred to as finishing. Surfaces that require finishing are indicated by placing a "check mark", as seen in Figure 7.26, on the edge view of the surface to be finished.

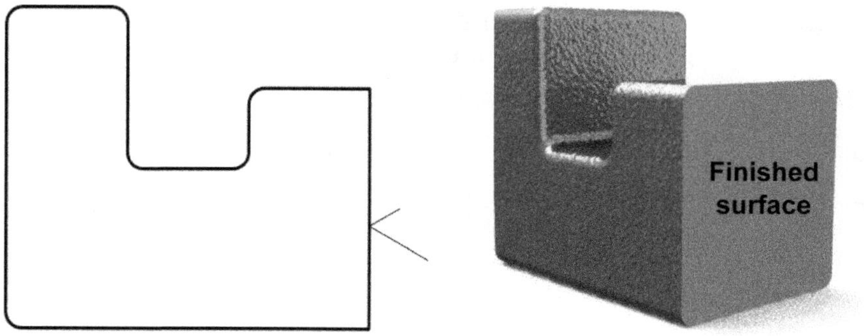

Figure 7.26 Check mark indicates surface to be finished.

3.22 Symmetrical Objects

Symmetry is normally shown by placing a note (OBJECT IS SYMMETRICAL) on the drawing. Dimensioning symmetrical objects is made easier because fewer dimensions need to be shown. Oftentimes, a center line can be used to indicate symmetry (see Figure 7.27 for an illustration). Although it is common to use centerlines to indicate symmetry, this practice is not considered to be in compliance with dimensioning standards. Using centerlines in this manner is acceptable early in the design process, but all manufacturing drawings should be dimensioned according to standards.

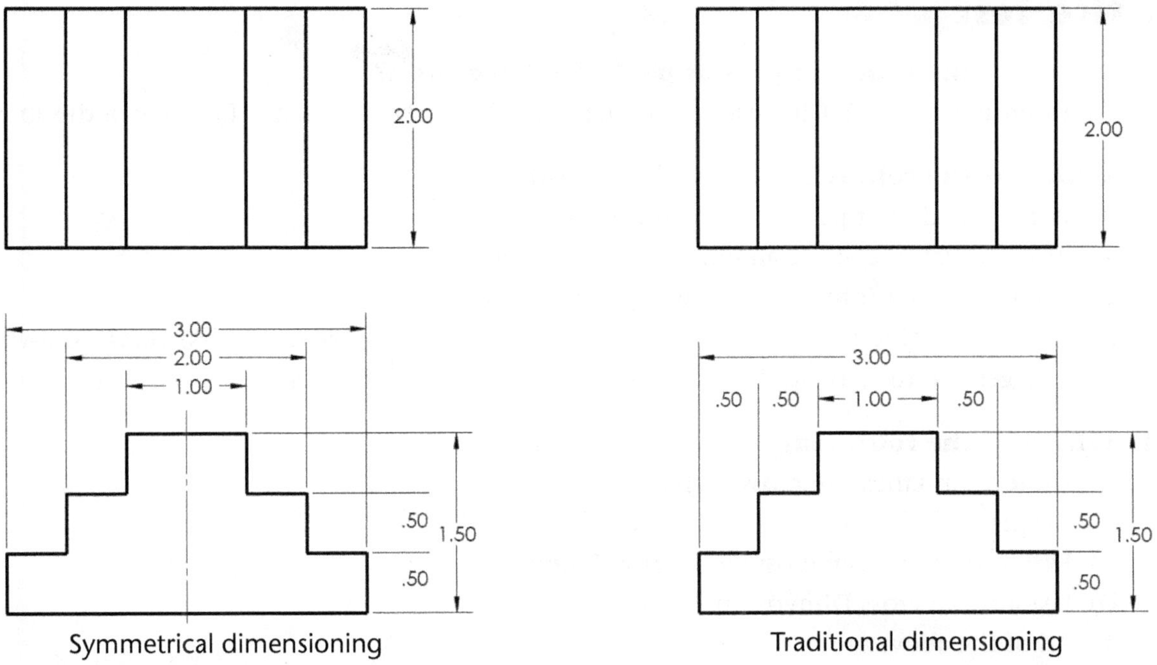

Figure 7.27 Dimensioning symmetrical objects

Practice Test

1. **Which of the following is not part of a dimension?**
   A) Extension Line B) Dimension Line C) Arrowhead D) Gap Line E) None of the above

2. **Which of the following statements is false?**
   A) If possible do not put any dimension inside any view
   B) As a general rule, place all dimensions between views
   C) Arrange dimensions so the drawing is clean and readable
   D) There is no need to fully dimension a drawing if the manufacturer works on the drawing.
   E) As a general rule, avoid duplicate dimensions in your drawings

3. **Which of the following statements is true?**
   A) Dimension Lines can cross each other
   B) Extension Lines can cross each other
   C) Extension Lines can cross Dimension Lines
   D) Leaders can cross Dimension Lines
   E) All of the above are true

4. **Holes are always dimensioned ____**
   A) Using diameter in the rectangular view
   B) Using radius in the rectangular view
   C) Using diameter in the circular view
   D) Using radius in the circular view
   E) None of the above

5. **Cylinders are always dimensioned ____**
   A) Using diameter in the rectangular view
   B) Using radius in the rectangular view
   C) Using diameter in the circular view
   D) Using radius in the circular view
   E) None of the above

6. **Fillets and rounds are always dimensioned ____**
   A) Using diameter in the rectangular view
   B) Using radius in the rectangular view
   C) Using diameter in the rounded view
   D) Using radius in the rounded view
   E) None of the above

7. **Which of the following is a proper dimensioning style?**
   A) Unidirectional
   B) General
   C) Aligned
   D) A and C are correct
   E) A, B, and C are correct

8. A duplicate dimension shown between parentheses and provided to avoid the need for calculations is called ____ dimension.

A) Secondary B) Auxiliary C) Derived D) Reference E) None of the above

9. Which of the following statements is true?

A) If possible, always dimension to T-Joints
B) Do not dimension to hidden lines
C) Never dimension to a center line
D) B and C are correct
E) A, B, and C are correct

10. Which of the following statements is false?

A) Leaders originate from the end or the beginning of the note
B) An elbow should always be provided when using leaders
C) Leaders should only have one arrowhead at one of its ends
D) Leaders should only be used when dimensioning cylindrical features
E) Leaders are used to direct a dimension to a particular feature in the object

DIMENSION THE OBJECTS SHOWN BELOW. NOTE THAT NOT ALL VIEWS WILL REQUIRE DIMENSIONS. DO NOT OVERDIMENSION. ASSUME THE GRID IS 0.1".

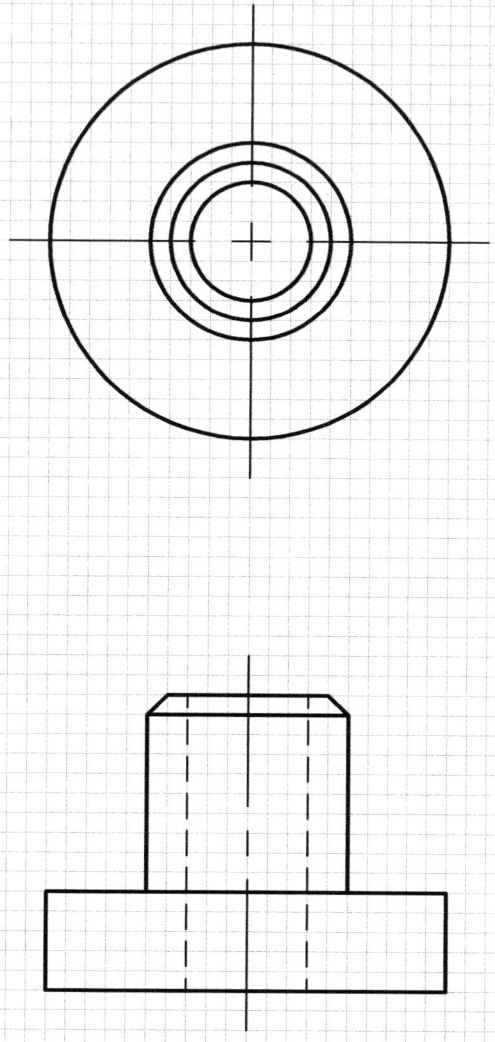

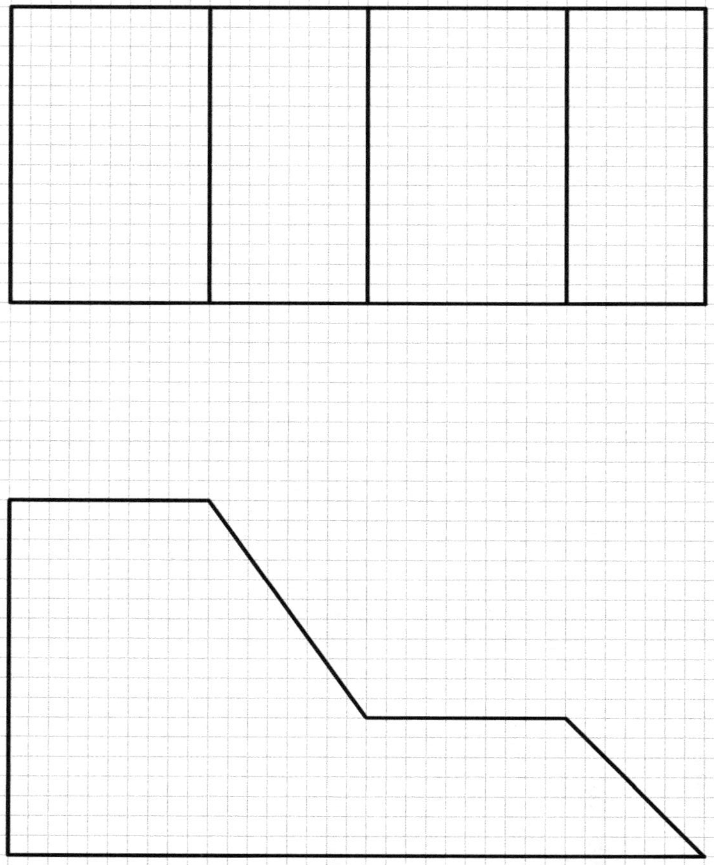

AUTHOR:

SCALE:

TEAM: SECTION:

DIM01

DIMENSION THE OBJECTS SHOWN BELOW. NOTE THAT NOT ALL VIEWS WILL REQUIRE DIMENSIONS. DO NOT OVERDIMENSION. ASSUME THE GRID IS 0.1".

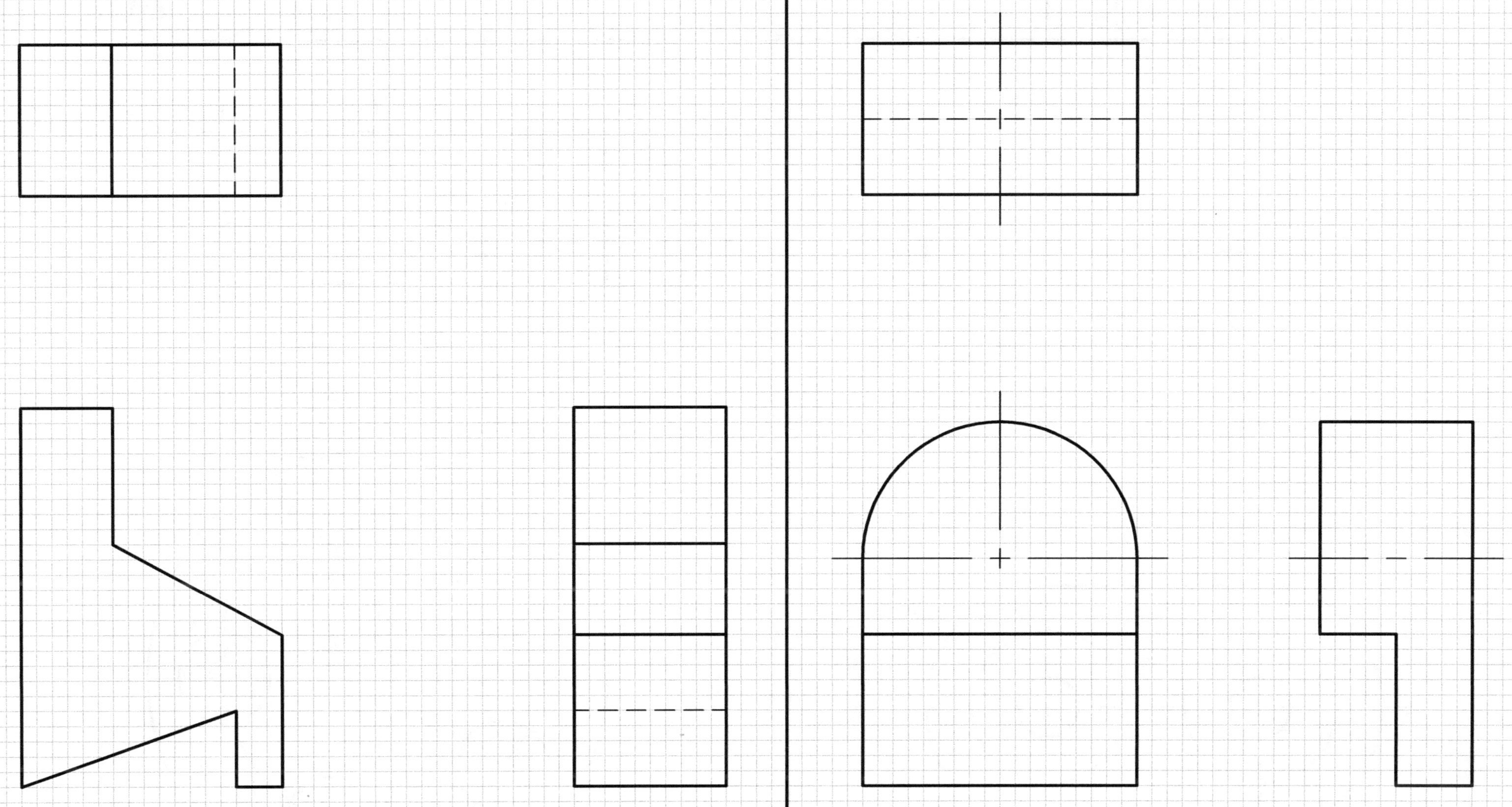

AUTHOR:		SCALE:
TEAM:	SECTION:	DIM02

DIMENSION THE OBJECTS SHOWN BELOW. NOTE THAT NOT ALL VIEWS WILL REQUIRE DIMENSIONS. DO NOT OVERDIMENSION. ASSUME THE GRID IS 0.1".

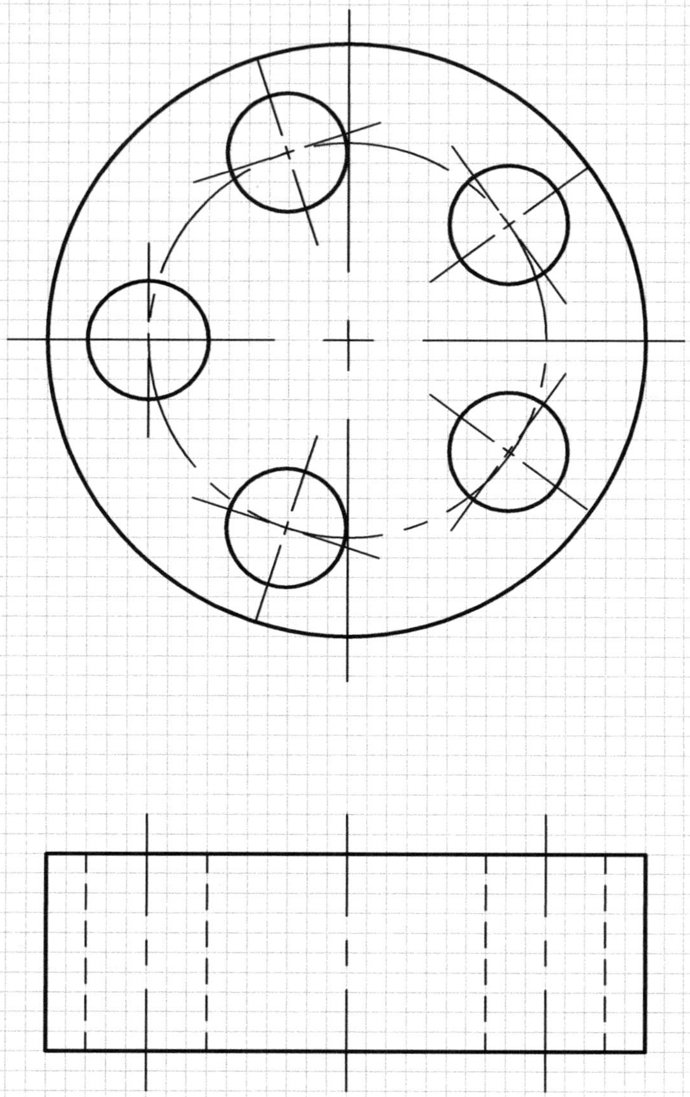

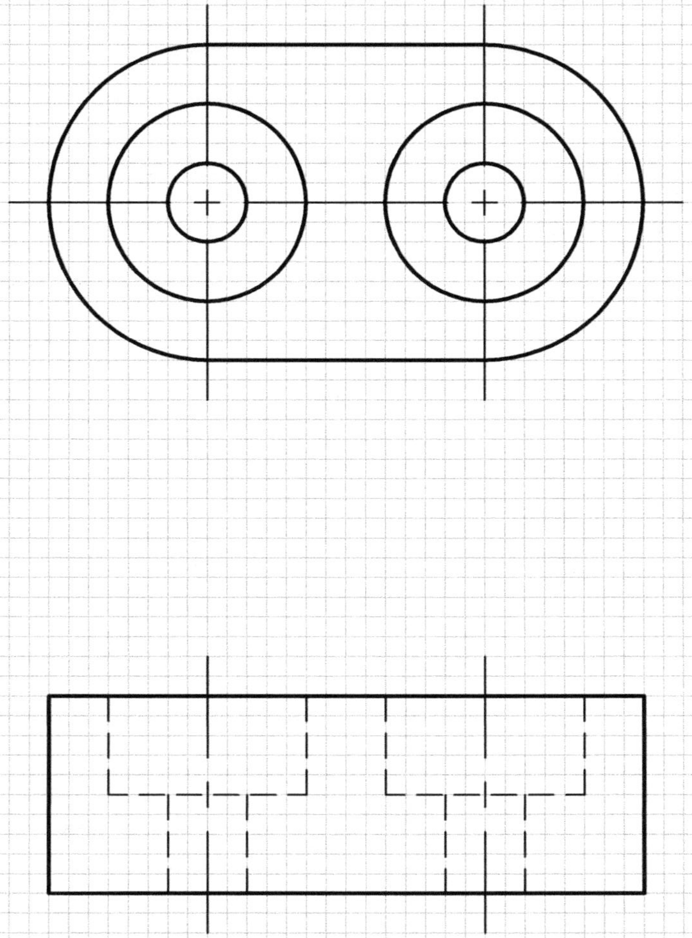

AUTHOR:		SCALE:
TEAM:	SECTION:	DIM03

DIMENSION THE OBJECTS SHOWN BELOW. NOTE THAT NOT ALL VIEWS WILL REQUIRE DIMENSIONS. DO NOT OVERDIMENSION. ASSUME THE GRID IS 0.1".

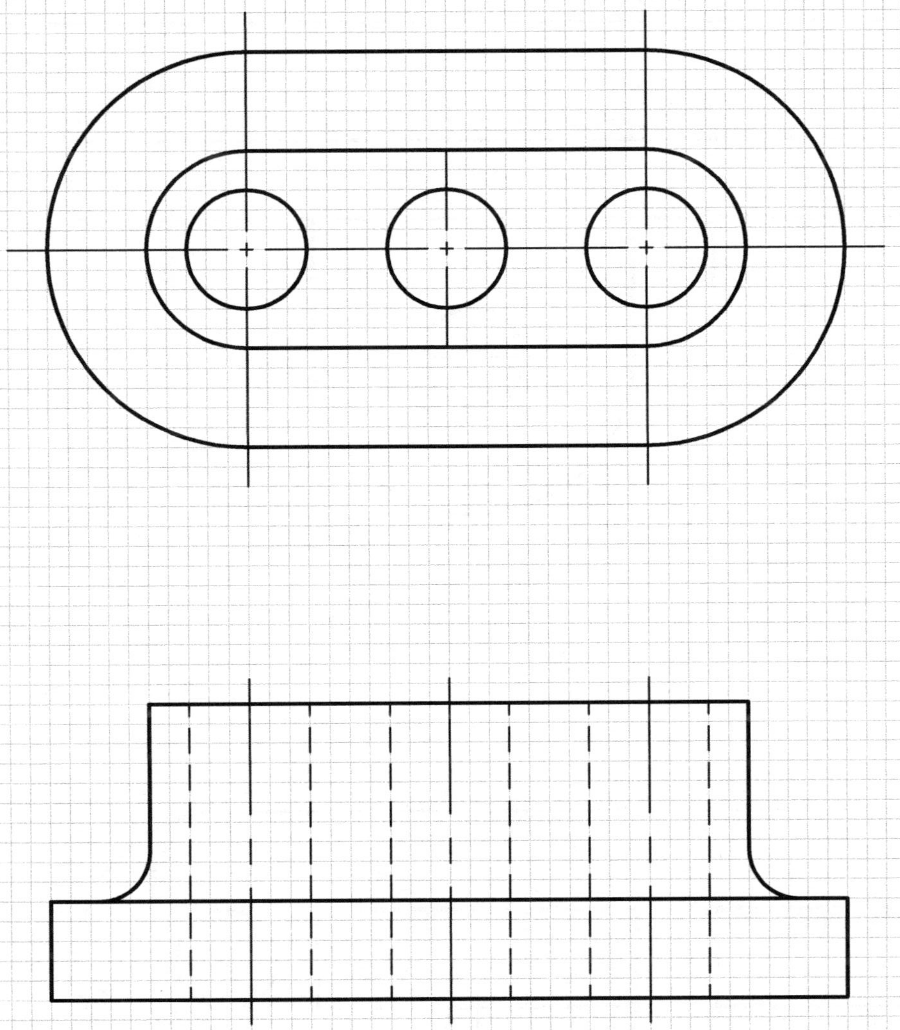

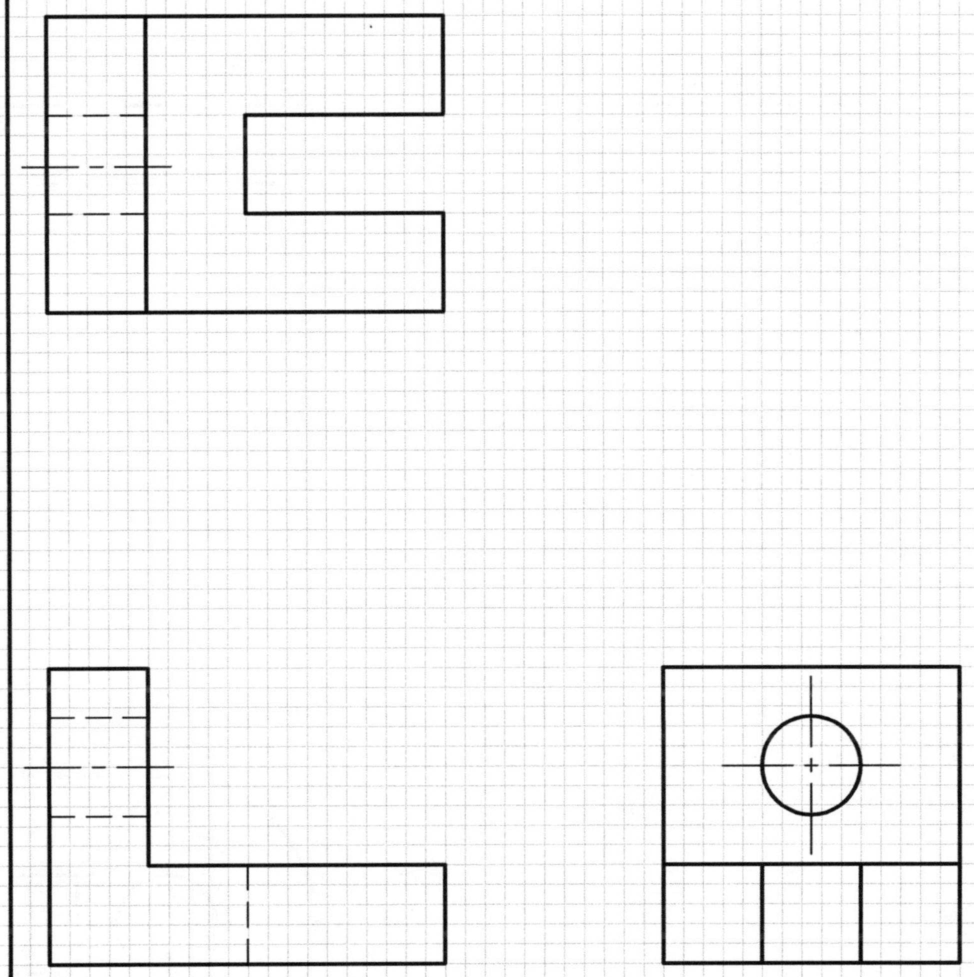

AUTHOR:	SCALE:
TEAM: SECTION:	DIM04

DIMENSION THE OBJECT SHOWN BELOW. NOTE THAT NOT ALL VIEWS WILL REQUIRE DIMENSIONS. DO NOT OVERDIMENSION. ASSUME THE GRID IS 0.1".

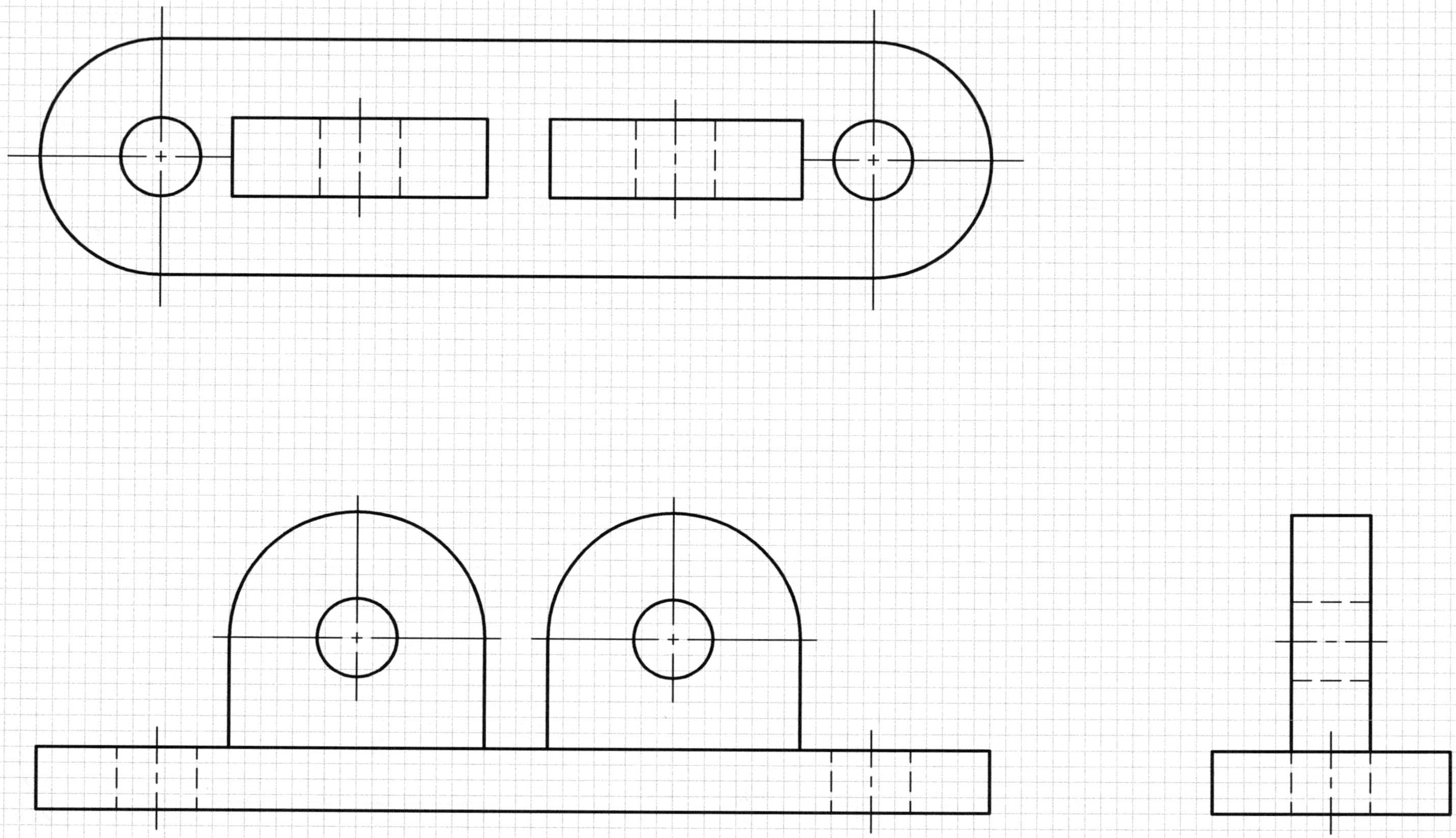

AUTHOR:		SCALE:
TEAM:	SECTION:	DIM05

DIMENSION THE OBJECT SHOWN BELOW. NOTE THAT NOT ALL VIEWS WILL REQUIRE DIMENSIONS. DO NOT OVERDIMENSION. ASSUME THE GRID IS 0.1". THE SCALE IS DOUBLE SIZED.

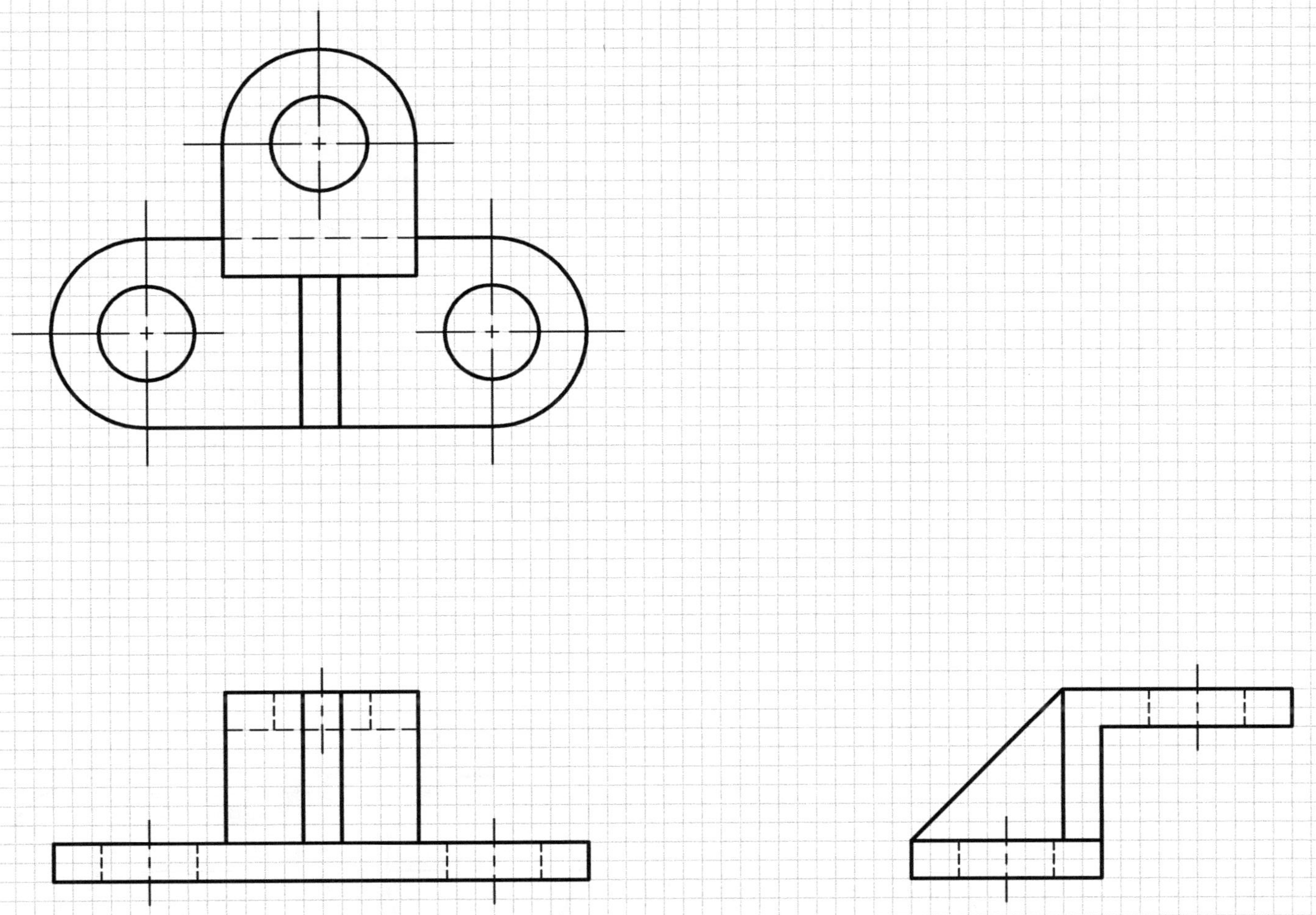

AUTHOR:		SCALE:
TEAM:	SECTION:	DIM06

DIMENSION THE OBJECT SHOWN BELOW. NOTE THAT NOT ALL VIEWS WILL REQUIRE DIMENSIONS. DO NOT OVERDIMENSION. ASSUME THE GRID IS 0.1".

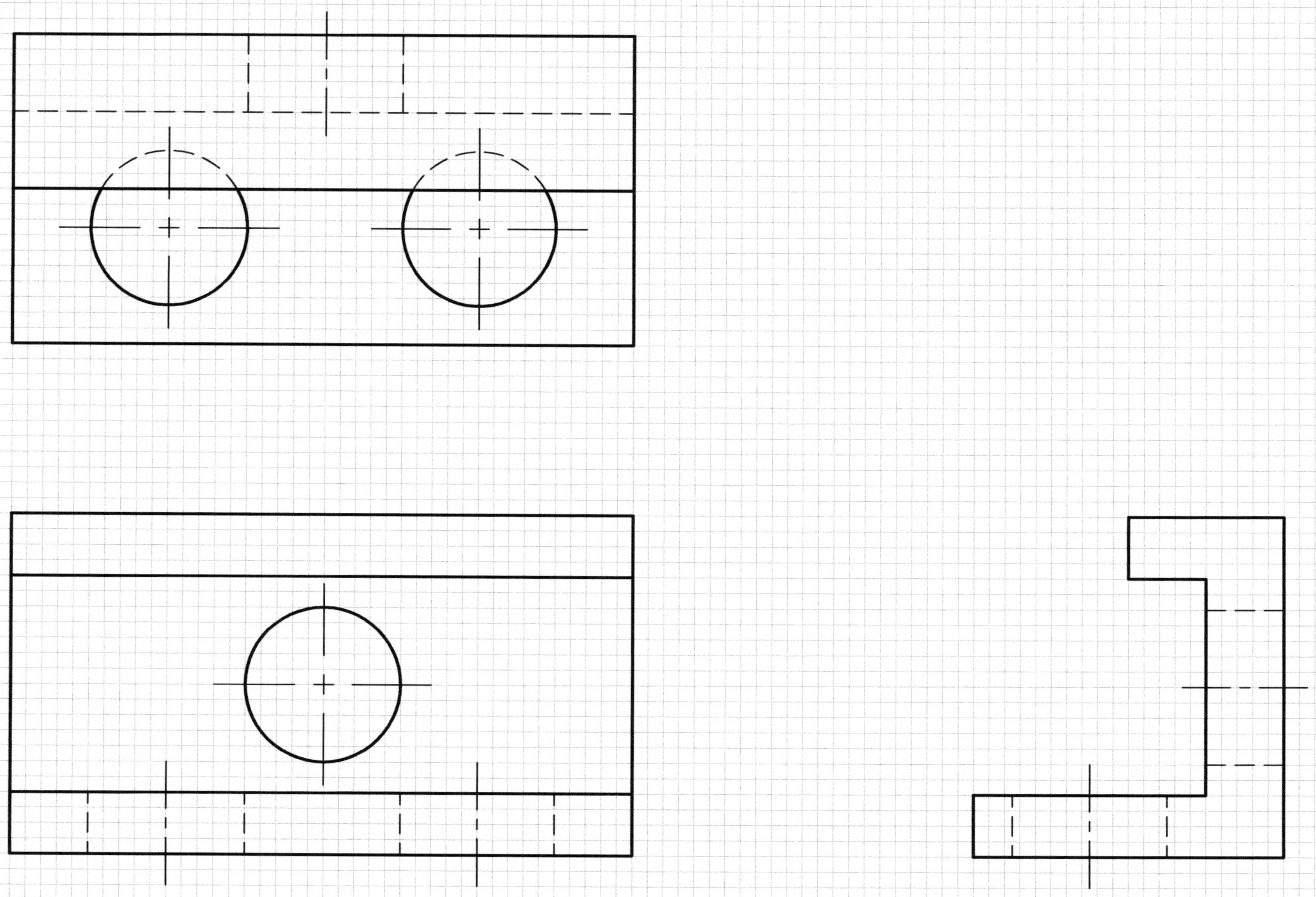

AUTHOR:

SCALE:

TEAM:

SECTION:

DIM07

DIMENSION THE OBJECT SHOWN BELOW. NOTE THAT NOT ALL VIEWS WILL REQUIRE DIMENSIONS. DO NOT OVERDIMENSION. ASSUME THE GRID IS 0.1".

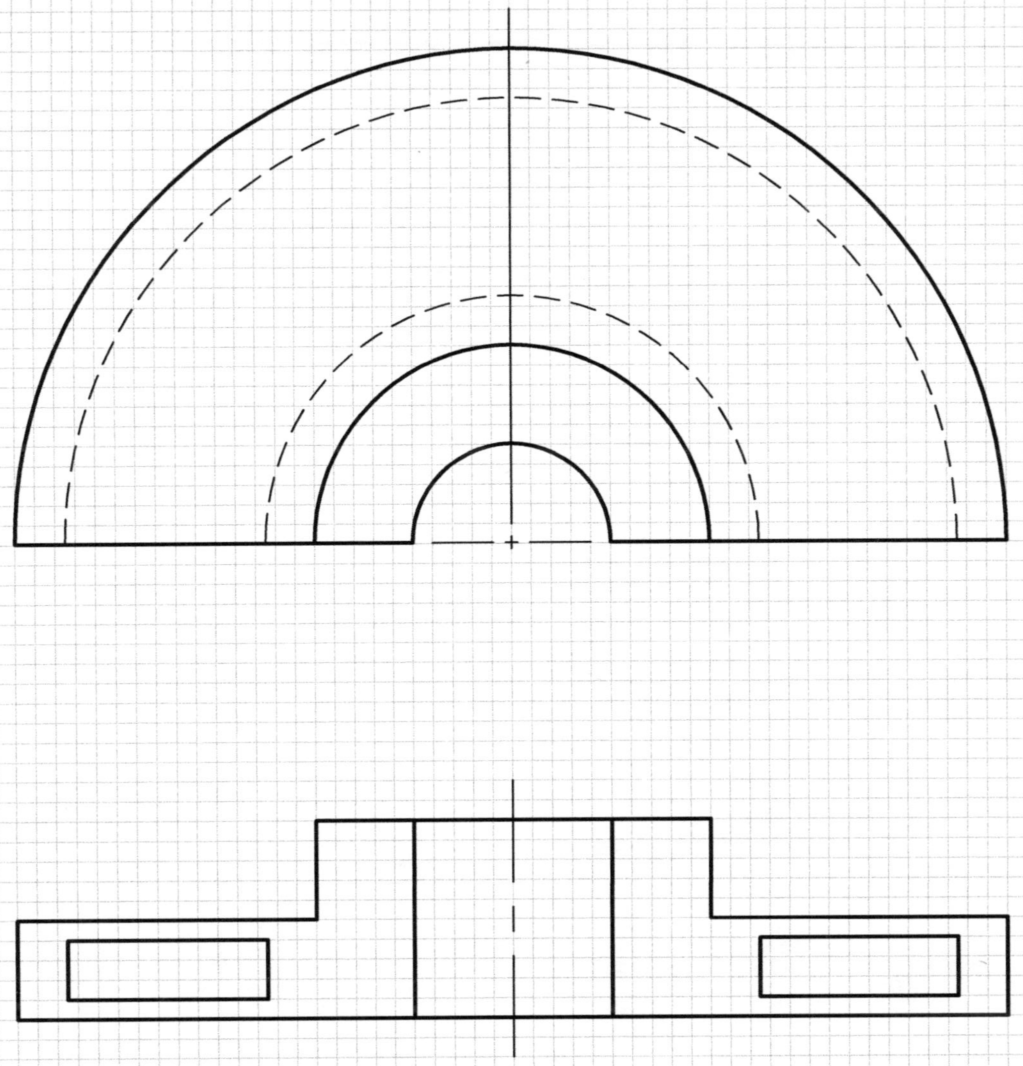

AUTHOR:		SCALE:
TEAM:	SECTION:	DIM08

CHAPTER 8

Tolerancing

1. Introduction

Features without any error are impossible to manufacture. If a hole has a design diameter of one inch, how close to one inch would it have to be to be considered acceptable? The greater the accuracy, the higher the cost of manufacturing. Incorrectly manufactured parts have to be discarded (either thrown away or re-melted so the material can be used again), more expensive equipment may have to be used (laser instead of a drill press), and expert craftsmen demand higher labor costs. Proper design must consider the relationship between end use and manufacturing cost. A bolt that is intended for a picnic table assembly, for example, could be made cheaper than a bolt intended for the space shuttle because its purpose does not require extreme strength and accuracy.

Tolerancing is a dimensioning method that allows for proper manufacturing and helps to ensure interchangeability between parts in an assembly. Standard tolerancing practices and their applications are discussed in this chapter.

2. Definition and Types of Tolerances

The term tolerance refers to the permissible level of error that a machinist is allowed on a specific dimension. A tolerance can be understood as a range of acceptable values for a particular dimension. For example, a drawing of a cylindrical part with a diameter of one inch and a tolerance of positive or negative .05 (specific details on tolerance representation will be explained later) will let a machinist know that the part to be manufactured is acceptable as long as its diameter is between .95 and 1.05 inches. Tolerances alert the machinist what equipment and methods are required to make a specific part.

There are three types of tolerances; general, linear, and geometric. General tolerances apply to all dimensions on a drawing whereas linear tolerances refer to specific features that require more accuracy than general tolerances provide. Geometric tolerances are concerned with a feature's shape or profile, not its size or dimensions.

3. General Tolerances

General tolerances are defined as the error limit that is acceptable for all dimensions on a drawing. This information is often found in the title block of all drawings intended for manufacturing or as a general note. An example is shown in Figure 8.1.

General tolerances are normally given in bilateral form, defining a symmetric limit above and below a dimension (forms of correct tolerancing will be explained shortly). A common general tolerance for English drawings is ±.005 inches. As an example, a hole that has a design diameter of 1.00 inches would be acceptable if it was actually drilled to a diameter of 0.995 inches, 1.002 inches, or 1.005 inches because these values fall between the allowed error defined by the tolerance.

BRACKET		MATERIAL: CAST IRON
DRAWN BY:	CREATION DATE:	SHEET: 1 OF 3
CHECKED BY:	REVISION DATE:	COMMENTS:
SCALE: 1 = 1	**TOLERANCES** **DEC ±.005** **ANG ±.005**	
DRAWING NO: 5105A01		

Figure 8.1 General tolerances indicated in a title block (DEC: Decimal, ANG: Angular)

4. Linear Tolerances

Linear tolerances are used when a specific feature requires greater accuracy than the one expressed by the general tolerance. Linear tolerances can be indicated by several methods. A tolerance shown in limit form provides a range of acceptable values. As long as the manufactured feature falls between these values, the feature is accepted.

A linear tolerance may also be shown in unilateral form providing a range of acceptable values with either an upper or lower value that cannot be crossed (the range of acceptable error extends only in one direction).

A tolerance shown in bilateral form provides a target diameter, with a symmetric limit above and below.

The examples shown in Figure 8.2 illustrate the different forms used to indicate linear tolerances.

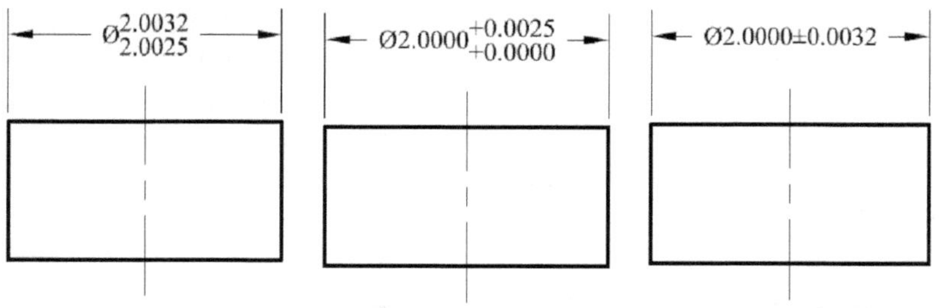

Limit form Unilateral form Bilateral form

Figure 8.2 Forms to represent linear tolerances

4.1 Linear Fit Relationships

When developing designs with mating parts, such as a shaft mating with a hole or a block sliding along a slot, tolerances become critical. Not only are the individual tolerances of each part important, but also the relationship between the two parts (how tightly or loosely they are going to fit) needs to be considered. From a design standpoint, there are four parameters of interest: the tolerance of the first mating part, the tolerance of the second mating part, the maximum clearance, and the allowance. A simple assembly will be used as an example to illustrate these concepts (see Figure 8.3), but the terms can be applied to any two mating parts. In this design, the two components should fit together like a puzzle piece, with the stud (shaft) from the top piece mating with the hole in the bottom piece

The tolerance is defined as the difference between the largest and smallest possible sizes of a feature. Therefore, the difference between the largest and smallest possible holes comprises the hole tolerance and the difference between the largest and smallest possible shafts comprises the shaft tolerance. The hole tolerance does not consider the shaft at all, but determines the cost of manufacturing the hole. The shaft tolerance does not consider the hole at all, but determines the cost of manufacturing the shaft.

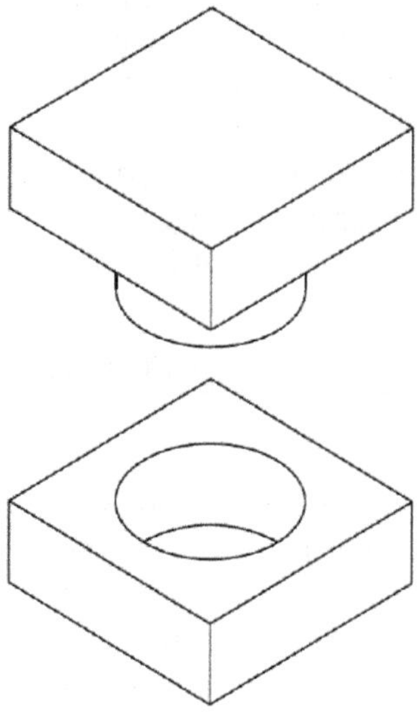

Figure 8.3 Basic assembly

The maximum clearance determines the maximum amount of space allowed between two mating parts, providing for the loosest fit between them. Allowance is defined as the minimum amount of space allowed between two mating parts, providing for the tightest fit. Allowance can also be defined as the minimum clearance between parts.

Tolerances, maximum clearance, and allowance are calculated using four equations. These equations are self-descriptive and should be easy to memorize. The equations using the hole and shaft example are shown below.

Hole Tolerance (HT)	=	Largest Possible Hole (LPH)	–	Smallest Possible Hole (SPH)
Shaft Tolerance (ST)	=	Largest Possible Shaft (LPS)	–	Smallest Possible Shaft (SPH)
Maximum Clearance (MC)	=	Largest Possible Hole (LPH)	–	Smallest Possible Shaft (SPS)
Allowance (A)	=	Smallest Possible Hole (SPH)	–	Largest Possible Shaft (LPS)

For convenience, these equations are listed again in simplified form:

HT = LPH – SPH	ST = LPS – SPS	MC = LPH – SPS	A = SPH – LPS

Using the hole/shaft assembly example, and given the following information:

Shaft: Ø $\begin{matrix} .4994 \\ .4990 \end{matrix}$ Hole: Ø $\begin{matrix} .5007 \\ .5000 \end{matrix}$

The hole and shaft tolerances, maximum clearance and allowance of the assembly can be easily calculated by substituting the values in the corresponding equations.

Hole Tolerance (HT)	=	.5007	–	.5000	=	**.0007**
Shaft Tolerance (ST)	=	.4994	–	.4990	=	**.0004**
Maximum Clearance (MC)	=	.5007	–	.4990	=	**.0017**
Allowance (A)	=	.5000	–	.4994	=	**.0006**

4.2 Types of Fits

Linear tolerances can be classified in four general categories or types of fits based on the interaction between the mating parts. A type of fit represents the degree of tightness between two mating parts. These types of fits are: clearance, interference, line, and transition. Some situations require tighter fits than others. The selection of the most appropriate type of fit for a particular assembly is based on design requirements and specifications.

In a clearance fit, the internal mating part (shaft) is always smaller than the external mating part (hole). Therefore, the parts will always fit together with room to spare. Some designs require parts that do not interfere with each other, such as pistons inside cylinders in an internal combustion engine.

In an interference fit (also called force fit), the internal mating part (shaft) will always be larger than the external mating part (hole). In order to assemble two parts with an interference fit, it is necessary to apply force to them. As an example, many travel mugs have lids that are smaller in diameter than the lip of the mug. The purpose of this design is to make a seal so the liquid does not leak out.

In a line fit, the internal mating part (shaft) could possibly be smaller or equal to the external mating part (hole). This type of fit may result in mating parts with either a positive or zero clearance. Line fits are used in assemblies with stationary parts, but which can be easily assembled and disassembled.

In a transition fit, the internal mating part (shaft) could be larger or smaller than the external mating part (hole). In this case, the two parts may either clear or interfere with each other.

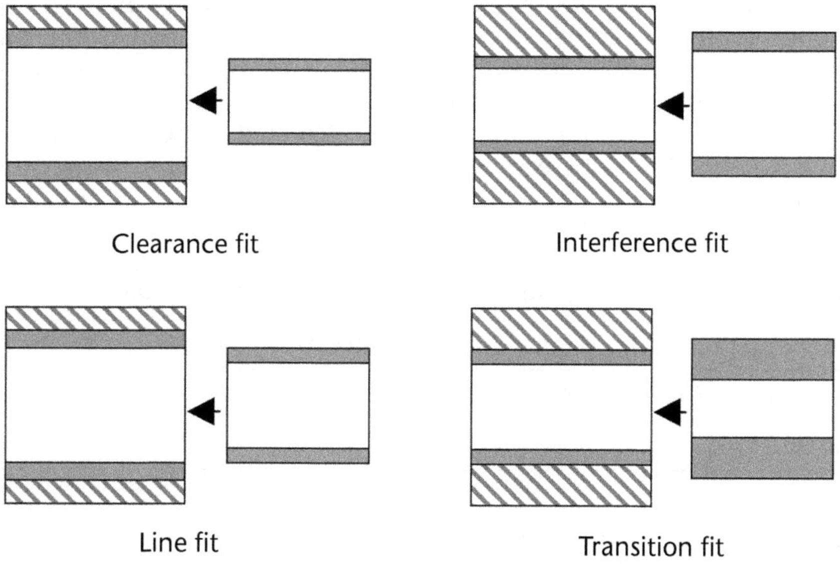

Figure 8.4 Types of fits

Transition fits are normally used for location of centers of holes, since it is the cheapest way to manufacture a part.

The sectional views shown in Figure 8.4 illustrate the different types of fits. The dark areas represent the largest possible diameters, whereas the white areas represent the smallest possible diameters.

For convenience, the relationship between positive and negative values of allowance and clearance depending upon the type of linear tolerance desired is shown in Figure 8.5.

	Allowance	Maximum Clearance
Clearance Fit	+	+
Interference Fit	–	–
Line Fit	0	+
Transition Fit	–	+

Figure 8.5 Linear Tolerance Fit Relationships

Examples of the different types of fits with proper dimensions are illustrated in Figure 8.6. Note that the larger value appears on top, with a diameter symbol (Ø) halfway between each value.

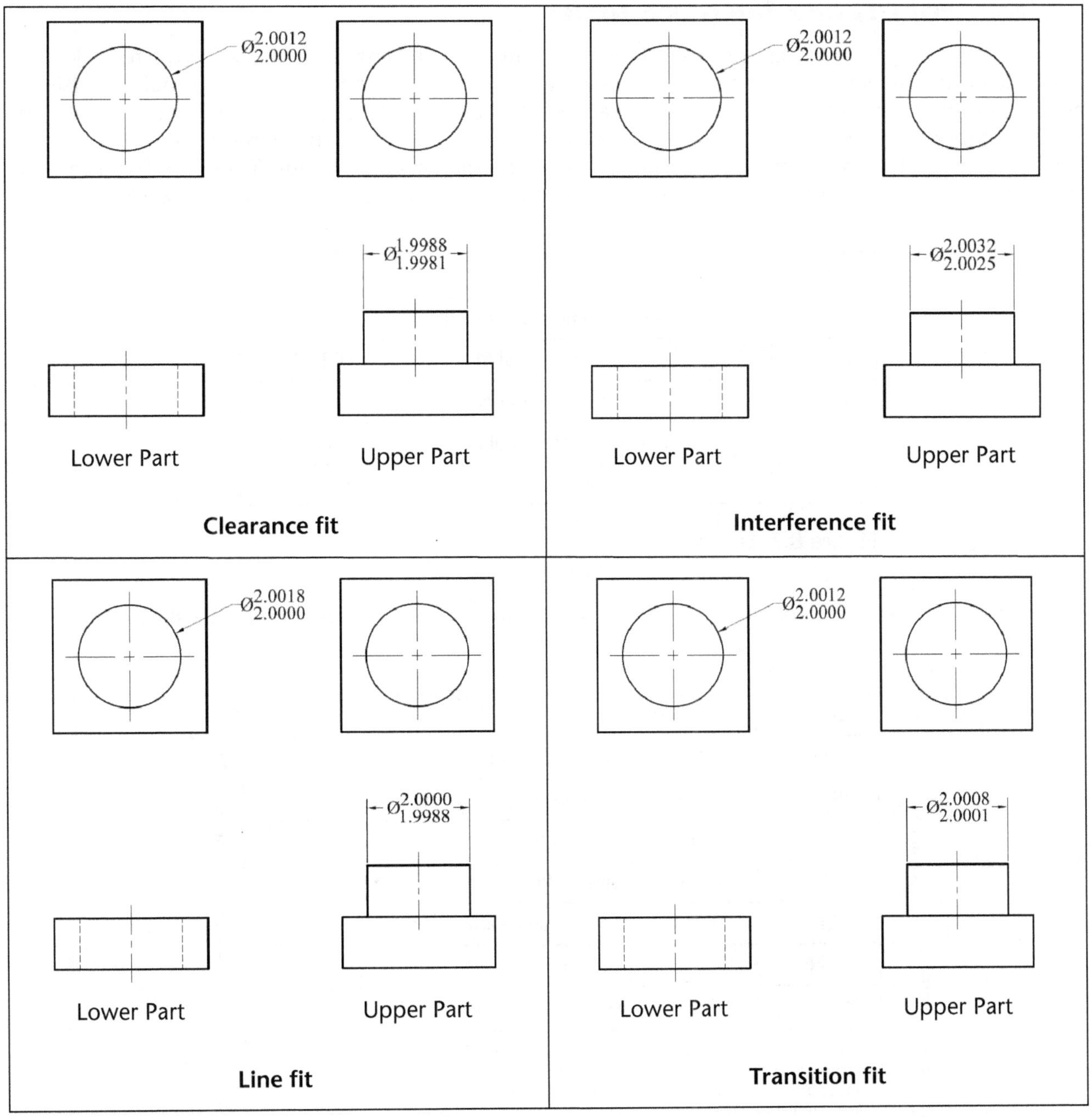

Figure 8.6 Types of fits

4.3 Standard Fits and Tolerance Tables

In order to make parts interchangeable, tolerance values have been standardized and recorded in tables for both English and metric systems. This way, a designer does not need to make decisions about specific tolerance values. Instead, tolerance tables are used to obtain specific values based on the type of fit and a basic size. The four general types of fits (clearance, interference, line, and transition) are further developed in both the English and metric tolerance tables. There are five types of fits (with different grades or classes) in the English system, depending on its tightness (see Figure 8.7).

Type	Description	Classes
RC:	Running or sliding clearance fit	RC1-RC9
LC:	Clearance locational fit	LC1-LC11
LT:	Transition locational fit	LT1-LT6
LN:	Interference locational fit	LN1-LN3
FN:	Force and shrink fit	FN1-FN5

Figure 8.7 English fits

The metric system defines ten types of fits, depending on its tightness (see Figure 8.8).

	Type		Description	
	Hole Basis	Shaft Basis		
Clearance Fits	H11/c11	C11/h11	Loose running fit	↑ More Clearance
	H9/d9	D9/h9	Free running fit	
	H8/f7	F8/h7	Close running fit	
	H7/g6	G7/h6	Sliding fit	
	H7/h6	H7/h6	Locational clearance fit	
Transition Fits	H7/k6	K7/h6	Locational transition fit	
	H7/n6	N7/h6	Locational transition fit (more accurate)	
Interference Fits	H7/p6	P7/h6	Locational interference fit	
	H7/s6	S7/h6	Medium drive	
	H7/u6	U7/h6	Force fit	More Interference ↓

Figure 8.8 Metric fits

Example (English):
Using a basic hole/shaft assembly and given the following information:

- The diameter is desired to be a clearance fit, in particular a RC3 fit (ANSI Running and Sliding Fit)
- The desired diameter for both the shaft and hole should be 2.00 inches

The hole and shaft tolerances can be calculated using the corresponding tolerance table. A portion of this table is shown in Figure 8.9. The nominal size range 1.97 – 3.15 is selected because 2.00 inches falls in between these values.

Nominal Size Range, Inches	Class RC 1			Class RC 2			**Class RC 3**		
		Standard Tolerance Limits			Standard Tolerance Limits			Standard Tolerance Limits	
Over To	Clear-ance	Hole H5	Shaft g4	Clear-ance	Hole H6	Shaft g5	Clear-ance	Hole H7	Shaft f6
...	Values shown below are in thousandths of an inch								
0.24 – 0.40	0.2 0.6	+0.25 0	-0.2 -0.35	0.2 0.85	+0.4 0	-0.2 -0.45	0.5 1.5	+0.6 0	-0.5 -0.9
0.40 – 0.71	0.25 0.75	+0.3 0	-0.25 -0.45	0.25 0.95	+0.4 0	-0.25 -0.5	0.6 1.75	+0.7 0	-0.6 -1.0
0.71 – 1.19	0.3 0.95	+0.4 0	-0.3 -0.55	0.3 1.2	+0.5 0	-0.3 -0.7	0.8 2.1	+0.8 0	-0.8 -1.3
1.19 – 1.97	0.4 1.1	+0.4 0	-0.4 -0.7	0.4 1.4	+0.6 0	-0.4 -0.8	1.0 2.6	+1.0 0	-1.0 -1.6
1.97 – 3.15	0.4 1.2	+0.5 0	-0.4 -0.7	0.4 1.6	+0.7 0	-0.4 -0.9	1.2 3.1	**+1.2** **0**	**-1.2** **-1.9**
3.15 – 4.73	0.5 1.5	+0.6 0	-0.5 -0.9	0.5 2.0	+0.9 0	-0.5 -1.1	1.4 3.7	+1.4 0	-1.4 -2.3
4.73 – 7.09	0.6 1.8	+0.7 0	-0.6 -1.1	0.6 2.3	+1.0 0	-0.6 -1.3	1.6 4.2	+1.6 0	-1.6 -2.6
7.09 – 9.85	0.6 2.0	+0.8 0	-0.6 -1.2	0.6 2.6	+1.2 0	-0.6 -1.4	2.0 5.0	+1.8 0	-2.0 -3.2
...	...	...	...	...	...	...	...	...	...

Figure 8.9 Portion of ANSI Running and Sliding Fits (RC) table

From the tolerance table, the hole limits are determined to be +1.2 and 0 and the shaft limits are -1.2 and -1.9. Note that these values are in thousandths of an inch, so in reality, the hole limits are 0.0012 and 0. The shaft limits are -0.0012 and -0.0019. These values will be added or subtracted from the target diameter (depending upon the sign) to obtain the proper acceptable range.

$$\text{Largest Possible Hole (LPH)} = 2.00 + 0.0012 = 2.0012$$
$$\text{Smallest Possible Hole (SPH)} = 2.00 + 0 = 2.0000$$
$$\text{Largest Possible Shaft (LPS)} = 2.00 - 0.0012 = 1.9988$$
$$\text{Smallest Possible Shaft (SPS)} = 2.00 - 0.0019 = 1.9981$$

From these values, the maximum clearance and allowance can be calculated.

$$MC = LPH - SPS = 2.0012 - 1.9981 = 0.0031$$
$$A = SPH - LPS = 2.0000 - 1.9988 = 0.0012$$

Example (Metric):
Using a basic hole/shaft assembly and given the following information:

- The diameter is desired to be a force fit, in particular an U7/h6 fit (ANSI Interference Fit)
- The desired diameter for both the shaft and hole should be 8 mm

The hole and shaft diameters are given directly in the corresponding tolerance table. A portion of this table is shown in Figure 8.10. Unlike English tolerances, metric tolerances give the exact diameter, not the range.

		...	Locational Interference			Medium Drive			Force		
Basic Size		...	Hole P7	Shaft h6	Fit	Hole S7	Shaft h6	Fit	**Hole U7**	**Shaft h6**	Fit
...	...	...	...	...	...	...	...	...	...	...	...
5	Max	...	4.992	5.000	0.000	4.985	5.000	-0.007	4.981	5.000	-0.011
	Min		4.980	4.992	-0.020	4.973	4.992	-0.027	4.969	4.992	-0.031
6	Max	...	5.992	6.000	0.000	5.985	6.000	-0.007	5.981	6.000	-0.011
	Min		5.980	5.992	-0.020	5.973	5.992	-0.027	5.969	5.992	-0.031
8	Max	...	7.991	8.000	0.000	7.983	8.000	-0.008	7.978	8.000	-0.013
	Min		7.976	7.991	-0.024	7.968	7.991	-0.032	7.963	7.991	-0.037
10	Max	...	9.991	10.000	0.000	9.983	10.000	-0.008	9.978	10.000	-0.013
	Min		9.976	9.991	-0.024	9.968	9.991	-0.032	9.963	9.991	-0.037
12	Max	...	11.989	12.000	0.000	11.979	12.000	-0.010	11.974	12.000	-0.015
	Min		11.971	11.989	-0.029	11.961	11.989	-0.039	11.956	11.989	-0.044
16	Max	...	15.989	16.000	0.000	15.979	16.000	-0.010	15.974	16.000	-0.015
	Min		15.971	15.989	-0.029	15.961	15.989	-0.039	15.956	15.989	-0.044
20	Max	...	19.986	20.000	-0.001	19.973	20.000	-0.014	19.967	20.000	-0.020
	Min		19.965	19.987	-0.035	19.952	19.987	-0.048	19.946	19.987	-0.054
...	...	...	...	...	...	...	...	...	...	...	...

Figure 8.10 Portion of ANSI Preferred Shaft Basis Metric Transition and Interference Fits table

The acceptable range of hole diameters is between 7.978 and 7.963 mm. The acceptable range of shaft diameters is between 8.000 and 7.991 mm.

Largest Possible Hole (LPH) = 7.978
Smallest Possible Hole (SPH) = 7.963
Largest Possible Shaft (LPS) = 8.000
Smallest Possible Shaft (SPS) = 7.991

From these values, the maximum clearance and allowance can be calculated.

$$MC = LPH - SPS = 7.978 - 7.991 = -0.013$$
$$A = SPH - LPS = 7.963 - 8.000 = -0.037$$

Note that both of these values are negative, which is expected for a force fit.

5. Geometric Tolerances

Geometric tolerances are used to define the shape of features. How parallel must two sides be? How concentric must a hole be with a semicircular arc? Geometric tolerancing is a system that controls the allowed level of error related to the geometry of features, not the size.

Geometric tolerances are indicated using symbols called feature control frames. Feature control frames (or call out boxes) are added to the end of a leader line pointing to the feature that the geometric tolerance is being applied. They consist of a standard geometric tolerance symbol, tolerance zone, and possible datum identifier (if needed). An example can be seen in Figure 8.11. In this example, the feature control frame is indicating a geometric tolerance calling for the top to be perpendicular to side "A" to within 0.1 inches.

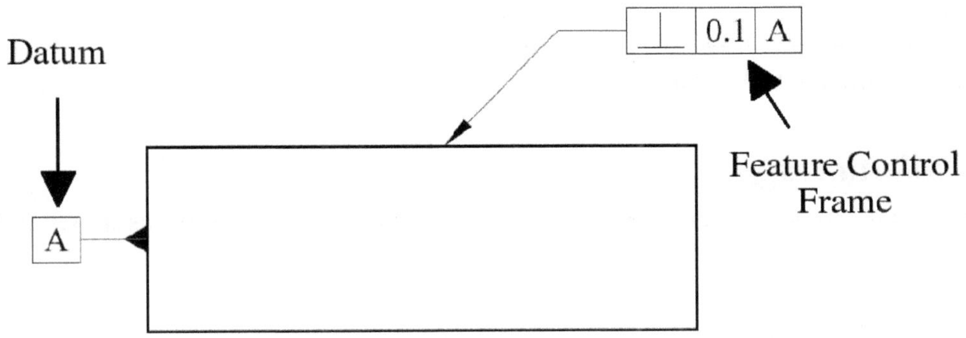

Figure 8.11 Feature control frame

Geometric tolerances can be grouped into several categories. These categories contain form, profile, orientation, location, and runout (see Figure 8.12).

Form	Profile	Orientation	Location	Runout
Flatness	Profile-Line	Angularity	Position	Circular Runout
Roundness	Profile-Surface	Parallelism	Concentricity	Total Runout
Straightness		Perpendicularity		
Cylindricity		Symmetry		

Figure 8.12 Feature control frame

5.1 Form Tolerances

Flatness Flatness is achieved when a surface exists within a flat tolerance zone, as shown in Figure 8.13.

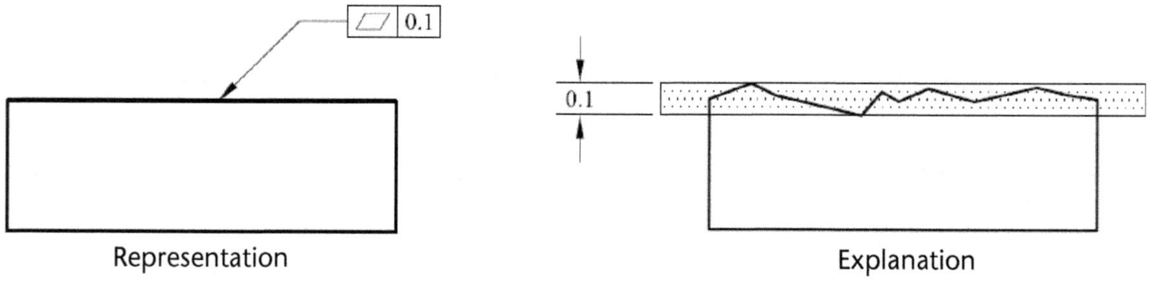

Figure 8.13 Flatness geometric tolerance

Roundness Roundness (also known as circularity) is a geometric tolerance term which defines a feature as being round if the circle exists inside the circular tolerance zone. This concept is illustrated in Figure 8.14.

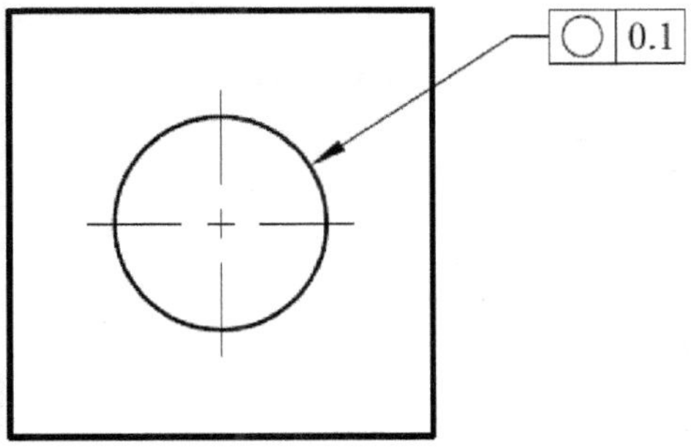

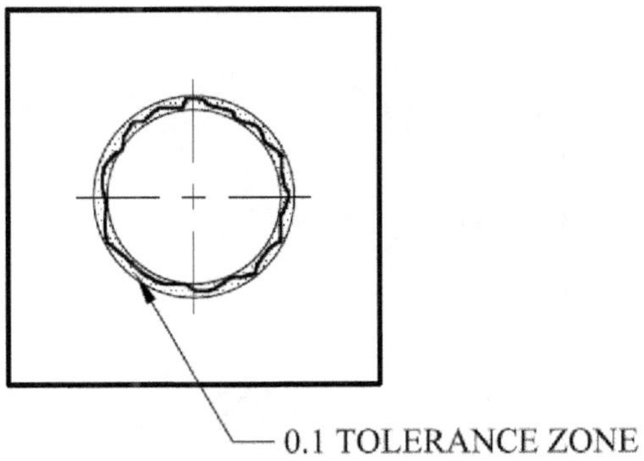

Representation

Explanation

Figure 8.14 Roundness geometric tolerance

Straightness Straightness is achieved when a line exists inside a straight tolerance zone, as illustrated in Figure 8.15.

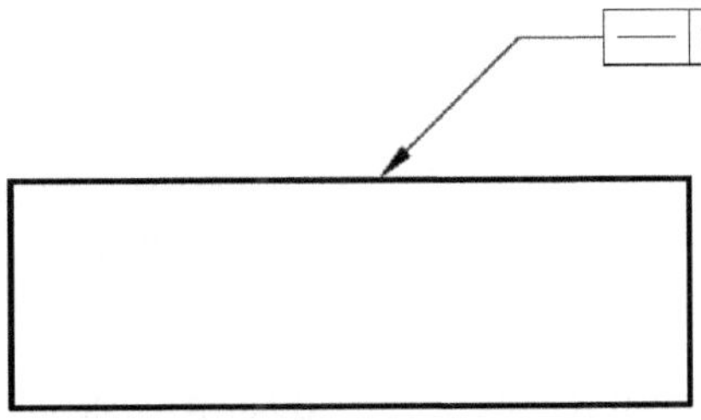

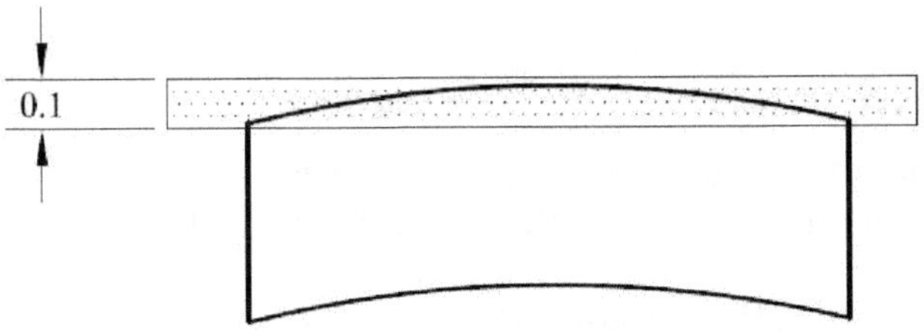

Representation

Explanation

Figure 8.15 Straightness geometric tolerance

Cylindricity Cylindricity is a combination of roundness and straightness. An example can be seen in Figure 8.16.

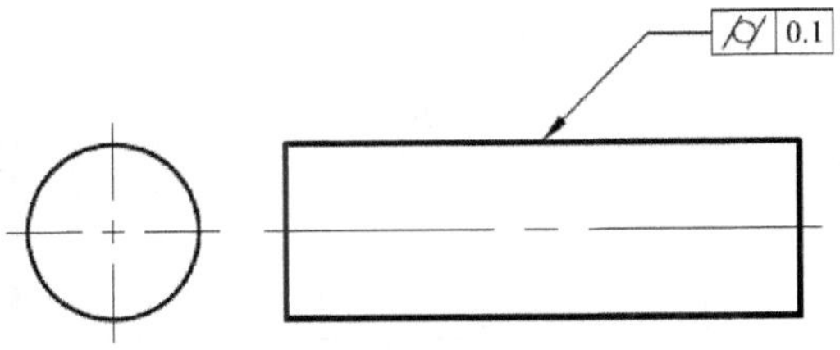

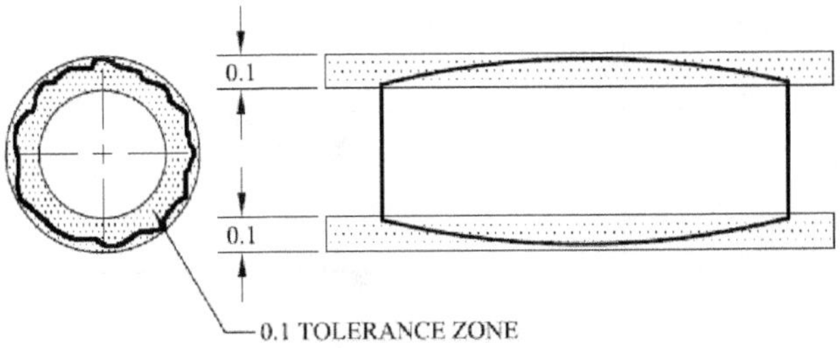

Representation

Explanation

Figure 8.16 Cylindricity geometric tolerance

5.2 Profile Tolerance

Profiles of lines and surfaces can be said to be within geometric tolerance if they lie within a profile tolerance zone. An example of a geometric tolerance for a profile of a line is shown in Figure 8.17.

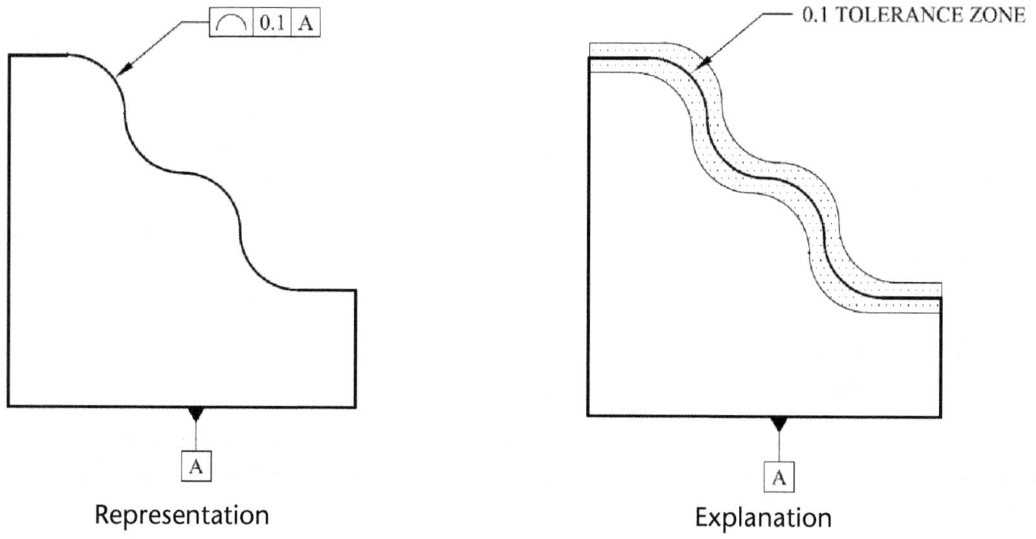

Figure 8.17 Profile of a line geometric tolerance

5.3 Orientation Tolerances

Geometric tolerances contained within the orientation category must contain a datum, which is added to the end of the feature control frame.

Angularity Angularity is achieved if an angle is within an angular tolerance zone, measured from a datum. Angularity can be seen in Figure 8.18.

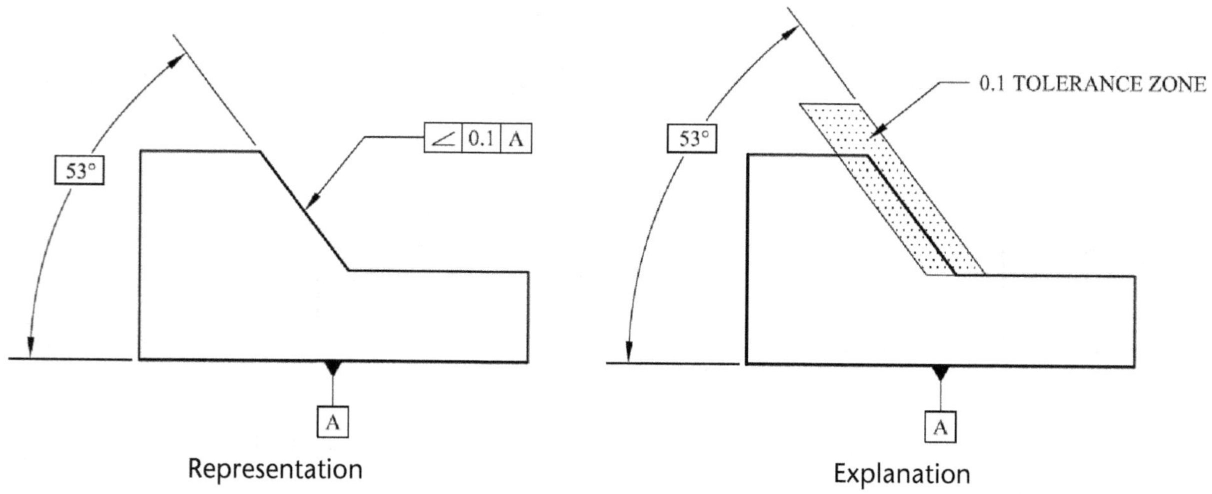

Figure 8.18 Angularity geometric tolerance

Parallelism Parallelism is achieved if one side of an object is parallel to another, within a tolerance zone, measured from a datum. Parallelism is illustrated in Figure 8.19.

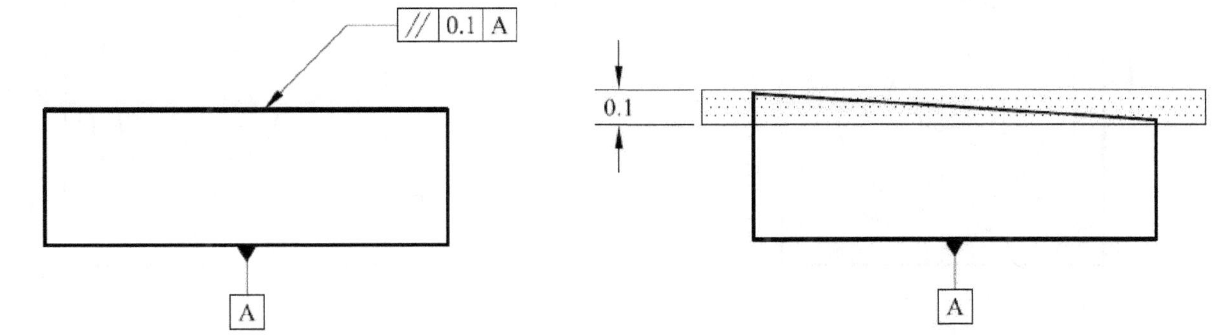

Representation   Explanation

Figure 8.19 Parallelism geometric tolerance

Perpendicularity Perpendicularity is achieved if one side of an object is perpendicular to another, within a tolerance zone, measured from a datum. Perpendicularity can be seen in Figure 8.20.

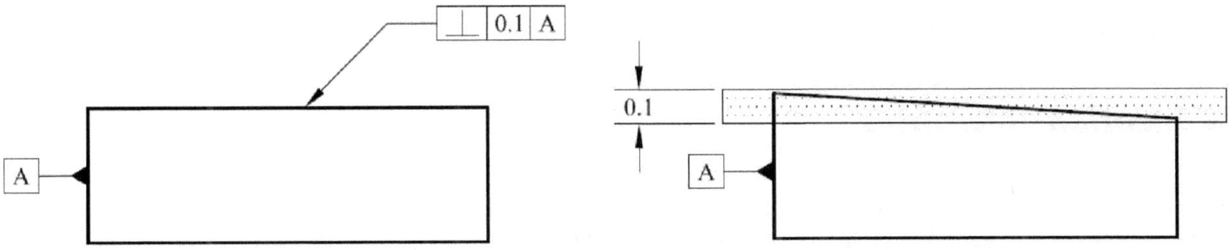

Representation   Explanation

Figure 8.20 Perpendicularity geometric tolerance

Symmetry An object is said to be symmetrical if one side of an object is symmetrical to another, within a tolerance zone, measured from a datum. Symmetry is illustrated in Figure 8.21.

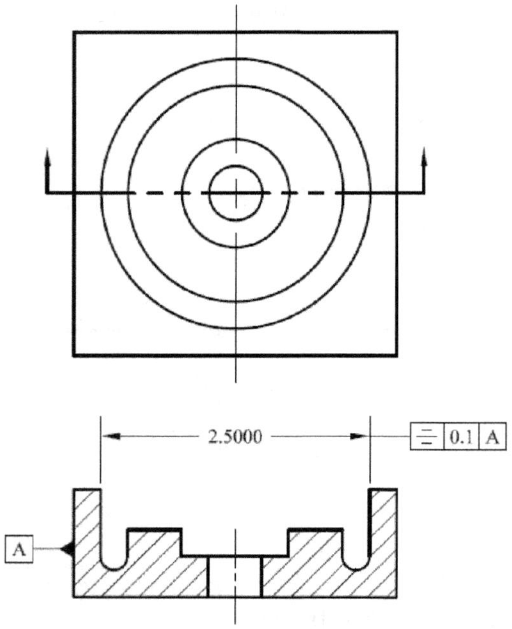

Representation

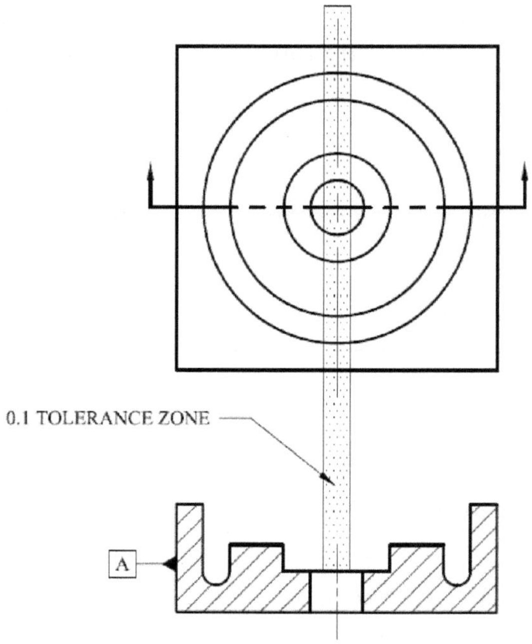

Explanation

Figure 8.21 Symmetry geometric tolerance

5.4 Location Tolerances

Position A hole's position can be dimensioned with position tolerancing, either by a position geometric tolerance or a linear tolerance in the horizontal and vertical direction. The position tolerance is more accurate, as is provides a circular zone, instead of a square zone. See Figure 8.22 for an illustration.

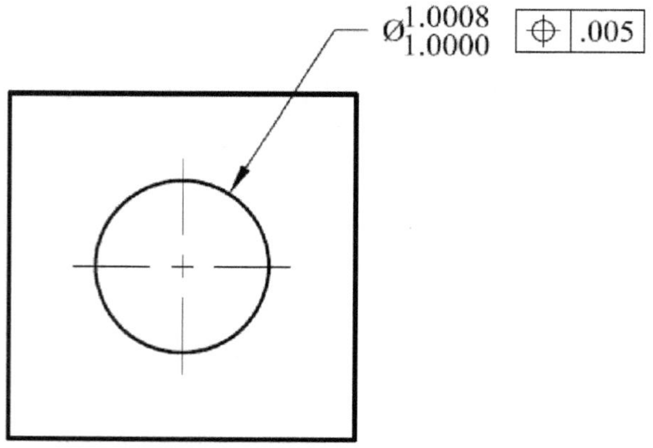

Representation

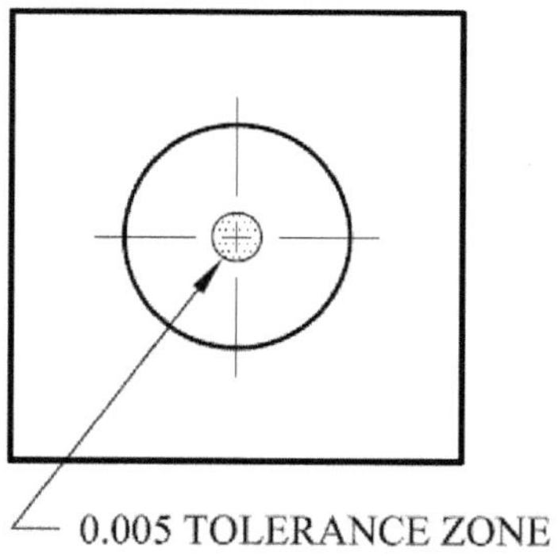

Explanation

Figure 8.22 Position geometric tolerance

Concentricity Concentricity is achieved when two cylinders share an identical axis, within a tolerance zone. Concentricity is shown in Figure 8.23.

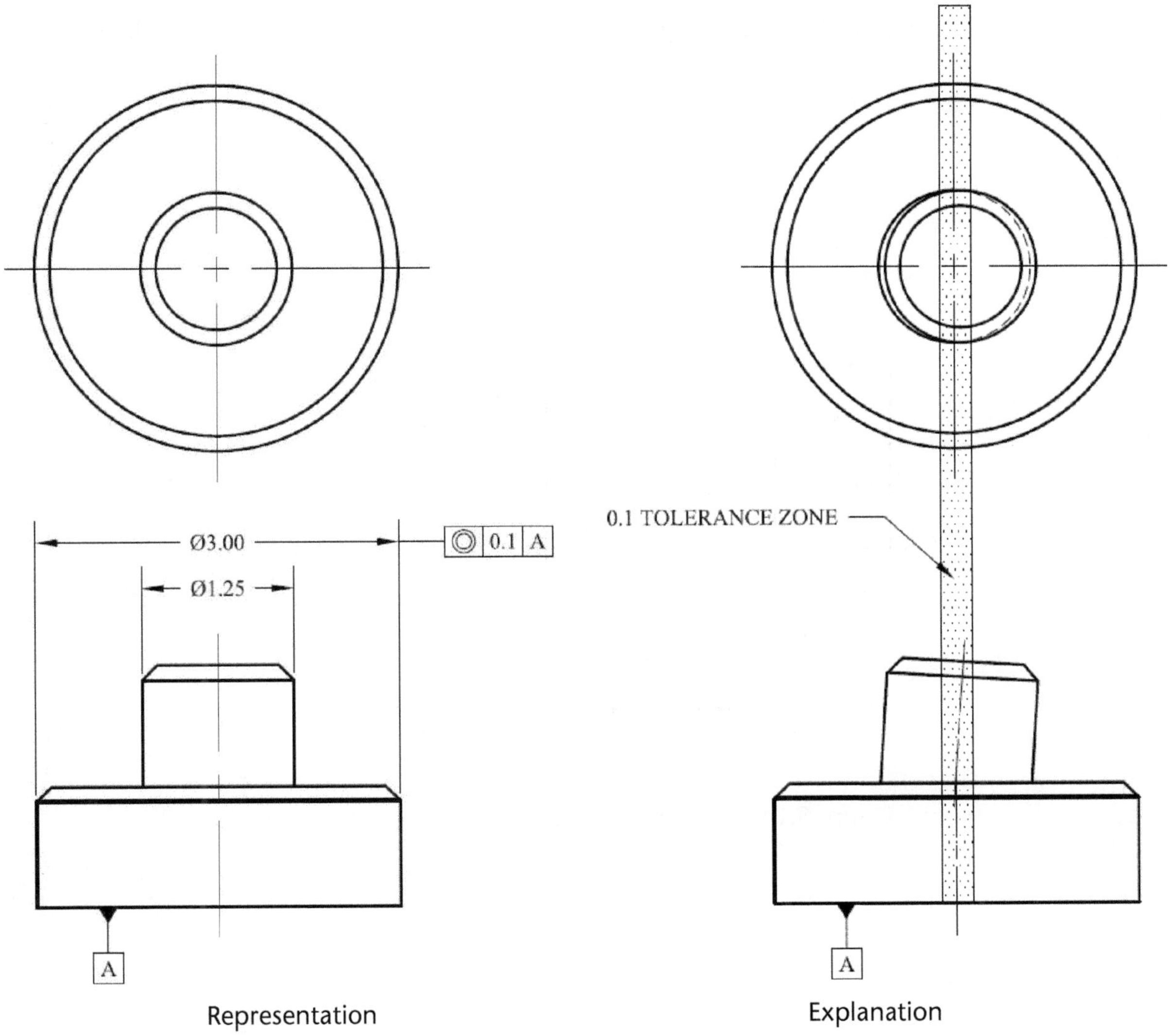

Figure 8.23 Concentricity geometric tolerance

5.5 Runout Tolerance

Runout geometric tolerances can be either circular or total. Runout is a combination of roundness, straightness, angularity, and profile. Runout tolerances always require a datum. An example of runout and the proper method to indicate it is shown in Figure 8.24. In this example, the right most cylindrical flange needs to be round, straight, angular, and reflect the profile within 0.05 inches to the datum.

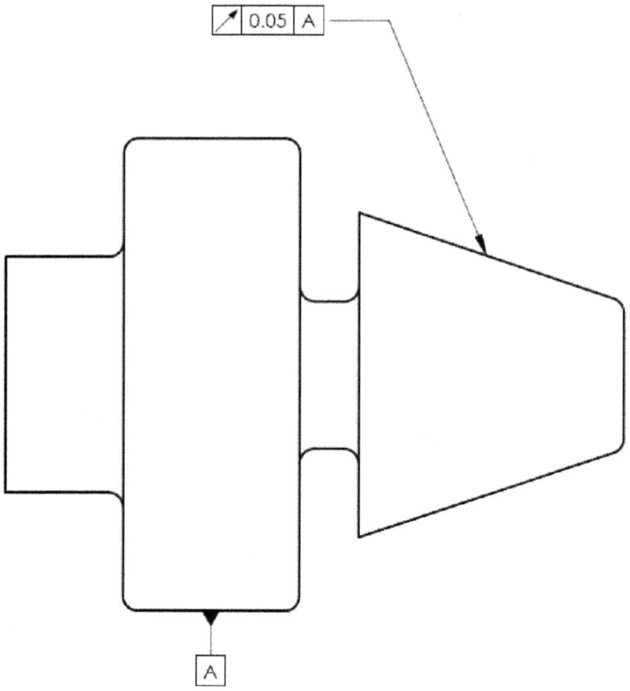

Figure 8.24 Circular runout geometric tolerance

Practice Test

1. **Roundness and Flatness can be used interchangeably when using geometric tolerances.**
   A) True
   B) False

2. **Linear tolerances always add to the cost of manufacturing.**
   A) True
   B) False

3. **An H11/c11 metric tolerance is the same as a C11/h11.**
   A) True
   B) False

4. **All geometric tolerance feature control frames require a datum.**
   A) True
   B) False

5. **The maximum clearance is defined as the loosest fit between mating parts.**
   A) True
   B) False

6. **RC4 hole and shaft limits are the same, regardless of the range of diameters.**
   A) True
   B) False

7. **Roundness and Circularity can be used interchangeably.**
   A) True
   B) False

8. **In some instances, the allowance can be greater than the maximum clearance.**
   A) True
   B) False

9. **The maximum clearance between a shaft that has acceptable diameters between 2.0032 and 2.0025 and a hole with acceptable diameters between 2.0012 and 2.000 is:**
   A) 0.0013.
   B) -0.0013.
   C) 0.0032.
   D) -0.0032.
   E) None of the above.

10. **The shaft tolerance for a shaft with acceptable diameters between 2.0032 and 2.0025 is:**
    A) -0.0007.
    B) 2.0000.
    C) 0.0007.
    D) -2.0000.
    E) None of the above.

Problems

For a classic shaft-hole assembly with the conditions listed below, calculate the acceptable range of diameters and the Hole Tolerance, Shaft Tolerance, Maximum Clearance, and Allowance.

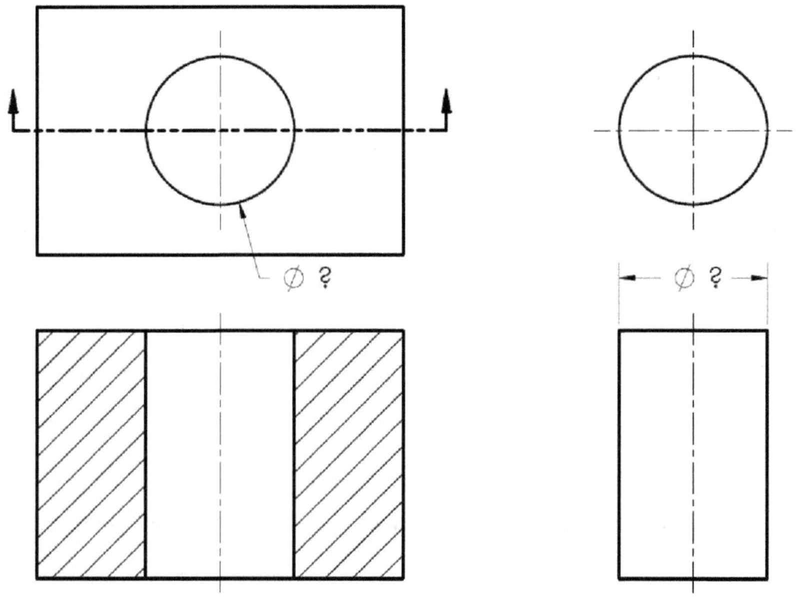

English	Metric
1. Nominal Size Ø 4.00, LC2	11. Nominal Size Ø 5.0, H11/c11
2. Nominal Size Ø 0.75, LT4	12. Nominal Size Ø 2.0, H8/f7
3. Nominal Size Ø 0.50, LN1	13. Nominal Size Ø 16.0, H7/h6
4. Nominal Size Ø 0.875, FN1	14. Nominal Size Ø 2.5, H7/k6
5. Nominal Size Ø 15.00, RC5	15. Nominal Size Ø 10.0, C11/h11
6. Nominal Size Ø 4.00, RC9	16. Nominal Size Ø 4.00, G7/h6
7. Nominal Size Ø 0.25, LC5	17. Nominal Size Ø 1.2, D9/h9
8. Nominal Size Ø 200, LT1	18. Nominal Size Ø 25.0, N7/h6
9. Nominal Size Ø 12.00, LN3	19. Nominal Size Ø 10.0, U7/h6
10. Nominal Size Ø 6.00, FN4	20. Nominal Size Ø 1.6, F8/h7

DIMENSION THE HOLES AND SHAFTS FOR THE REQUIRED LINEAR TOLERANCE. ALSO CALCULATE THE HOLE TOLERANCE, SHAFT TOLERANCE, MAXIMUM CLEARANCE, AND ALLOWANCE.

NOMINAL SIZE: 2.000 INCHES
TYPE OF TOLERANCE: RC2

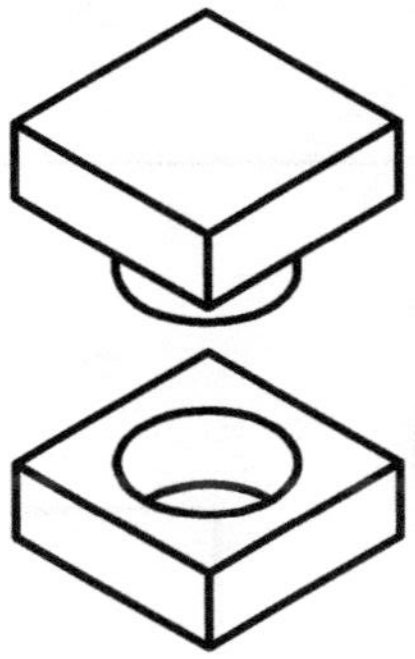

HOLE TOLERANCE: . . . .

SHAFT TOLERANCE: . . . .

MAXIMUM CLEARANCE: . . . .

ALLOWANCE: . . . .

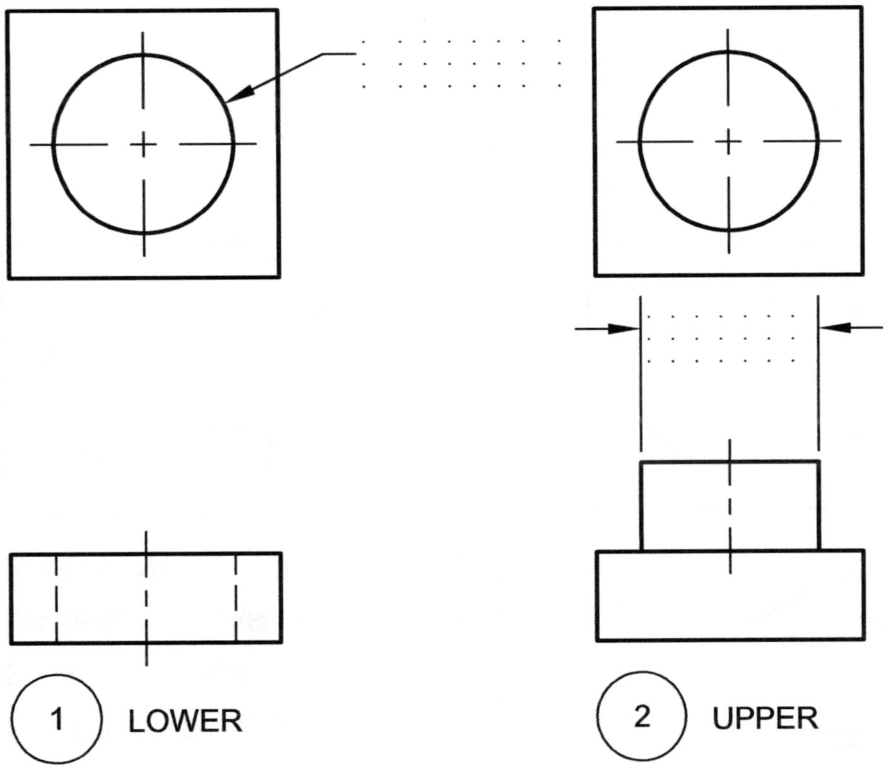

NOMINAL SIZE: 0.5000 INCHES
TYPE OF TOLERANCE: LC5

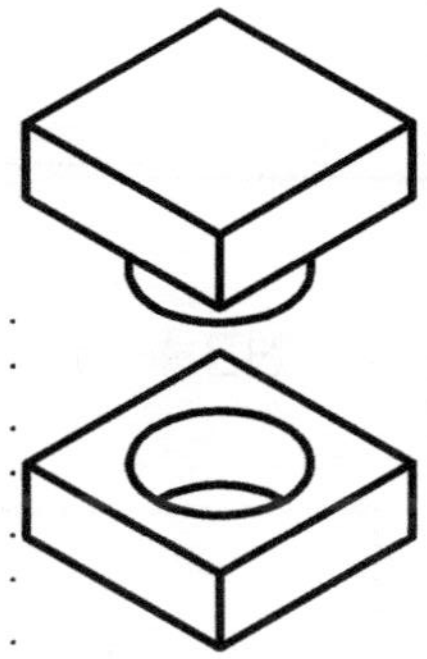

HOLE TOLERANCE: . . . .

SHAFT TOLERANCE: . . . .

MAXIMUM CLEARANCE: . . . .

ALLOWANCE: . . . .

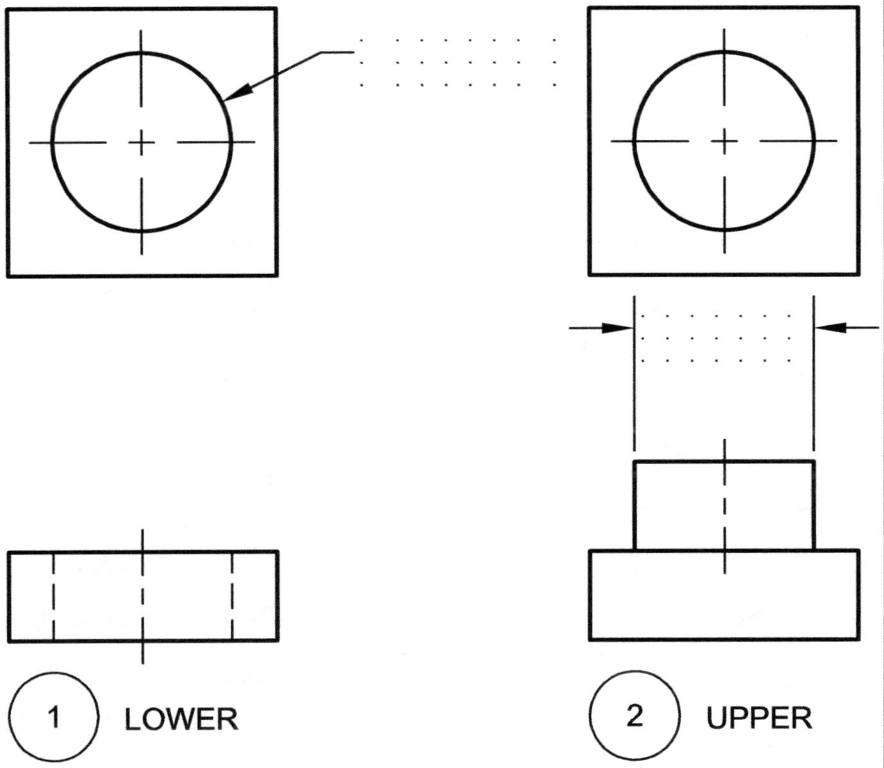

AUTHOR: . . . . SCALE: . . . .

TEAM: . . . . SECTION: . . . . DATE: . . . .

DIMENSION THE HOLES AND SHAFTS FOR THE REQUIRED LINEAR TOLERANCE. ALSO CALCULATE THE HOLE TOLERANCE, SHAFT TOLERANCE, MAXIMUM CLEARANCE, AND ALLOWANCE.

NOMINAL SIZE: 0.7500 INCHES
TYPE OF TOLERANCE: FN1

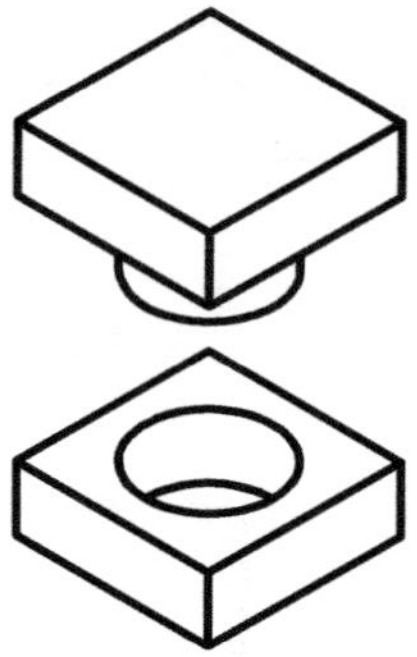

HOLE TOLERANCE: ....

SHAFT TOLERANCE: ....

MAXIMUM CLEARANCE: ....

ALLOWANCE: ....

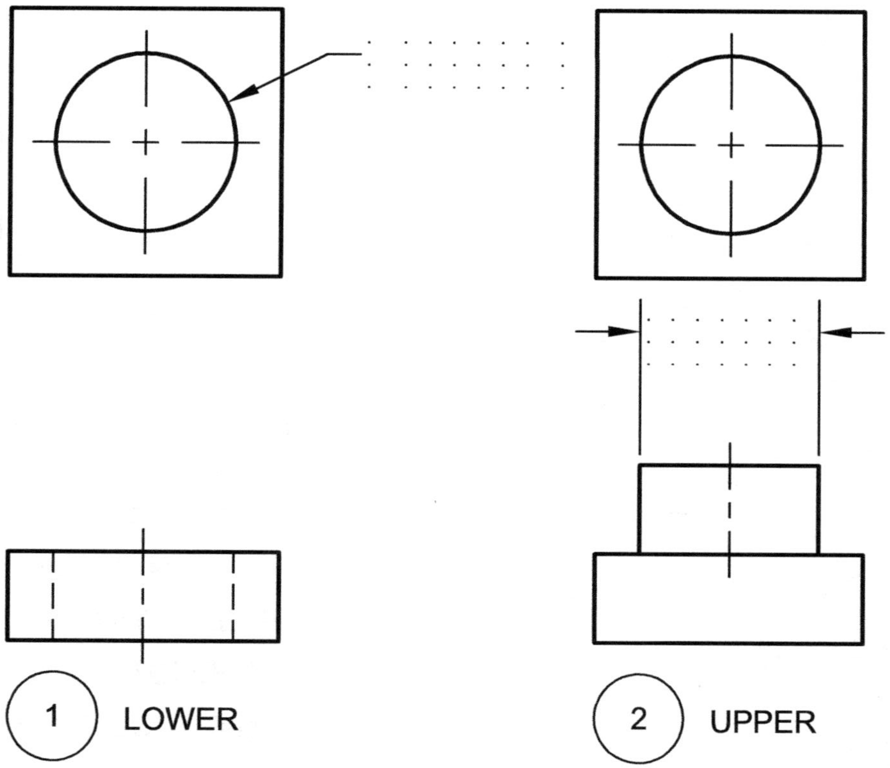

NOMINAL SIZE: 1.7500 INCHES
TYPE OF TOLERANCE: LT3

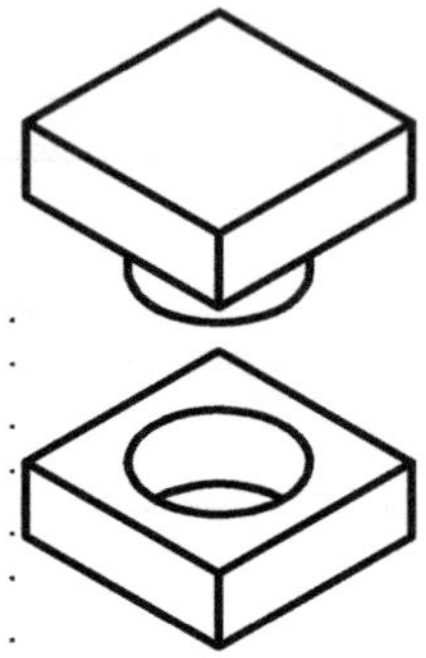

HOLE TOLERANCE: ....

SHAFT TOLERANCE: ....

MAXIMUM CLEARANCE: ....

ALLOWANCE: ....

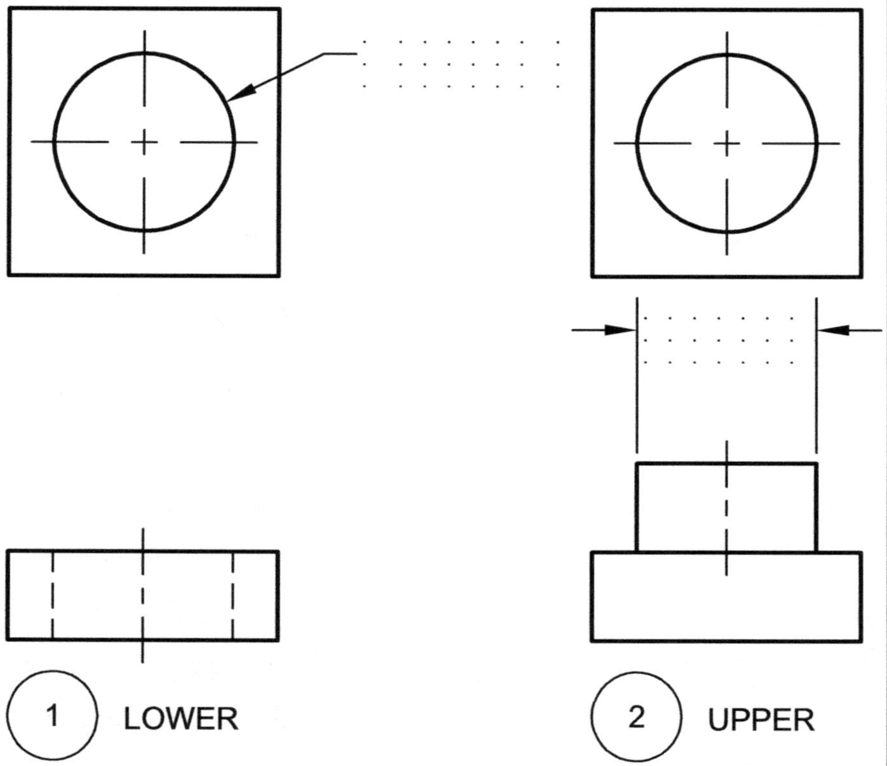

AUTHOR:	SCALE:
TEAM: SECTION:	DATE:

DIMENSION THE HOLES AND SHAFTS FOR THE REQUIRED LINEAR TOLERANCE. ALSO CALCULATE THE HOLE TOLERANCE, SHAFT TOLERANCE, MAXIMUM CLEARANCE, AND ALLOWANCE.

NOMINAL SIZE: 12 mm
TYPE OF TOLERANCE: H9/d9

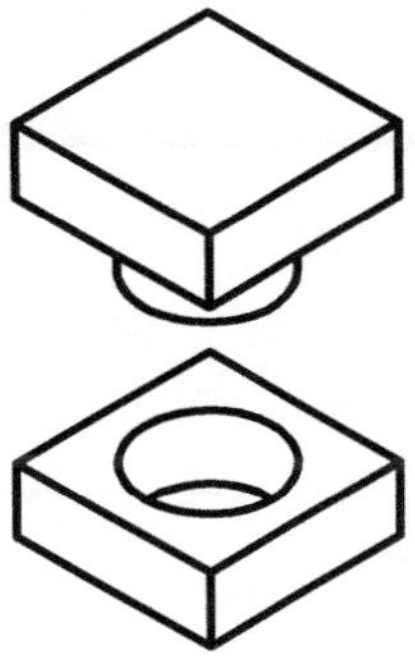

HOLE TOLERANCE: . . . .

SHAFT TOLERANCE: . . . .

MAXIMUM CLEARANCE: . . . .

ALLOWANCE: . . . .

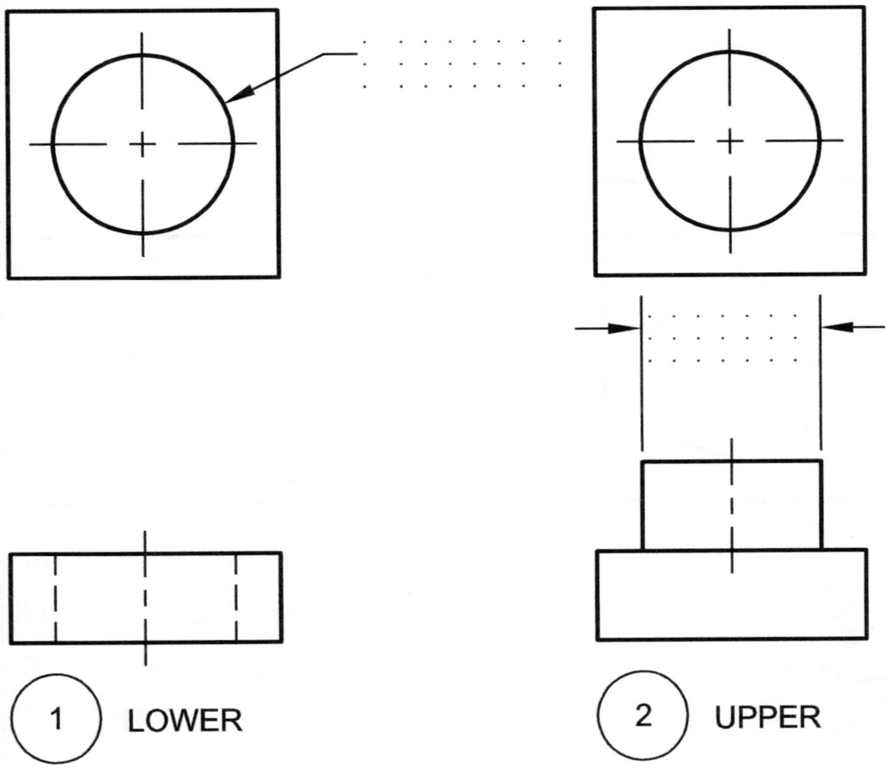

NOMINAL SIZE: 4 mm
TYPE OF TOLERANCE: H7/p6

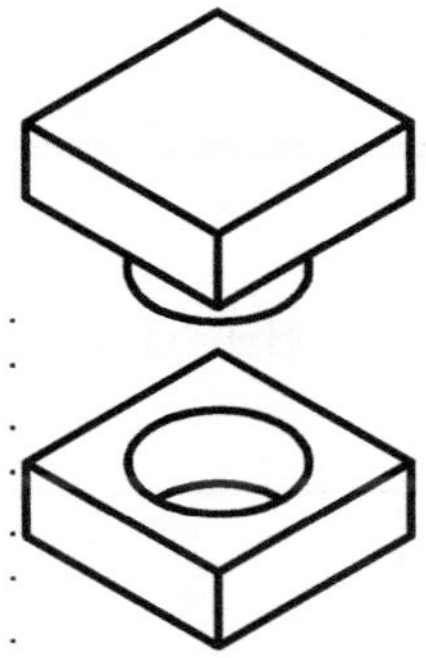

HOLE TOLERANCE: . . . .

SHAFT TOLERANCE: . . . .

MAXIMUM CLEARANCE: . . . .

ALLOWANCE: . . . .

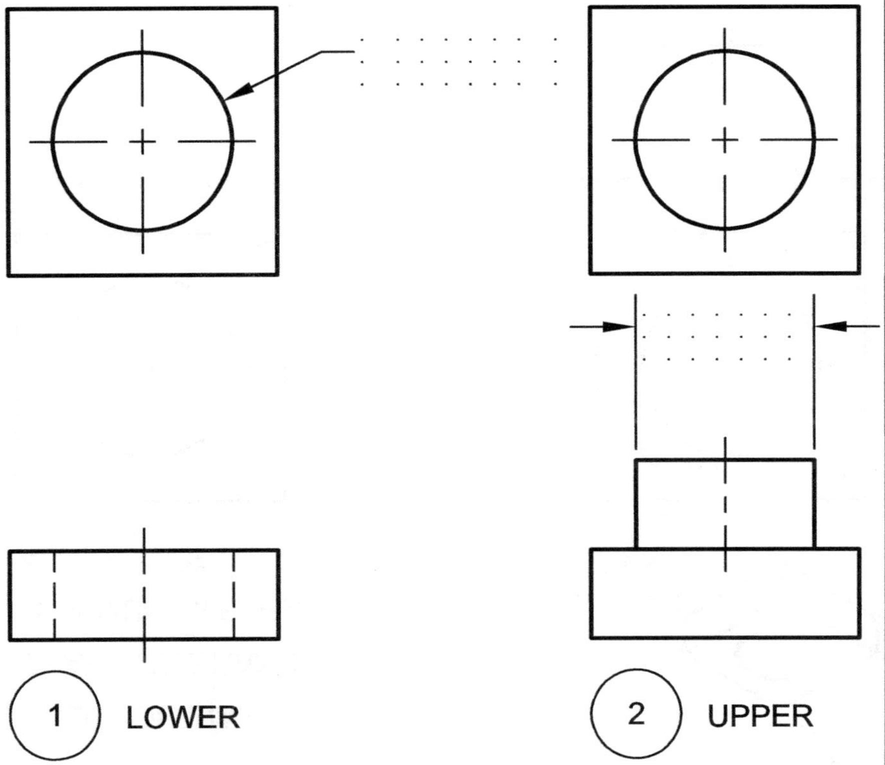

AUTHOR: . . . . SCALE: . . . .

TEAM: . . . . SECTION: . . . . DATE: . . . .

DIMENSION THE HOLES AND SHAFTS FOR THE REQUIRED LINEAR TOLERANCE. ALSO CALCULATE THE HOLE TOLERANCE, SHAFT TOLERANCE, MAXIMUM CLEARANCE, AND ALLOWANCE.

NOMINAL SIZE: 16 mm
TYPE OF TOLERANCE: U7/h6

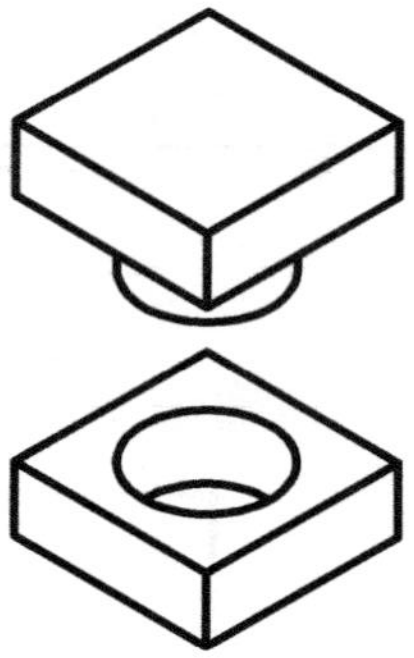

HOLE TOLERANCE: . . . .

SHAFT TOLERANCE: . . . .

MAXIMUM CLEARANCE: . . . .

ALLOWANCE: . . . .

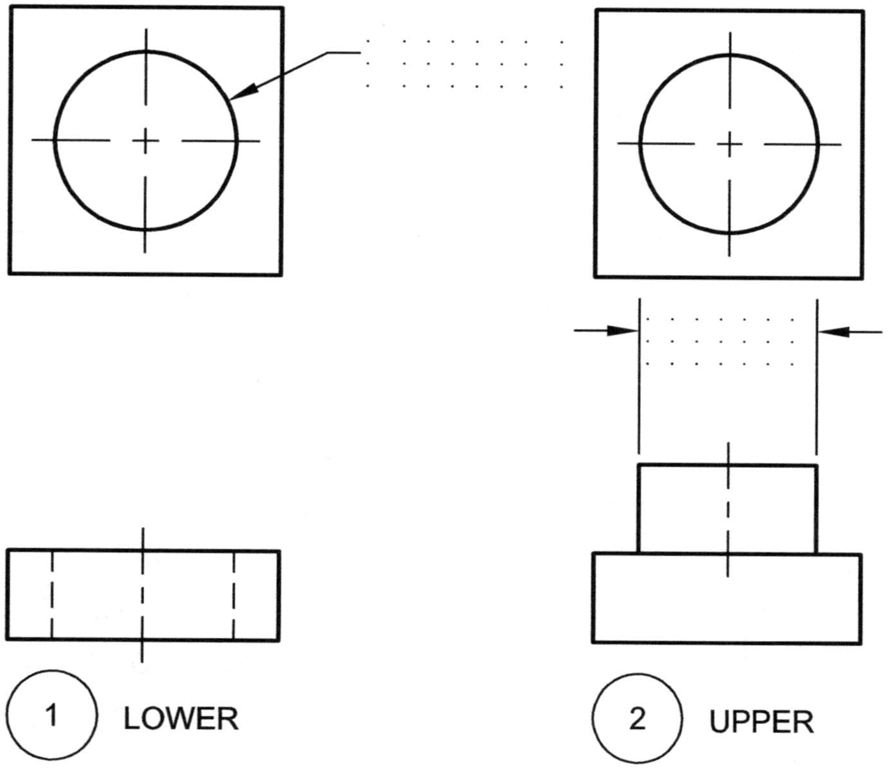

NOMINAL SIZE: 20 mm
TYPE OF TOLERANCE: F8/h7

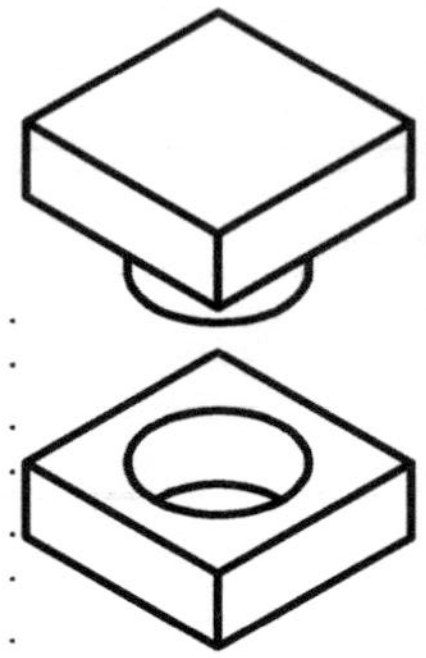

HOLE TOLERANCE: . . . .

SHAFT TOLERANCE: . . . .

MAXIMUM CLEARANCE: . . . .

ALLOWANCE: . . . .

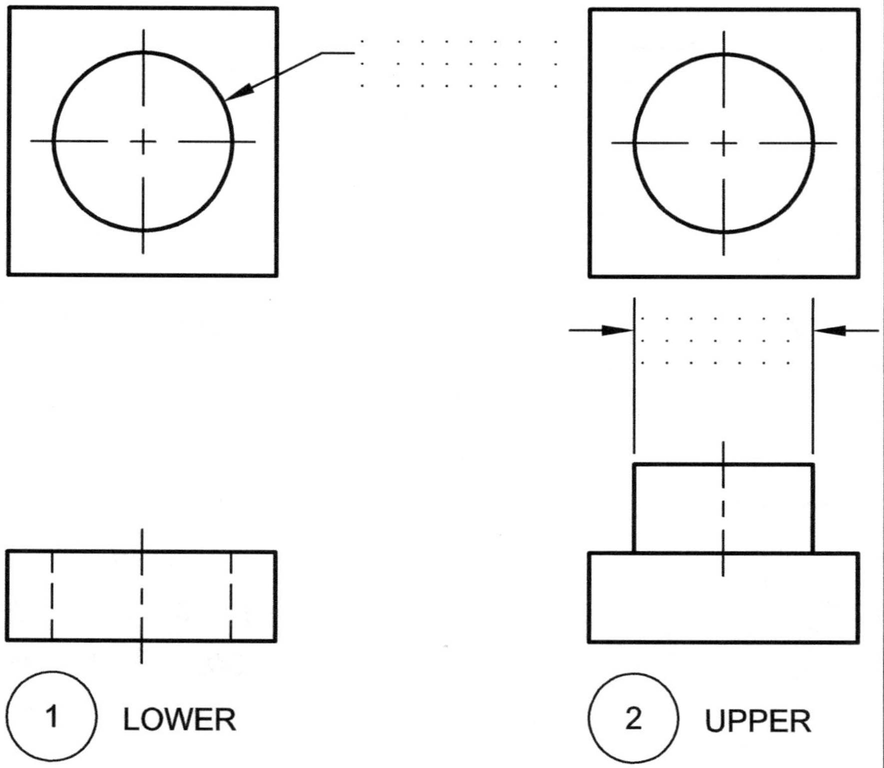

AUTHOR:	SCALE: . . . .
TEAM: . . . . SECTION:	DATE: . . . .

Apply the geometric tolerances as noted.

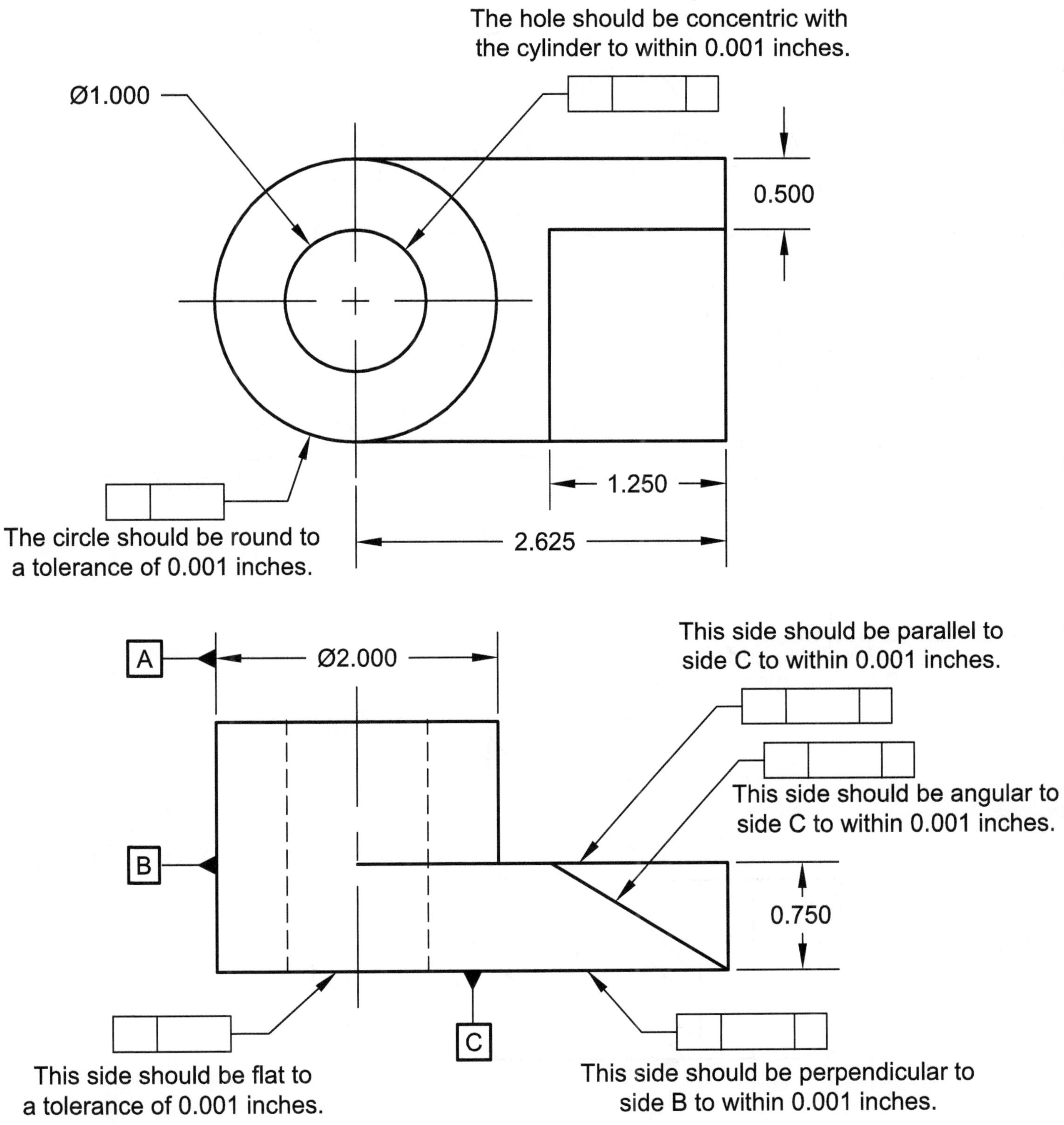

AUTHOR:	SCALE:
TEAM: SECTION:	DATE:

Apply the geometric tolerances as noted

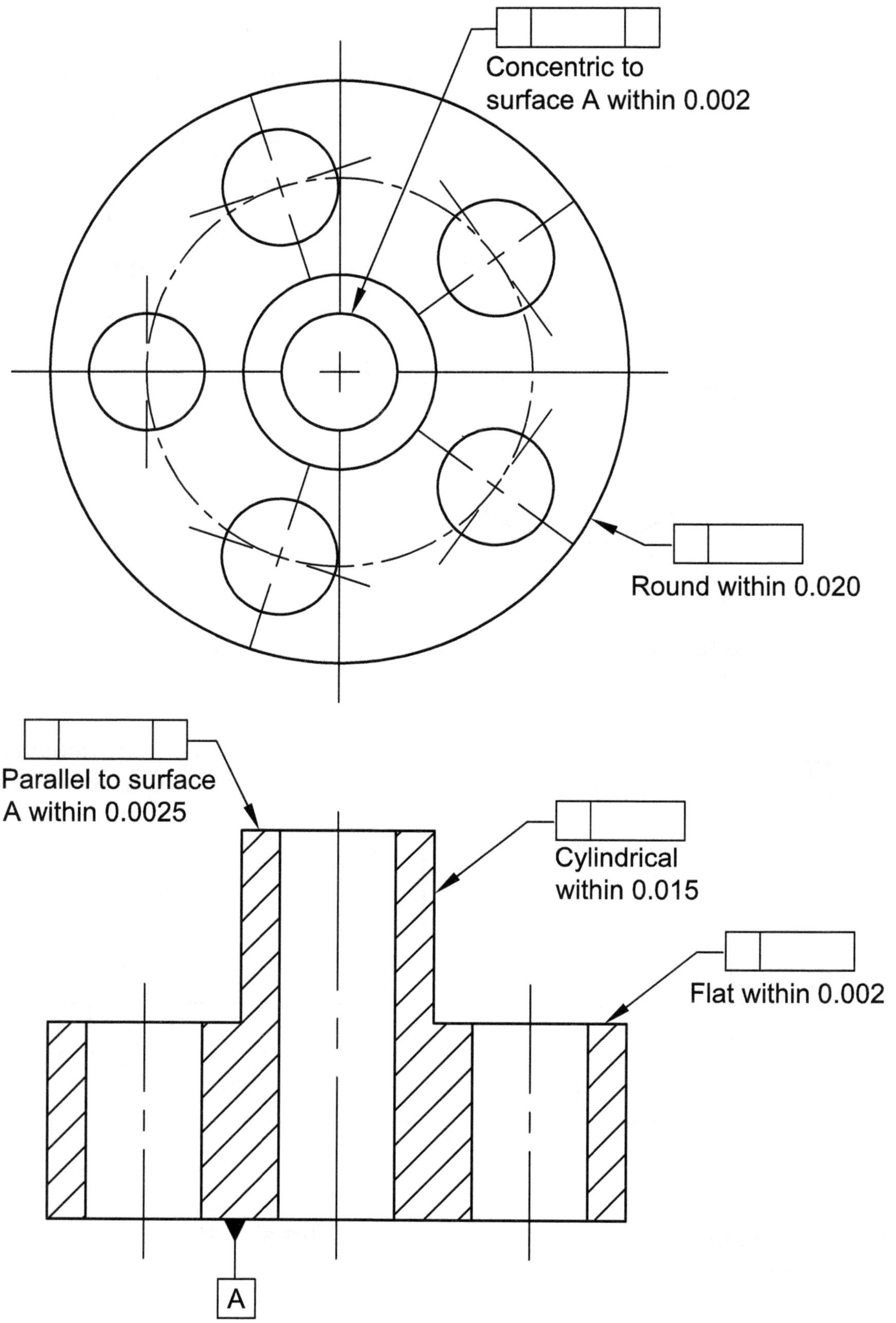

AUTHOR:	SCALE:
TEAM: SECTION:	DATE:

Fasteners and Other Standard Parts

9 CHAPTER

1. Introduction

Most, if not all, engineering designs require the use of certain parts and features that serve common purposes. For example, most assemblies require bolts and screws to join two or more components together. Many designs involve the use of springs to resist and react to forces. Others require devices to transmit force and motion, such as gears and cams.

The specifications of commonly used parts have been standardized for both the US and the metric systems. Such parts are widely available in numerous forms, shapes, and sizes. They are interchangeable and relatively inexpensive. As a result, standard parts are normally purchased from vendors instead of being designed and manufactured from scratch for every new assembly. For example: any designer that specifies a 1/2-20UNF-2A Hex Cap Screw will receive an identical product by any manufacturer accepting the order. Some specific situations may exist, however, that require the use of specially engineered bolts or springs or modified versions of a particular standard.

In engineering drawings, time is not wasted drawing and dimensioning standard parts in detail. Instead, standard representations and notations have been developed for such purposes so the information is communicated clearly and unambiguously.

2. Fasteners

Most engineering designs consist of multiple parts working together as single functional units. Determining how parts are assembled and held together in an assembly is a major concern. The process by which different components of a design are held or joined together is called fastening. A fastener is a mechanical device designed for this purpose. Based on its application, fasteners often receive varying degrees of precision and capabilities. They directly affect the design's durability, size, and weight, and ensure adequate functioning of the product under pre-established conditions.

Some fastening methods are intended to be permanent, such as welding, nailing, stapling, or gluing. Others allow for disassembly and reassembly of the design so parts can be adjusted, serviced or replaced easily. Examples of non-permanent fasteners include bolts, nuts, screws, and pins. In this section, the types of non-permanent fasteners commonly used in industry and defined by current standards (both ANSI and ISO) will be discussed.

2.1 Threads

Some fasteners are threaded while others are unthreaded. A thread is a spiral profile wrapped around a cylindrical or conical surface (external thread) or inside a hole (internal thread) designed to convert between rotational and linear motion and/or secure two or more parts together (see Figure 9.1).

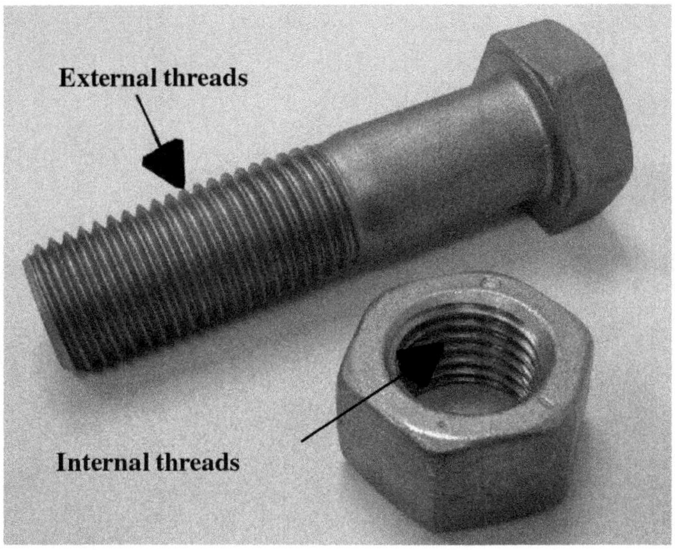

Figure 9.1 External and internal threads

Knowledge of threads begins with understanding the terminology associated with them. The illustration of the bolt shown in Figure 9.2 lists the most important terms used to describe threads

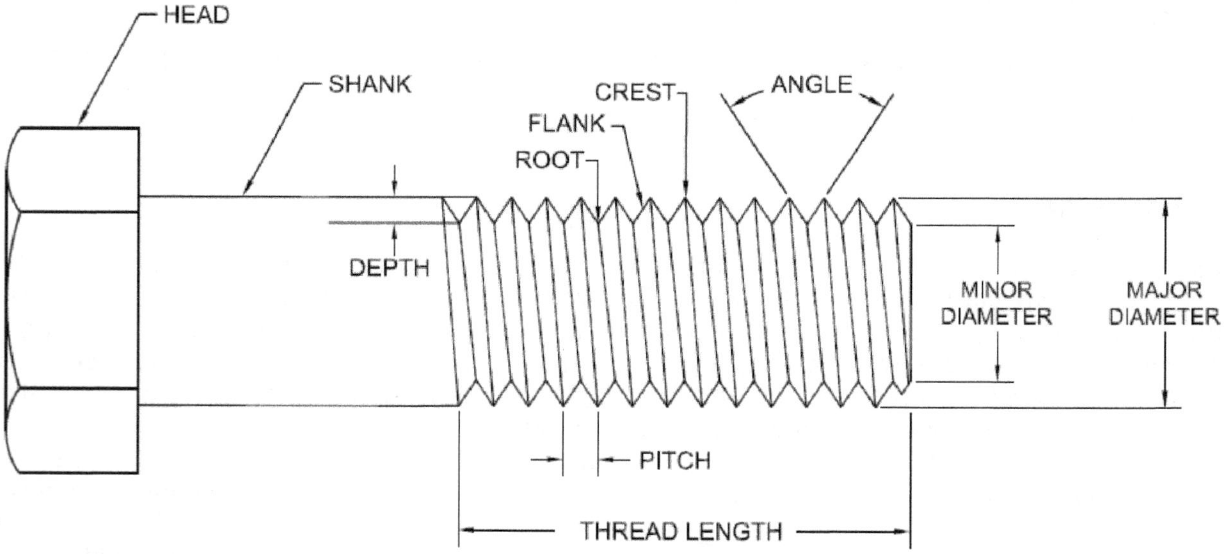

Figure 9.2 Thread terminology

Basic Terminology

- Crest: peak or prominent point of a thread.
- Root: bottom point at which the sides of a thread meet.
- Flank: thread surface that connects a crest and a root.
- Angle: degrees of the angle between two flanks.
- Pitch: distance between two adjacent crests or two adjacent roots.
- Major Diameter: largest diameter of a thread, whether external or internal.
- Minor Diameter: smallest diameter of a thread, whether external or internal.
- Thread Length: length of the threaded portion of a shaft measured longitudinally.
- Depth: difference between a crest and a root measured perpendicularly to the longitudinal axis of the thread.
- Shank: unthreaded portion between the head and the threads.
- Threads per inch: number of threads that occur per every inch of axial length.

Lead The lead is defined as the distance a threaded part travels forward or backward when rotated one complete revolution. Most parts are single threaded, which means that the lead equals the pitch (the part advances one pitch distance when rotated one revolution).

In certain situations, multiple threads (normally double or triple) may be required. The lead of a double thread equals twice the pitch. Therefore, a double threaded part will advance two pitch distances when rotated one revolution. The lead of a triple thread equals three times the pitch. In this case, the threaded part will advance three pitch distances when rotated one revolution. Multiple threads do not provide as much resistance as single threads. They are used almost exclusively in scenarios that require quick assembly and disassembly of parts rather than tightness.

Direction The majority of threaded fasteners are right handed, which means that they advance forward as they are rotated clockwise (when viewed toward the mating thread) and backward when rotated counterclockwise. However, certain situations require the use of left handed threads, or threads that advance forward as they are rotated counterclockwise when viewed toward the mating part (see Figure 9.3). Examples of these situations include keeping counterclockwise rotating assemblies from becoming loose (the left hand pedal on a bicycle or some lawn mower blades), or preventing hazardous misconnections on gas pipes and supply links, especially for flammable gases. Common turnbuckles used to adjust the tension of ropes and cables also use left handed threads (see Figure 9.4).

Figure 9.3 Right handed threads (top) and left handed threads (bottom)

Figure 9.4 Turnbuckles use right and left handed threads

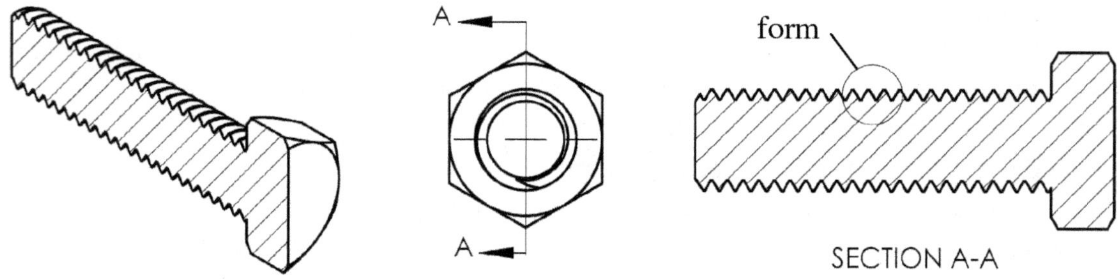

Figure 9.5 Section view of a fastener illustrating its thread form

Form The form of a thread refers to the shape of its profile (see Figure 9.5). There is a wide variety of thread forms defined by different standards. Standardization of threads was developed to facilitate compatability and interchangeability of parts between different manufacturers. The standardization process of threads still continues today.

Thread forms are identified by short letter codes based on the specific standard that defines them. The most common thread forms and their codes are illustrated in Figure 9.6.

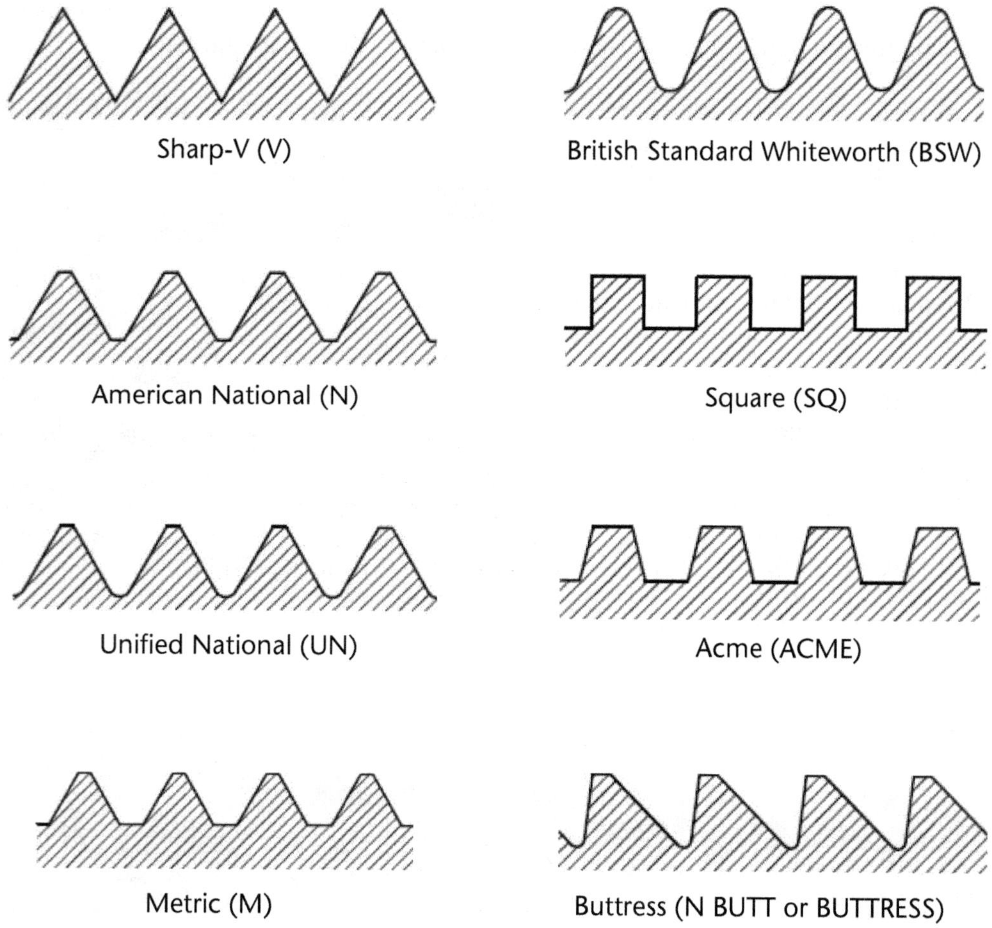

Figure 9.6 Common thread forms and letter codes

The most common thread forms are the Unified National (defined in the standard ANSI Y14.6) and the Metric (defined in the standard ISO 261). Other standards and variations still remain in use in certain areas and/or for some specific applications (medical technology, microscopy, aeronautics, etc).

The Sharp-V profile was the first thread standard in the United States. Its form increased the friction of the thread and allowed very tight seals. However, because of the difficulty of manufacturing such a thread (the crests are often damaged in the process) the Sharp-V profile was soon replaced by the American National Form. The American National profile is still widely used in North America and the United Kingdom. It is based on the Sharp-V form, but with the crests and roots flattened. The Unified National thread is the current standard in the US, Canada, and UK. The metric thread is the international standard.

Square, Acme, and Buttress threads are used mainly in machinery applications for power transmission. The lack of a thread angle in square threads eliminates radial pressure, which makes this thread very efficient. Its main disadvantage is the difficulty in machining such a thread. Acme threads are normally found where large loads or high accuracy are required. They are also easier to manufacture than square threads. Buttress threads are designed to have very low friction and handle extremely high axial loads in one direction.

Series The series of a thread determines the number of threads per inch for a given diameter. Different standards define different series. For example, some of the series defined in the Unified National (UN) form include: C (Coarse pitch series), F (Fine pitch series), EF (Extra fine pitch series), JC, JF, and JEF (Coarse, Fine, and Extra fine series with external thread controlled root radius), M (Miniature thread series), and S (Special series).

Most standards define coarse, fine, and extra fine series. A coarse thread has a larger pitch (fewer threads per inch) than a fine thread (more threads per inch) (see Figure 9.7).

Figure 9.7 Difference in threads per inch between coarse (top) and fine threads (bottom)

Coarse threads are the most common. They are used in a wide variety of environments and with different materials. Finely threaded fasteners are used when additional force is required to hold the components of the assembly together because of extreme pressure during machine operation. Extra fine threads are normally used for short and small diameter shanks.

Thread Fit (Tolerance) The term thread fit refers to the tightness or tolerance between two mating threaded parts. Different standards define different levels of tolerance, ranging from loose fits, for assemblies where looseness between threaded parts is acceptable and/or desired, to very close and tight fits, for assemblies required to handle very high levels of pressure and stress. Fits are generally represented with a number (the larger the number, the tighter the fit).

2.2 Thread Representation

Drawing threads in a realistic manner can be a difficult and time consuming process. Instead, three conventional symbols have been developed: detailed, schematic, and simplified (see Figure 9.8). Any of these three methods is acceptable in engineering drawings.

Detailed symbols are the most realistic, but schematic and simplified symbols are easier and quicker to draw. Simplified thread symbols use dashed lines (similar to hidden lines) to represent the minor diameter of the thread. If the simplified thread is drawn in a section view, the hidden lines are still shown (see Figure 9.9). Schematic thread symbols use lines drawn perpendicular to the longitudinal axis of the thread to approximate the thread pitch and the minor diameter. The shorter line that approximates the minor diameter are referred to as root lines (and are drawn thicker than visible lines) and the longer lines that represent the major diameter are called crest lines.

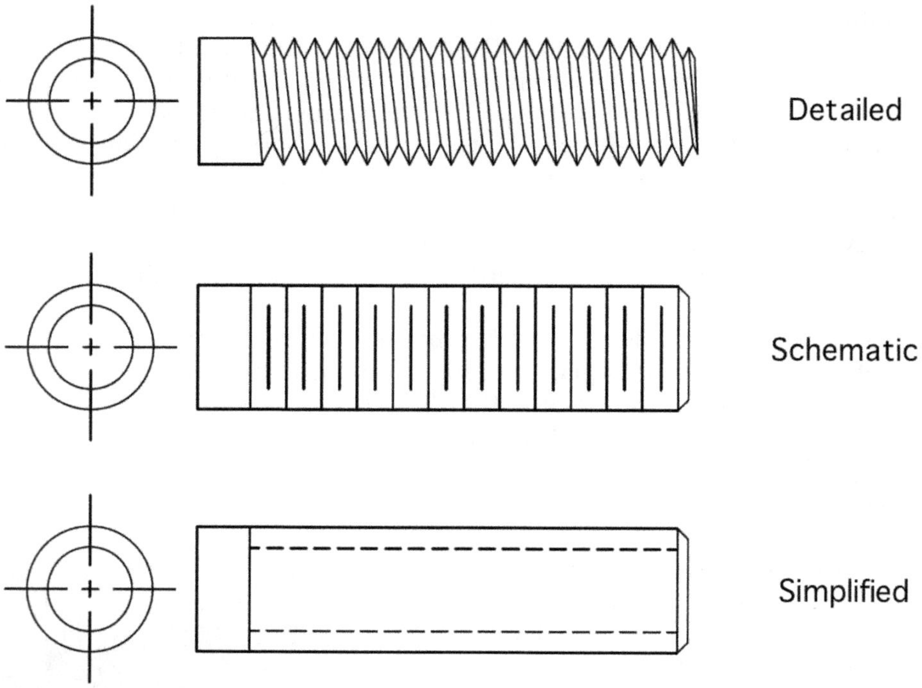

Figure 9.8 External thread representations

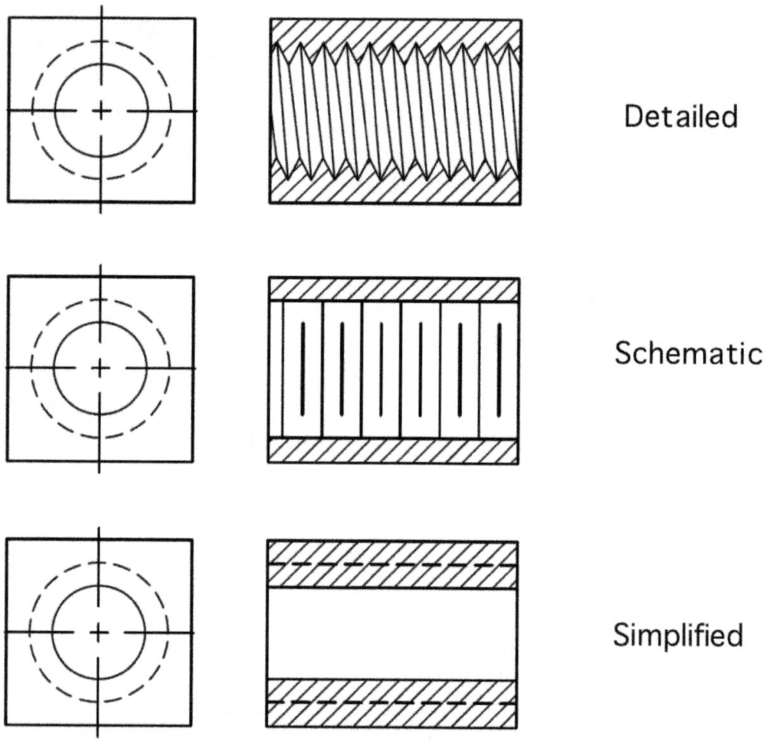

Figure 9.9 Internal thread representation (sectional views)

2.3 English Thread Notes

Thread representations require standard notes to provide complete and accurate information about the specific type of thread being used in a particular design. Thread notes are given with a leader and an elbow to the thread symbol, if the thread is external, or to the circular view, if the thread is internal. English thread notes are used to describe American National and Unified National Series. The format of an English thread note and an example are illustrated in Figures 9.10 and 9.11.

Major Diameter (in inches)	–	**Threads Per inch**	**Form**	**Series**	–	**Fit**	**External/Internal**	Additional Information (if required)

Figure 9.10 Format of an English thread note

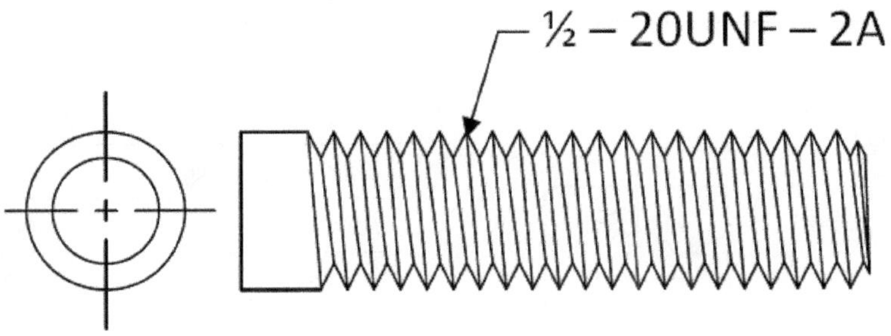

Figure 9.11 Use of an English thread note

The first part of the note is the major diameter followed by a dash. Although decimal numbers are acceptable, the major diameter is normally given as a fraction. Next, threads per inch followed by the thread form (N for American National or UN for Unified National) and series (most often C for Coarse, F for Fine or EF for Extra Fine). After that, another dash followed by a number that indicates the type of fit (1 for Loose, 2 for Regular, and 3 for Tight) and a letter (A if the thread is external or B if it is internal).

Additional information at the end of the note may be required in some cases. If a thread is left handed the letters LH must appear at the end of the note. Also, for double and triple threads, the words DOUBLE or TRIPLE (or the abbreviations DBL or TPL) must be included. Some examples of thread notes for American and Unified National threads are shown in Figure 9.12.

ANSI Thread Note	Description	
¼ – 20UNC – 3A – LH 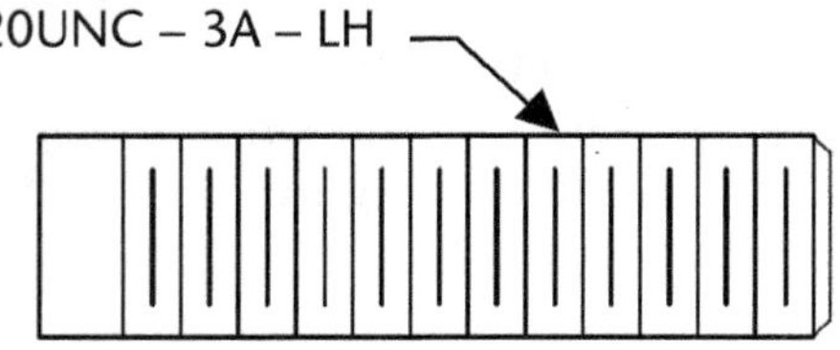	*Major diameter:*	¼″ (.25 inches)
	Threads per inch:	20
	Form:	UN (Unified National)
	Series:	C (Coarse)
	Fit:	3 (Tight)
	External/Internal:	A (External)
	Additional:	LH (Left Handed)
¾ – 16NF – 1A – DOUBLE 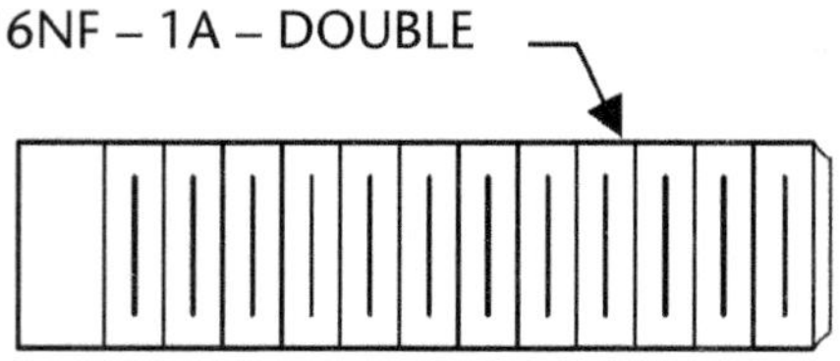	*Major diameter:*	¾″ (.75 inches)
	Threads per inch:	16
	Form:	N (American National)
	Series:	F (Fine)
	Fit:	1 (Loose)
	External/Internal:	A (External)
	Additional:	DOUBLE (Double Thread)
1.00 – 20UNEF – 2B 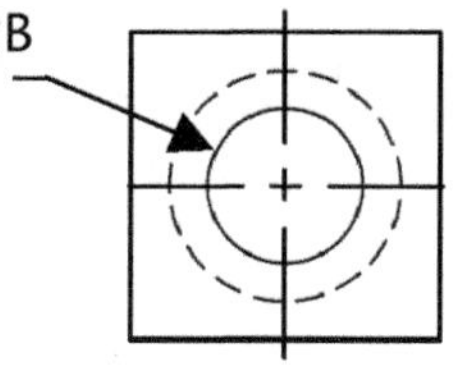	*Major diameter:*	1.00″
	Threads per inch:	20
	Form:	UN (Unified National)
	Series:	EF (Extra Fine)
	Fit:	2 (Regular)
	External/Internal:	B (Internal)
	Additional:	–
3.00 – 4UNC – 3B – LH 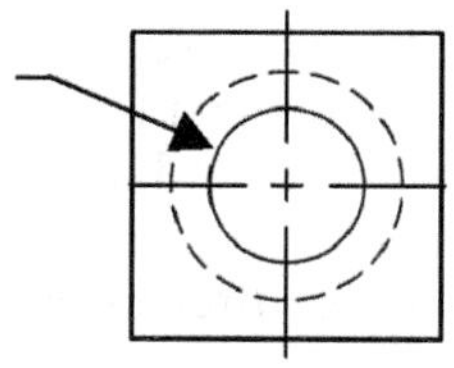	*Major diameter:*	3.00″
	Threads per inch:	4
	Form:	UN (Unified National)
	Series:	C (Coarse)
	Fit:	3 (Tight)
	External/Internal:	B (Internal)
	Additional:	LH (Left Handed)

Figure 9.12 Examples of English thread notes

Specific information about thread notes is obtained from standard thread tables. A portion of those tables is shown in Figure 9.13. As an example, the threads per inch of a coarse thread (UNC) with a major diameter of 3/4″ can be located on the table by intersecting the appropriate row and column (see Figure 9.13).

ANSI – UNIFIED STANDARD SCREW THREAD SERIES

Sizes		Basic Major Diameter	Threads per Inch											Sizes
			Series with graded pitches			Series with constant pitches								
Primary	Secondary		Coarse UNC	Fine UNF	Extra Fine UNEF	4UN	6UN	8UN	12UN	16UN	20UN	28UN	32UN	
...	...	...	↓	...	...	...	...	...	...	...	...	...	...	...
1/2		0.500	13	20	28	-	-	-	-	16	UNF	UNEF	32	1/2
9/16		0.5265	12	18	24	-	-	-	UNC	16	20	28	32	9/16
5/8		0.625	11	18	24	-	-	-	12	16	20	28	32	5/8
	11/16	0.6875	-	-	24	-	-	-	12	16	20	28	32	11/16
3/4	→	0.750	**10**	16	20	-	-	-	12	UNF	UNEF	28	32	3/4
	13/16	0.8125	-	-	20	-	-	-	12	16	UNEF	28	32	13/16
...	...	...	...	...	...	...	...	...	...	...	...	...	...	...

Figure 9.13 Portion of ANSI table for UN threads illustrating how to find the threads per inch in a 3/4" coarse thread

2.4 Metric Thread Notes

Metric thread specifications are defined in the ANSI standard B1.13, which is based on ISO references. The format of a basic metric thread note and an actual example are illustrated in Figures 9.14 and 9.15.

Form (M)	**Nominal size (in mm)**	X	**Pitch**	Additional information (if required)

Figure 9.14 Format of a metric thread note

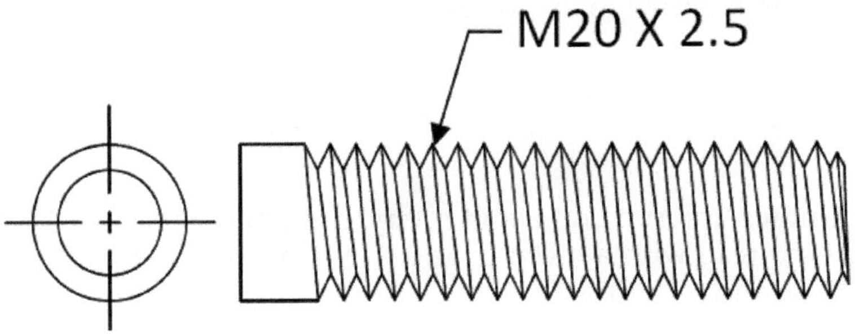

Figure 9.15 Use of metric thread note

Metric thread specifications are defined for coarse and fine threads. The first element of the note is the capital letter M, indicating metric form, followed by the nominal major diameter of the thread in millimeters. Next, the symbol "X" (read as "by") and the thread pitch, also in millimeters.

For general applications, the basic format of a metric thread note (nominal size X pitch) is sufficient. Some situations require additional information to be provided at the end of the note. This information may include the tolerance class and the thread length. Also, the capital letter LH must be provided if the thread is left handed.

The tolerance class is expressed with a number ranging from, and including, 3 to 9, and a letter, which can be E, G or H. The number represents the tolerance grade: the larger the number, the larger the tolerance. In most cases, a medium tolerance of 6 is used). The letter represents the type of fit: E (loose fit), G (regular fit), and H (tight fit). Lowercase letters are used for external threads, and capital letters for internal threads.

The thread length can be given as a number, in millimeters, or as a capital letter (S for short, N for normal, or L for long). Some examples of metric thread notes are shown in Figure 9.16.

Metric Thread Note	**Description**	
M24 X 3 – 8gS 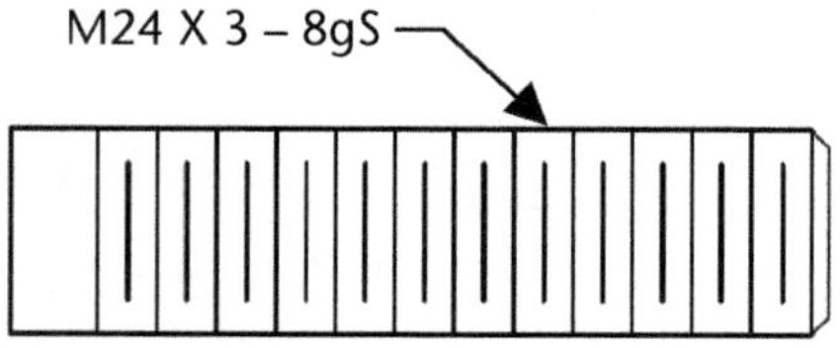	*Form:*	M (Metric)
	Major diameter:	24 mm
	Pitch:	3 mm
	Tolerance class:	8g (Coarse – Regular Fit)
	External/Internal:	g in lower case (External)
	Length:	S (Short)
M20 X 2.5 – 6g – LH 20 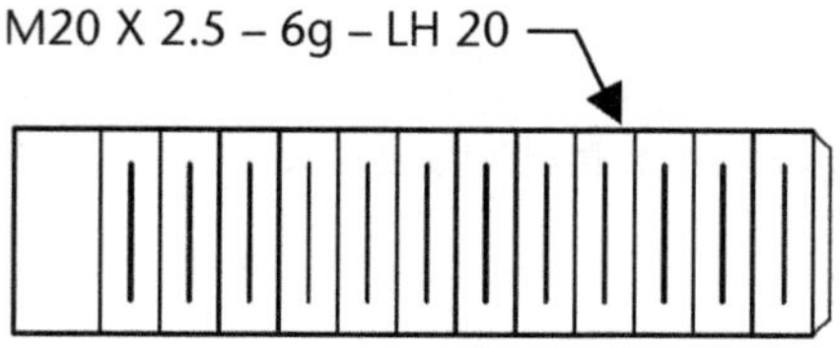	*Form:*	M (Metric)
	Major diameter:	20 mm
	Pitch:	2.5 mm
	Tolerance class:	6g (Medium – Regular Fit)
	External/Internal:	g in lower case (External)
	Length:	20 mm
	Additional:	LH (Left Handed)
M8 X 1.25 – 4H – 15 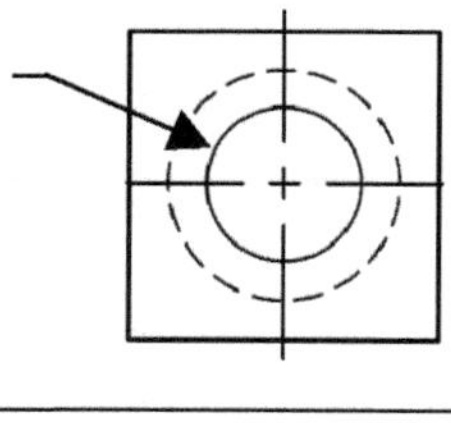	*Form:*	M (Metric)
	Major diameter:	8 mm
	Pitch:	1.25 mm
	Tolerance class:	4H (Fine – Tight Fit)
	External/Internal:	H in uppercase (Internal)
	Length:	15 mm

Figure 9.16 Examples of metric thread notes

Like English notes, information about metric threads is also available from standard tables. The pitch of a specific thread with a given diameter can be located on the metric table by intersecting the appropriate row and column (see Figure 9.17).

ANSI—UNIFIED Standard Screw Thread Series (Metric)														
Nominal Size		Diameter				Tap Drill (for 75% thread)			Threads per Inch		Pitch (mm)		Threads per Inch (Approx.)	
		Major		Minor										
Inch	mm	Inch	mm	Inch	mm	Drill	Inch	mm	UNC	UNF	Coarse	Fine	Coarse	Fine
...	...	...	...	...	...	...	...	...	...	...	↓	...	...	...
-	M16	0.630	16.002	-	-	-	-	14	-	-	2	1.5	12.5	17
-	M18 →	0.709	18.008	-	-	-	-	15.50	-	-	2.5	1.5	10	17
3/4	-	0.750	19.050	0.6201	15.748	21/32	0.6562	16.668	10	-	-	-	-	-
3/4	-	0.750	19.050	0.6688	16.967	11/16	0.6875	17.462	-	16	-	-	-	-
-	M20	0.787	19.990	-	-	-	-	17.50	-	-	2.5	1.5	10	17
-	M22	0.866	21.996	-	-	-	-	19.50	-	-	2.5	1.5	10	17
...	...	...	...	...	...	...	...	...	...	...	...	...	...	...

Figure 9.17 Portion of ANSI Metric table for UN threads illustrating how to find the pitch of an 18 mm coarse thread

2.5 Threaded Parts: Bolts, Nuts, and Screws

General use threaded fasteners can be divided in three different groups: bolts, nuts, and screws (see Figure 9.18). The difference between a bolt and a screw is somewhat ambiguous. In fact, an official and universal distinction between a bolt and a screw does not exist.

In general, a bolt is a headed and externally threaded cylindrical surface designed to be used in combination with a nut to hold multiple parts of an assembly together or to be fastened into a threaded hole. The seal between the bolt and the nut is tightened or loosened when either the nut or the bolt is rotated, depending upon the design.

A screw is a headed and externally threaded cylindrical surface that normally creates its own internal thread in the mating part as it is fastened. Some screws, however, also work in combination with a nut.

A nut is an internally threaded part used in combination with a bolt or a screw to fasten different parts of an assembly together.

There is a wide variety of shapes, sizes, and forms of bolts, nuts, and screws depending on the specific purpose and the area of application. In this section, the most common types and their corresponding specification notes are described.

Bolt and Nut Screw

Figure 9.18 Example of bolt, nut, and screw

Head and Drive The head is the top part of a bolt or a screw designed to make the fastener turn using tools such as a wrench or a screwdriver. Most bolts and screws have a head, although some exceptions exist (some head types are not designed to drive the fastener and there exist special screws that are headless).

The drive is a recessed shape designed on the head of some bolts and screws that matches with specific tools to make the fasteners rotate. There are hundreds of different head and drive combinations used for bolts and screws. The most common ones are shown in Figures 9.19 and 9.20. Some bolts are shown in Figure 9.21 to illustrate some of the head-drive combinations.

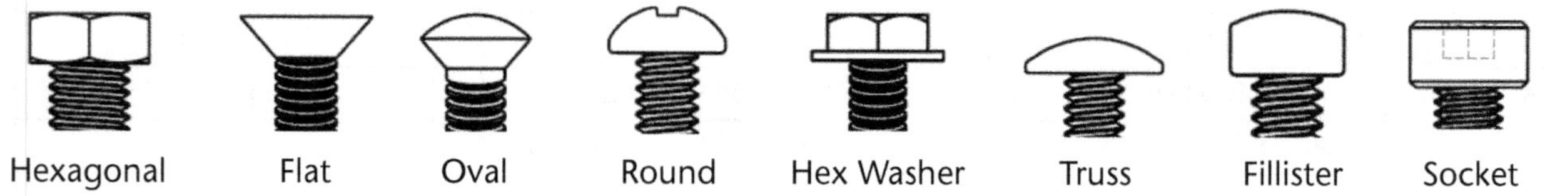

Figure 9.19 Common head types for bolts and screws

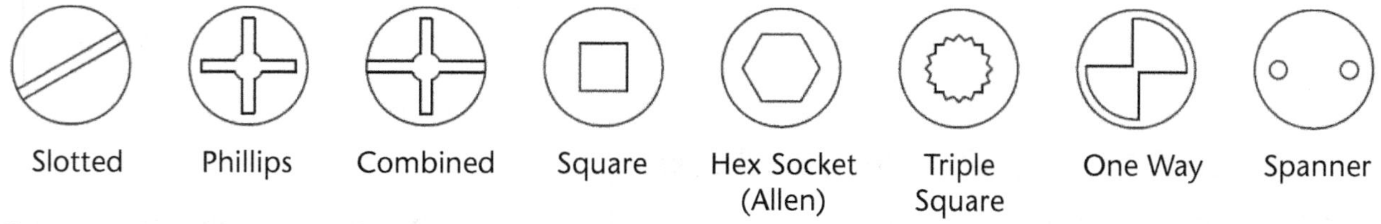

Figure 9.20 Common drives for bolts and screws

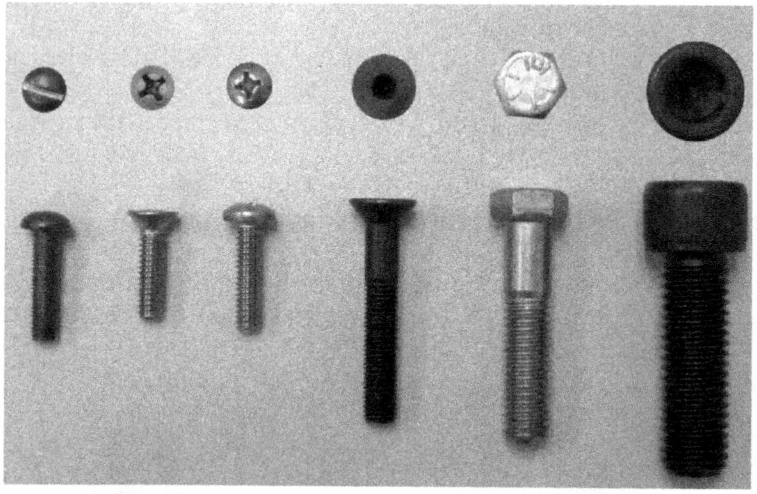

Figure 9.21 Examples of different head-drive combinations on bolts

Bolts In engineering drawings, bolts are defined with a note. This note includes the thread specification (English or metric), the length of the bolt (not including the head), and the head type (HD indicates "head"). Other information such as the material designation is optional. The examples shown in Figure 9.22 illustrate proper ways to designate bolts. When a bolt is not threaded all the way to the head, the length of the threaded portion can be specified with a linear dimension. This information can also be placed in a part note in a working drawing, which will be discussed in a later chapter.

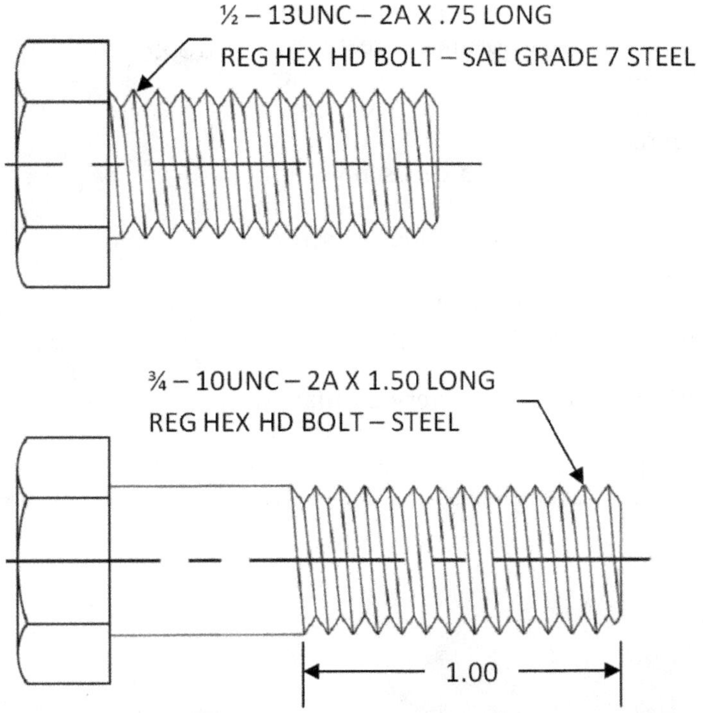

Figure 9.22 Examples of properly designated bolts

Nuts The variety of nut sizes, shapes, and forms is as wide as it is for bolts and screws. Selecting a particular type of nut depends on the application, although square and hex nuts are the most common ones. Some common nut types are shown in Figure 9.23. Nuts are specified with notes, similar to the notes used with bolts: thread note first and nut type (see Figure 9.24).

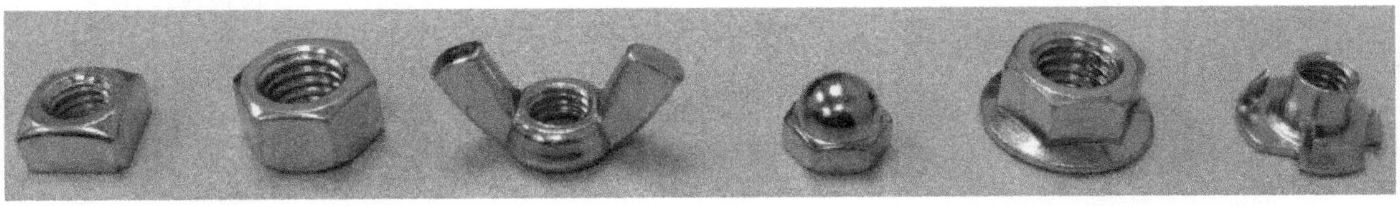

Square Hex Wing Acorn Flange Tee

Figure 9.23 Common types of nuts

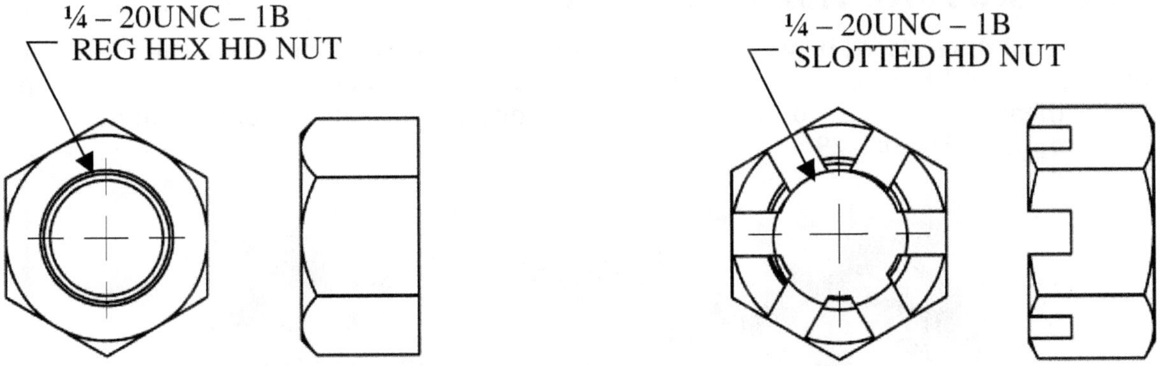

Figure 9.24 Examples of properly designated nuts

Screws Screws can be found in many different forms and lengths (see Figure 9.25). Since screws normally create their own internal threads on the mating part as they go in, there is also a variety of end points, which facilitate easier insertion depending on the design.

Most common screws are divided into cap screws and machine screws. Machine screws are normally smaller than cap screws. Some special types of screws include set screws (see Figure 9.26), which have no head and are intended to be inserted entirely inside a hole, wood screws, specifically designed to fasten wooden parts, and self-tapping screws (see Figure 9.27).

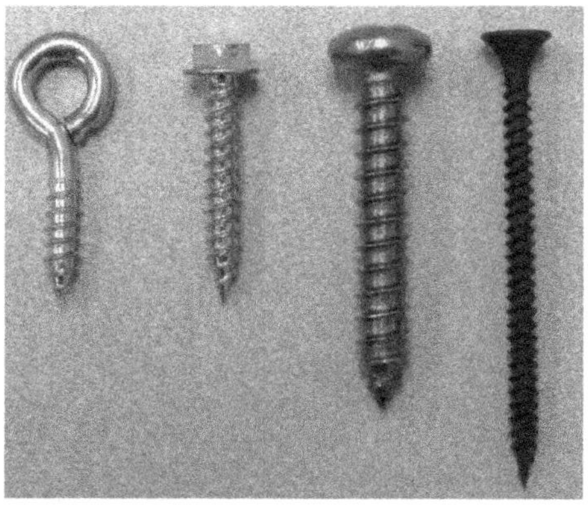

Figure 9.25 Different types of screws

The notes used to designate screws are very similar to those used for bolts. First, the thread is described. The length of the screw must be given as well as the head type. Some cases require the point type of the screw (PT indicates "point"). The examples shown in Figure 9.28 illustrate proper ways to designate screws.

Figure 9.26 Set screw

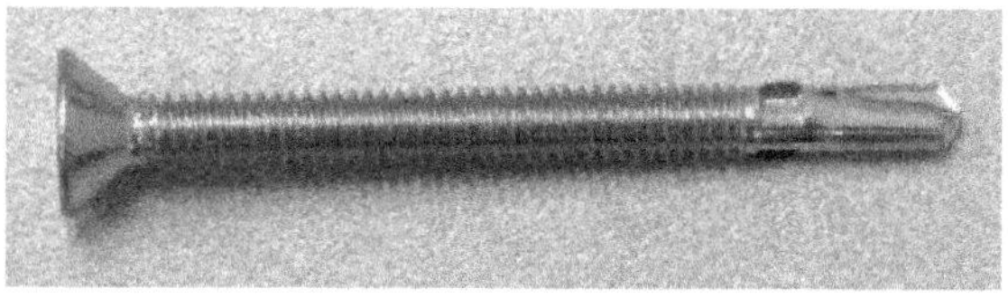

Figure 9.27 Self-tapping screw

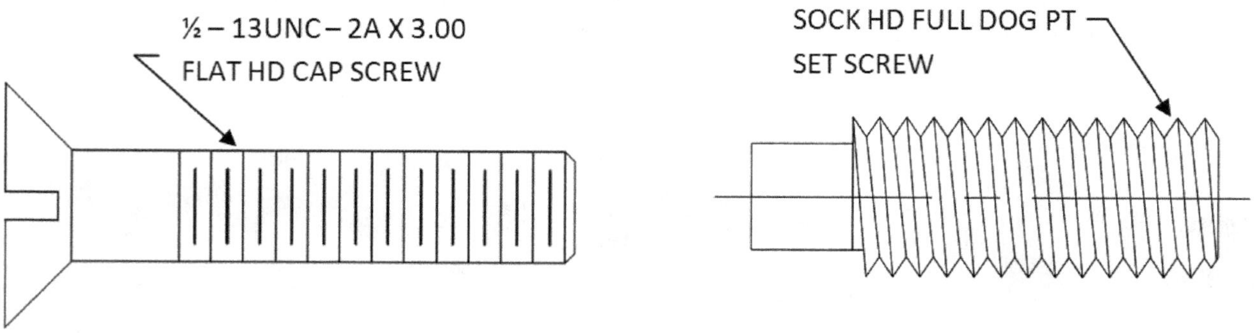

Figure 9.28 Examples of properly designated screws

2.6 Unthreaded Parts: Washers and Pins

Washers are small discs placed between a bolt and a nut used to evenly distribute the pressure over the fastening area and to smooth the assembling surface. Lock washers also protect threaded parts and prevent them from coming loose if the assembly undergoes vibration. Some common types of washers are shown in Figure 9.29.

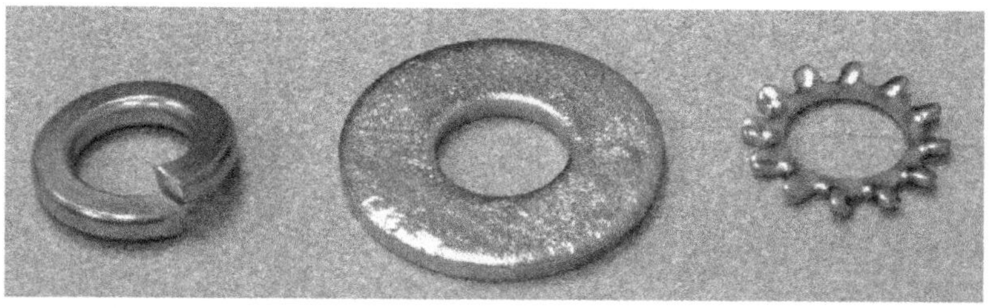

Lock Plain Tooth lock

Figure 9.29 Examples of different types of washers

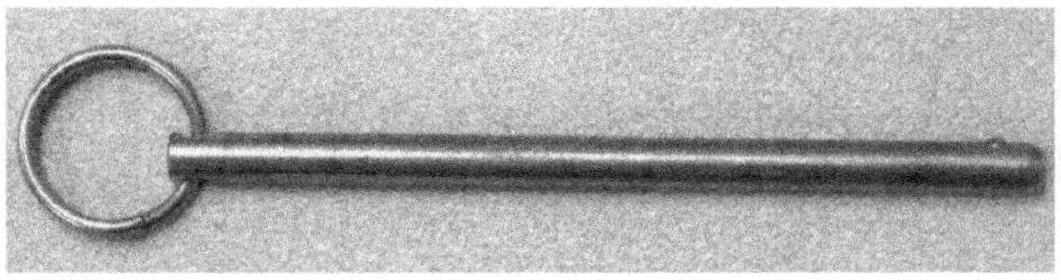

Figure 9.30 Dowel pin

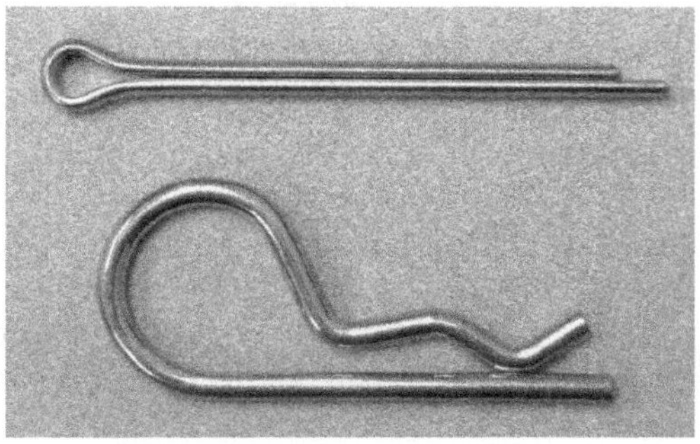

Figure 9.31 Standard (top) and hairpin (bottom) cotter pins

Plain washers are designated by giving their inside and outside diameters and their thickness. This information is based on the bolt or screw size and can be found in ANSI standard tables. Lock and tooth lock washer notes require the inside diameter to be given.

The following are examples of valid notes for washers:

"1.50 × 2.75 × .165 TYPE A PLAIN WASHER" (Inside dia = 1.50, Outside dia = 2.75, and thickness = .165)

"EXTERNAL TOOTH LOCK WASHER .625 – TYPE A" (Inside dia = .625)

Pins are unthreaded locking devices used to maintain different parts together in a fixed position or orientation. Selecting a specific pin depends on the precision desired in the assembly and the strength, among other factors. Most common types of pins include dowel pins (see Figure 9.30), cotter pins (see Figure 9.31), and clevis pins.

3. Other Parts and Features

Other common parts and features have also been standardized. Specifications for standard components such as retaining-rings, U-bolts, anchors, gears, keys, or springs are commonly available. Features such as slots, grooves, holes, and knurls have also been normalized.

A knurl is a rolled pattern cut into a cylindrical part with the purpose of improving gripping, or simply for decoration. They are very common in certain types of bolt heads, thumb screws and nuts (see Figure 9.32). An example of a properly designated knurl is shown in Figure 9.33.

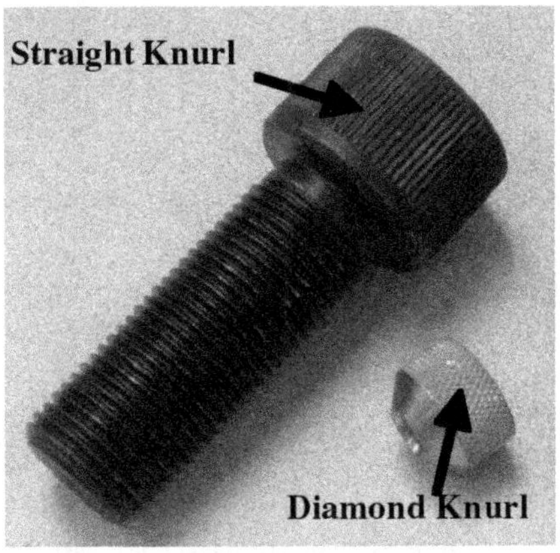

Figure 9.32 Examples of knurls

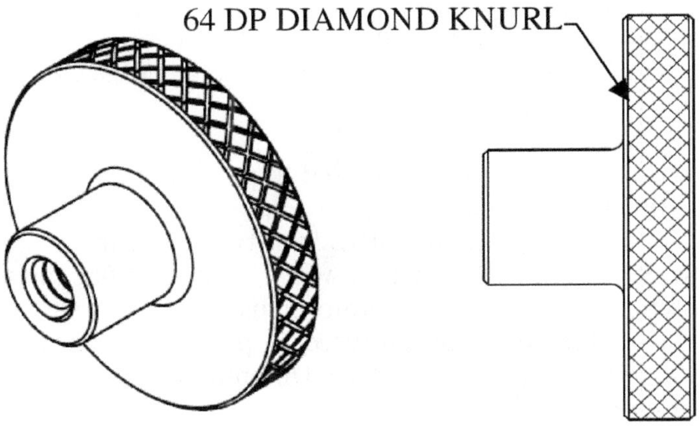

Figure 9.33 Example of properly defined knurl (DP= Diametral Pitch, number of crests per inch of diameter)

4. Machined Holes

Holes are one of the most common features in engineering designs. They are placed on parts and assemblies for a variety of reasons: to make a part lighter (or use less material), to put other objects into or through a part, or to fasten or secure two or more parts together. Holes are frequently circular. They can be blind (at a specific depth) with flat or conical bottoms (most common), or pass entirely through the part. An example is illustrated in Figure 9.34.

Holes have different profiles depending on their intended purpose and the manufacturing techniques used to create them. Common machined hole types are shown Figure 9.35.

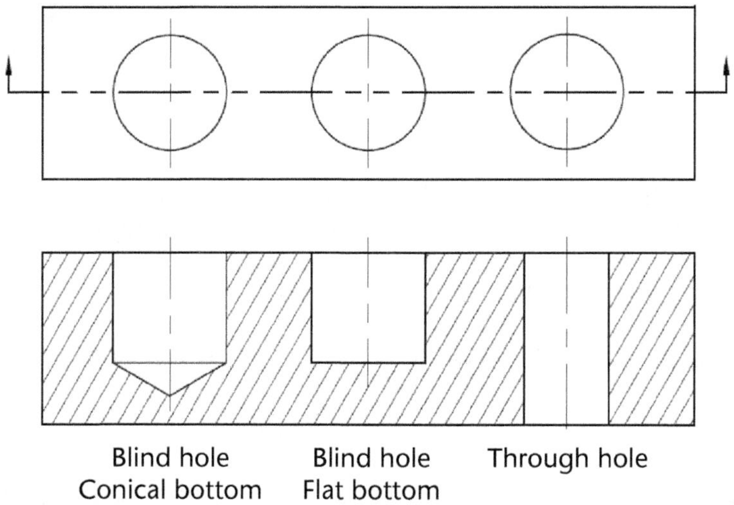

Figure 9.34 Example of blind holes with flat and conical bottoms, and a through hole.

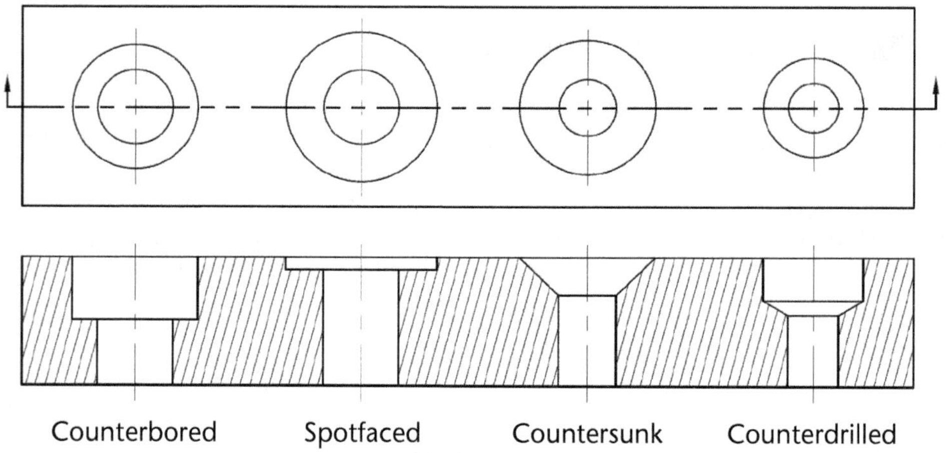

Figure 9.35 Types of machined holes

Counterbored holes are used in parts that require a bolt or a cap screw to fit flush with or below the finished surface.

Spotfaced holes are a particular type of counterbored hole where the bored surface is very shallow. Spotfaced holes are used when a finished flat surface is required to provide better contact for a fastener head. They are commonly used with cast surfaces.

Countersunk holes are used in parts that require a countersunk flat head bolt or screw to fit flush with the finished surface. Different angles can be used, but 82°, 90°, and 92° are the ones most commonly used.

Counterdrilled holes are normally used in parts that require a bolt or a screw head to sit below the finished surface. The space above the screw or bolt head is then covered with a plug to give the assembly a fine finish.

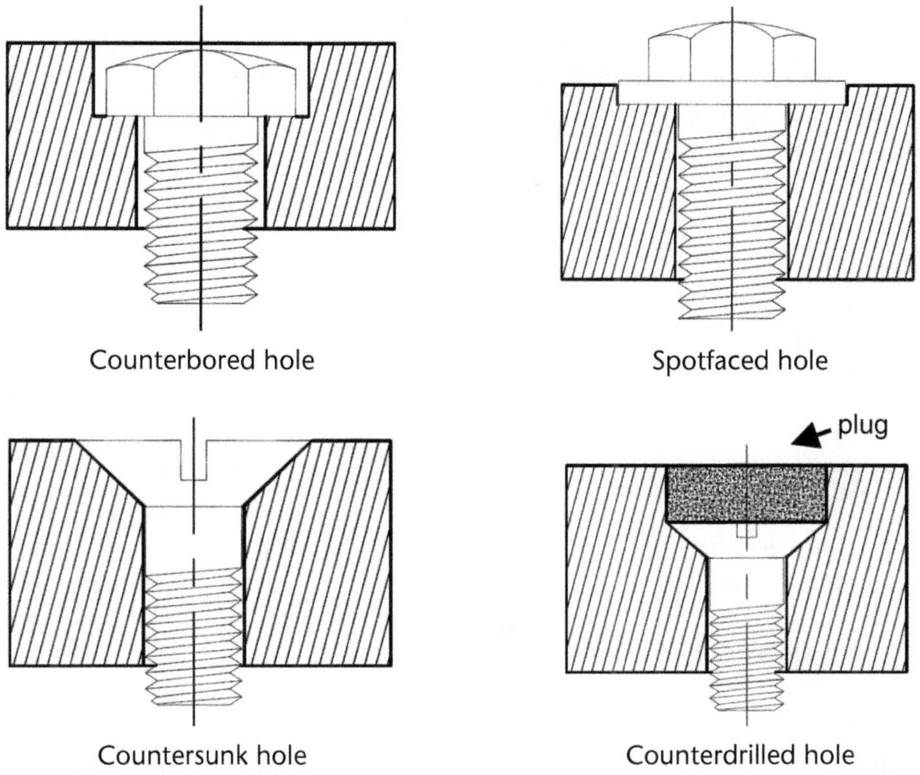

Figure 9.36 Machined holes and fasteners

The examples shown in Figure 9.36 illustrate section views of different holes with fasteners passing entirely through.

4.1 Dimensioning Machined Holes

Machined holes require specific notes and designations. Holes are always dimensioned in their circular view with a leader pointing to the center of the hole. The information required in the dimension includes the diameter and depth of the actual hole, a symbol or abbreviation indicating the hole type, and the dimensions of the bored, drilled, or sunk part.

The most common symbols and abbreviations are shown in Figure 9.37. Although abbreviations are still in use, they may be subject to interpretation, especially when translated into different languages. For this reason, the use of symbols is highly recommended.

When the depth of a hole is not given, it is assumed that it passes entirely through the part. The abbreviation THRU can also be used.

The example shown in Figure 9.38 illustrates how to properly dimension blind holes. The diameter of the hole is 1.00 inch and the depth is also 1.00 inch. The note "Ø1.00 × 1.00 DP" is also acceptable. If a hole is threaded, the thread note must be given as part of the general note. For example, the note "Ø.4375 × .75 DP FLAT BTM, Ø.50 – 13UNC – 2B, .50 DP" designates a flat bottomed hole with a diameter of .4375 inches and a depth of .75 inches. The hole is threaded .50 inches deep.

Description	Abbreviation	Symbol
Diameter	DIA	Ø
Depth	DP	↧
Counterbored	CBORE or CB	⌴
Spotfaced	SFACE or SF	⌴
Countersunk	CSINK or CSK	⌵
Counterdrilled	CDRILL or CD	*No symbol*

Figure 9.37 Abbreviations and symbols

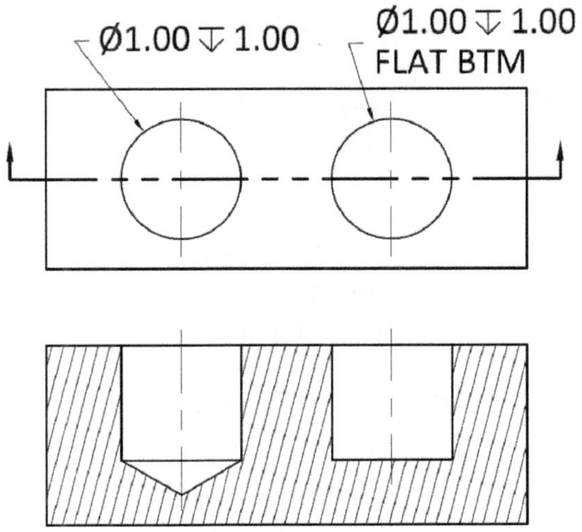

Figure 9.38 Properly dimensioned blind holes

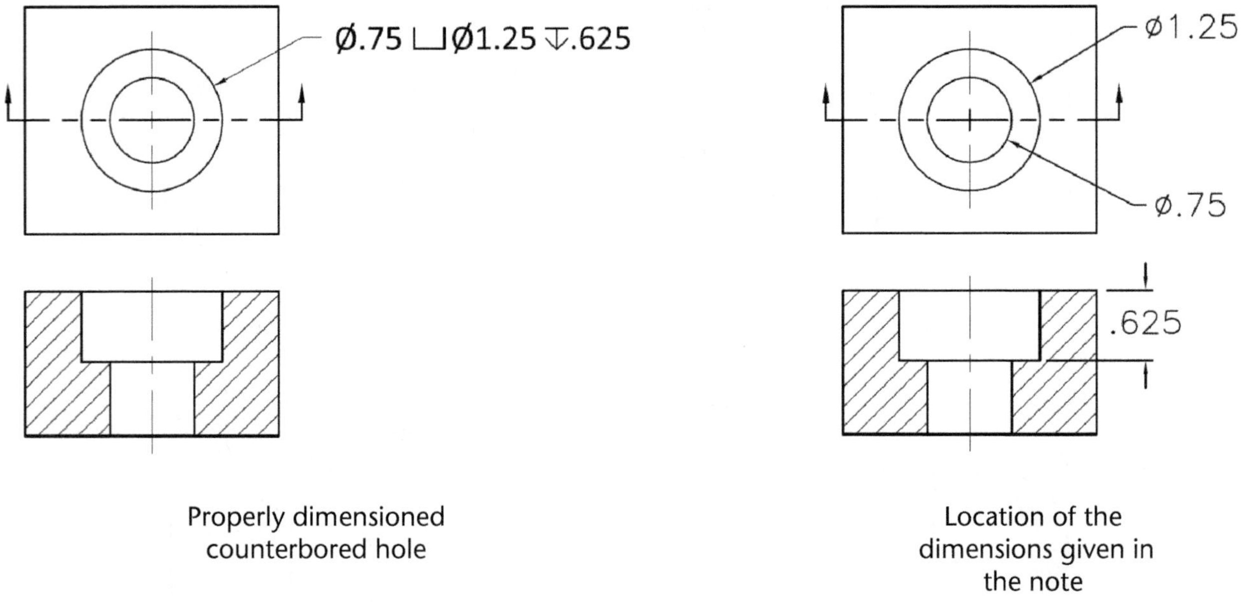

Figure 9.39 Properly dimensioned counterbored hole

The example shown in Figure 9.39 illustrates how to properly dimension a counterbored hole. The diameter of the hole is .75 inches. Since no depth is given, a through hole is assumed. The symbol indicates a counterbored hole with a diameter of 1.25 inches and a depth of .625 inches. The note "Ø.75 THRU, CB Ø1.25 × .625 DP" is also acceptable.

The example shown in Figure 9.40 illustrates how to properly dimension a spotfaced hole. The diameter of the hole is .75 inches. Since no depth is given, a through hole is assumed. The symbol indicates a spotfaced hole with a diameter of 1.25 inches. No depth is required for the spotfaced part of the hole. This determination is normally left to the machinist. The note "Ø.75 THRU, SF Ø1.25" is also acceptable.

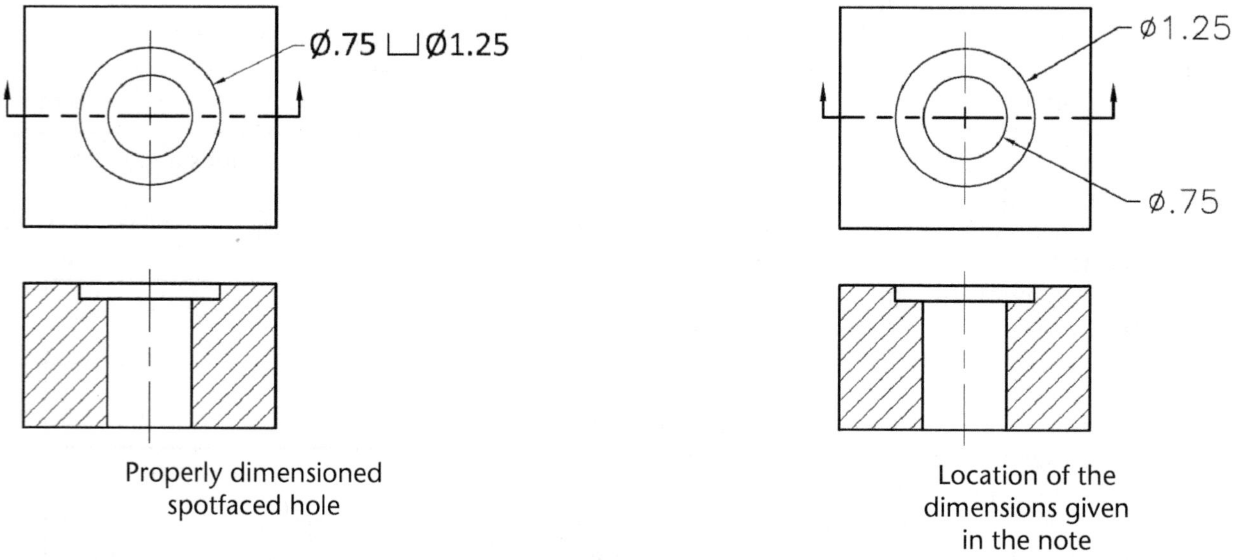

Figure 9.40 Properly dimensioned spotfaced hole

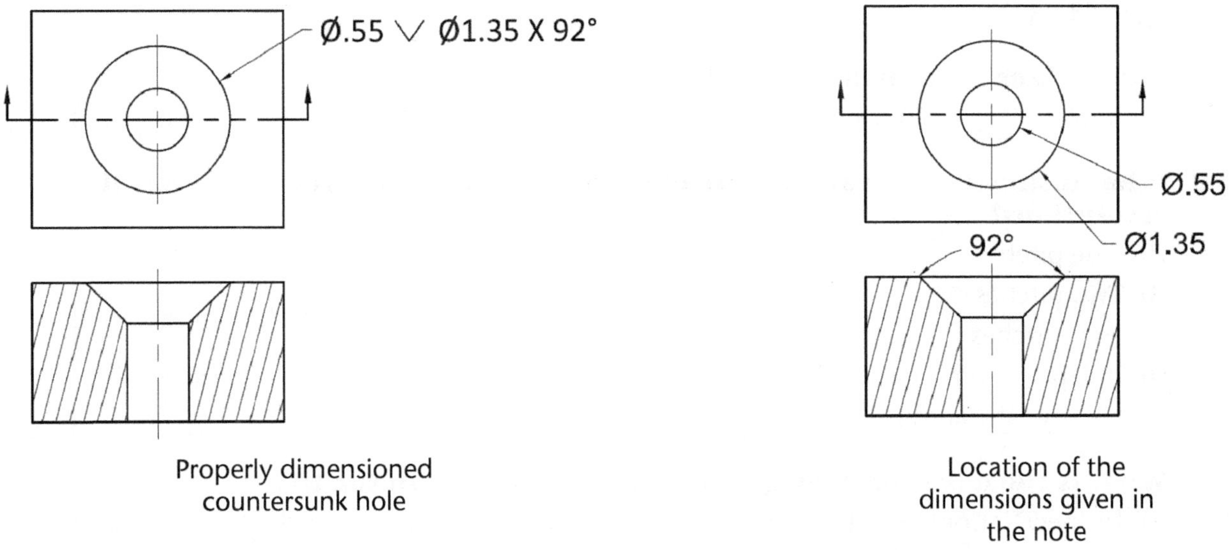

Figure 9.41 Properly dimensioned countersunk hole

The example shown in Figure 9.41 illustrates how to properly dimension a countersunk hole. The diameter of the hole is .55 inches. Since no depth is given, a through hole is assumed. The symbol indicates a countersunk hole with a diameter of 1.35 inches and an angle of 92°. No depth is required. The note "Ø.55 THRU, CSK Ø1.35 X 92°" is also acceptable.

The example shown in Figure 9.42 illustrates how to properly dimension a counterdrilled hole. The diameter of the hole is .50 inches. Since no depth is given, a through hole is assumed. The letters CD indicate a counterdrilled hole with a diameter of 1.00 inch and a depth of .425 inches. The note "Ø.50 THRU, CD Ø1.00 × .425 DP" is also acceptable.

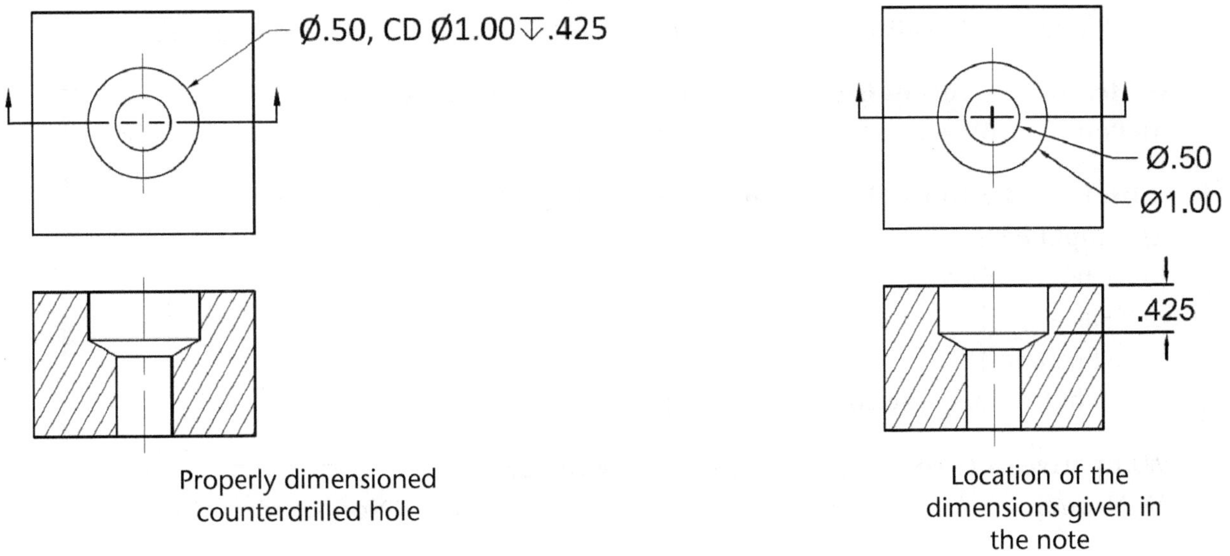

Figure 9.42 Properly dimensioned counterdrilled hole

Practice Test

1. **The peak or prominent point of a thread is called ____**

   A) Root B) Flank C) Head D) Crest E) None of the above

2. **What distance does a single threaded bolt advance when rotated two full revolutions?**

   A) One pitch
   B) Two pitches
   C) Four pitches
   D) It depends on the thread form
   E) None of the above

3. **What is the most common general use thread form in the US?**

   A) Unified National B) Acme C) Buttress D) Metric E) None of the above

4. **The thread note "¾ – 10UNC – 2A – LH" designates a/an ____ fit.**

   A) Tight B) Average C) Loose D) Internal E) None of the above

5. **How many threads per inch does the thread note "½ – 28UNEF – 1A" indicate?**

   A) .5 B) 28 C) 1 D) Cannot be determined E) None of the above

6. **What is the pitch of the thread designated by the note "M12 × 1.75 – 6g – 20" indicate?**

   A) 12 B) 1.75 C) 6 D) 20 E) None of the above

7. **Which of the following is not a type of bolt and screw drive?**

   A) Slotted B) Fillister C) Philips D) Triple square E) Spanner

8. **Which of the following is not a type of threaded fastener?**

   A) Bolt B) Acorn Nut C) Set screw D) Cotter Pin E) Cap Screw

9. **Which of the following is not a type of thread representation?**

   A) Simplified
   B) Detailed
   C) Coarse
   D) Schematic
   E) All of the above are correct thread representations

10. **What thread form is represented by the letter "N"?**

    A) Neutral B) American National C) Unified National D) Non-standard E) Normal

Problems

Give the proper thread note for the following specifications:

Thread Note	Description

....

....

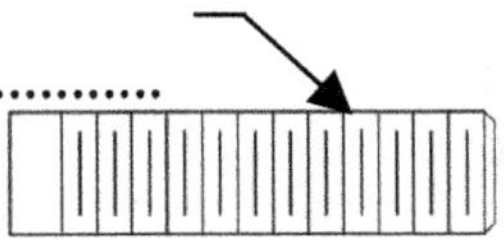

Major diameter:	.625 inches
Form:	Unified National
Series:	Coarse
Fit:	Regular
Additional:	Right Handed

....

....

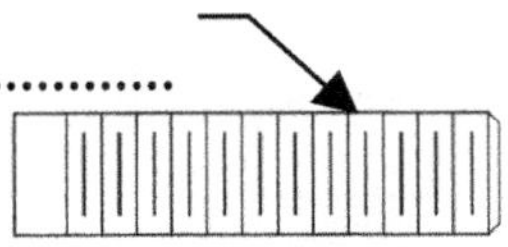

Major diameter:	.75 inches
Form:	Unified National
Series:	Fine
Fit:	Tight
Additional:	Double Thread, Right handed

....

....

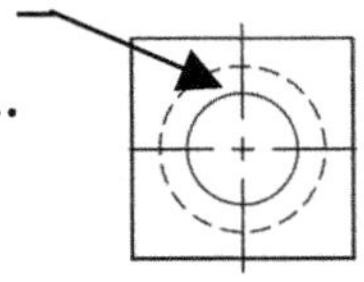

Major diameter:	1.125 inch
Form:	Unified National
Series:	Extra Fine
Fit:	Regular
Additional:	Left Handed

....

....

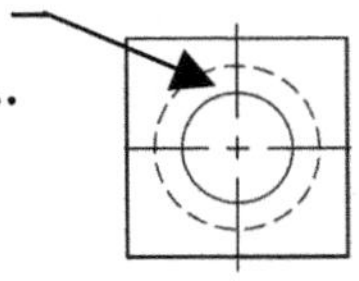

Major diameter:	4.00 inches
Form:	Unified National
Series:	Coarse
Fit:	Loose
Additional:	Left Handed

....

....

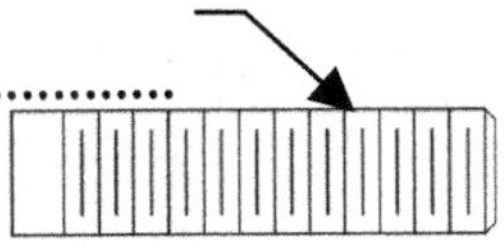

Major diameter:	2.5 mm
Form:	ISO Metric
Series:	Coarse
Fit:	Regular
Additional:	Right Handed

....

....

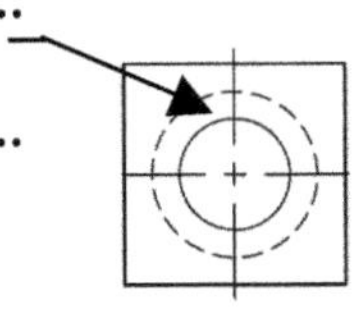

Major diameter:	20 mm
Form:	ISO Metric
Series:	Coarse
Fit:	Regular
Additional:	Left Handed

Identify the head type and drive of the following fasteners:

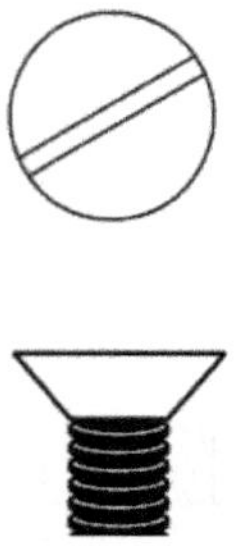 1. **Head:** **Drive:**	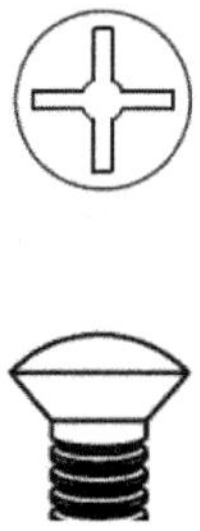 2. **Head:** **Drive:**	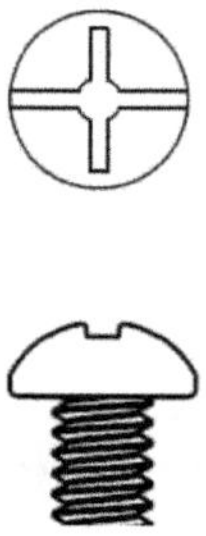 3. **Head:** **Drive:**
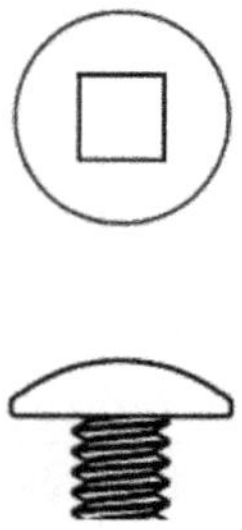 4. **Head:** **Drive:**	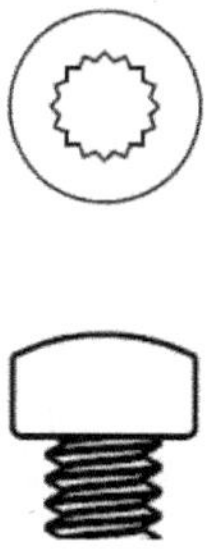 5. **Head:** **Drive:**	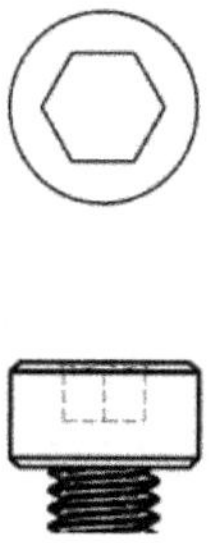 6. **Head:** **Drive:**

Identify the following hole types:

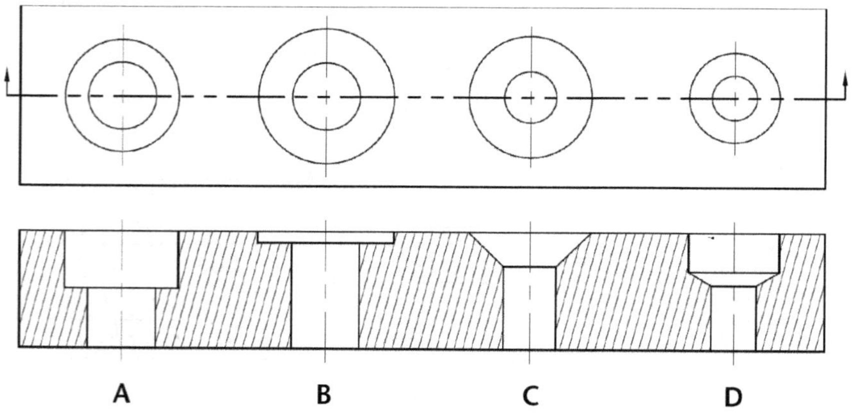

A. ____________________

B. ____________________

C. ____________________

D. ____________________

Give the proper note for the following holes. Each grid square is .10 inches.

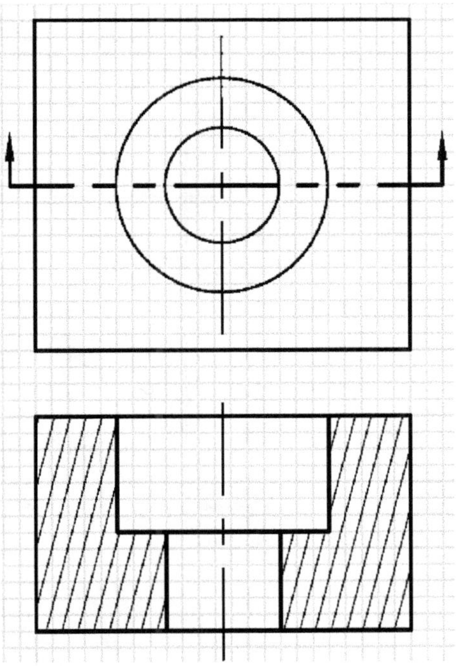

1. ....

Note: ....

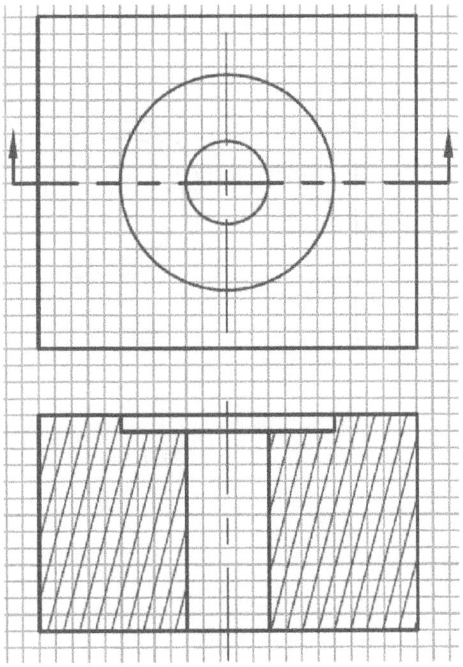

2. ....

Note: ....

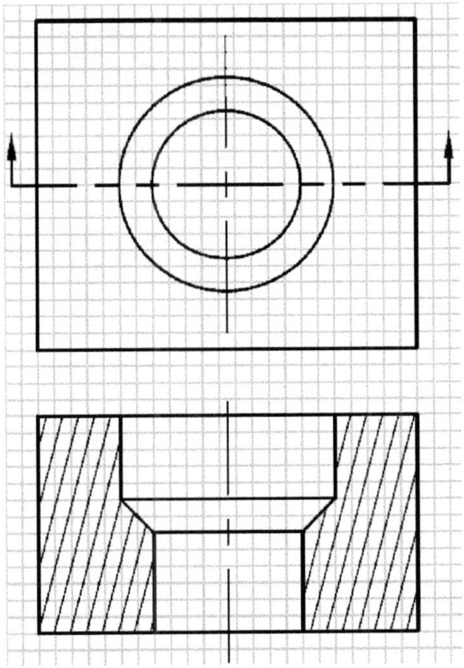

3. ....

Note: ....

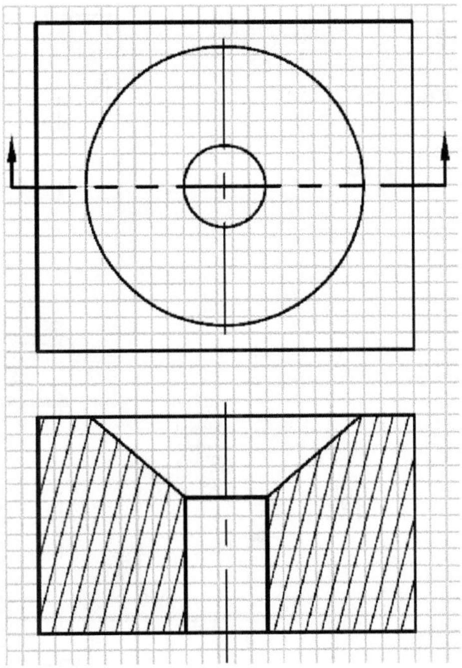

4. ....

Note: ....

Apply Thread Notes to the Drawing

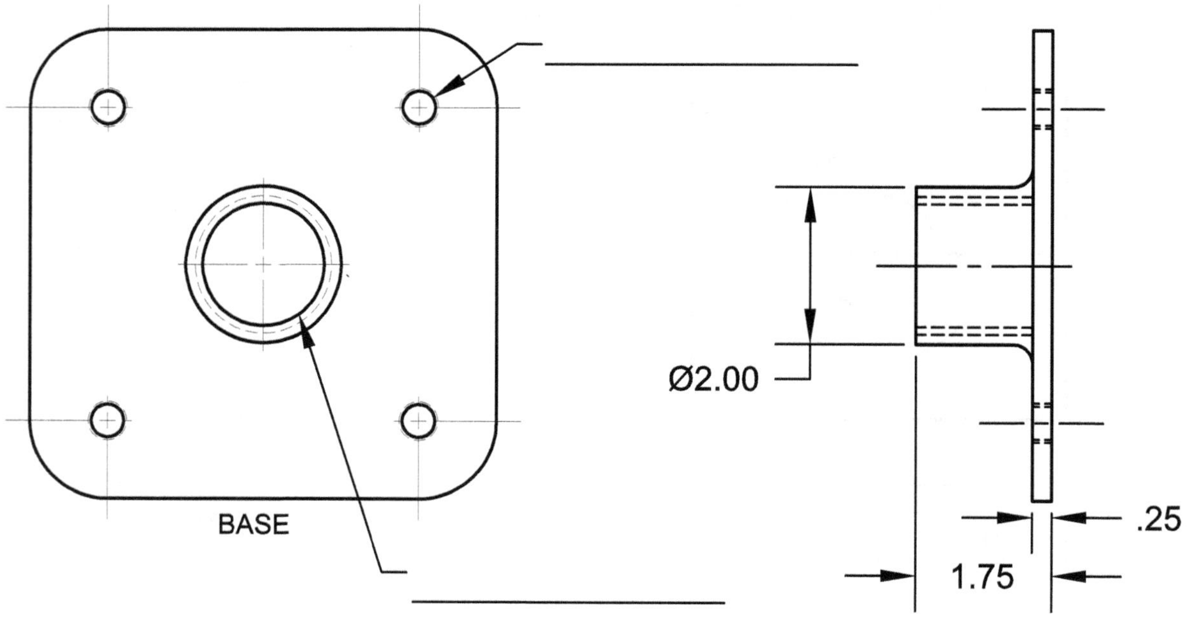

BASE AND AXLE:
UNIFIED NATIONAL
COARSE SERIES
REGULAR FIT

BASE AND SCREWS:
UNIFIED NATIONAL
FINE SERIES
TIGHT FIT
DOUBLE

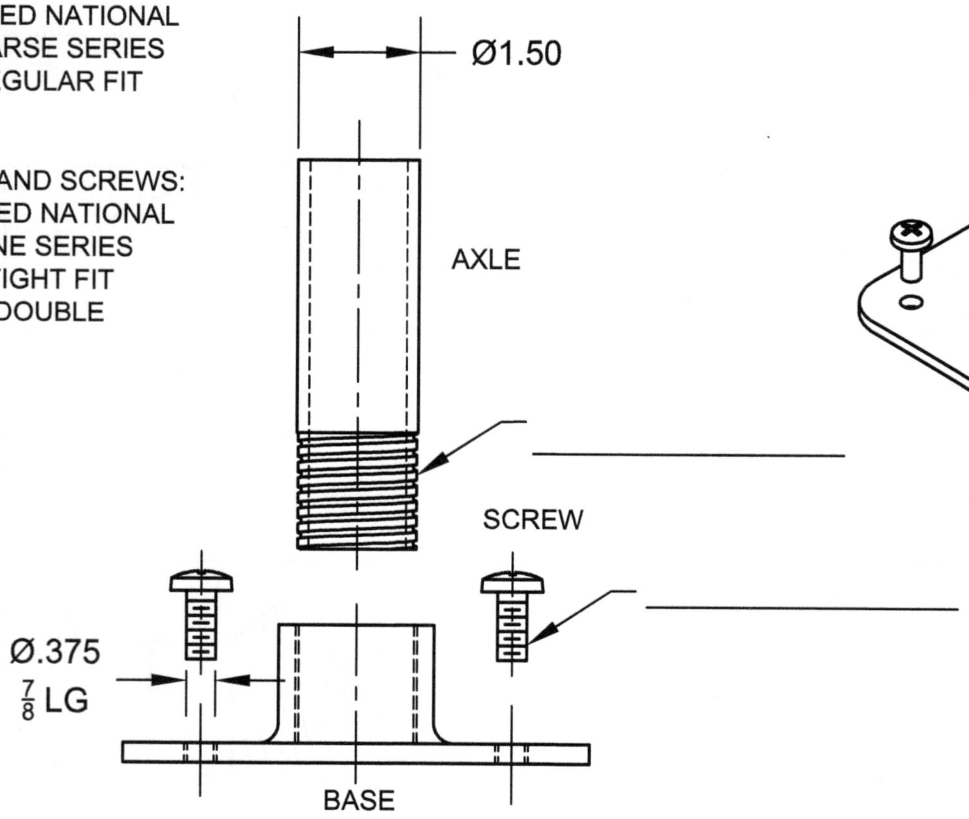

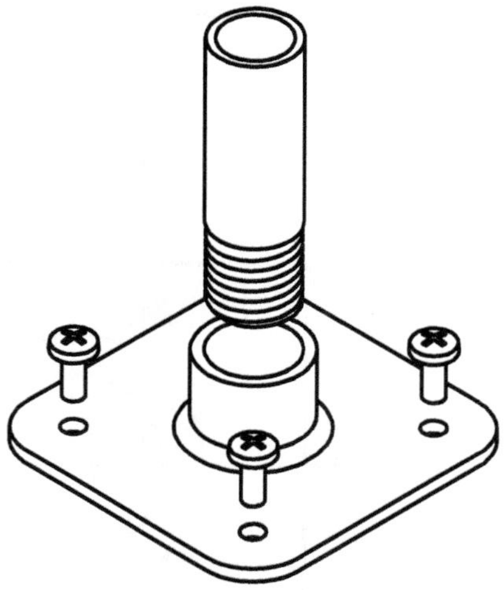

AUTHOR:	SCALE:
TEAM: SECTION:	DATE:

Apply Thread Notes to the Drawings

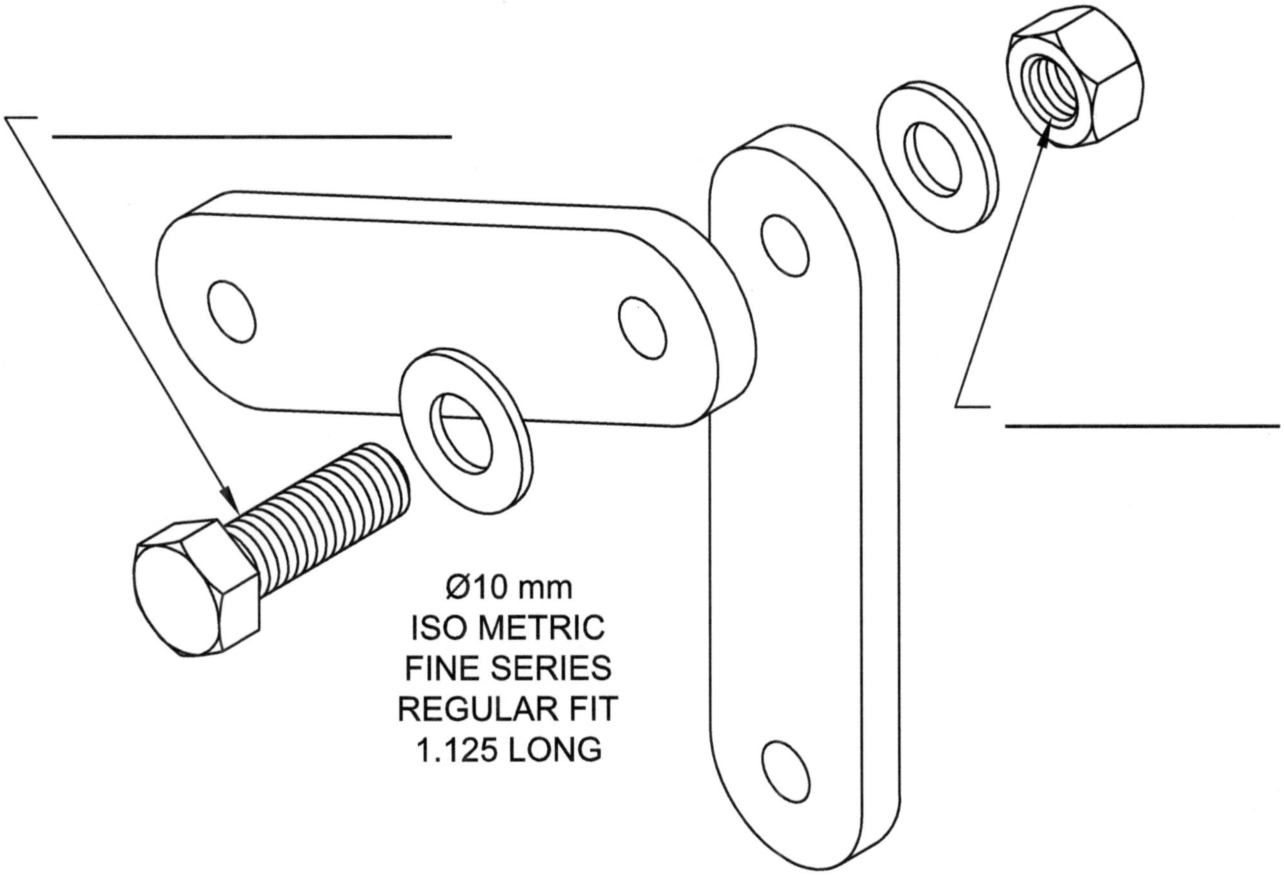

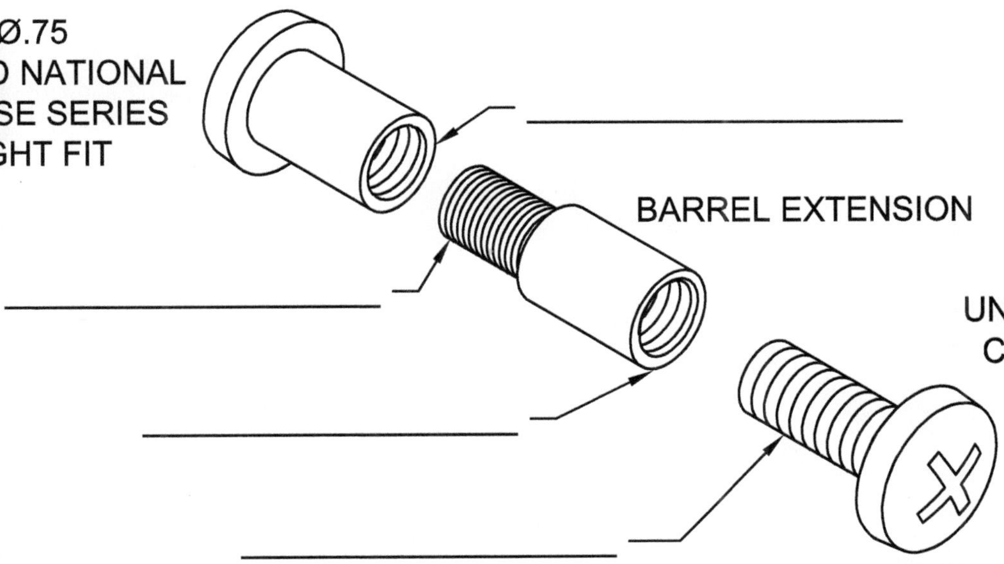

AUTHOR:	SCALE:
TEAM: SECTION:	DATE:

10

CHAPTER

Working Drawings

1. Introduction

Working drawings, also called design drawings, are a complete and accurate set of drawings and specifications that allow a final design to be manufactured. If an engineer has placed their seal on the drawing, it is considered a legal document and is used to communicate all the details and fabrication procedures of the design to the manufacturer. The complexity of a product determines the number and complexity of the corresponding working drawings.

Working drawings rely on the drawing practices introduced in previous chapters (orthographic projection, pictorials, sectioning, dimensioning, etc). In many instances, a client only knows or provides some basic sketches and dimensions that are required for a specific design. It is up to the designer to define the working dimensions that are missing.

To avoid error and misinterpretation, many aspects of working drawings have been standardized. Many industries, however, use their own customized formats. In this chapter, standard and conventional practices, as well as the different parts of the working drawings, are discussed.

2. General Considerations

2.1 Paper Size

Selecting an appropriate paper size and orientation (portrait vs. landscape) for working drawings depends on several factors: the availability of printers and plotters, the desired scale of the drawings, the drawing layout, or even the designer's preference.

The standard paper sizes used in engineering drawings are defined by both ANSI and ISO (see Figure 10.1).

2.2 Title Block

Every page of a set of working drawings must have a border and a title block. The border defines the drawing area on the sheet and the title block provides administrative and technical information about the drawing. Although both ANSI and ISO define standard title blocks, with precise sizes and contents, most companies use their own formats. Title blocks are normally located in the lower right hand corner of each page. The information given in

ANSI		ISO	
Format	**Size (inches)**	**Format**	**Size (mm)**
A	8.5 × 11	A4	210 × 297
B	11 × 17	A3	297 × 420
C	17 × 22	A2	420 × 594
D	22 × 34	A1	594 × 841
E	34 × 44	A0	841 × 1189

Figure 10.1 Standard paper sizes

a title block varies depending on the purpose of the drawing and the target audience, but it should always contain basic information, such as the following:

- Name of the designer or company who created the drawing and the date the drawing was created.
- Part name, number, scale, and general tolerances.
- Sheet number and total number of sheets (i.e. "SHEET 2 OF 6").

Other information that can be given in the title block includes the drawing number (for archiving), material of the part, surface finish, and projection method.

In many cases, especially when subcontractors are used, it is necessary to provide space in the title block for signatures, revision and approval dates, and comments. For large projects, a revision table is useful. The example shown in Figure 10.2 illustrates a common title block.

MOUNTING FRAME		MATERIAL: 1020 STEEL
DRAWN BY:	CREATION DATE:	SHEET: 1 OF 3
CHECKED BY:	REVISION DATE:	COMMENTS:
SCALE: 1 = 2	TOLERANCES DEC ±.005 ANG ±.005	
DRAWING NO: 1162R02		

Figure 10.2 Title block

3. Structure of Working Drawings

A typical set of working drawings includes the following elements (see Figure 10.3):

- One or multiple assembly and subassembly drawings. These drawings are used to identify all the parts of the design and show how the parts are assembled together. Assembly drawings are commonly found first in a design drawing document.
- Detail drawings of non standard (custom) parts. These drawings are used to describe each custom part of the assembly, both geometrically and dimensionally, for manufacturing. Manufacturing notes can also be found in detail drawings (ex. is a part forged, milled, etc.).
- Specifications of standard parts, providing a written description about the parts of the assembly that can be purchased from a vendor. Standard part notes are usually placed in unused space in the detail drawings. A separate page(s) is not necessary.

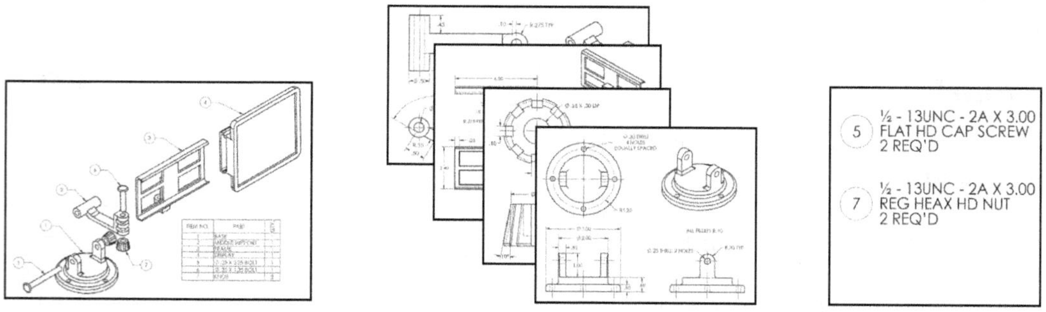

Assembly drawing Detail drawings Standard parts

Figure 10.3 Elements of a set of working drawings

3.1 Assembly Drawings

Assembly drawings provide a comprehensive view of the overall design by showing all the parts of the assembly in their functional location. The assembly drawing is normally the first sheet in a set of working drawings.

There are three components in an assembly drawing: the assembly view, leader lines with balloons, and a parts list or Bill of Materials (BOM). The example shown in Figure 10.4 illustrates the different components of an assembly drawing.

The assembly view is a drawing of the entire assembly. A variety of views can be used: pictorial, orthographic, sectioned, or a combination of different views. They can also be represented as exploded (with the components slightly separated by distance illustrating the order of assembly) or assembled (collapsed). Determination of the best views for an assembly drawing depends on the complexity of the assembly and the number of components, among other factors. Assembly drawings are only used to show how the design is assembled, not how the design is manufactured.

Several conventions must be followed when creating assembly drawings:

- The number of views in the assembly drawing must be kept to the minimum. Only the minimum number of views required to fully describe the assembly must be included.
- All parts in the assembly must be visible in the drawing, if possible. If an exploded assembly is used, the components must appear in their correct location, clearly illustrating the order of assembly.

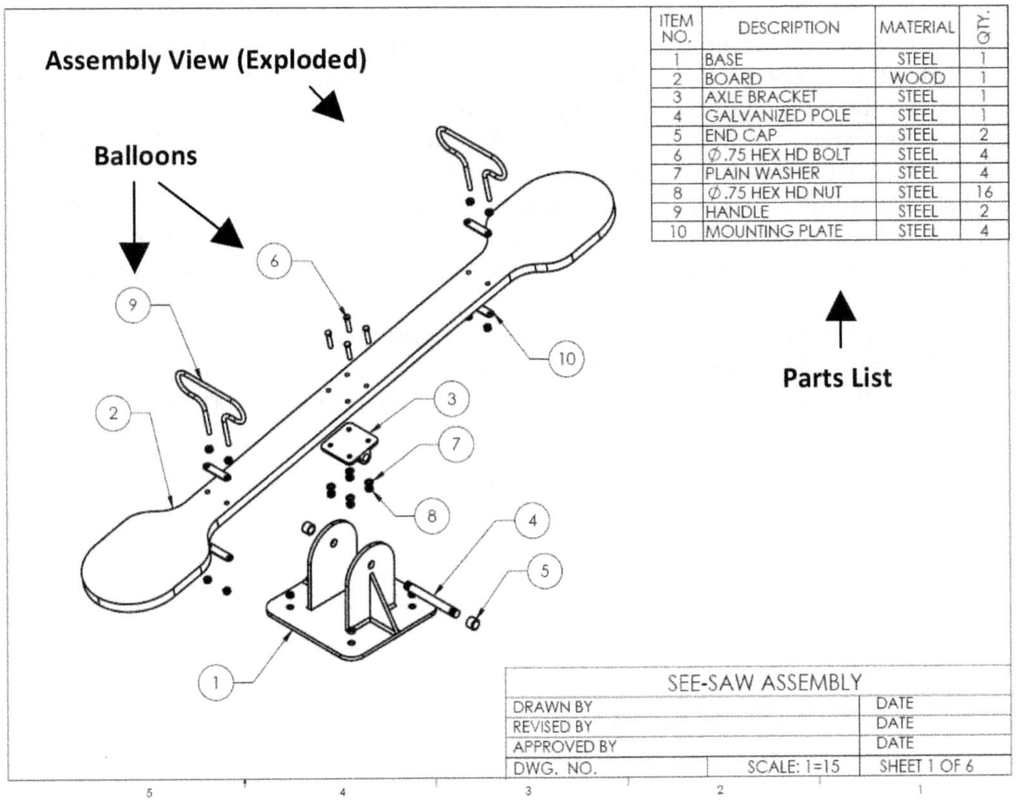

Figure 10.4 Assembly drawing

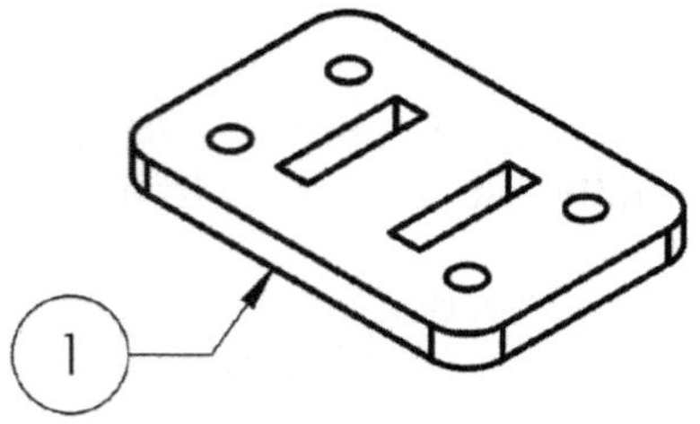

Figure 10.5 Leader line with balloon

NO.	DESCRIPTION	MAT	REQ
1	BASE PLATE	CAST IRON	1
2	BRACKET	CAST IRON	2
3	BUSHING	RUBBER	2
4	COTTER PIN	STEEL	2
5	HEX HD BOLT	STEEL	4
6	ACORN NUT	STEEL	4

Figure 10.6 Parts list

- Some parts may be simplified. For example, it is not necessary to represent detailed threads of bolts.
- Hidden lines, dimensions, and tolerances must not be included, unless they are essential for clarity.

Each component of the assembly must be identified using a part number. This number will appear centered inside a circle, known as balloon, connected to a leader line pointing to the designated part (see Figure 10.5). Most leaders are shown with elbows. The size of the balloons is three to four times the standard letter height. The balloons must be properly located on the page without interfering with the drawing. If a component is used multiple times (ex. an assembly that uses four bolts), only one component has a part number balloon attached.

Each balloon number in the assembly drawing is associated to a description shown on a table known as parts list or bill of materials (BOM). The information provided in this description varies, but it must include at least the part number, the name of the part, and the quantity required. Other information such as the part material or supplier may also be incorporated (see Figure 10.6). Part numbers must be consistent throughout a design. If a flat washer is part number 1 in the standard notes, it must also be part number 1 in the bill of materials.

The parts list is normally placed either at the bottom right hand corner of the page with part numbers shown in descending order or at the top right hand corner with part numbers in ascending order (see Figure 10.7). This arrangement allows easy addition of new items to the list without recreating the whole table.

For large and complex assemblies, it may be useful to provide separate drawings of the different functional units (subassemblies) of the global design. In this case, each drawing is annotated in a similar manner, providing the subassembly name and identifying its components.

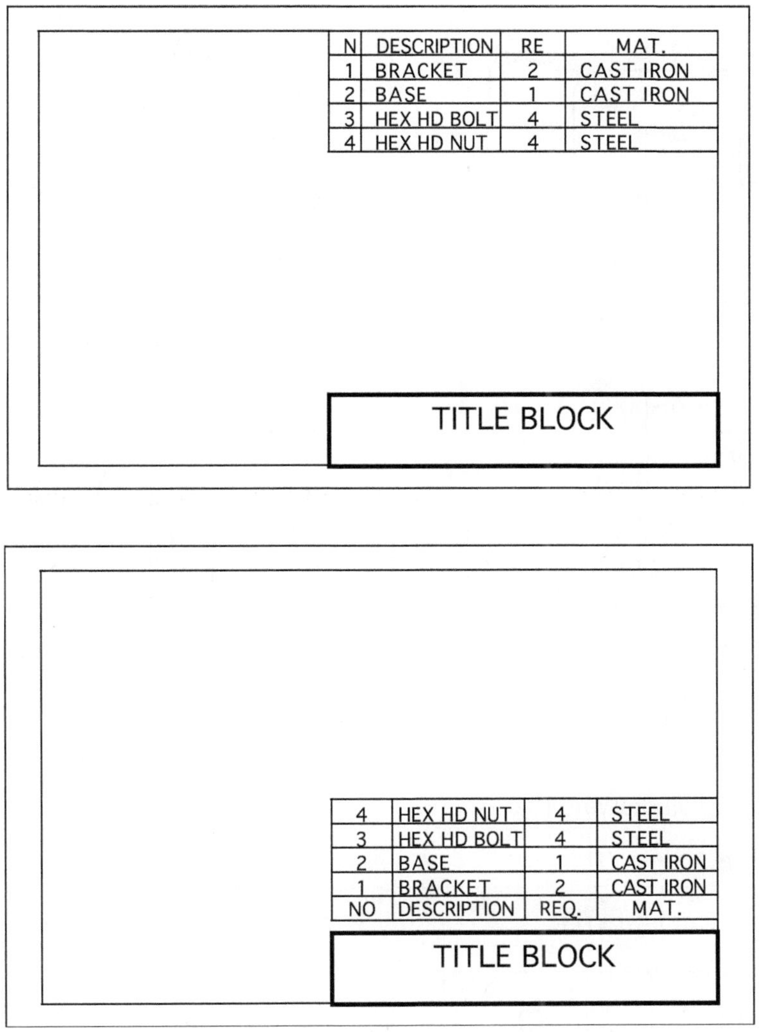

Figure 10.7 Proper locations of the parts list on a drawing

3.2 Detail Drawings

Each custom part of the assembly requires detail drawings. Detail drawings include sufficient orthographic drawings of the part with dimensions and tolerances. There is no required order of the parts, but the information provided in every detail drawing (part name and number) must agree with the information given in the parts list and balloons (see Figures 10.8 and 10.9).

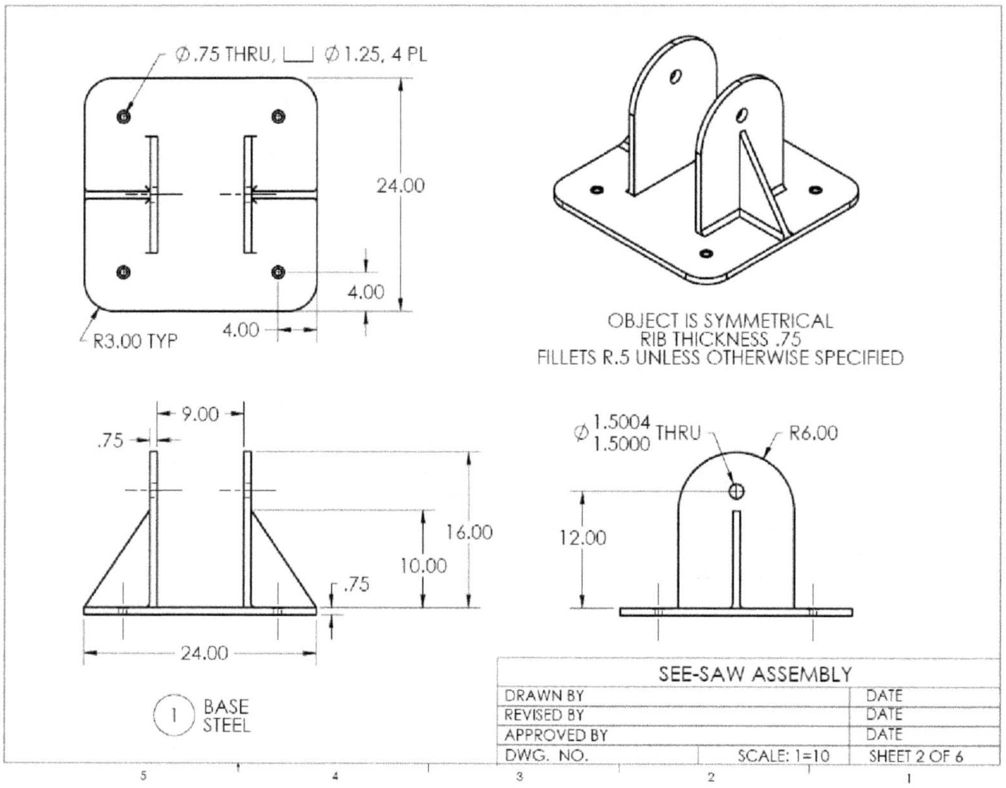

Figure 10.8 Detail drawing (base)

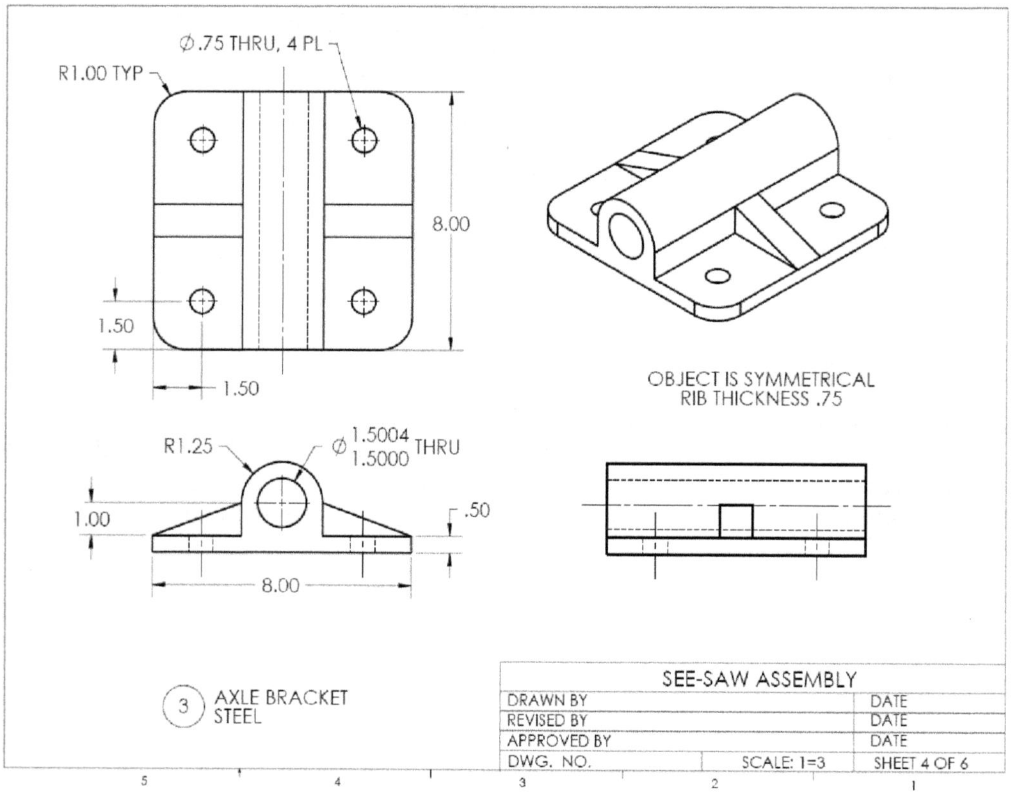

Figure 10.9 Detail drawing (axle bracket)

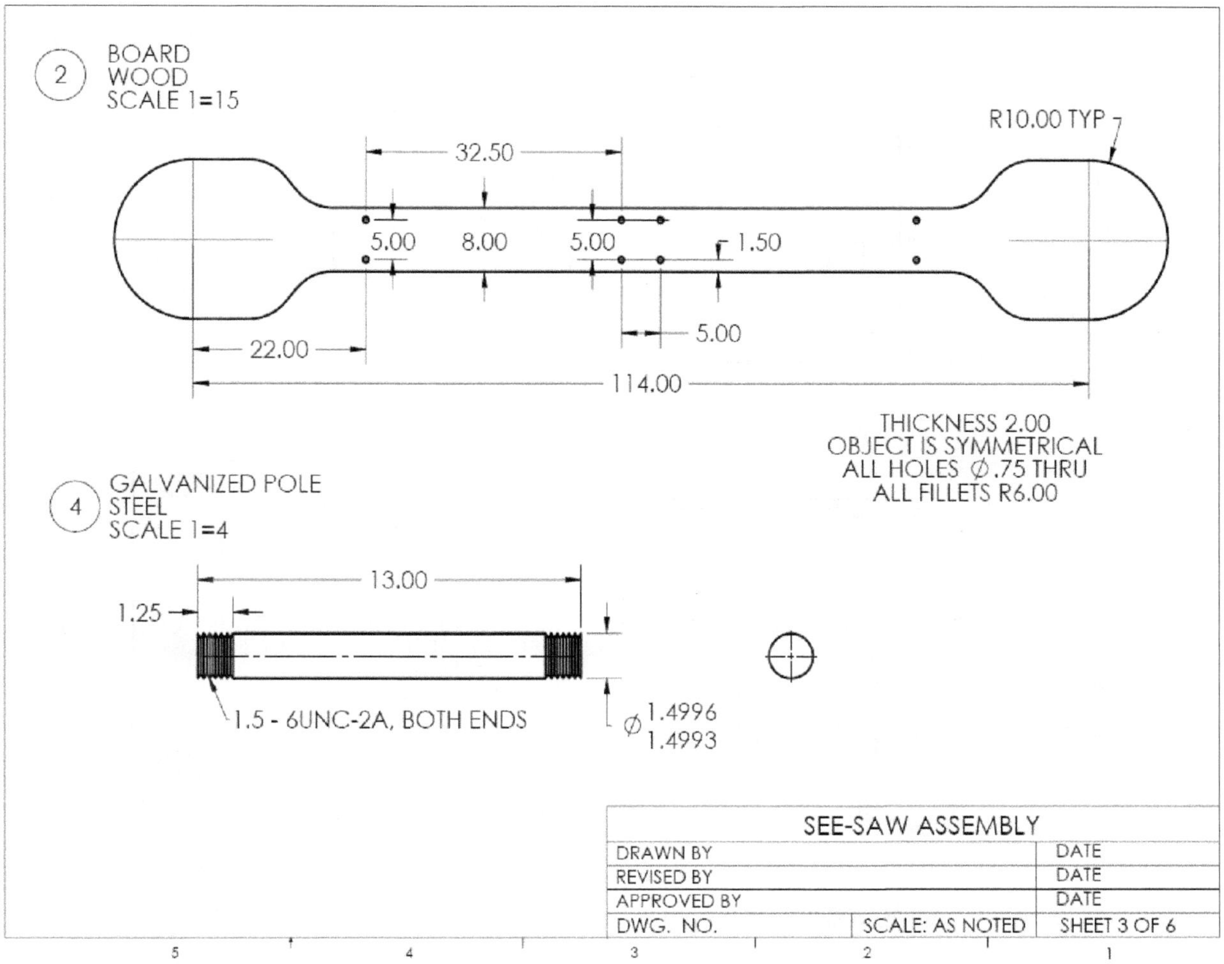

Figure 10.10 Detail drawing with two parts in the same sheet

Multiple parts can be drawn on the same sheet if space allows, but each part must be properly identified. Auxiliary and section views may be used if necessary. Only the minimum number of views required to fully describe the object must be used (see Figures 10.10 and 10.11). The scale of the detail drawings may vary from part to part, but not from view to view, and must be clearly indicated in the part name tag. If two parts drawn on the same page require different scales, "AS NOTED" should be placed in the scale portion of the title block. Identical parts are drawn only once, with the number required shown in the part name tag. Only the name of the part is given on the top line of the tag. The material, number required, etc. is usually given on separate lines. If only one instance of the part is required, the number required is omitted from the part name tag.

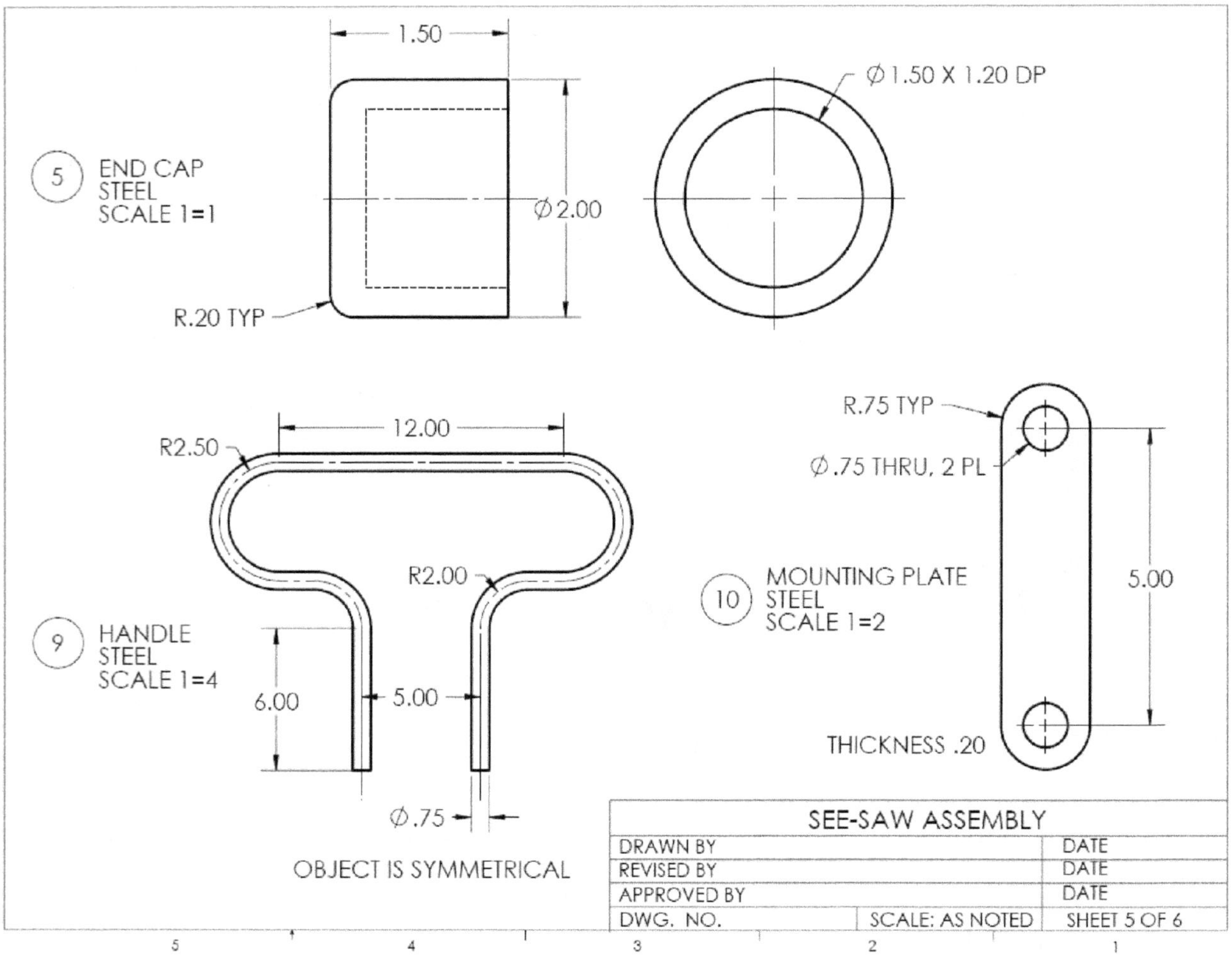

Figure 10.11 Detail drawing with three parts in the same sheet

3.3 Standard Parts

Standard parts that can be purchased from a vendor (nuts, bolts, washers, bearings, gaskets, etc) do not need to be drawn. Instead, a written specification using standard notes is provided. Standard specifications are normally included in any unused space on the detail drawings, if space permits. In large construction projects, however, a separate sheet may be used for the notes and attached to the set of drawings (see Figure 10.12).

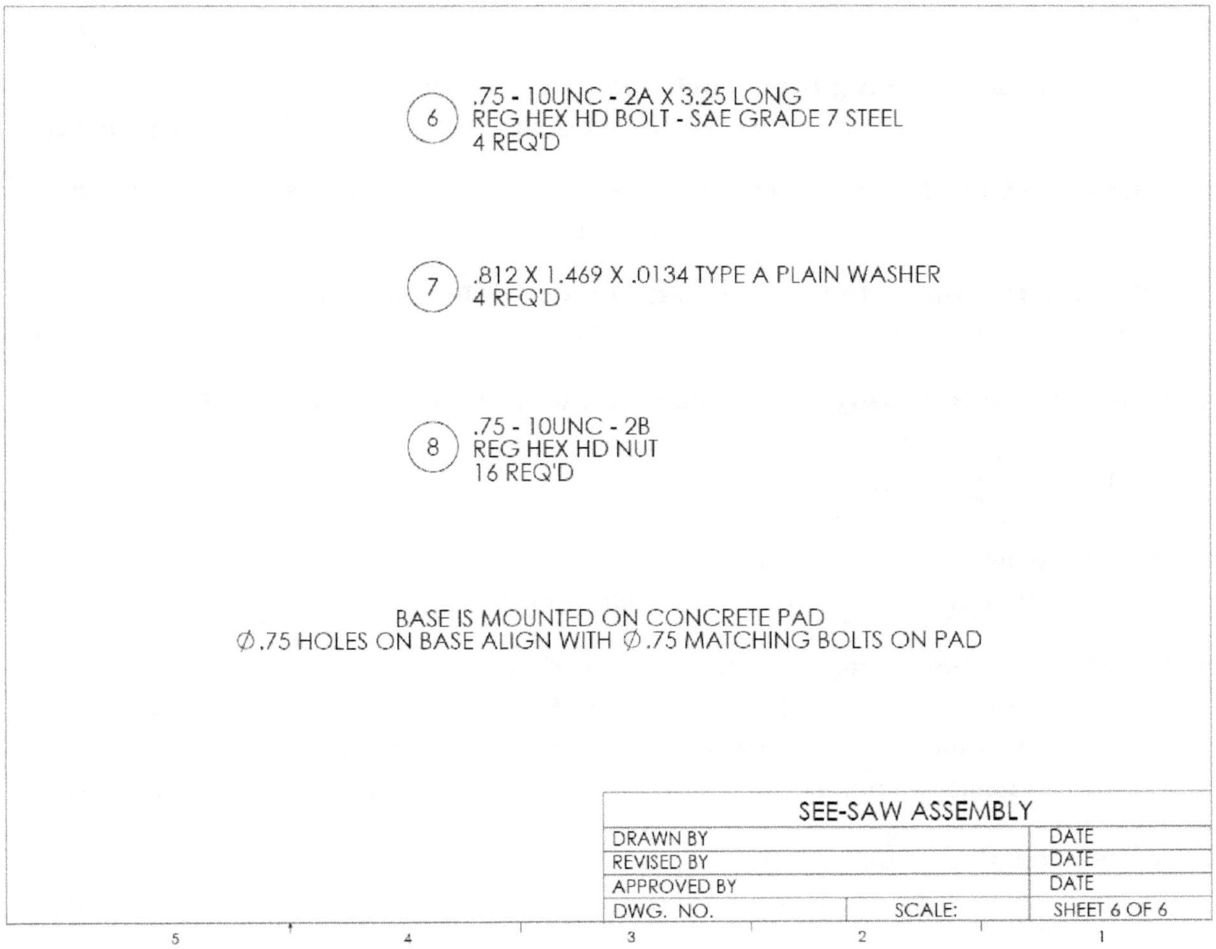

Figure 10.12 Standard parts specifications provided on a separate sheet

Practice Test

1. **Which of the following is not included in a title block?**
   A) Name of the drafter B) Revision date C) Scale D) Parts list E) Drawing number

2. **Which of the following paper designations represents the largest paper size?**
   A) A B) B C) A4 D) D E) A3

3. **What are the dimensions (in inches) of a size B sheet of paper?**
   A) 11×17 B) 8.5×11 C) 22×34 D) 11×22 E) 10×20

4. **Which of the following is not a part of a set of working drawings?**
   A) Assembly drawing
   B) Detail drawings of custom parts
   C) Standard parts notes
   D) Detail drawings of standard parts
   E) All of the above are parts of a set of working drawings

5. **Which of the following statements about detail drawings is false?**
   A) Detail drawings provide orthographic views of the parts that need to be manufactured
   B) Dimensions and tolerances must be included in the detail drawings
   C) The part name and number on a detail drawing must agree with the information on the parts list
   D) Detail drawings of different parts can be included in the same sheet
   E) In a detail drawing, the scale of the orthographic drawings of a part may vary from view to view

6. **Which of the following statements about assembly drawings is false?**
   A) Assembly drawings show how the components of an assembly are put together
   B) Assembly drawings may include sectional views of the assembly
   C) Assembly drawings must be always fully dimensioned
   D) It is acceptable to simplify certain components in an assembly drawing
   E) Information about individual components of the assembly is provided in the parts list

7. **What type of assembly drawing shows the components slightly separated by distance illustrating the order of assembly?**
   A) Isometric B) Exploded C) Section D) Auxiliary E) None of the above

8. **Which of the following statements about balloons is false?**
   A) Balloons appear in assembly drawings connected to leader lines
   B) Balloons provide the part number information
   C) The number within a balloon must agree with the part number in the bill of materials
   D) The balloon size is three to four times the standard letter height
   E) Balloons are sometimes used as a reference to the tolerance value for a specific part

Problems

Provide a complete set of working drawings of the following designs. Some of the dimensions may have to be inferred.

WIRE CHEESE SLICER ASSEMBLY

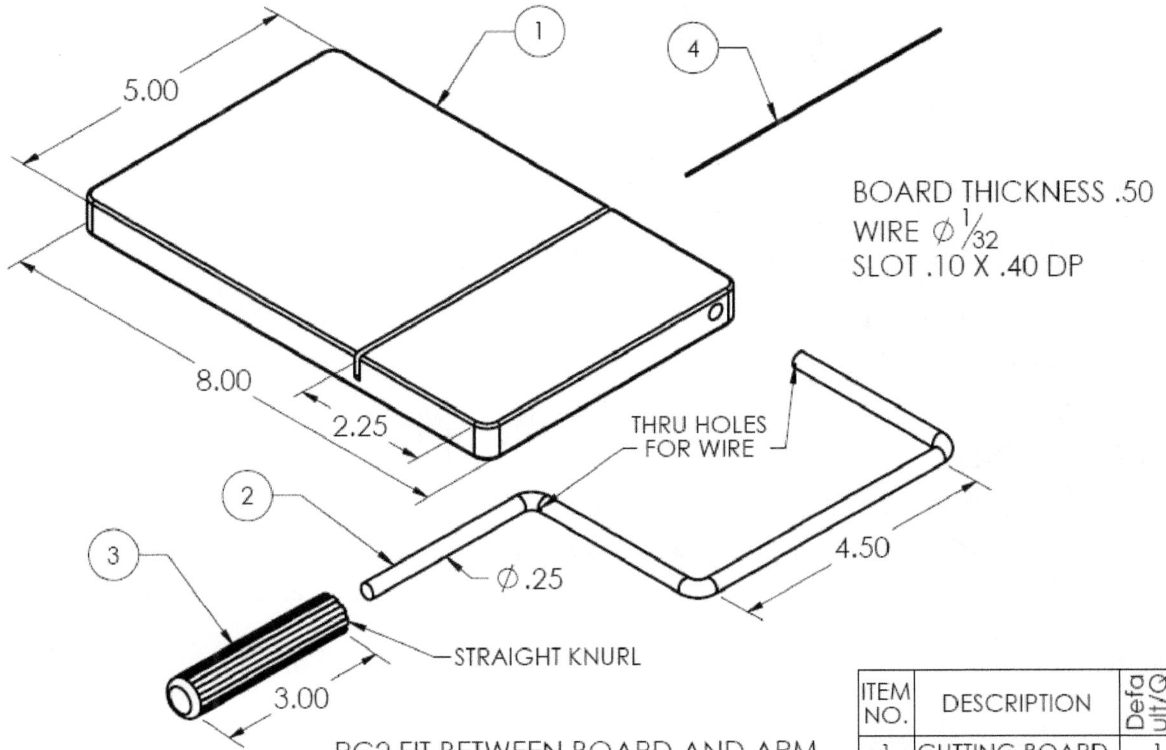

RC2 FIT BETWEEN BOARD AND ARM

FN1 FIT BEWEEN HANDLE AND ARM

ITEM NO.	DESCRIPTION	Default/QTY.
1	CUTTING BOARD	1
2	ARM	1
3	HANDLE	1
4	WIRE	1

Optional Design Improvements

Redesign the wire cheese slicer so it can be adjusted for left-handed users.

TOY CANNON ASSEMBLY

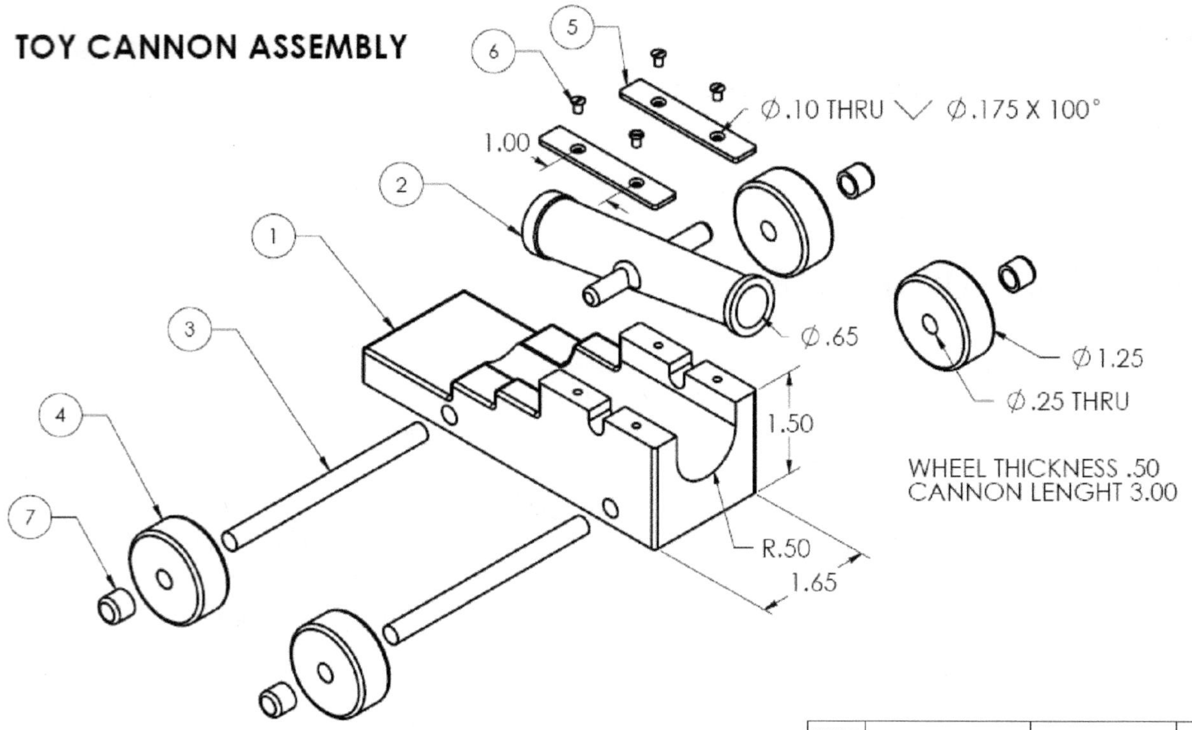

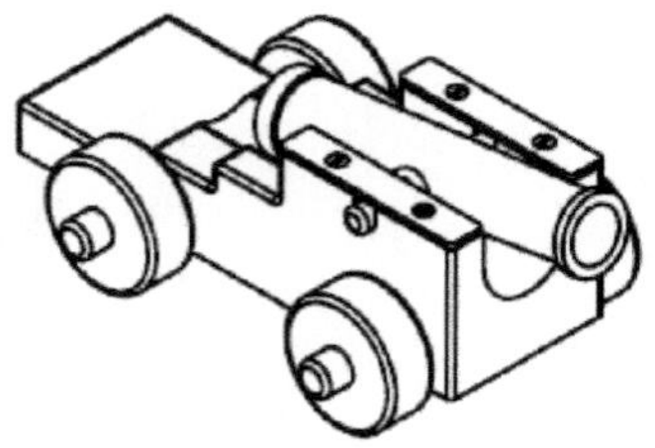

RC1 FIT BETWEEN CANNON AND BASE

FN1 FIT BEWEEN END CAPS AND AXLES

RC1 FIT BETWEEN WHEELS AND AXLES

LC4 FIT BETWEEN BASE AND AXLES

ITEM NO.	DESCRIPTION	MATERIAL	QTY.
1	BASE	PLASTIC	1
2	CANNON	CAST IRON	1
3	AXLE	PLASTIC	2
4	WHEEL	PLASTIC	4
5	PLATE	STEEL	2
6	FLAT HD SCREW	STEEL	4
7	END CAP	PLASTIC	4

Optional Design Improvements

Redesign the toy cannon in order to improve its mobility and increase its angle of fire.

BASIC CRANKSHAFT ASSEMBLY

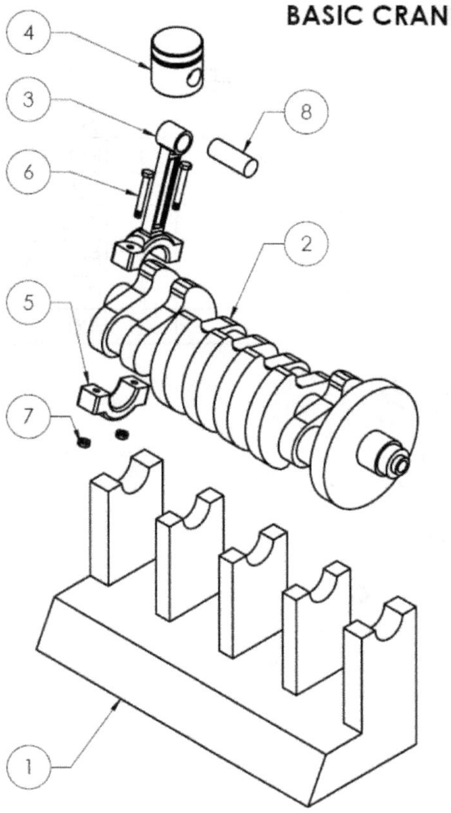

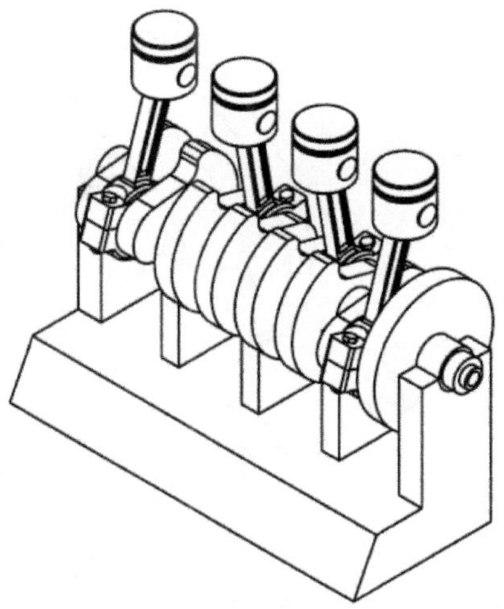

ITEM NO.	PART NUMBER	QTY.
1	BASE	1
2	CRANKSHAFT	1
3	UPPER CONNECTING ROD	4
4	PISTON	4
5	LOWER CONNECTING ROD	4
6	HEX HD BOLT	8
7	FLANGE HEX NUT	8
8	PIN	4

(2) CRANKSHAFT

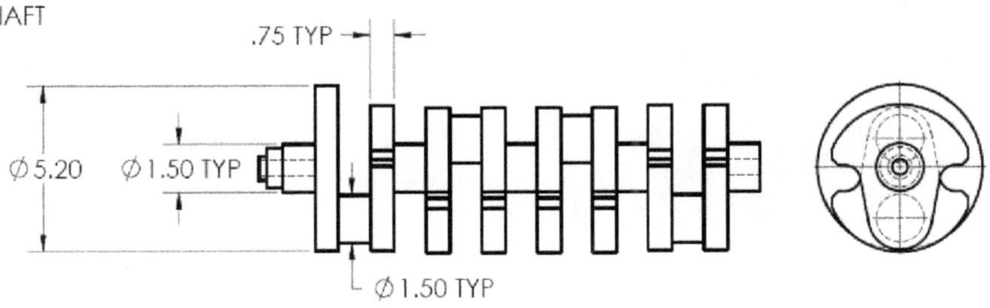

(3) UPPER CONNECTING ROD

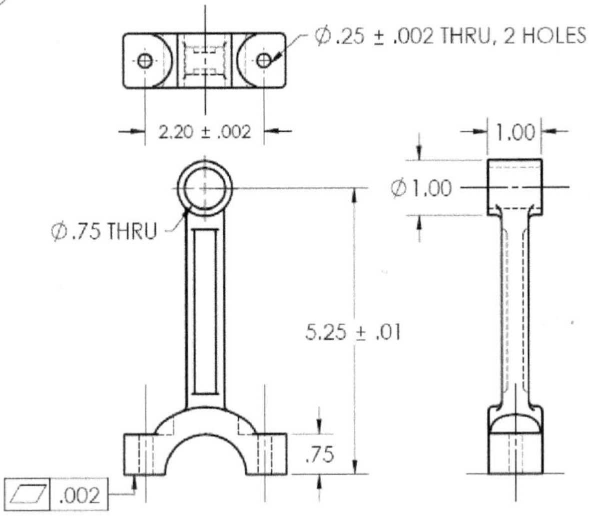

(4) PISTON

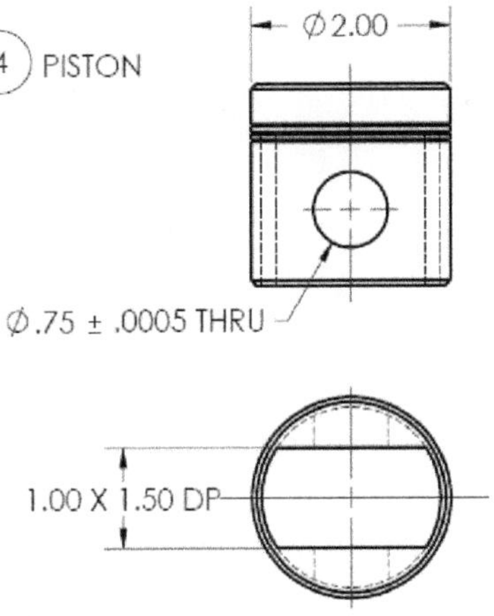

INDUSTRIAL HOOK ASSEMBLY

PARTS LIST		
PART NUMBER	QTY	DESCRIPTION
1	1	BODY
2	1	HOOK
3	1	RING
4	1	BEARINGS
5	1	PULLEY
6	1	CAP
7	1	AXLE

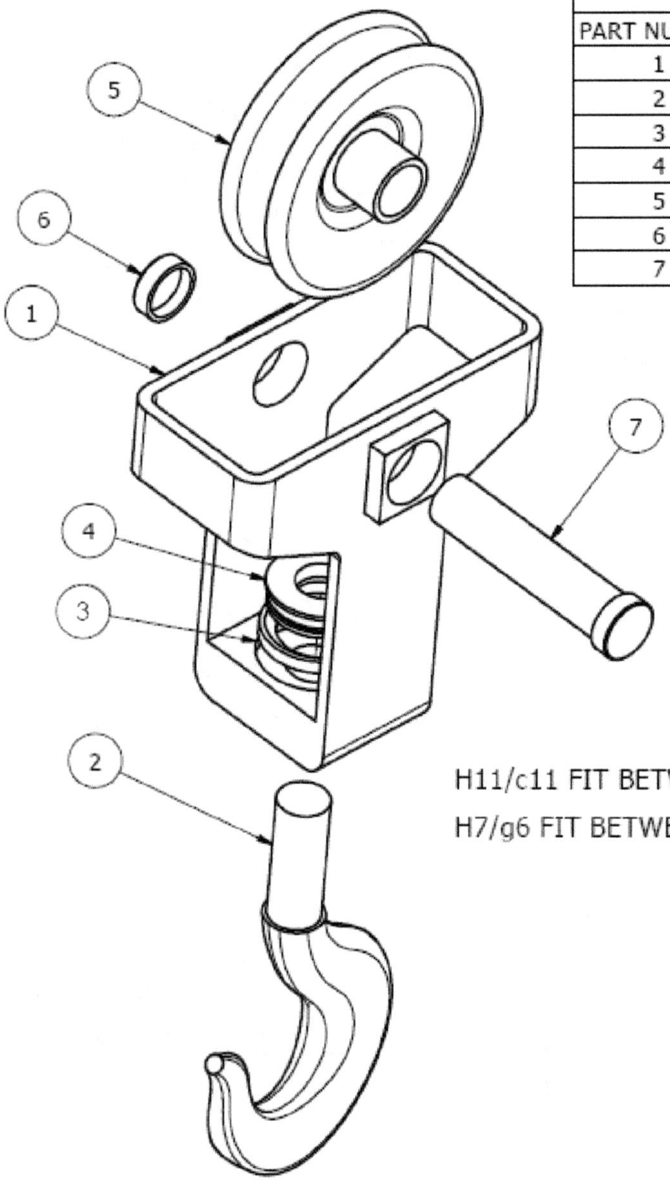

H11/c11 FIT BETWEEN PULLEY AND AXLE

H7/g6 FIT BETWEEN BODY AND HOOK

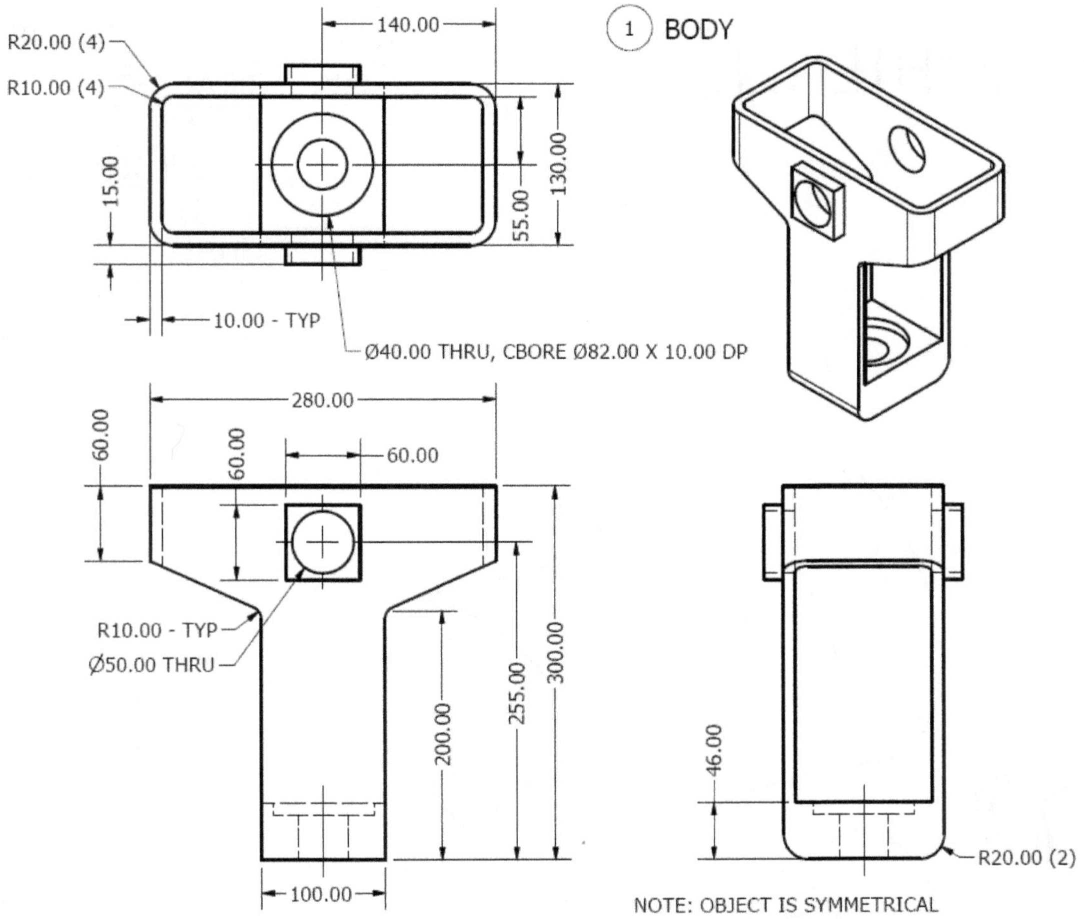

1 BODY
R20.00 (4)
R10.00 (4)
140.00
15.00
55.00
130.00
10.00 - TYP
Ø40.00 THRU, CBORE Ø82.00 X 10.00 DP
280.00
60.00
60.00
60.00
R10.00 - TYP
Ø50.00 THRU
200.00
255.00
300.00
100.00
46.00
R20.00 (2)
NOTE: OBJECT IS SYMMETRICAL

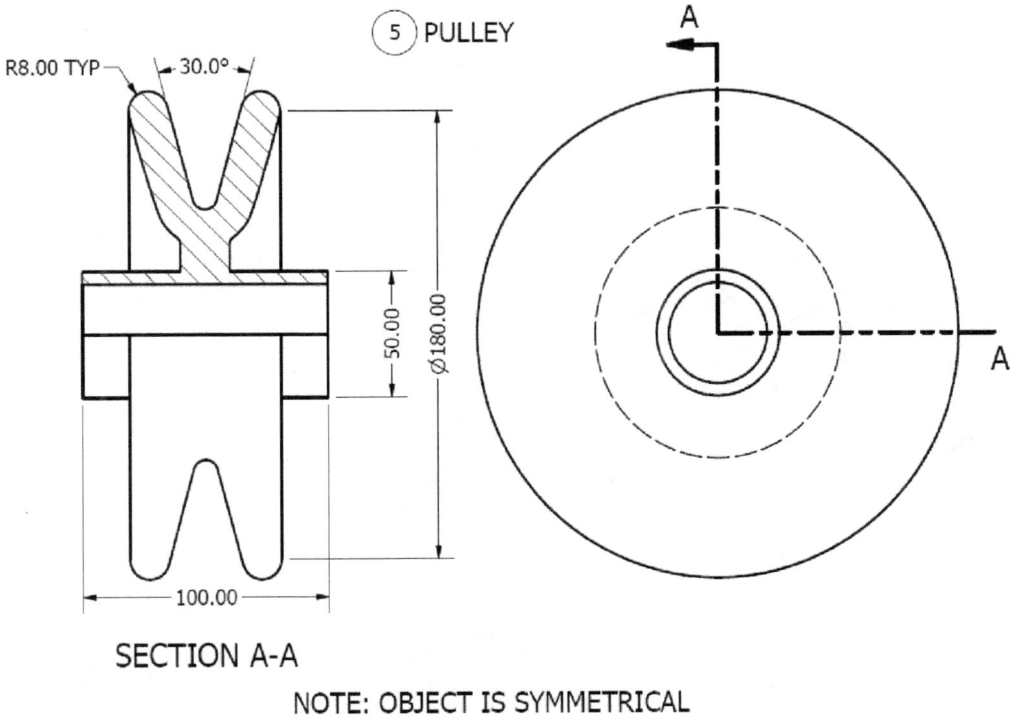

5 PULLEY
R8.00 TYP
30.0°
50.00
Ø180.00
100.00
A
A
SECTION A-A
NOTE: OBJECT IS SYMMETRICAL

BELT TIGHTENER

ITEM NO.	DESCRIPTION	QTY.
1	BRACKET	1
2	FRAME	1
3	SHOULDER BOLT	2
4	PULLEY	1
5	SHAFT	1
6	3/16 X 3/16 X 1.50 KEY	1
7	.875 REG HEX NUT	1

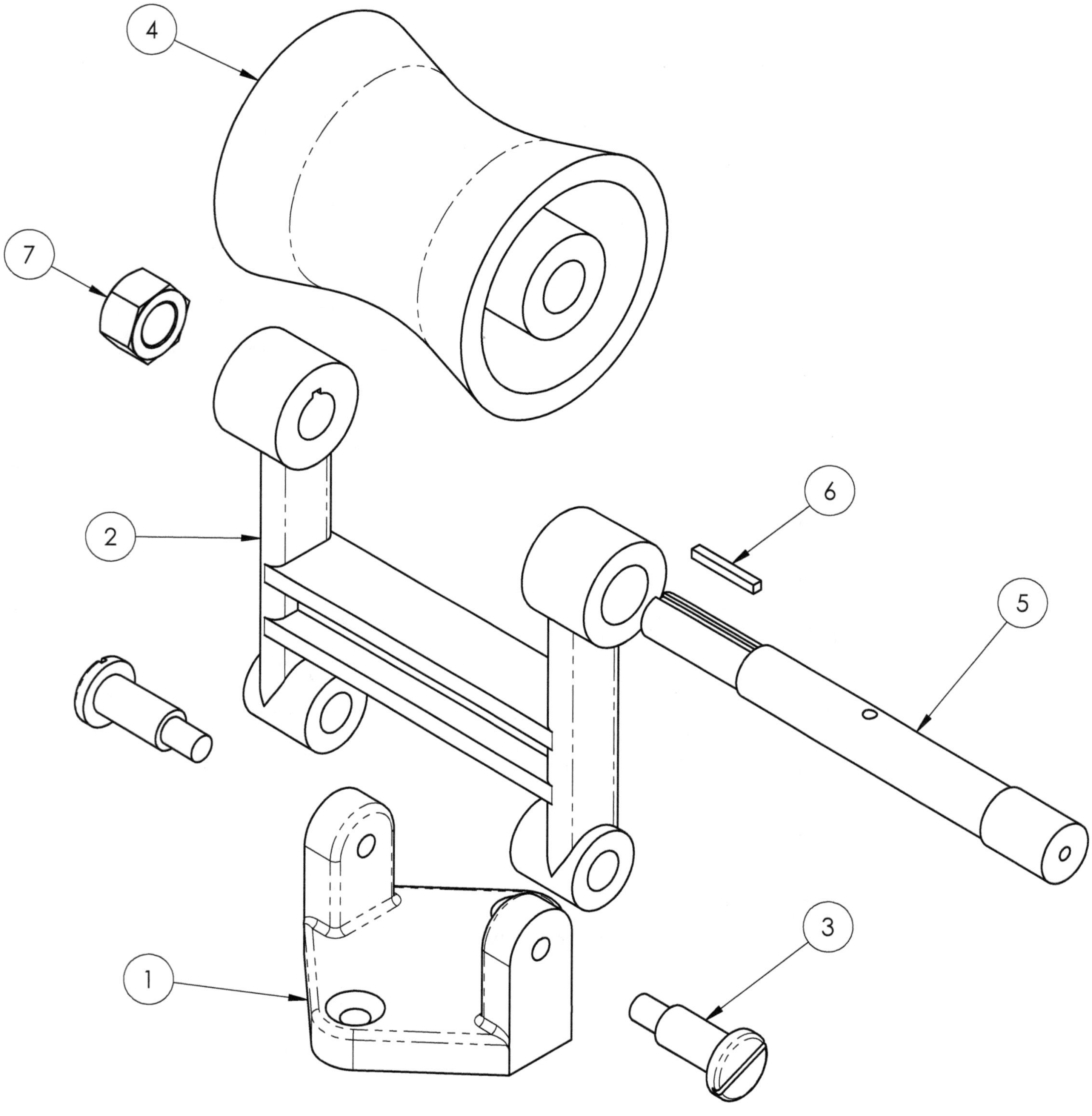

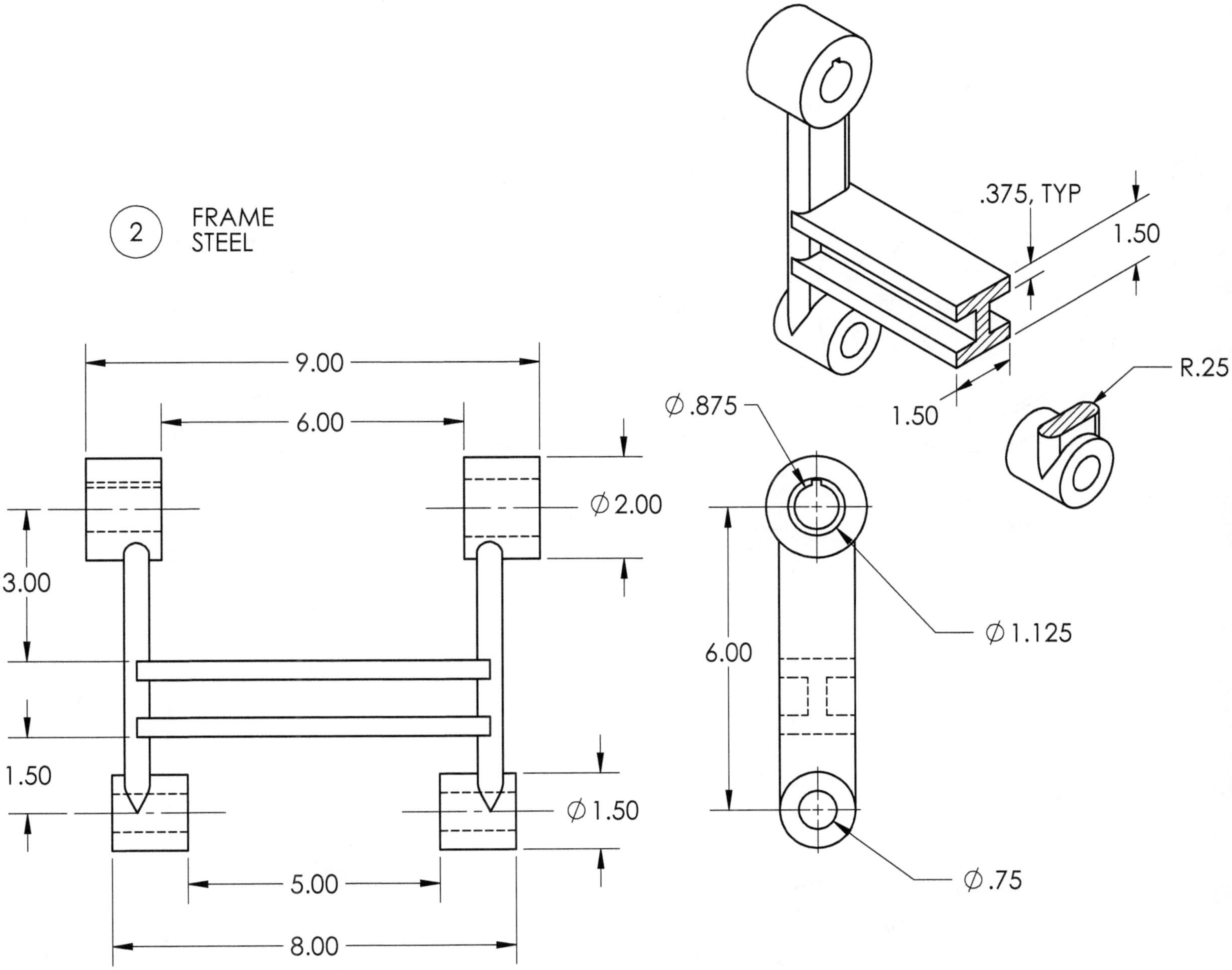
2
FRAME
STEEL
.375, TYP
1.50
1.50
R.25
9.00
6.00
Ø 2.00
3.00
Ø .875
Ø 1.125
6.00
1.50
Ø 1.50
5.00
8.00
Ø .75

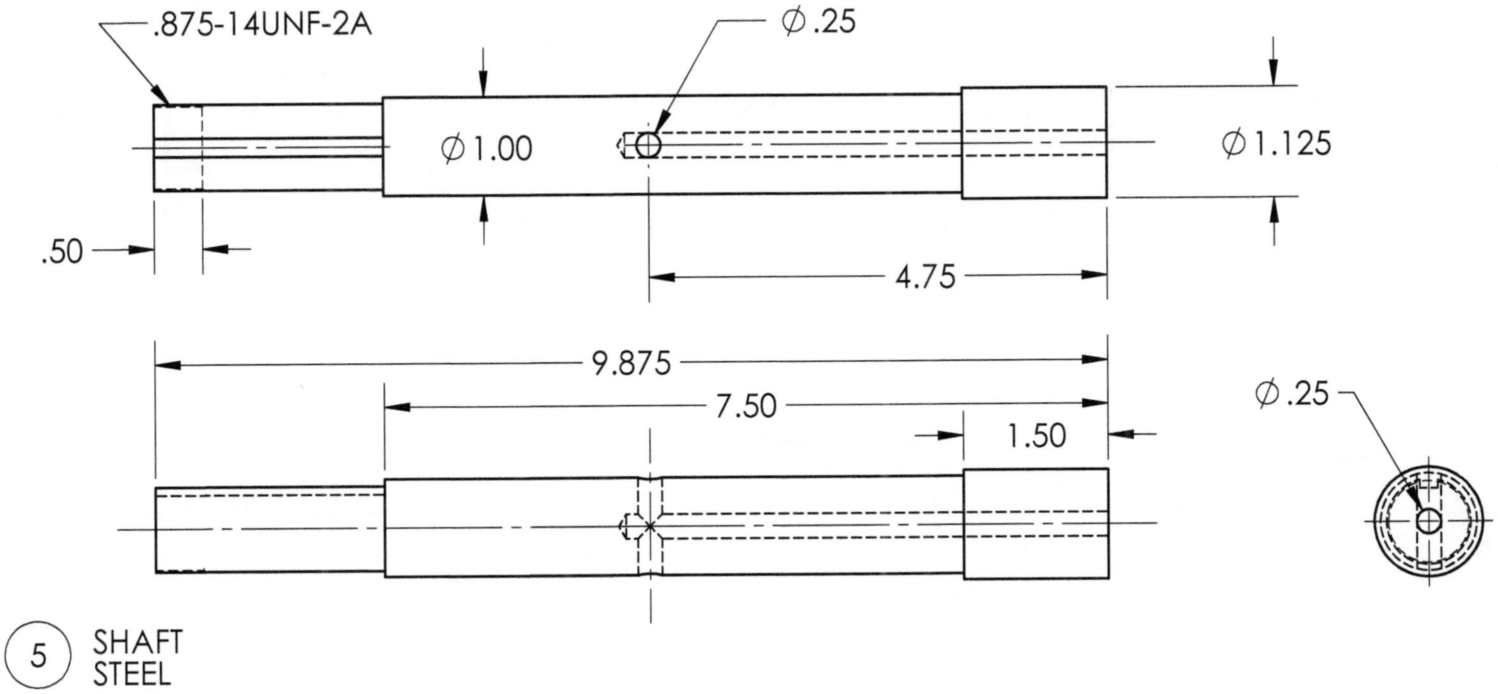
.875-14UNF-2A
Ø .25
Ø 1.00
Ø 1.125
.50
4.75
9.875
7.50
1.50
Ø .25
5
SHAFT
STEEL

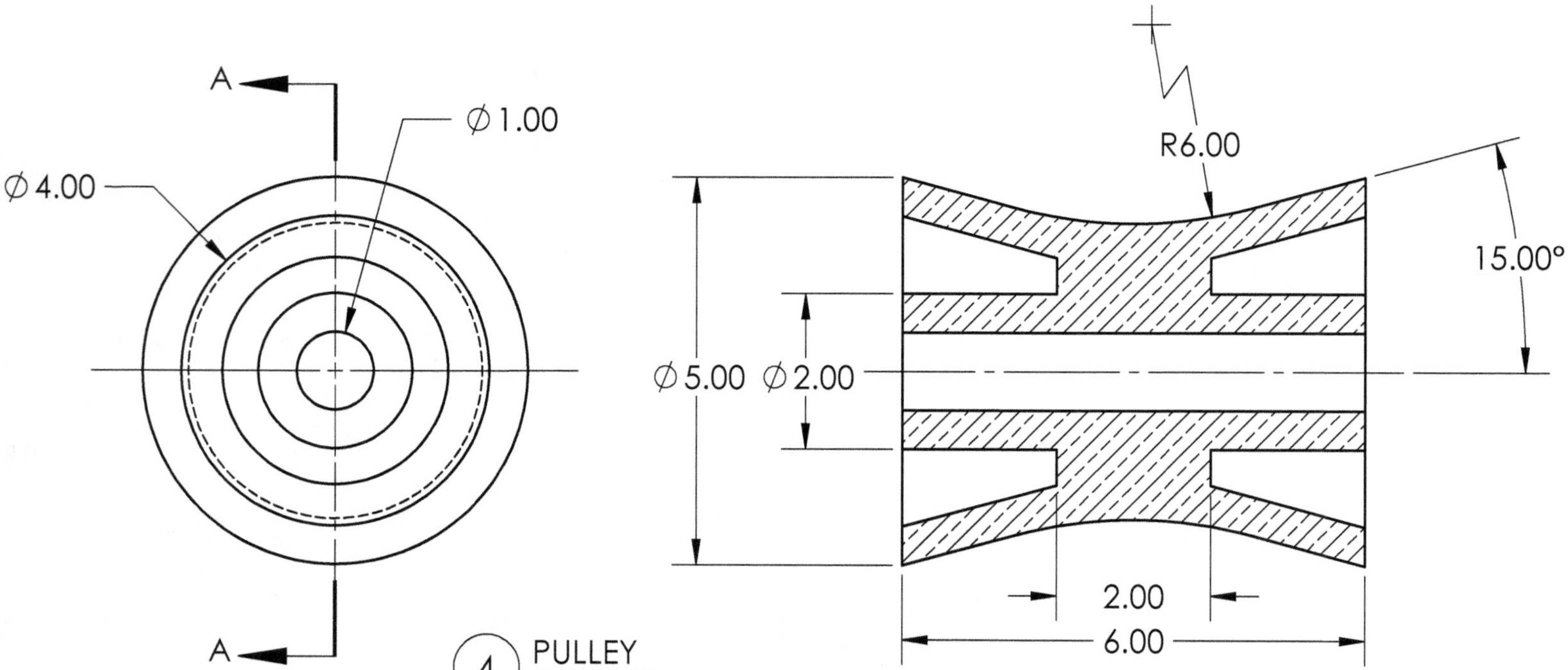

(4) PULLEY
BRONZE

SECTION A-A

(1) BRACKET
ALUMINUM

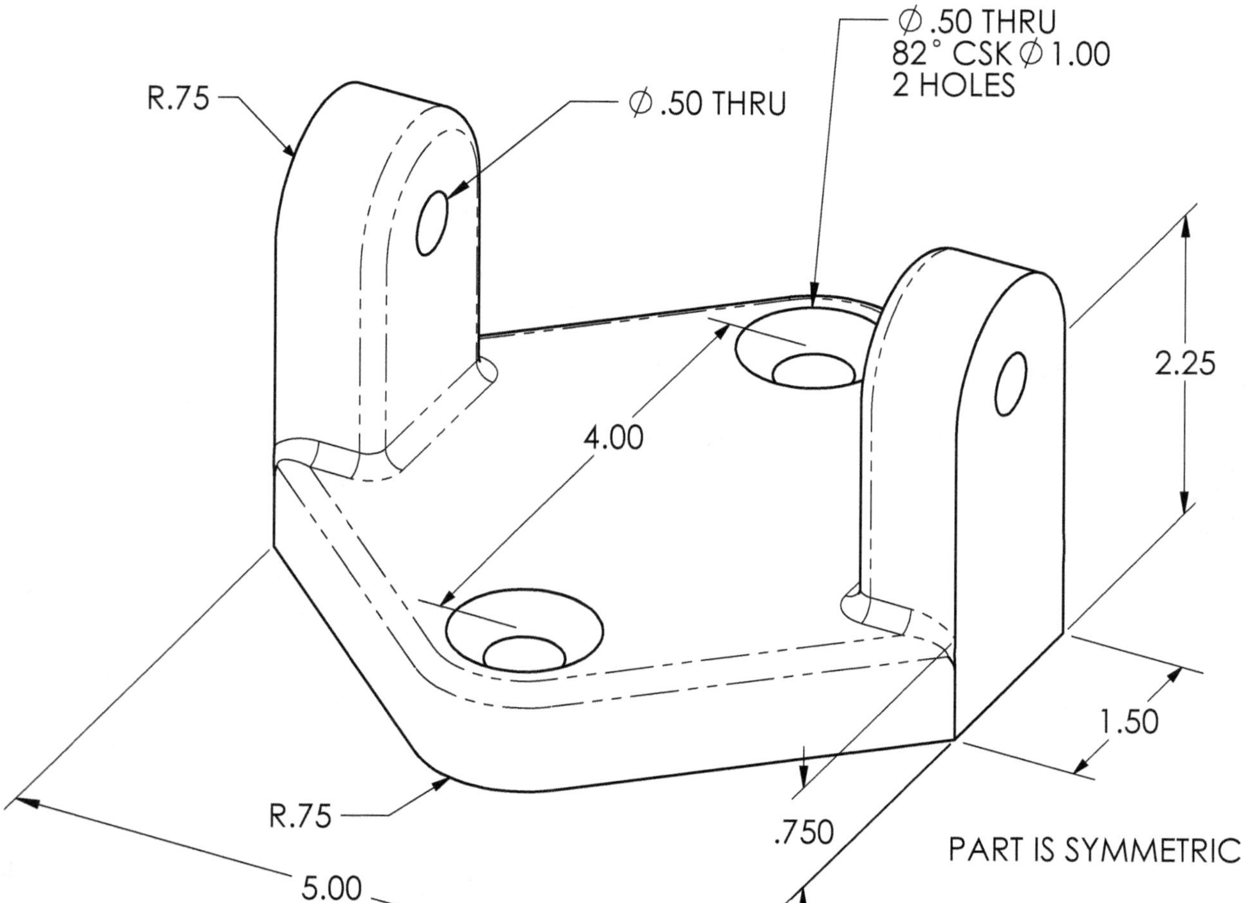

FOLD-UP TABLE

ITEM NO.	DESCRIPTION	QTY.
1	BOARD	1
2	SIDE SUPPORT	2
3	GUIDE SUPPORT	2
4	GUIDE	1
5	BASIC LEG	2
6	LEG STRUCTURE	1
7	SPACER	4
8	FLAT HD SCREW ∅.25	8
9	ROUND HD SCREW ∅.164	2
10	BINDING POST BARREL	4
11	BINDING POST SCREW	4

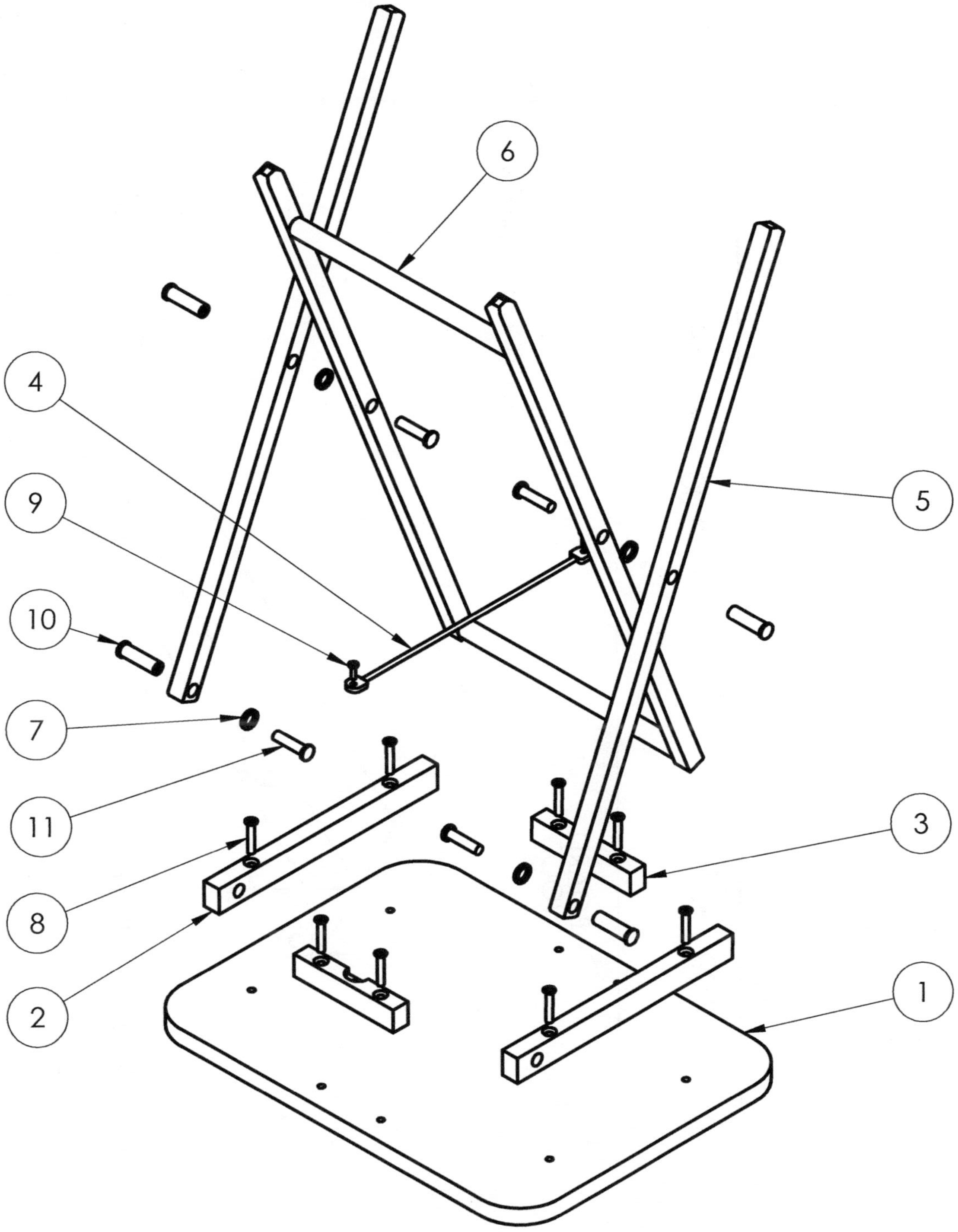

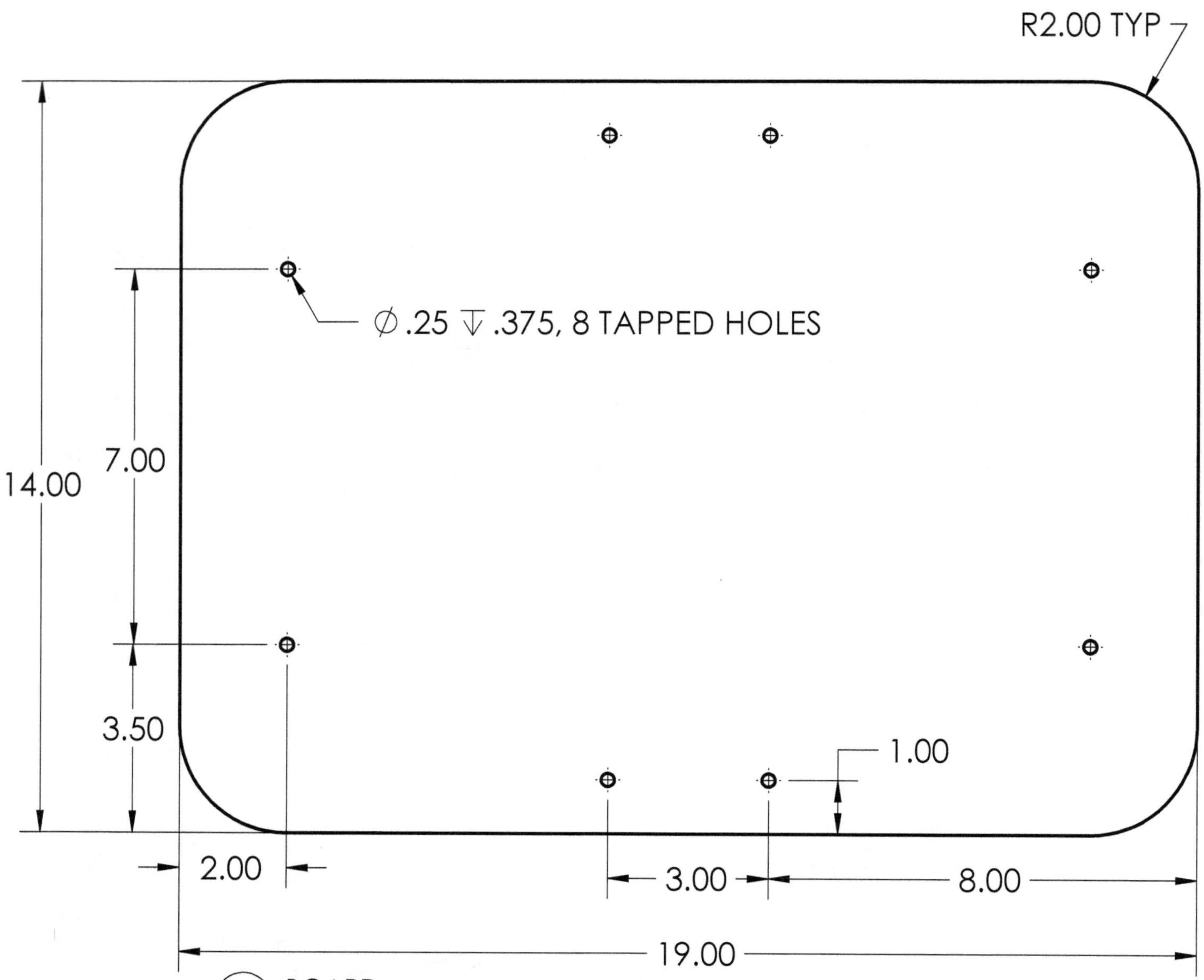

1 BOARD
WOOD

OBJECT IS SYMMETRICAL
PART THICKNESS .75

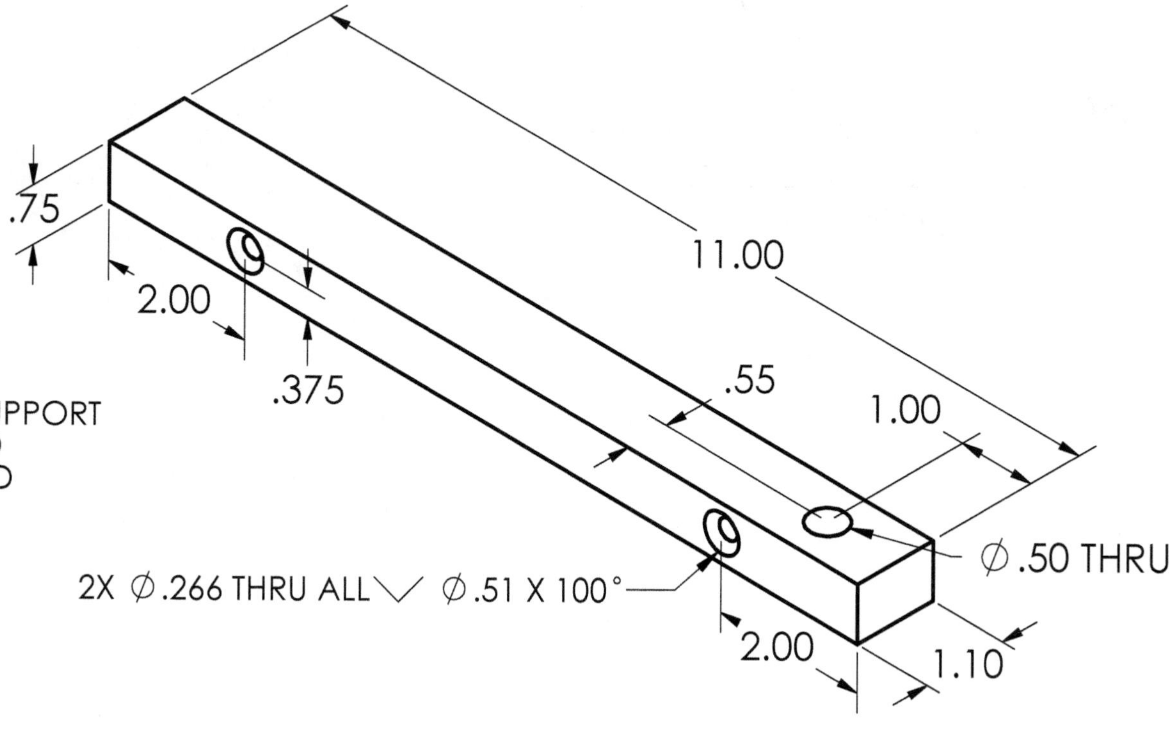

2 SIDE SUPPORT
WOOD
2 REQ'D

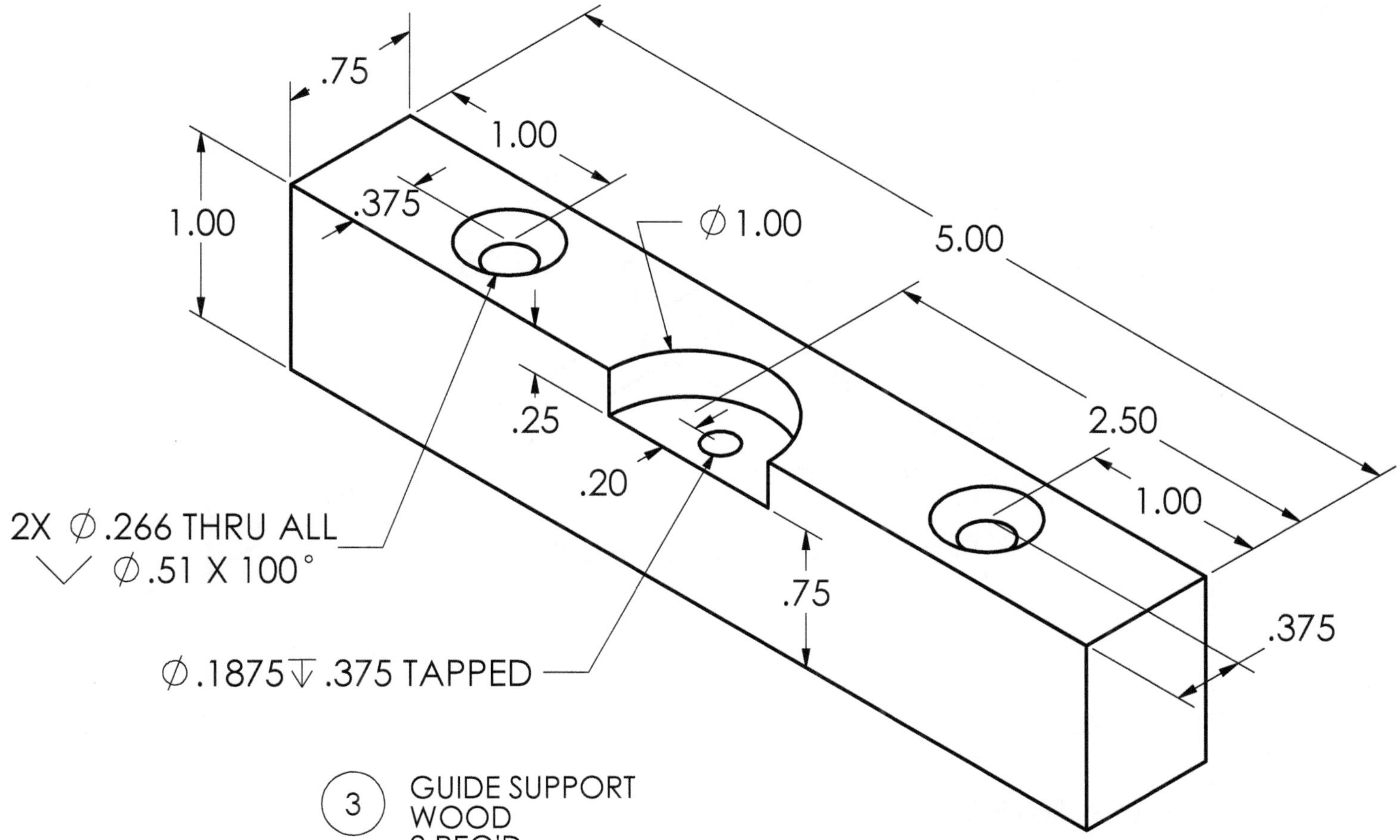

3 GUIDE SUPPORT
WOOD
2 REQ'D

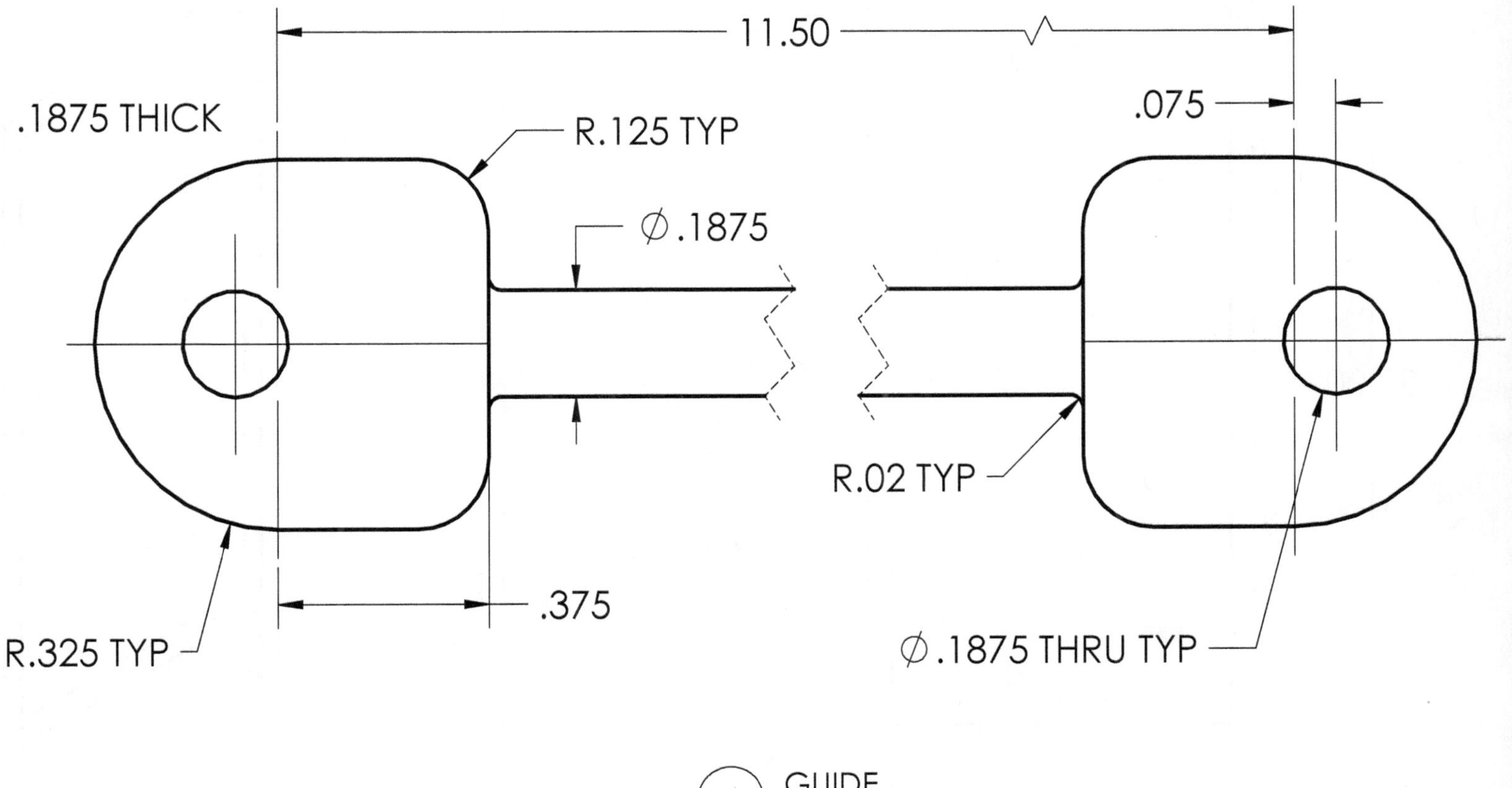

4 GUIDE
ALUMINUM

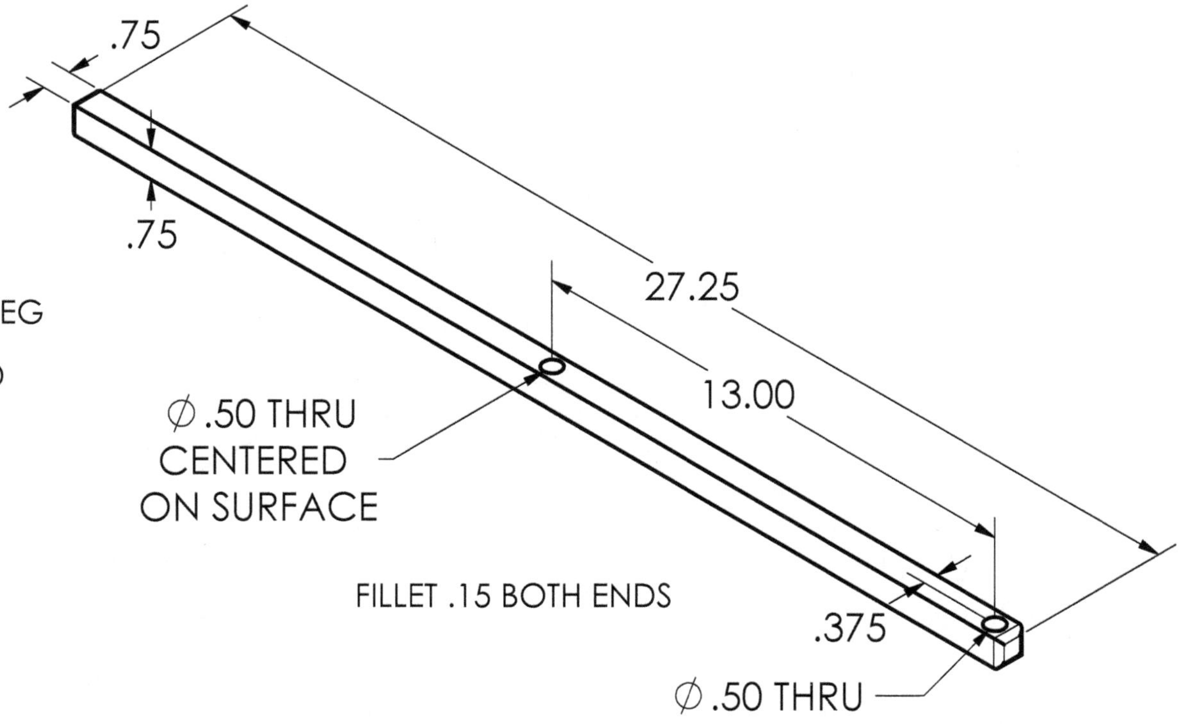

(5) BASIC LEG
WOOD
2 REQ'D

(6) LEG STRUCTURE
WOOD

BUILD AS SUBASSEMBLY

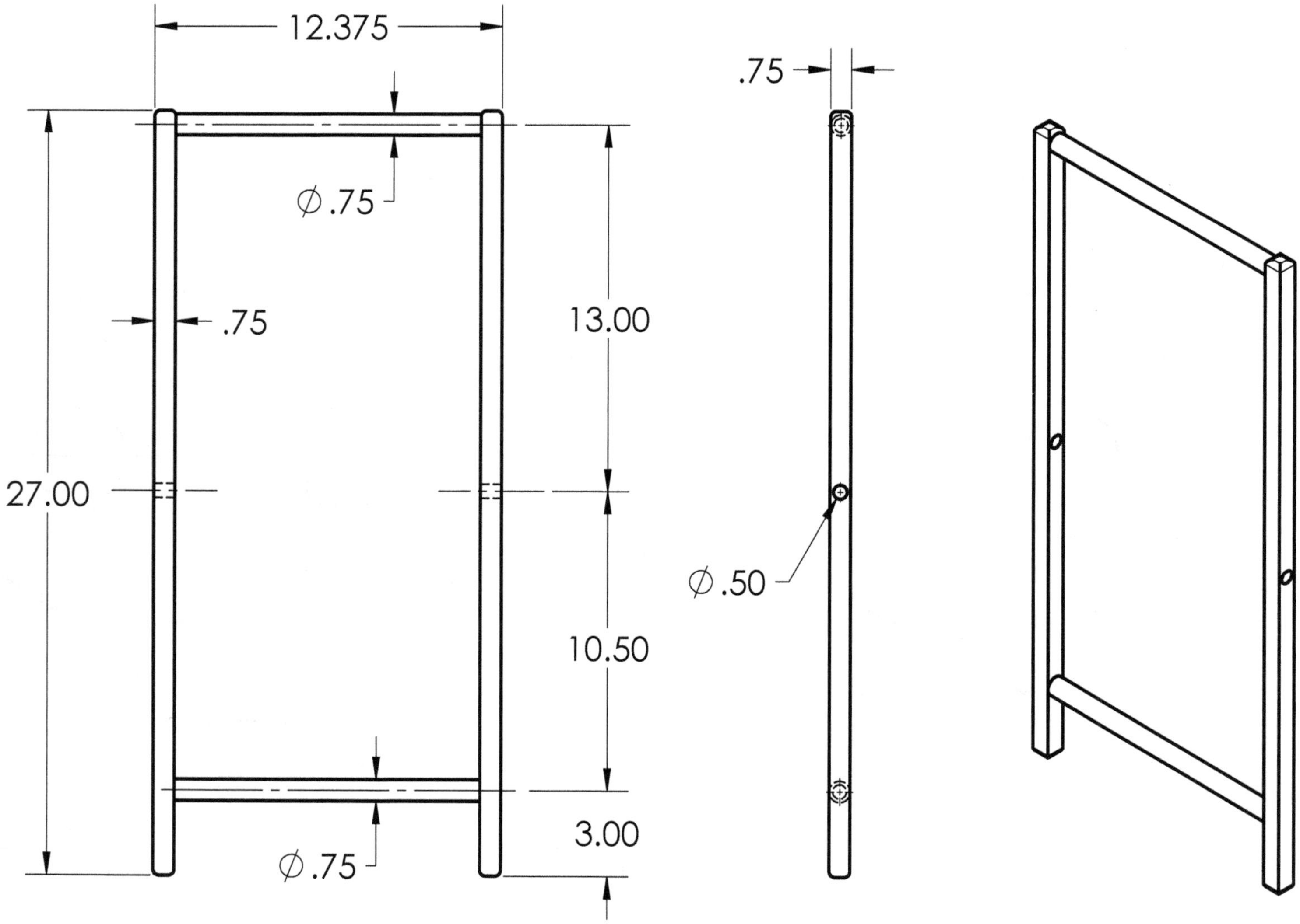

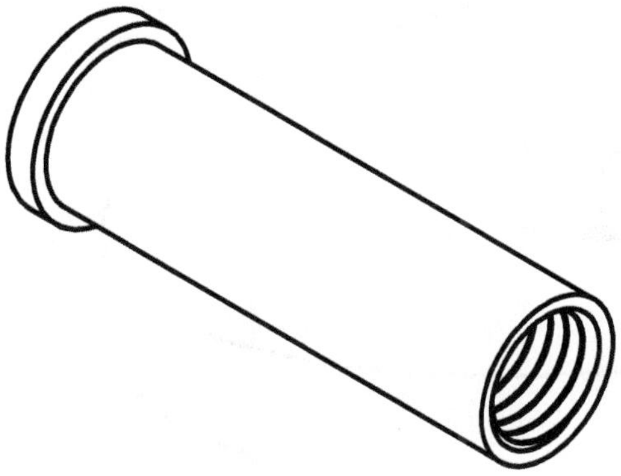

(10) ⌀3/8-16UNC-2B
BINDING POST BARREL 1.65 LG
4 REQ'D

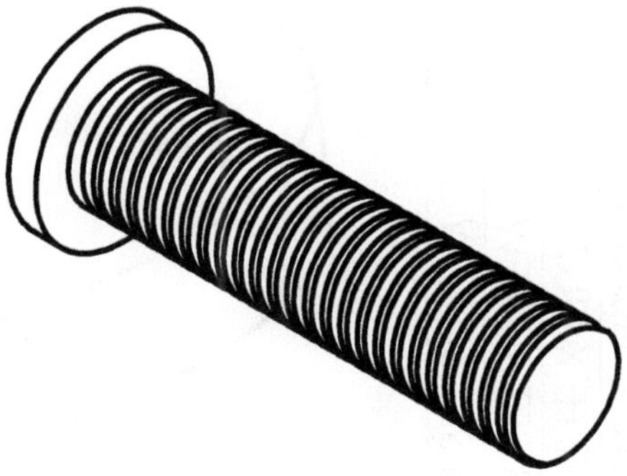

(11) ⌀3/8-16UNC-2A
BINDING POST SCREW1.50 LG
4 REQ'D

(7) SPACER (ID .50 OD .75 X 3/32)
PLASTIC
4 REQ'D

(8) ⌀1/4-20UNC-2A
FLAT HD SCREW 1.2675 LG
8 REQ'D

(9) ⌀.164-32UNC-2A
ROUND HD SCREW .50 LG
2 REQ'D

GRINDER ASSEMBLY

ITEM NO.	DESCRIPTION	QTY.
1	SUPPORT BASE	2
2	SHAFT	1
3	BELT PULLEY	1
4	COLLAR	2
5	GRINDING WHEEL	2
6	SLEEVE	2
7	ARBOR MOUNT	4
8	M12 HEX NUT	2
9	M8 SET SCREW	2
10	M5 SET SCREW	1

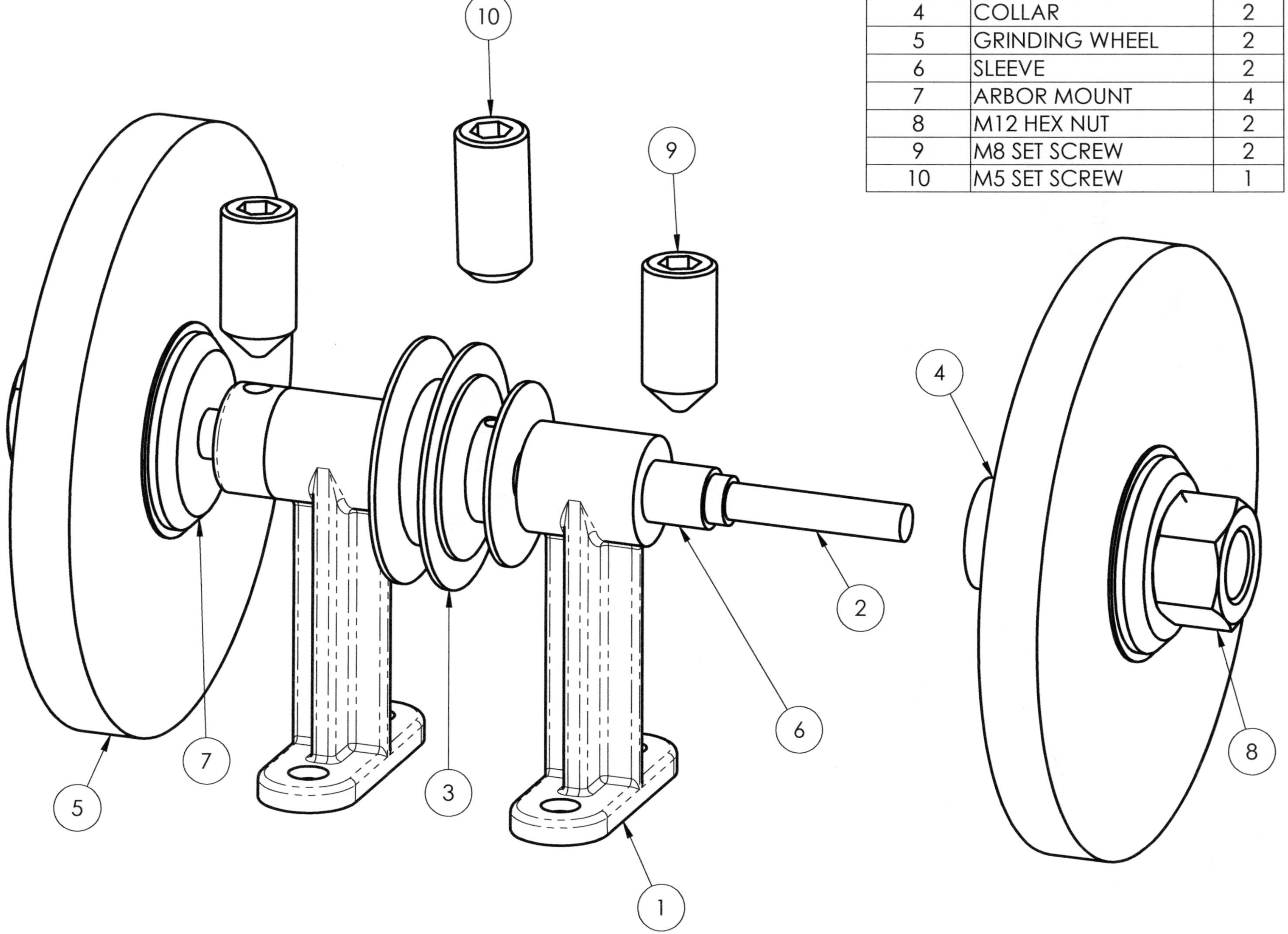

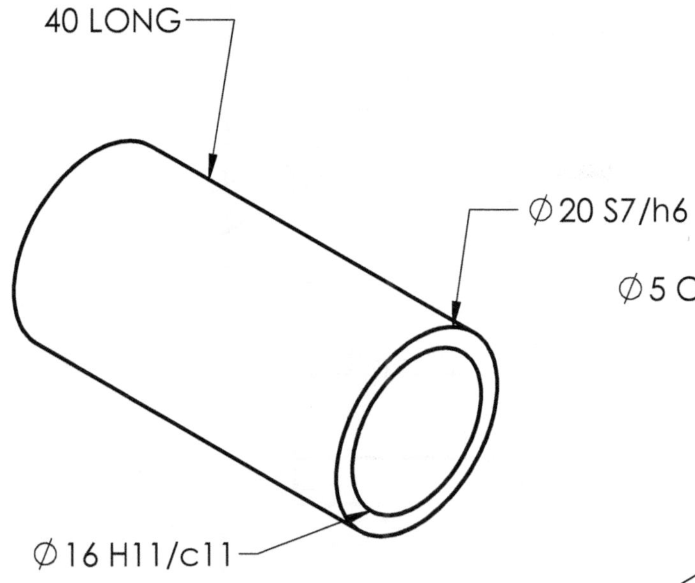

6 SLEEVE
BRASS

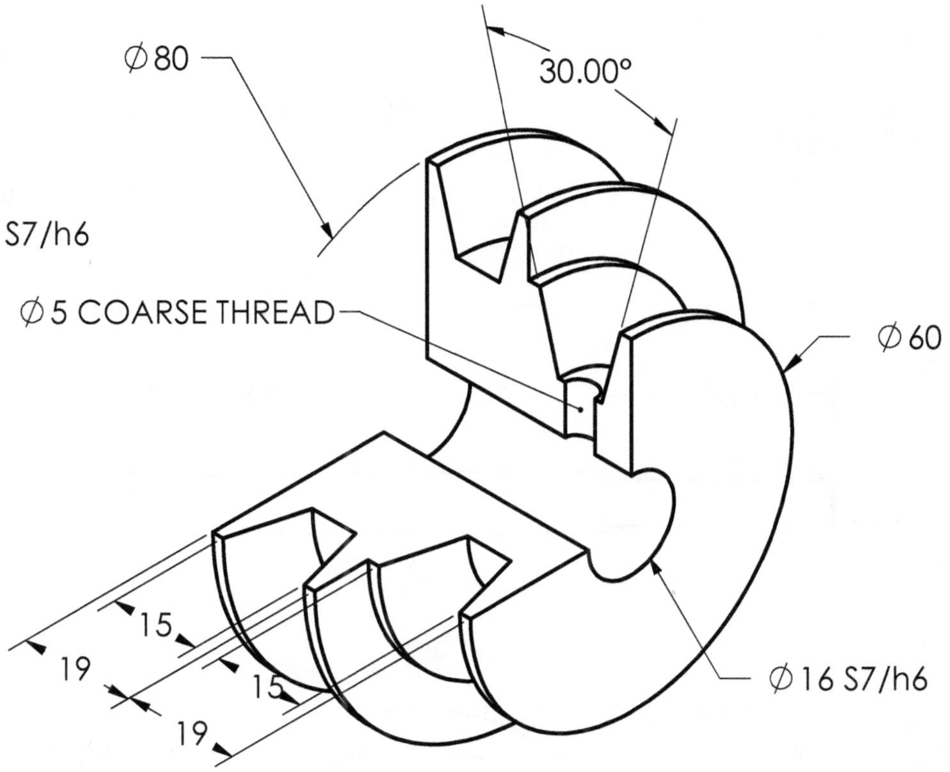

3 BELT PULLEY
STEEL

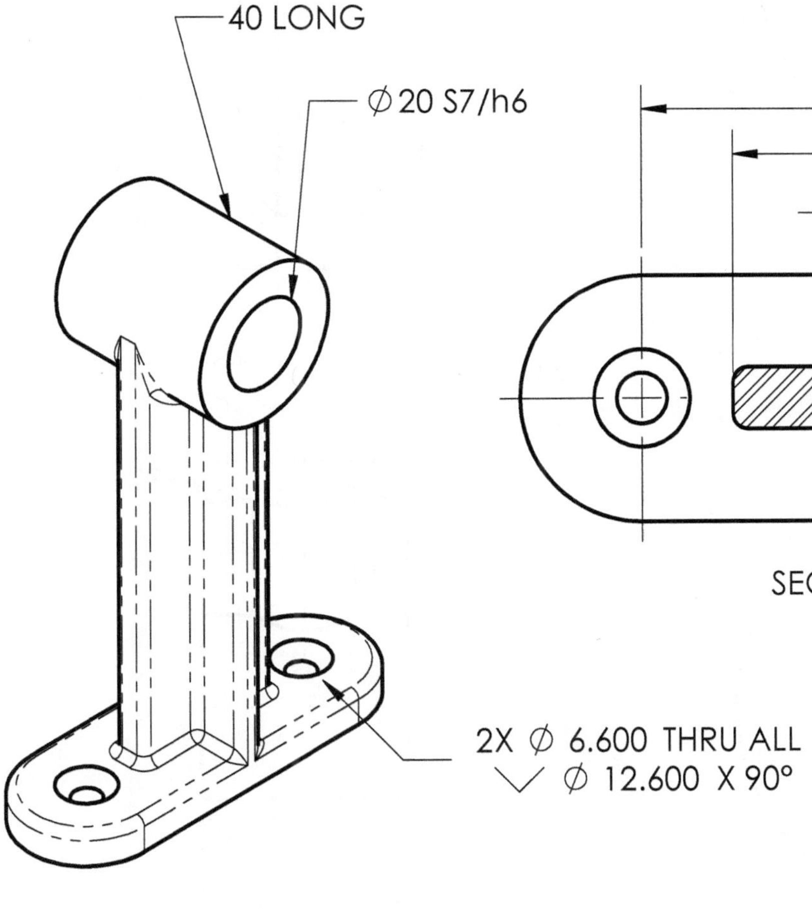

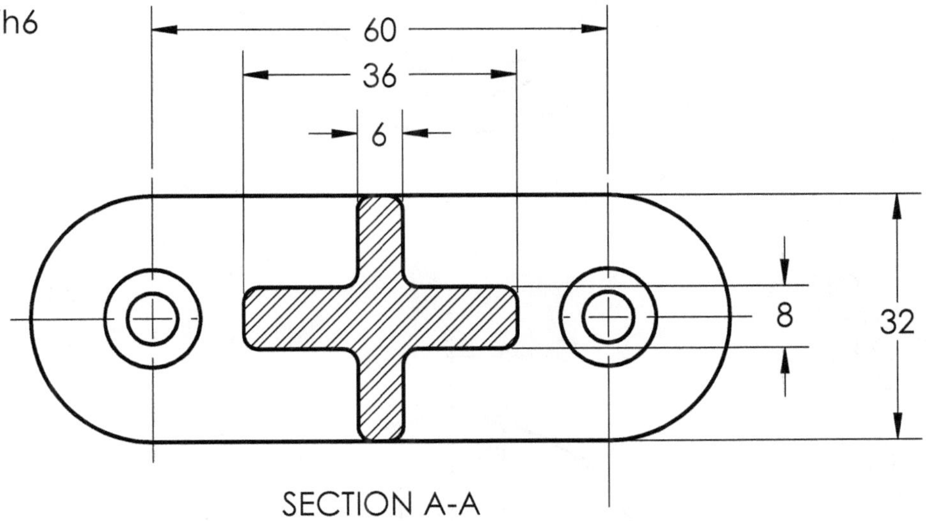

SECTION A-A

1 SUPPORT BASE
STEEL

2 SHAFT
STEEL

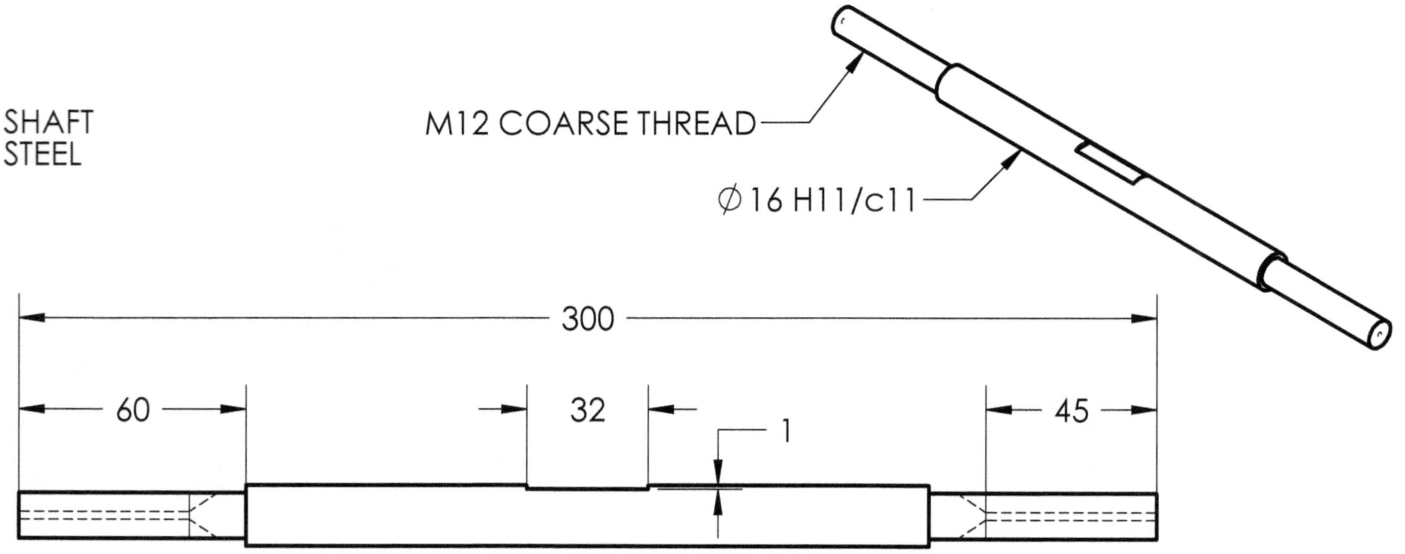

7 ARBOR MOUNT
STEEL
4 REQ.

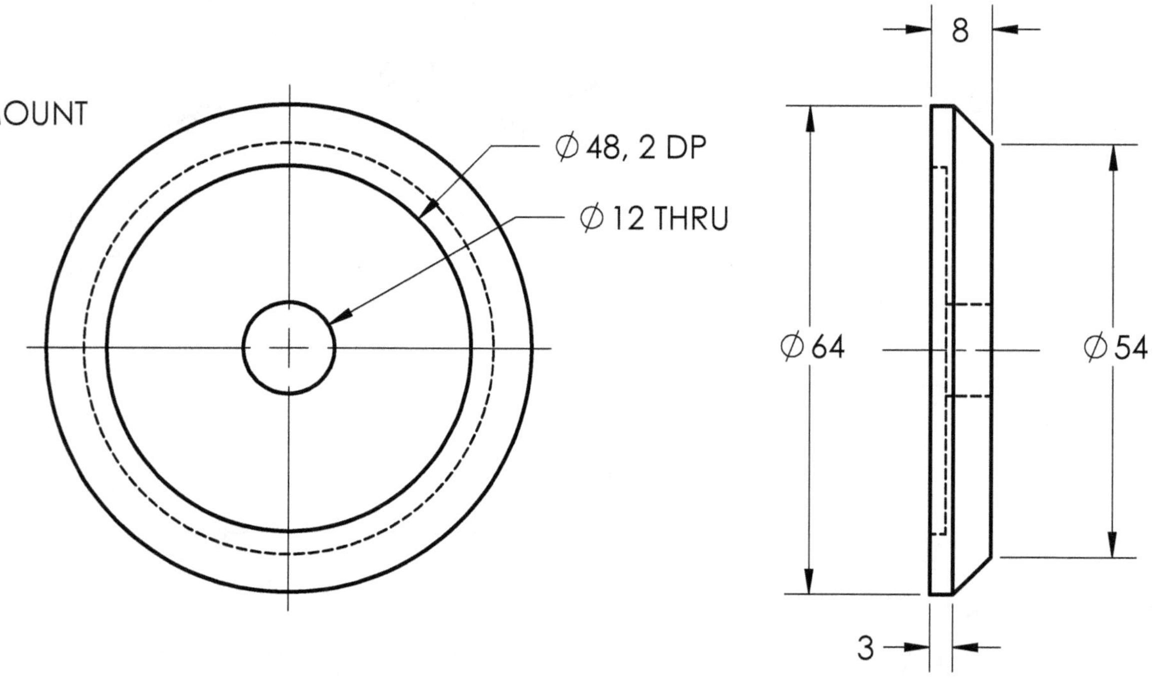

4 COLLAR
STEEL
2 REQ.

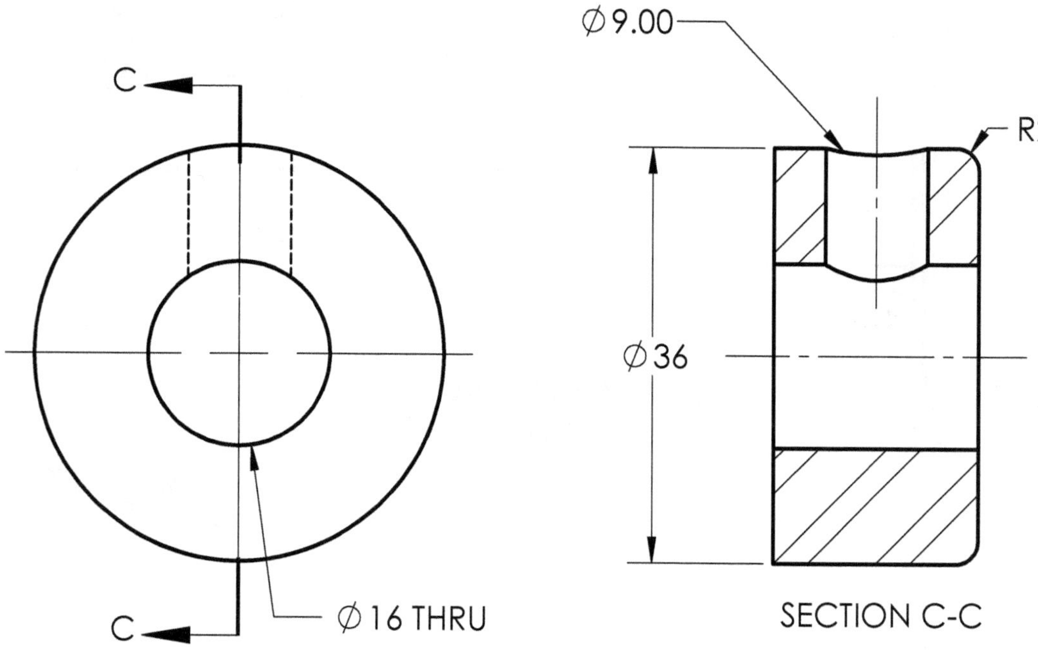

HANGING ROLLER

ITEM NO.	DESCRIPTION	QTY.
1	WHEEL	1
2	AXLE	1
3	ARM	2
4	MOUNTING PLATE	1
5	HEX HD BOLT	4
6	HEX NUT	4

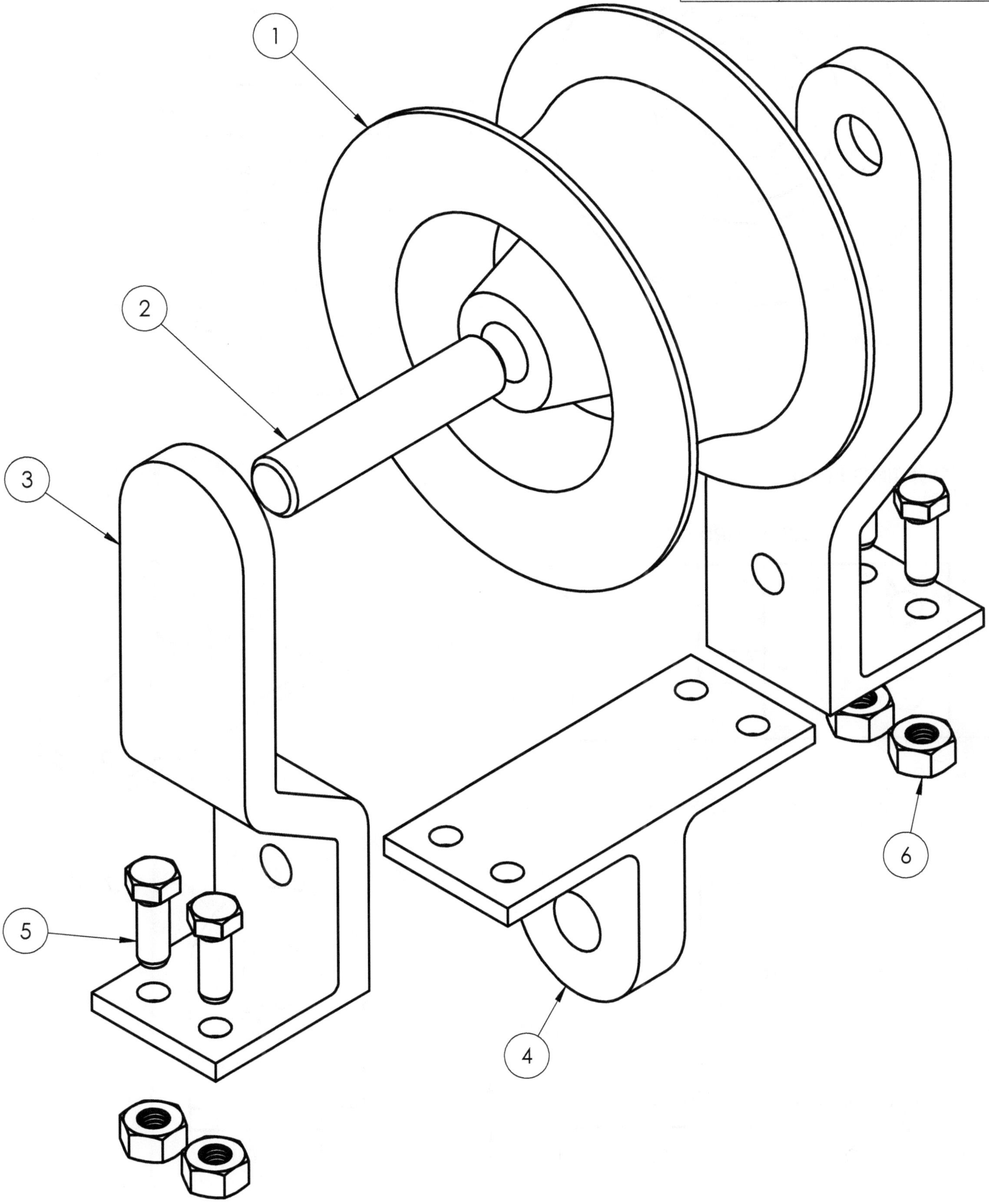

1 WHEEL
BRONZE

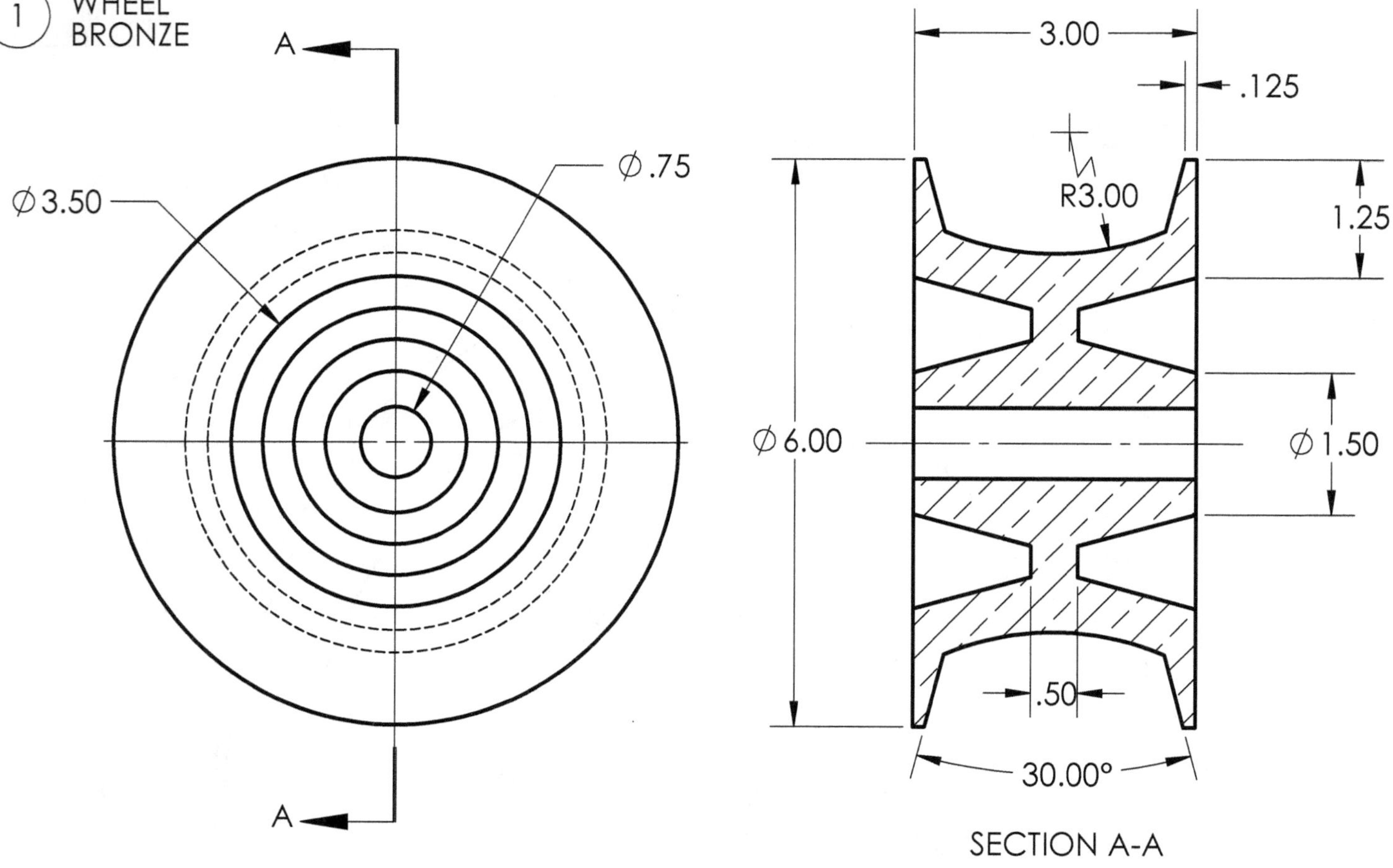

4 MOUNTING PLATE
STEEL

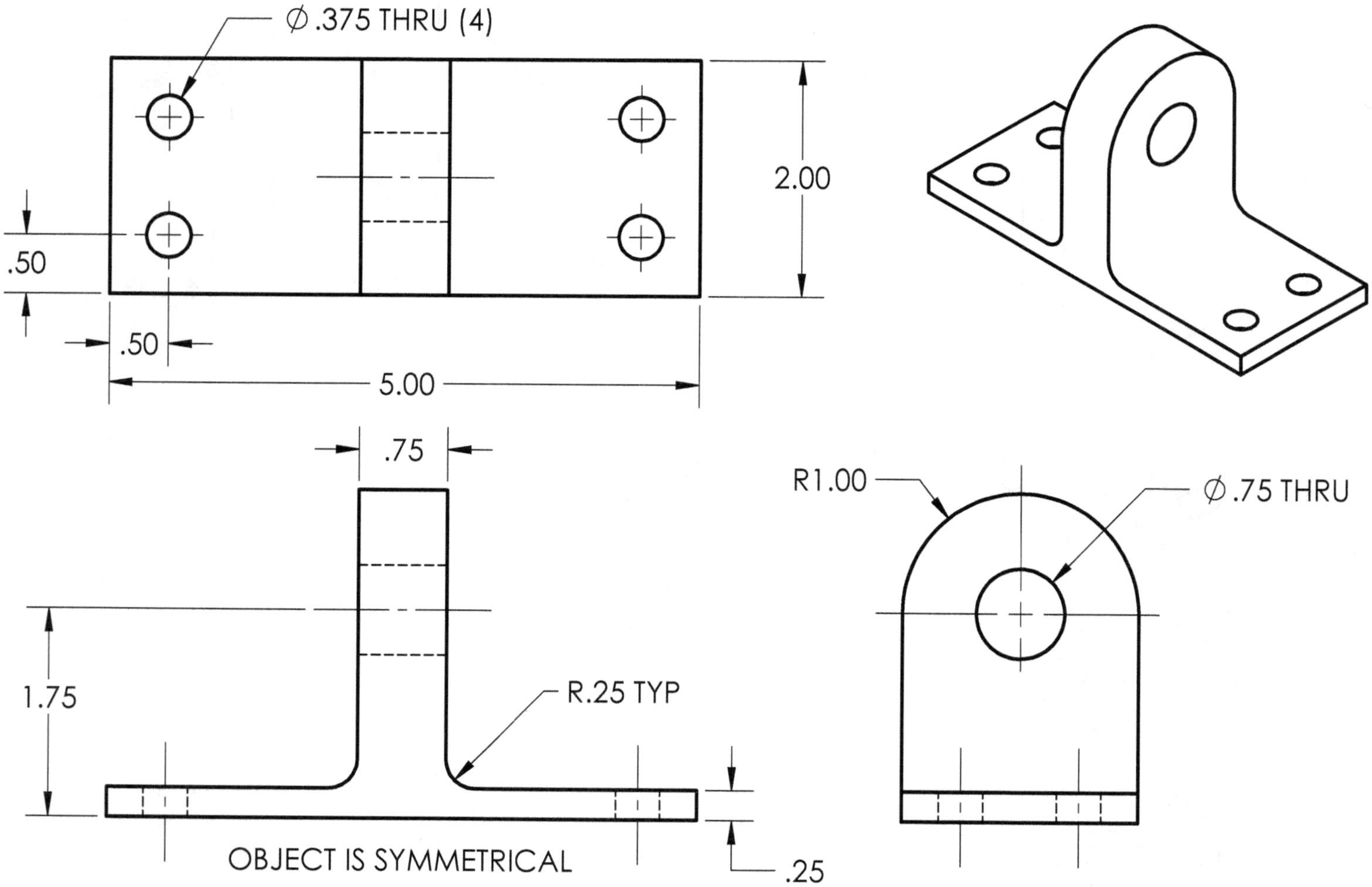

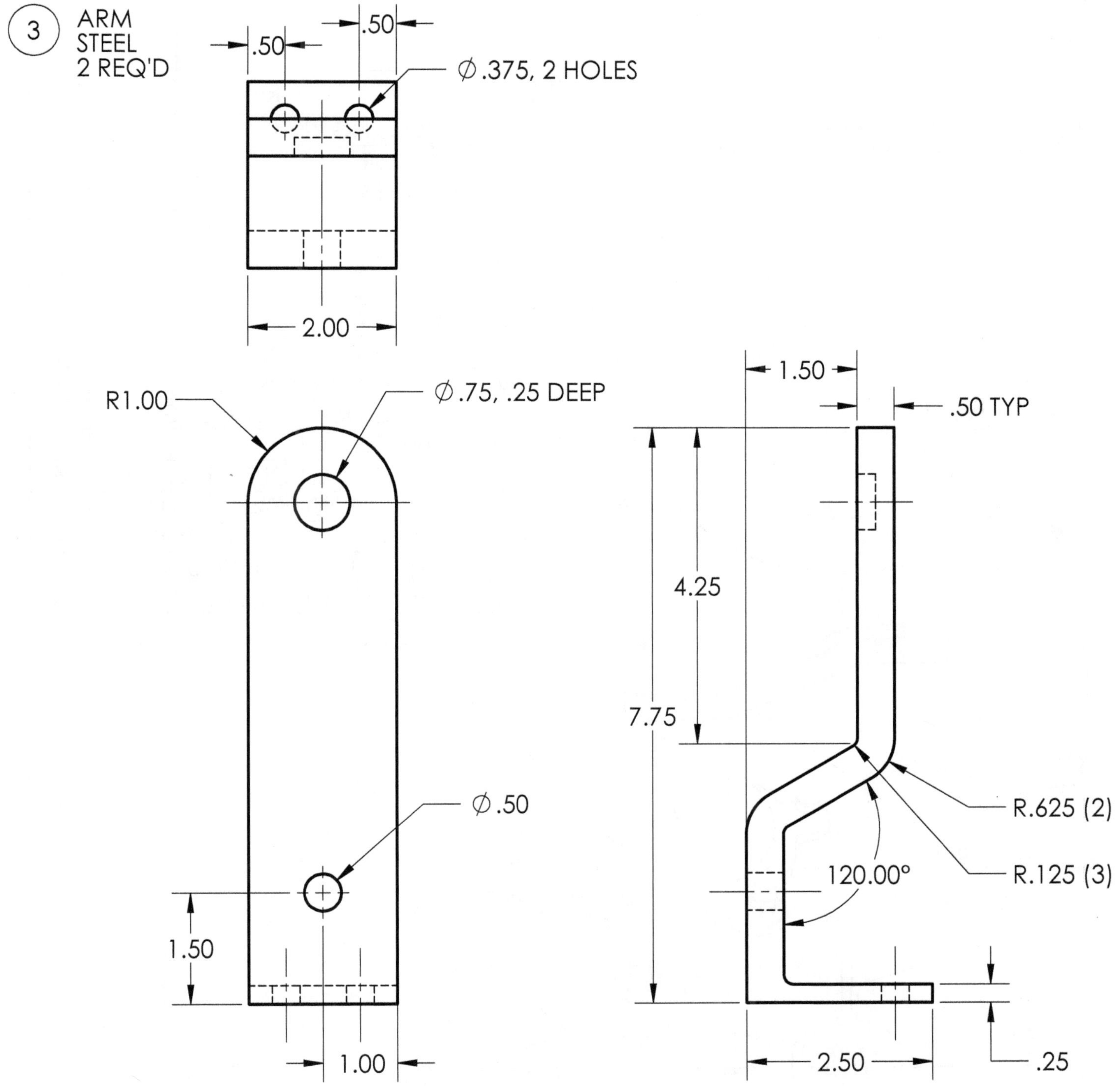

3
ARM
STEEL
2 REQ'D
.50
.50
Ø.375, 2 HOLES
2.00
R1.00
Ø.75, .25 DEEP
Ø.50
1.50
1.00
1.50
.50 TYP
4.25
7.75
R.625 (2)
R.125 (3)
120.00°
2.50
.25

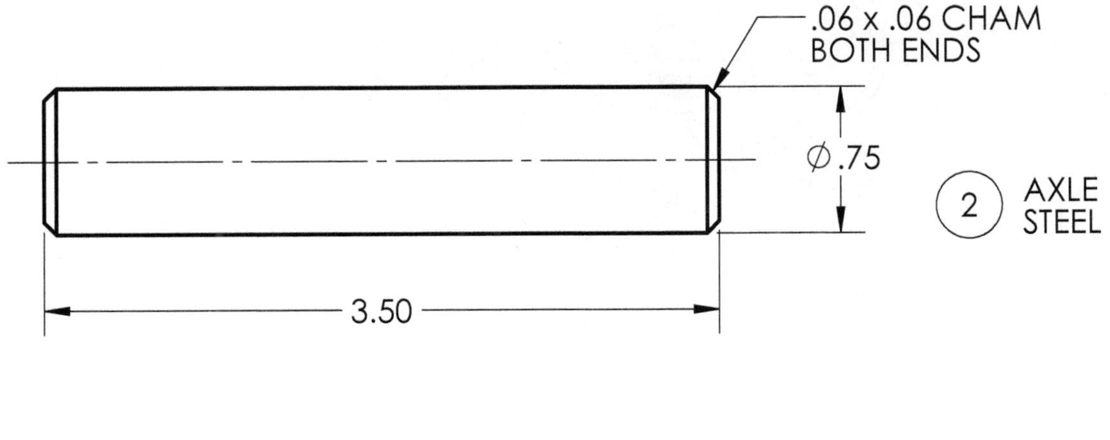

.06 x .06 CHAM
BOTH ENDS
Ø.75
2
AXLE
STEEL
3.50

HOSE PUMP

ITEM NO.	DESCRIPTION	QTY.
1	HOUSING	1
2	ROTOR	1
3	ROLLER	2
4	AXLE	2
5	SHAFT	1
6	BALL BEARING	2
7	COVER	1
8	RETAINING PLATE	2
9	RADIAL SUPPORT	1
10	SHOULDER BOLT, 120 LG	1
11	HANDLE	1
12	M8 HEX HD BOLT, 16 LG	6
13	M4 FLAT HD SCREW, 16 LG	1
14	M8 FLAT HD SCREW, 35 LG	4
15	M3 FLAT HD SCREW, 8 LG	8
16	M4 FLAT HD SCREW, 30 LG	1
17	TUBE	1

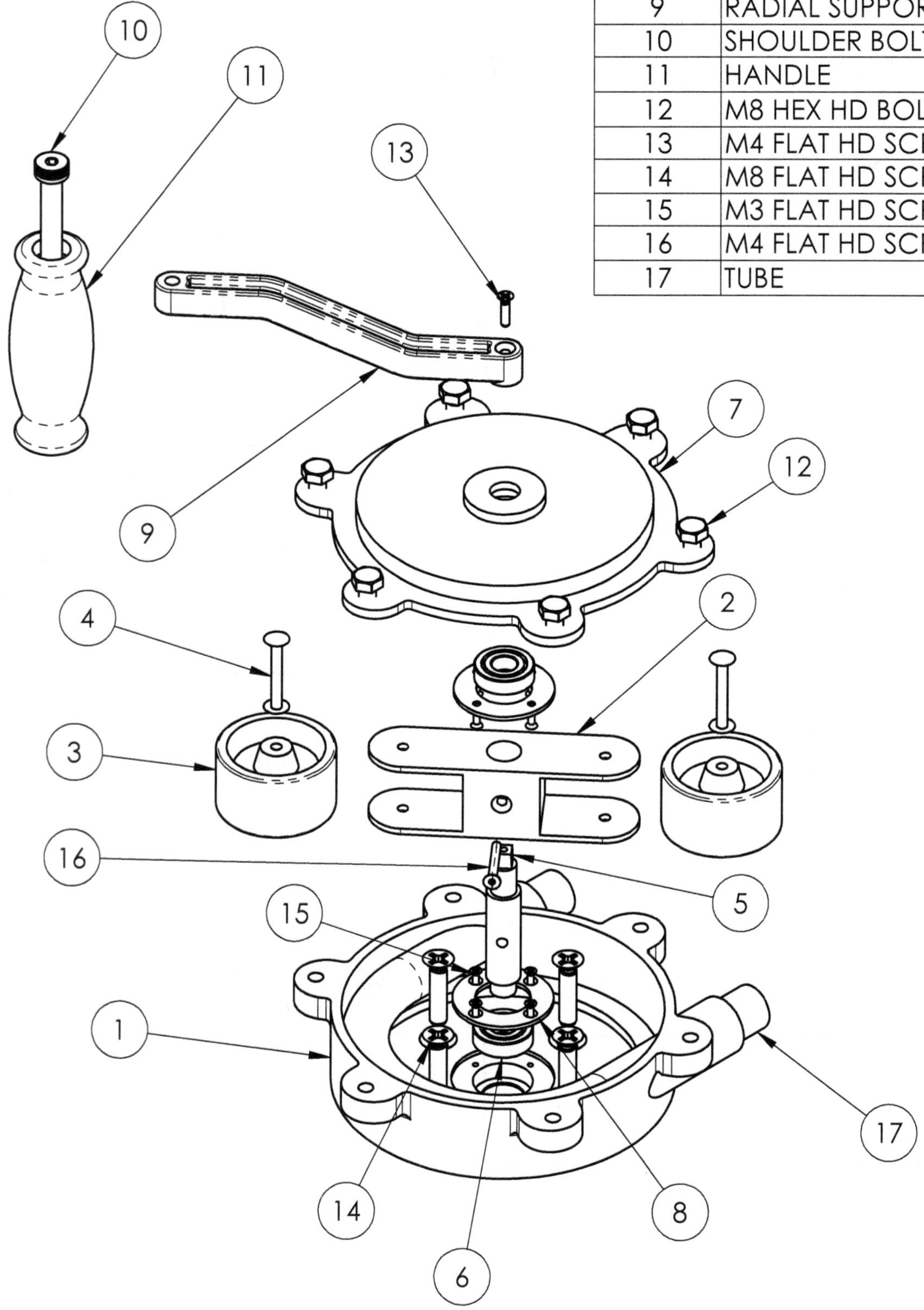

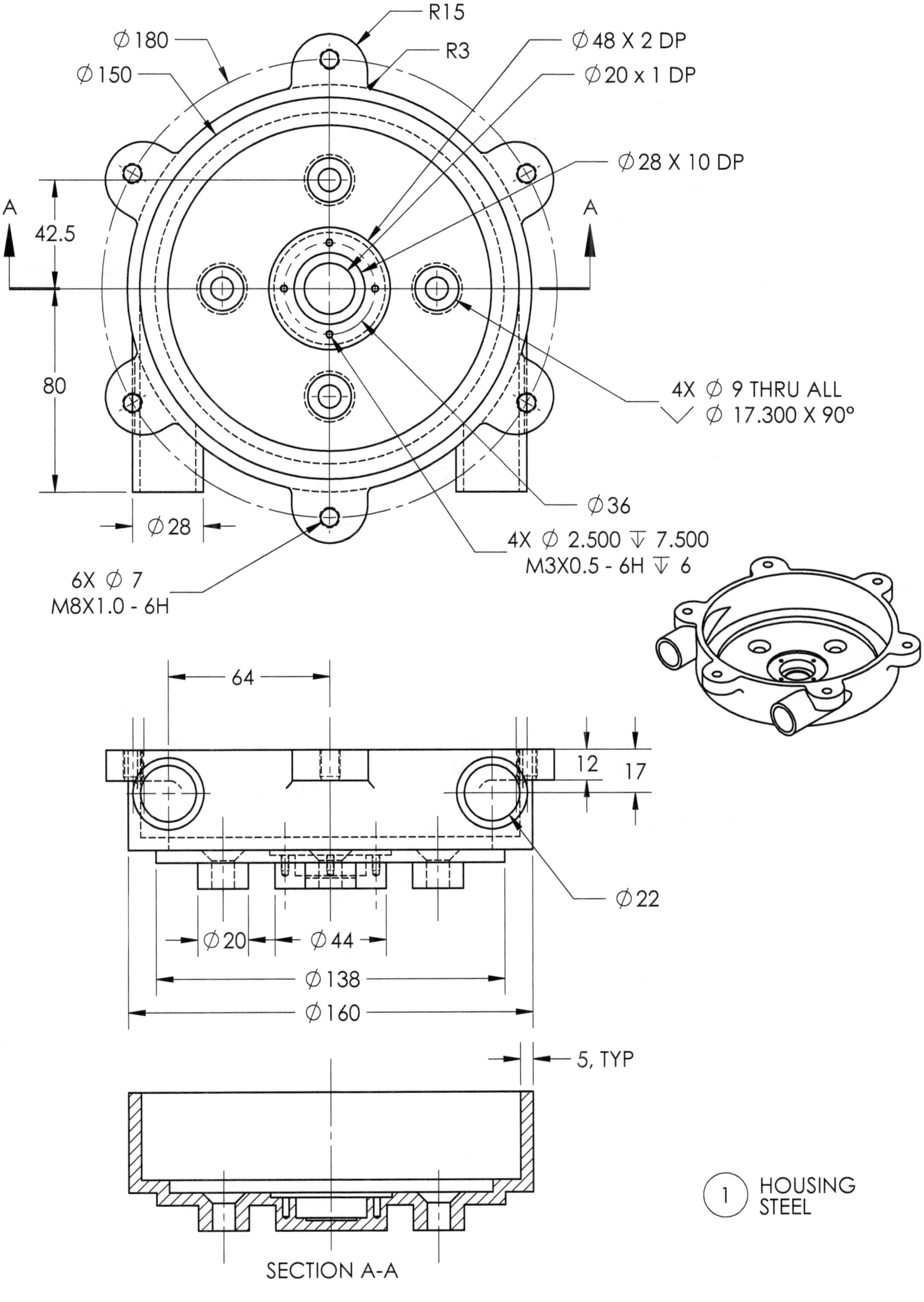
R15
Ø180
R3
Ø150
Ø48 X 2 DP
Ø20 x 1 DP
Ø28 X 10 DP
A
A
42.5
80
4X Ø 9 THRU ALL
⌵ Ø 17.300 X 90°
Ø36
Ø28
4X Ø 2.500 ↧ 7.500
M3X0.5 - 6H ↧ 6
6X Ø 7
M8X1.0 - 6H
64
12
17
Ø22
Ø20
Ø44
Ø138
Ø160
5, TYP
SECTION A-A
1
HOUSING
STEEL

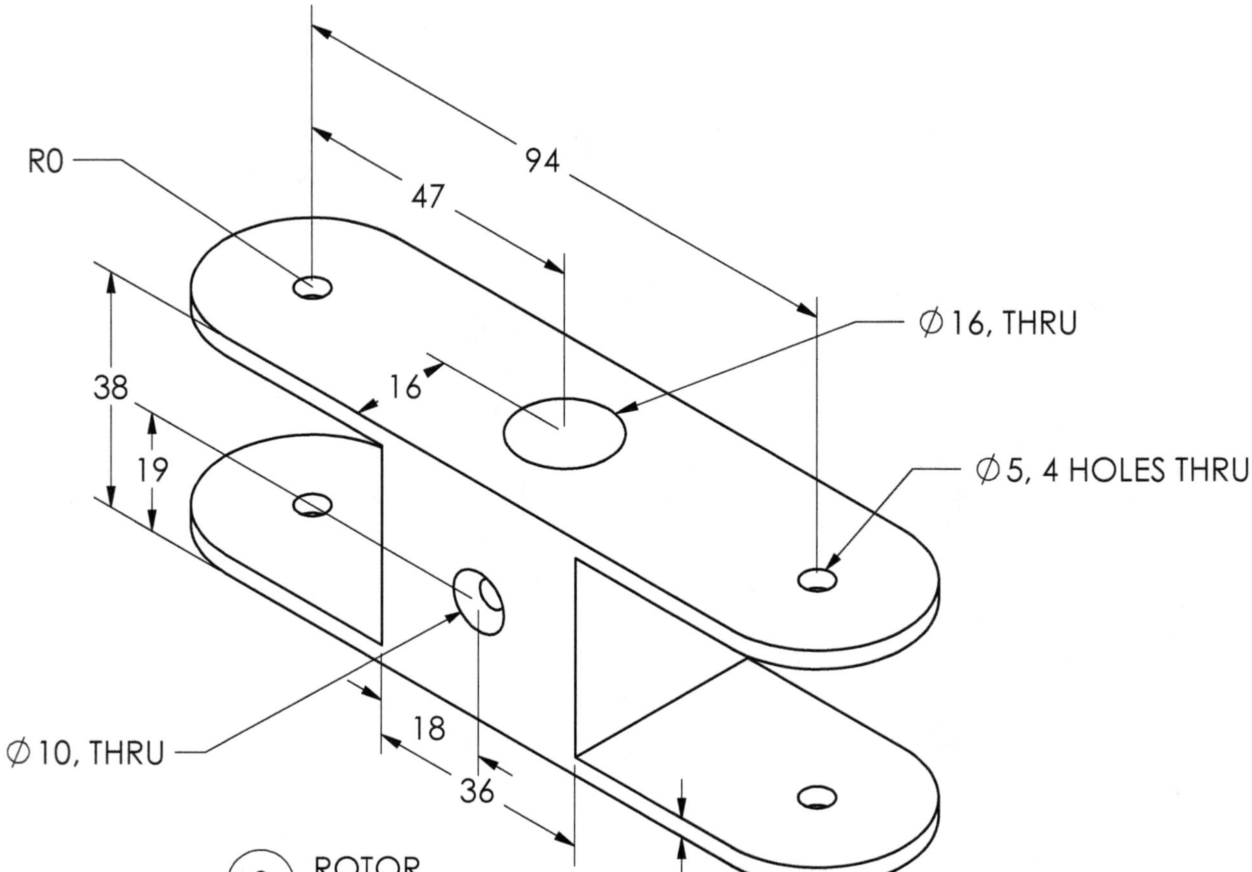

2 ROTOR
STEEL

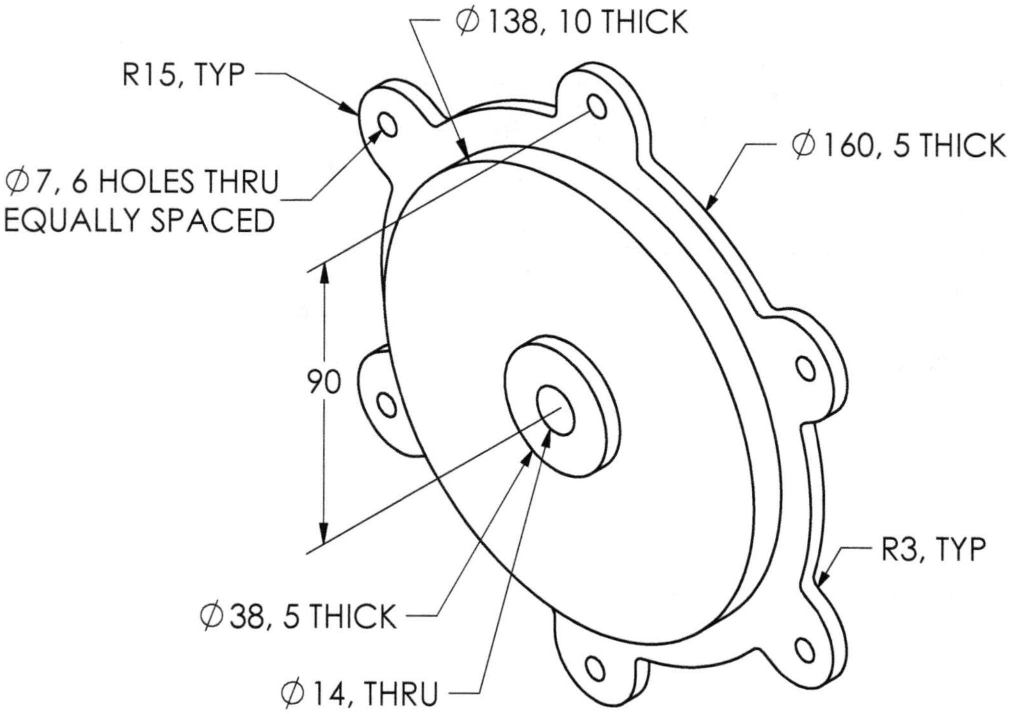

7 COVER
CAST IRON

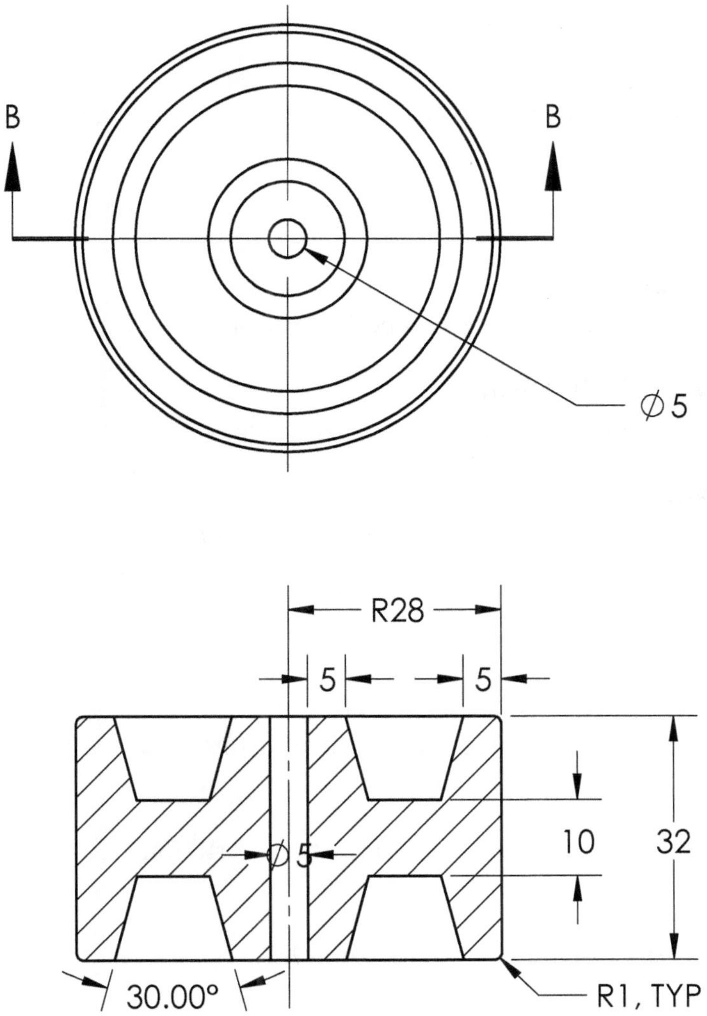

SECTION B-B

(3) ROLLER
HARD RUBBER
2 REQ'D

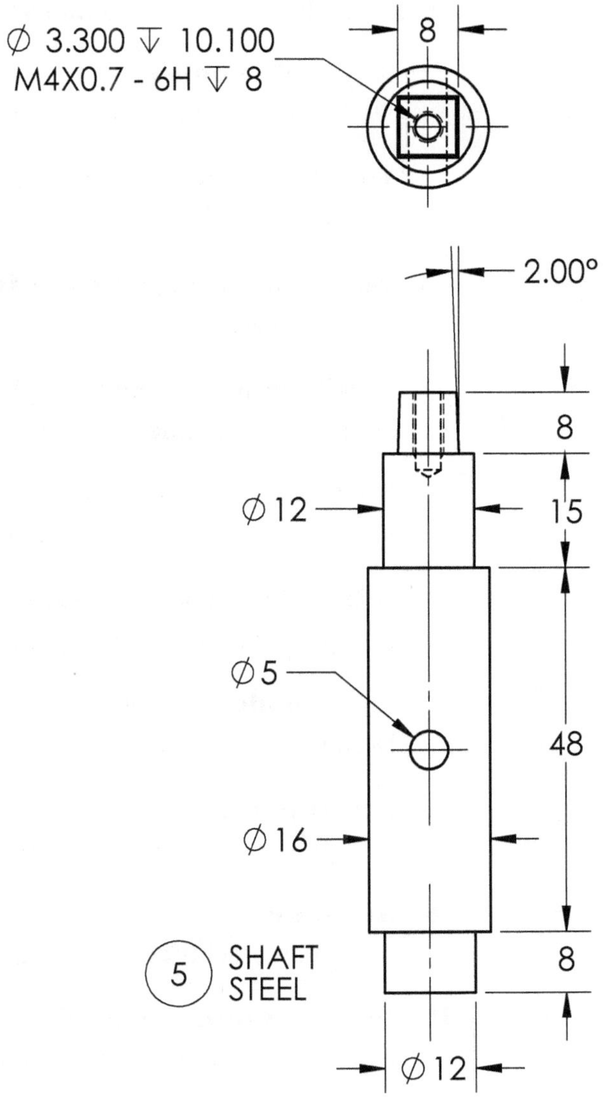

(5) SHAFT
STEEL

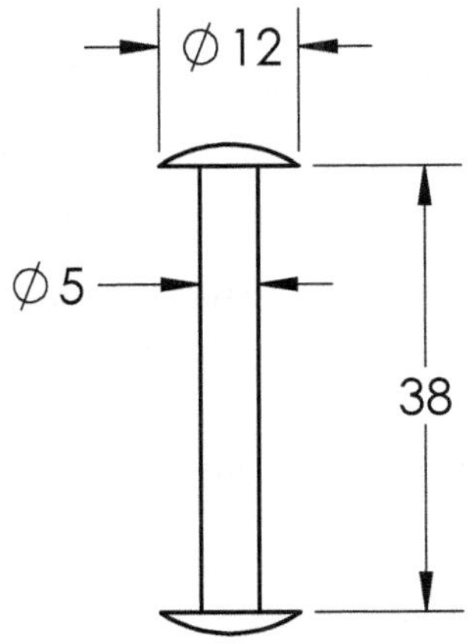

(4) AXLE
STEEL
2 REQ'D

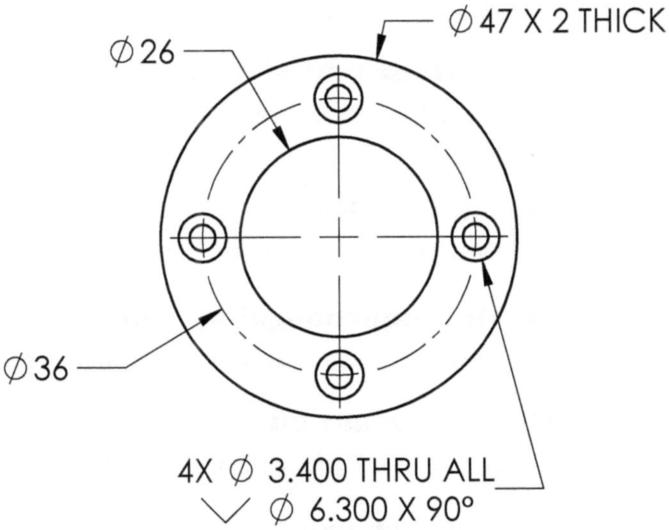

(8) RETAINING PLATE
STEEL
2 REQ'D

Design Problems

The following list provides suggestions and ideas for developing design projects and working drawings.

1. **Adjustable hole punch**
   Design a hole punch that can be adjusted for different paper sizes.

2. **Foot holder for sit-ups/crunches**
   The device must adapt to different types of crunches and body positions.

3. **Book holder/page turner for use in bathtub**
   Design an adjustable device to avoid getting the book or magazine wet.

4. **Fertilizer attachment for lawn mowers**
   Design an attachment for a lawn mower that will allow fertilizer distribution while mowing.

5. **Portable cement mixer**
   Design a portable cement mixer that can be operated manually.

6. **Adjustable tv wall mount**
   Design a wall mount to support flat TV sets allowing maximum adjustments and rotations.

7. **Fold-up picnic table and chairs set**
   Design a picnic table and chairs set that can be fold up for easy transportation.

8. **Porch swing**
   Design a porch swing that can be easily assembled and disassembled.

9. **Artist's easel**
   Design an easel that can be adjusted to different heights and it is easy to transport.

10. **Outdoors ping-pong table**
    Design a ping-pong table with an attachment for storing ping pong balls and paddles.

11. **Playground slide**
    Design a children's slide set for an outdoors playground.

12. **Multiuse vehicle ramp**
    Design a versatile ramp for the loading of recreational vehicles (motorcycles, ATVs, etc) into trucks.

13. **Outdoor shower**
    Design an adaptable shower head with a mounting system for its use on the outside.

14. **Multi-function pocket knife with silverware blades**
    Design a pocket knife set with blades that can be easily assembled and disassembled.

15. **Toy catapult kit**
    Design a fully functional medieval style toy catapult or trebuchet.

16. **Camera tripod**
    Design an adjustable and expandable tripod for a photographic camera.

17. **Paddle boat**
Design a recreational human-powered boat for two people for use in areas with mild currents.

18. **Recliner seat with leg rest mechanism**
Design an individual seat with a reclining back and an elevating leg rest support.

19. **Folding ladder**
Design an a-frame ladder with extensible legs for use on uneven ground.

20. **Children's tricycle**
Design a traditional tricycle for pre-school age kids.

21. **Carpenter's sawhorse**
Design an adjustable sawhorse that can be fold up for easy storage.

22. **Manual meat grinder**
Design a hand-powered meat grinder that can be disassembled easily for cleaning.

23. **Garden hose reel cart**
Design a garden hose holder with a storage tray to hold small watering accessories.

24. **Steering wheel lock**
Design a device to prevent rotation of the steering wheel of a vehicle to protect it from theft.

25. **Wine bottle corker**
Design a floor corking machine with a locking support to secure the wine bottle.

26. **Hand caulking gun**
Design a caulking gun to apply one-component sealants (such as silicone) sold in cartridges.

27. **Ceramic tile cutter**
Design a manual device to cut tiles to a required size or shape.

28. **Airport luggage cart**
Design an adjustable and stackable luggage cart to be used in airports and train stations.

29. **Ceiling suspended basketball system**
Design a folding frame for a ceiling suspended basketball system.

30. **Fishing reel**
Design a functional fishing reel to be used in conjunction with a fishing rod.

31. **Hand-help packaging tape dispenser**
Design an industrial packing tape dispenser that can be easily loaded and operated.

32. **Swing style dog door**
Design a dog door and mounting frame that can be installed on a regular door.

33. **Unicycle**
Design a one-wheel vehicle with an adjustable seat.

34. **Baby high chair**
Design a classic high chair with a detachable tray.

35. **"Spin and win" wheel**
Design a customizable wheel to play "spin and win" type of games.

Computer-Aided Design

11 CHAPTER

1. Introduction

Computer-aided design (CAD) or Computer-aided design and drafting (CADD) is the process of using computer applications to create construction and manufacturing drawings. CAD has revolutionized the engineering profession by replacing manual drafting techniques, therefore accelerating production and increasing the quality of technical drawings. Modern CAD packages are used throughout the entire design process for generating drawings, storing and tracking changes, and performing analysis.

The capabilities of CAD applications have increased gradually with the power of modern computers. In the past, generic CAD packages were used across all industries, regardless of whether the specific program fit the project, because discipline specific software did not exist. Now, CAD software has evolved into different, more powerful programs designed for specific applications. Road layout and construction, manufacturing process control, electrical systems, and automotive design are some examples of custom CAD packages targeted to specific industries that share a common set of tools and features. Most of these packages provide three dimensional tools and include powerful rendering engines to create realistic images of the designs.

In this chapter, the general concepts and tools used across different CAD applications are discussed. Common techniques in two dimensional drafting and drawing organization are explained.

2. Basic Concepts and General CAD Techniques

In its basic core, a CAD package is an electronic equivalent of a sheet of paper and a set of drawing tools. CAD offers many advantages over manual drafting. For example, straight lines and geometric figures such as circles or arcs can be created easier and more accurately. Many repetitive and time-consuming tasks in drafting such as hatching, dimensioning, or lettering, are automatic or semi-automatic in CAD. The true power of CAD, however, lies in its editing capabilities. Significant changes or mistakes that often required a traditional drawing to be manually redone from the beginning can be corrected quickly in an electronic drawing. Updated drawings can be shared without difficulty among team members working simultaneously on a project. CAD systems also provide tools that resemble drafting instruments. Traditional drawing aids such as protractors, T-Squares, and compasses have their electronic version on most CAD programs.

When working with CAD packages, there is a series of general techniques or practices that should always be followed in order to obtain optimal results and performance. First, drawings should always be created full scale (1:1) and only scaled when ready for printing. This practice makes drawings easier to read, analyze, and maintain. Second, the concept of repeatability should always be applied for efficiency. The concept of repeatability states that every object should only be created once. After creation, objects can be copied, mirrored, or patterned, and many new objects can be created by using existing items. Furthermore, in most CAD packages, specific objects or parts of a drawing can be exported and shared across different projects. In fact, most design firms often keep libraries of commonly used CAD components so they can be reused in multiple drawings. Third, measurements must be precise. There is no guessing or eye-ball dimensioning in technical drafting. CAD packages provide visual aids that guarantee accuracy when used properly. In fact, many commands will not perform properly if the program does not understand an object's geometry caused by inaccurate measurement. For this reason, CAD users must learn to use these tools in their specific package. Finally, CAD drawings must be well organized. Layers (which will be discussed later) must be used properly and efficiently and good naming conventions must be used to facilitate the maintenance of the drawings.

3. Drawing Settings

When using a CAD package, the largest portion of the computer screen is reserved for the drawing area. It is in this area where all geometric forms are created. The drawing area in all CAD systems is based on a two-dimensional Cartesian coordinate system. This coordinate system can be understood as an X-axis, running horizontally, and a Y-axis, running vertically. The intersection of these two axes (the origin 0,0) is typically located at the bottom left corner of the drawing area. Positive X values progress in the right direction, and positive Y values progress upwards. To locate any point in this 2D space, both its X and Y coordinates must be given, separated by a comma. See Figure 11.1 for an example. In some CAD packages, elevation can also be applied (in the Z direction), but with the advent of parametric modeling software, this feature is commonly unused. Also, it is possible to use the Polar coordinate system when creating objects that refer to previously selected points.

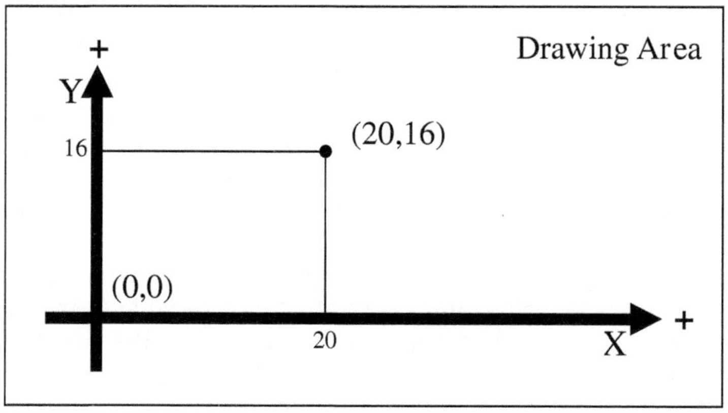

Figure 11.1 Cartesian coordinate system

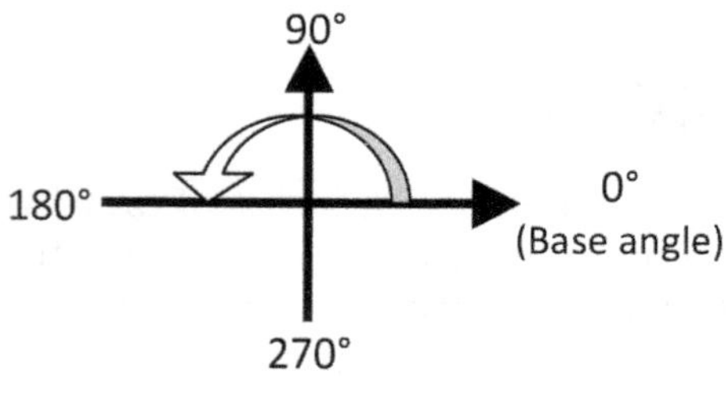

Figure 11.2 Angle measurement

Although angles are normally measured in a counterclockwise direction, from 0 degrees right of the positive X-axis (see Figure 11.2), both the direction and the base angle can be changed to any other settings in most CAD packages.

The drawing area extends infinitely in all directions. To navigate within this theoretically infinite amount of 2D space, CAD packages provide display tools that allow the user to zoom in and out of the drawing, and pan to translate the point of view in the X and Y directions. There are several settings that must be checked and established in a CAD system before a drawing is created. It is important to realize that the object remains in the same location, only the position of the observer changes. Also, pan and zoom capabilities do not change the actual size of the drawing.

Workspace

Modern CAD applications are fully customizable. Users can define and arrange toolbars, menus, palettes, colors, shortcuts, and what and where graphical information will be displayed on the screen. A useful CAD workspace should have adequate room on the screen to draw, without having to relocate palettes and toolbars.

Although workspace customization can be changed any time, it typically occurs after the installation of the CAD package, when the application is first used. Users tend not to change an application interface once they are familiar with it. Most CAD systems today allow multiple customizations, for cases when the computer is shared by multiple users, or if a single user decides to use a different interface arrangement for different projects. Predefined workspaces also exist for using CAD on multiple screens.

Many CAD systems provide multiple ways to input commands. Commands can be entered by typing in the command name (or abbreviation) in a text field, through pull down menus, by toolbars or palettes, or by user programmed digitizers.

Units

It is important to set the units of the drawing beforehand to avoid errors and scaling problems. Both English and metric systems are supported by virtually all CAD packages. Typically, linear units (metric and imperial), angular units (degrees and radians) and precision (number of decimal places displayed) need to be established.

Layers

Many technical drawings are complex and contain a large amount of objects, which often makes them difficult to understand and the user is prone to errors when making modifications. Traditionally, technical drawings created manually were drafted on a set of clear acetate sheets to separate different parts of the drawings in multiple layers. These sheets were then stacked on top of each other to form the complete design. For example, the drawing of a house can be separated into the following layers: Floor plan, electrical system, plumbing, and dimensions. In this way, a person can view the drawing using only the elements or the combination of elements that is needed.

Layering is a method used by most modern CAD systems that simulates the stacking of classic acetate sheets. Layers provide a way to distribute graphical entities of a drawing onto different levels. The majority of the layers in a drawing are created before any drafting is initiated. In many cases, design firms use templates with well-defined sets of layers, so no time is wasted setting up each drawing from scratch.

The functionality of layers in CAD packages extends beyond the traditional organization of drawings. Most CAD systems allow the user to set up multiple parameters that affect all the entities contained in a particular layer. Common parameters include visibility, which determines if the objects drawn on a layer are visible or hidden; the color, line type, and thickness of the entities drawn on the layer; and parameters that determine if a layer will print or if the objects drawn on that layer are selectable. The example shown in Figure 11.3 illustrates the parameters of a set of layers for a drawing of a house.

	PARAMETERS					
	Visible	Printable	Selectable	Color	Line type	Line weight
LAYERS						
Floor Plan	Yes	Yes	No	Black	_____	0.4 mm
Electrical	Yes	Yes	Yes	Red	_____	0.2 mm
Plumbing	Yes	No	No	Green	_____	0.2 mm
Dimensions	Yes	No	No	Blue	_____	0.1 mm

Figure 11.3 Layer definition example for a house drawing

For mechanical drawings it is helpful to define layers with different line types. See Figure 11.4.

	PARAMETERS					
	Visible	Printable	Selectable	Color	Line type	Line weight
LAYERS						
Visible	Yes	Yes	Yes	Black	_____	0.4 mm
Hidden	Yes	Yes	Yes	Red	- - - -	0.2 mm
Center	Yes	Yes	Yes	Green	— - —	0.2 mm
Section	Yes	Yes	Yes	Yellow	— — — —	0.4 mm
Dimensions	Yes	Yes	No	Blue	_____	0.1 mm

Figure 11.4 Layer definition example for a mechanical drawing

Grid

The grid is a common visual aid found in most CAD packages to help users in their drawing efforts. It is simply a set of points displayed in the drawing area used as a visual reference to assist the user in locating precise coordinates with ease. It can be turned on and off at any time during the drawing process and is helpful when the user is in need of quick visual locations in the drawing. Most CAD packages allow some level of customization of the grid such as grid size, distance between points, and type (orthogonal, isometric, etc). See Figure 11.5 for examples of grids.

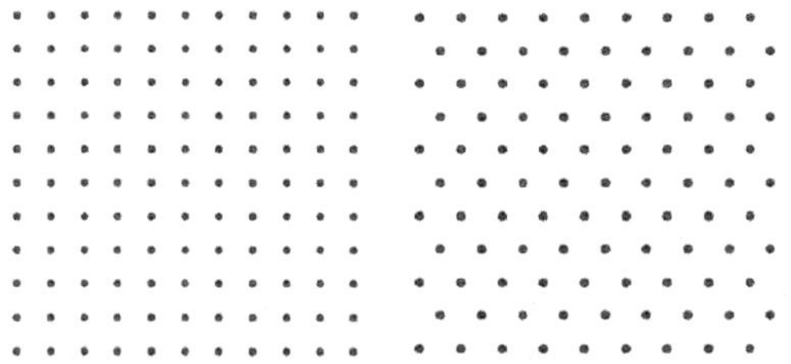

Figure 11.5 Orthogonal and isometric grids

Snap

Snapping tools give CAD the capability of easily drawing objects at specific points. Snaps are so important, that most users agree that it would be difficult and time consuming to draw accurately without them. Snapping tools are used in conjunction with drawing tools to force the user to select points located at specific coordinate locations without having to calculate distances or enter them manually. Snapping tools are normally enabled when needed, at any time during the drawing process, and disabled in situations where they hinder production. Learning to use the snapping tools wisely is required in order to create precise drawings quickly and efficiently.

Different snapping tools are found in different packages. For this reason, users should develop a good understanding of what specific tools are provided by the CAD package of choice and how they work.

Generally, a grid snapping tool restricts the movement of the cursor in the drawing area so that only points located at specific X and Y increments can be selected. This snapping tool is normally used in combination with the grid explained in the previous section, although the grid and snapping tools can be used independently.

Feature-sensitive snapping tools are a very powerful feature in most CAD packages. These tools allow the user to locate points precisely on existing objects. For example, it is possible to select the midpoint of a line (see Figure 11.6), the center of a circle, or the intersection point between two lines, among others. Basic feature-sensitive snaps include endpoints of lines, midpoints, intersections, center points of circles, tangent points, and quadrants.

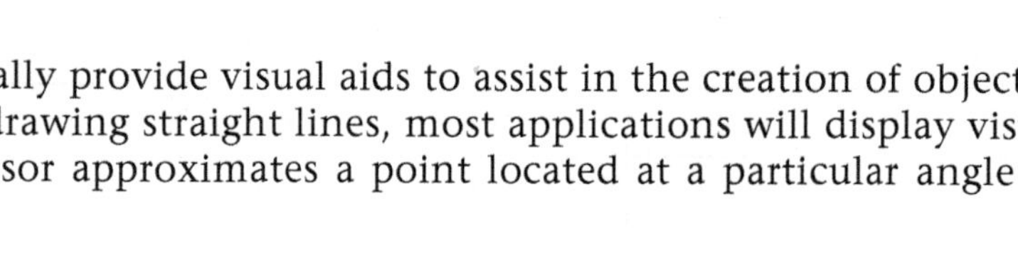

Figure 11.6 Snapping to the midpoint of a line

Angular Modes

CAD packages normally provide visual aids to assist in the creation of objects at common angles. For example, when drawing straight lines, most applications will display visual indicators on the screen when the cursor approximates a point located at a particular angle from the previously selected point (see Figure 11.7). Common angles include 0, 30, 45, 60, and 90 degrees but most packages allow the user to customize the angular indicator. In some cases, the angular mode can be a forced feature, locking the cursor at specific angles, so that users can only draw lines at the configured angles. This feature is especially useful when creating isometric drawings.

The 90 degree forced angular mode, also called orthogonal mode, is a common setting useful when drawing orthographic views and architectural floor plans. This mode ensures that all the lines are drawn horizontally or vertically.

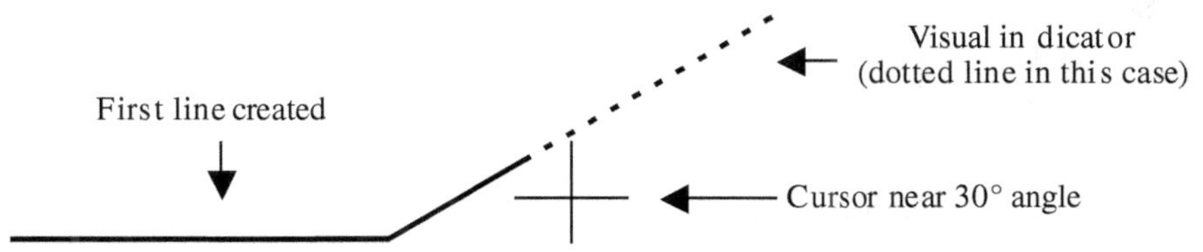

Figure 11.7 Visual indicator for angular mode found in a common CAD package

4. Drawing Tools

All drawings can be broken down into a series of basic geometric shapes and all CAD systems provide tools to draw basic geometric entities such as lines, circles, arcs, and polygons. All information needed to create different geometry can be entered visually on the screen or numerically using the keyboard. CAD packages provide multiple methods to define the same geometric objects.

Lines

Lines (really, line segments) are the most fundamental objects in CAD. Two endpoints are required to create a line. Construction lines or rays can also be drawn and are useful for guide lines. Other properties such as line type (solid, dashed, etc), color, and thickness can also be defined, but are usually differentiated by layer properties.

Circles

Another basic geometric object is the circle. Circles are very common elements in drafting when creating holes and shafts. Even with an object as simple as a circle, CAD packages generally provide multiple methods for its creation. The simplest method requires a center point and the circle's radius or diameter. A circle can also be created by defining three points on the circumference or using tangent information. See Figure 11.8.

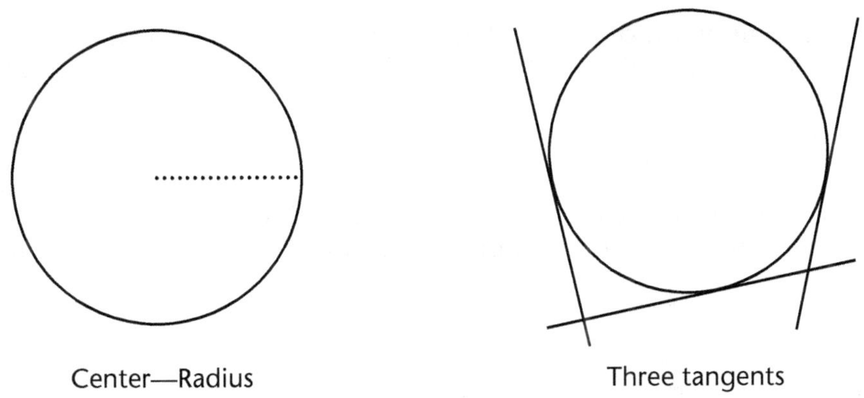

Figure 11.8 Common methods to define a circle

Arcs and Curves

An arc is defined as a segment on the circumference of a circle. The most common method used to create an arc is by defining its starting point, a second point on the arc, and its endpoint (see Figure 11.9). Other methods include start point-end point-radius and start point-end point-angle.

Figure 11.9 Basic creation of an arc

Exceptionally complex curves can be created with the arc tool. However, numerous CAD users find arc tools very difficult to control and master. For this reason, many users prefer to create circles and trim them accordingly to accomplish the same results. Basic trimming tools will be explained in the following section. Semicircular arcs can be easily created by using a fillet command and defining the arc radius to be half the distance.

Polygons

Although polygons can be created by connecting multiple lines, CAD packages normally provide tools to draw regular polygons by indicating the number of sides and the endpoint of an edge. Other common methods include defining a circle about which the polygon will be circumscribed, or in which the polygon will be inscribed. See Figure 11.10 for an example.

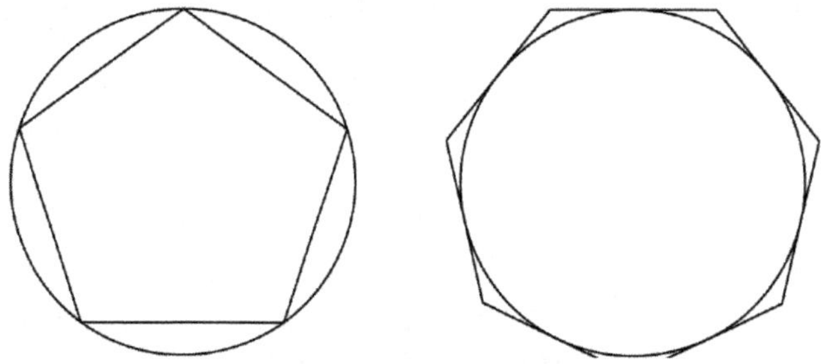

Figure 11.10 Inscribed and circumscribed polygons

Text

Text in drafting is used to add notes and enter information in title blocks. Text tools in CAD are typically provided in the form of a text editor, where settings such as font, text size, justification, and distance between letters can be modified. Oftentimes, text can be copied and pasted from word processing software.

Hatching

Hatching is a method to fill in a closed (no gaps) and well-defined area with a pattern. It is a common practice in sectional views of parts or as a shading method. Hatch patterns are groups of parallel lines in one or more different directions to create darker tones. Standard patterns for common materials are typically predefined in most CAD applications. These hatch patterns are generally grouped as a single object. In order for them to be edited, they have to be first broken up into their constituent components.

Common parameters associated with hatch patterns include scale (to determine the distance between the lines in the pattern) and rotation angle (to set up the orientation). See Figure 11.11 for examples of hatch patterns.

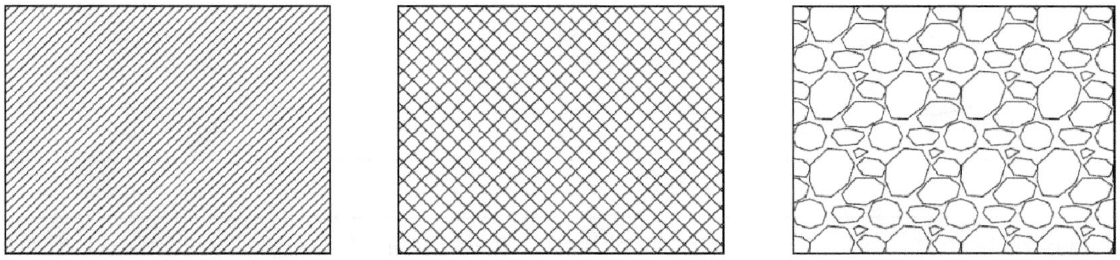

Figure 11.11 Examples of hatch patterns

5. Editing Tools

The true power of CAD lies in its editing capabilities. It is inefficient to draw every object of a drawing from scratch. Likewise, there is no need to erase and rebuild an object every time it needs modification. Instead, CAD packages provide an extensive set of editing tools that allow users to quickly modify existing objects and automate some of the most common processes.

Basic transformation tools such as move (translating objects to a different X-Y location), rotate (rotating objects a number of degrees around a pivot point), scale (resizing objects based on a scaling factor), erase (deleting objects from a drawing), and copy (making duplicates of objects) are frequently used.

Many CAD packages provide multiple methods to select objects for editing. Selection is primarily accomplished by clicking on each object individually. If groups of objects need to be selected, a window can be used to create the selection set by clicking in space and drawing a box around all of the desired objects. If all objects desired for selection are on a common layer, other layers can be temporarily turned off to facilitate easier selection.

Other common editing tools are listed below.

Offset	Description: It creates a copy of the selected object(s) at a defined distance.	Parameters: Objects to offset, offset distance, offset direction

Trim	Description: It deletes a portion of an object where it intersects with another object (trimming object).	Parameters: trimming object, objects to trim

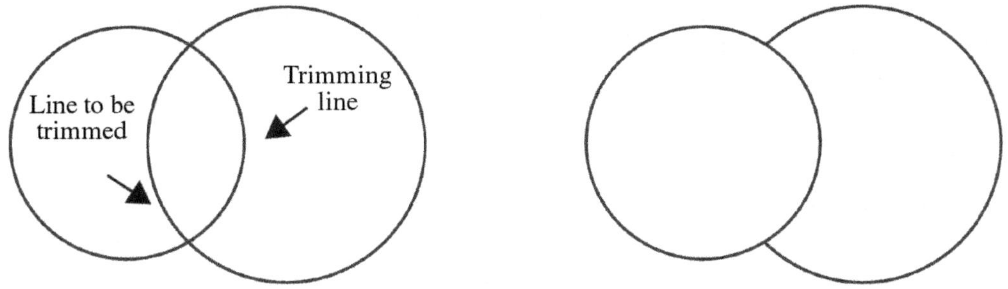

Extend	Description: It shortens or lengthens objects to meet the edges of another object (boundary).	Parameters: boundary, objects to extend

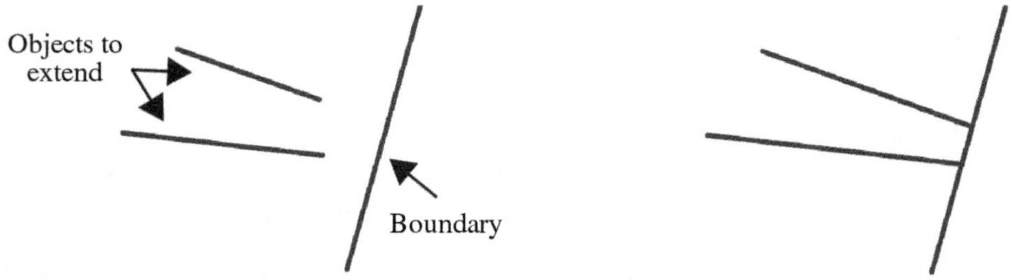

Fillet	Description: It rounds a corner.	Parameters: Edges to fillet, fillet radius
Chamfer	Description: It creates a beveled edge at a corner.	Parameters: Edges to chamfer, angle, chamfer dimension

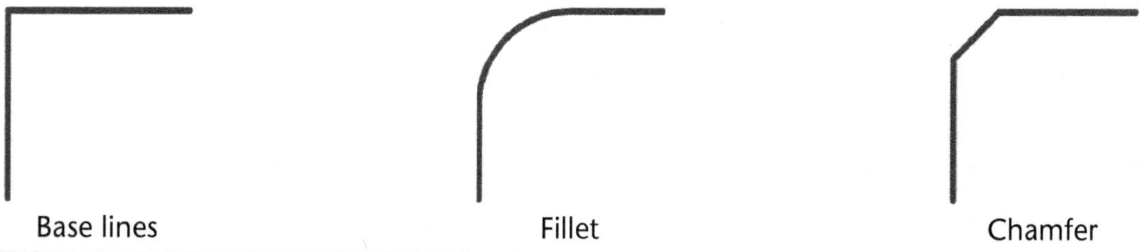

Mirror	Description: It creates the symmetrical pattern of the given objects.
	Parameters: Objects to mirror, and mirror line

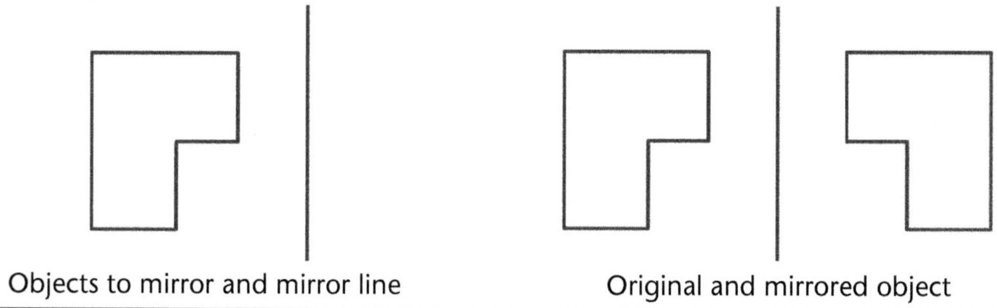

Pattern	Description: It creates copies of objects along one or two directions, or around a reference circle.
	Parameters: Objects to copy, number of copies, copies to skip. For linear patterns: distance between copies, and directions. For circular patterns: reference circle, degrees to cover

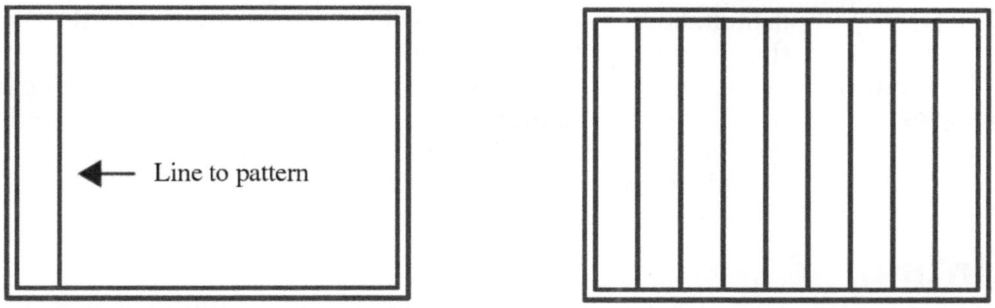

Settings: 8 instances along X axis

Linear pattern

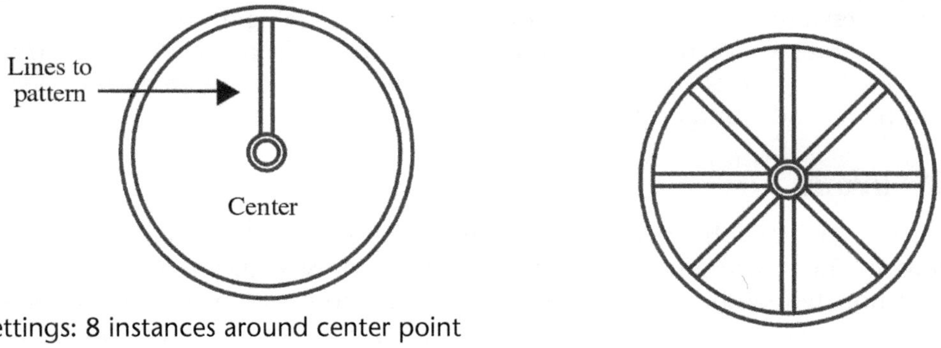

Settings: 8 instances around center point

Circular pattern

6. Dimensioning

From a drafter's standpoint, dimensioning is a repetitive and time consuming process. Traditionally, all lines of every dimension were drafted by hand, even to the smallest details such as gaps and text height. The dimensioning process in CAD has been reduced to the point of being semi-automatic. CAD users are still responsible for following proper dimensioning standards and practices, but the creation of dimension lines, leaders, and extension lines has been remarkably simplified.

In general, CAD packages provide a set of tools designed to assist in dimensioning objects in a drawing. These tools include linear dimension, baseline dimension, or leader dimension (to dimension arcs and circles). There are tools to automatically space stacked dimensions as well as tools that create jog lines in linear dimensions when a feature is too long to fit on a specific paper size.

Dimensioning a specific object requires the user to select the feature, the type of dimension to be applied, and the position of the dimension. All extension lines, leaders, elbows, and arrowheads are automatically created by the CAD application based on a set of customizable parameters. See Figure 11.12 for an example.

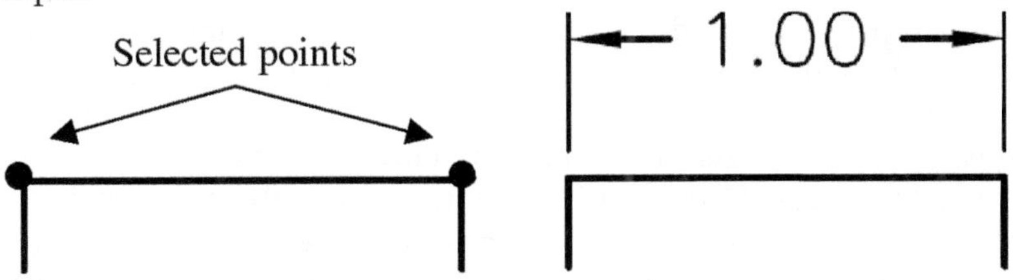

Object to dimension

Automatic creation of lines and measurement

Figure 11.12 Dimensioning using a CAD application

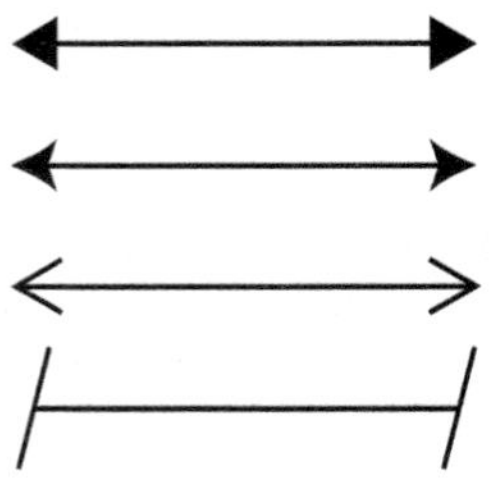

Figure 11.13 Arrowhead examples

Typically, CAD packages allow users to customize every part of a dimension. The dimension text size, the precision, the thickness of the dimension lines, the gap between the extension lines and the drawing, the gap between the dimension value and the dimension lines, or the type of arrowhead and its size (see Figure 11.13 for examples) are examples of common customizable parameters. Standard styles such as ANSI and ISO are usually defined by default. Some CAD packages will automatically update the dimension should the feature's distance be edited. Toleranced dimensions can be created for every dimension in a drawing through customization of the dimension style (though not recommended), or can be individually created by editing the specific dimension that requires a linear tolerance. Many CAD packages also provide for geometric tolerancing.

7. Printing

One of the essential techniques in CAD is that drawings should always be created full scale. This practice is reasonable due to the theoretically infinite amount of space provided as drawing area. However, when drawings are be prepared for printing, certain procedures must be followed in order to plot them successfully in the limited amount of space available on the sheet. CAD packages normally offer a separate set of tools and parameters specifically designed to manage the printing process. There are two general steps involved in printing a drawing: defining the sheet layout on the desired size of paper and configuring the printer settings.

7.1 Sheet Layout

Defining a layout includes the creation of a border and a title block, and the incorporation of the different views of the drawing (if more than one), properly scaled. Borders and title blocks are elements that do not frequently ever change from drawing to drawing. In general, design firms have title blocks already defined in templates, so they are normally imported to new drawings rather than created from scratch each time.

The creation of drawing views is a simple step in most CAD packages. It requires selecting the finished drawing from the drawing area and setting up a scale, so it fits within the limits of the page. Although users have the freedom to apply any scale to their drawings, standard engineering, metric, or architectural scales should be always be employed.

It is possible to define multiple sheet layouts for the same drawing. In fact, this practice is common in industry when multiple copies of a drawing need to be printed on different paper sizes at different scales, or when specific details of the drawing need to be emphasized and printed on a separate page.

7.2 Printer Settings

Printer settings are parameters required to interface directly with the printer. They are similar to setting up a page in a word processing program. CAD packages, however, require the user to be much more specific about how the drawing is going to be printed.

First, a printer must be selected and the paper size determined. The paper size setting is directly related to the printers installed on the system. A large plotter capable of printing E-size paper (34″ × 44″) will have many more paper size options available that a regular A-size printer (8.5″ × 11″).

Second, color assignments need to be set up. Color assignments determine how each color used in the drawing will appear on the paper. Although not a common practice, most CAD packages allow the user to define different color tables for the same drawing when printing in color. For example, it is possible to make red, orange, and yellow objects in the drawing print in plain red, or make blue objects print black. Default color assignments such as grayscale or monochrome (all lines print black) are typically predefined. It is important to select monochrome settings when using a black and white printer to print a colored drawing, otherwise, some colors will appear very light, making visibility difficult.

Third, the position of the drawing on the paper must be indicated. This step includes setting up the drawing orientation (portrait or landscape), the printable area (normally the extents of the layout defined earlier), and the position of the drawing with respect to the paper (centered, top left corner, etc).

8. Advanced Topics: Blocks and Reference Files

Portions of drawings used repeatedly (symbols, company logos, etc.) or entire drawings can be inserted into a current drawing. These items are commonly referred to as blocks and have many customizable features. Blocks can be inserted at specific locations, scaled, rotated, or broken up into their individual components during the insertion process. Blocks can be nested, be external files, or can even be composed of objects previously drawn in the current drawing. Blocks are a method of making the drawing process more efficient. Blocks can even be assigned text or numeric information (part number, price, etc.) which can be extracted later into a spreadsheet for further analysis.

Previously created drawing files can be linked electronically to a larger, project file, instead of insertion as blocks. If alterations are made to the linked file, when the project file is printed, the edited version of the linked file will be shown. Reference files are useful in multiple user environments, when a team is working on different portions of the overall design. If reference files are used, it is imperative that all files be grouped in a common location, since the linked file does not exist as an independent entity in the main project file.

Practice Test

1. **The drawing area in all CAD systems is based on a ____ coordinate system.**
   A) Parabolic B) Cartesian C) Polar D) Non-Euclidean E) B and C are correct

2. **General CAD techniques include**
   A) Always draw all objects full scale
   B) Repeatability (every object is drawn only once)
   C) All measurements must be precise
   D) All of the above
   E) None of the above

3. **What is the purpose of separating the objects of a drawing in multiple layers?**
   A) To keep the drawing organized
   B) To be able to associate different settings, such as the line type, to a group of objects
   C) To be able to turn on and off the visibility of certain objects for printing purposes
   D) All of the above
   E) None of the above

4. **How many layers should a drawing have?**
   A) One
   B) One for each object
   C) As many as needed to clearly organize the objects
   D) At least three
   E) None of the above

5. **What tool allows the user to create multiple copies of objects around a center point?**
   A) Linear pattern B) Circular pattern C) Mirror D) Extend E) None of the above

6. **What tool allows the user to create a beveled edge at a corner?**
   A) Fillet B) Round C) Chamfer D) Trim E) None of the above

7. **What tool allows the user to create a copy of an object at a defined distance?**
   A) Polygon B) Offset C) Mirror D) Move E) None of the above

8. **Why is it important to use templates when working in a CAD environment?**
   A) To save time
   B) To make the drawings more consistent
   C) To be able to reuse common objects in multiple drawings
   D) All of the above
   E) None of the above

9. **Which of the following is not a drawing aid in most CAD packages?**
   A) Grid B) Snap C) Dimension D) Angular mode E) None of the above

10. **Which of the following dimensioning tools are normally included in most CAD packages?**
    A) Linear dimensions B) Linear tolerance C) Angular dimension
    D) Leader E) None of the above

Problems

Use your preferred CAD application to draw the following objects.

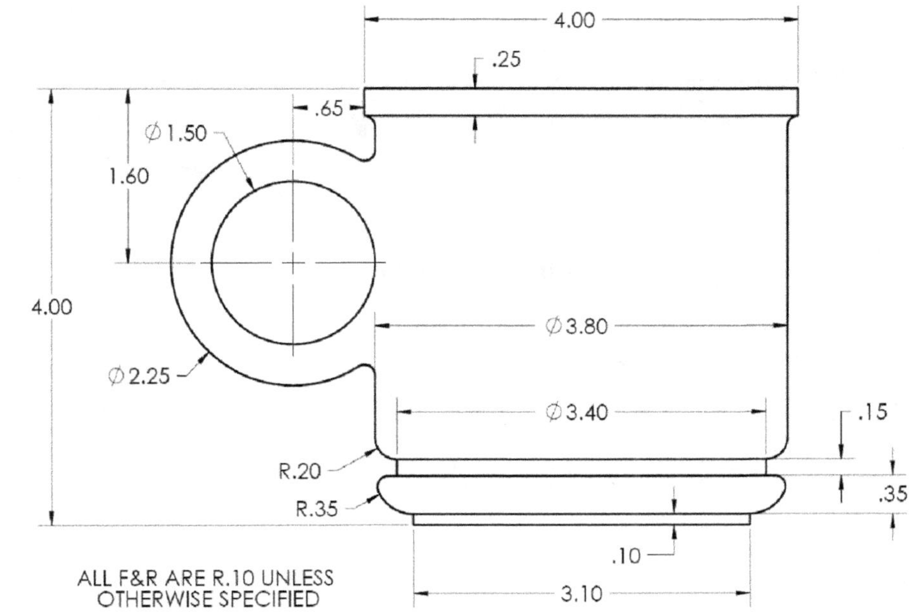

Problem 1—Coffee Mug

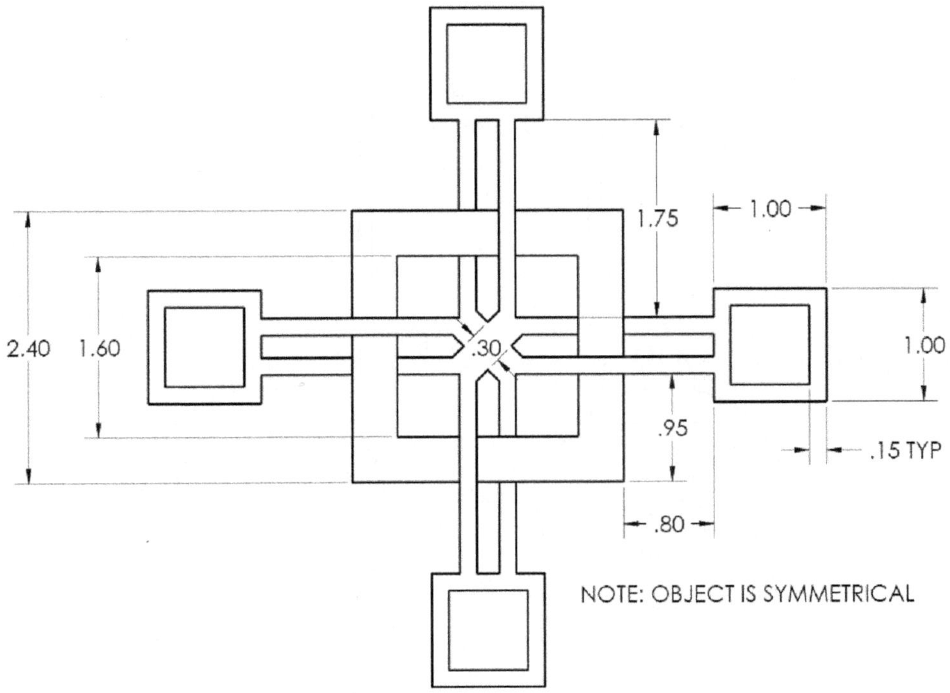

Problem 2—Engraved Detail

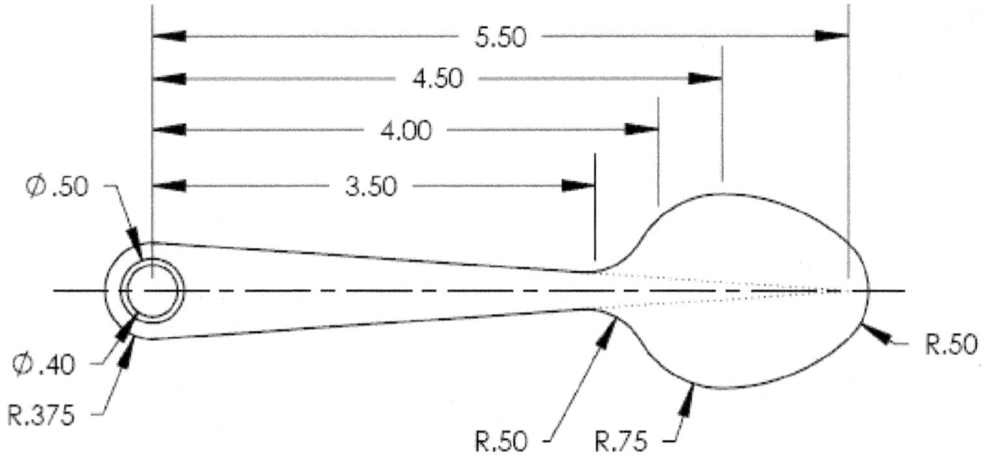

Problem 3—Spoon

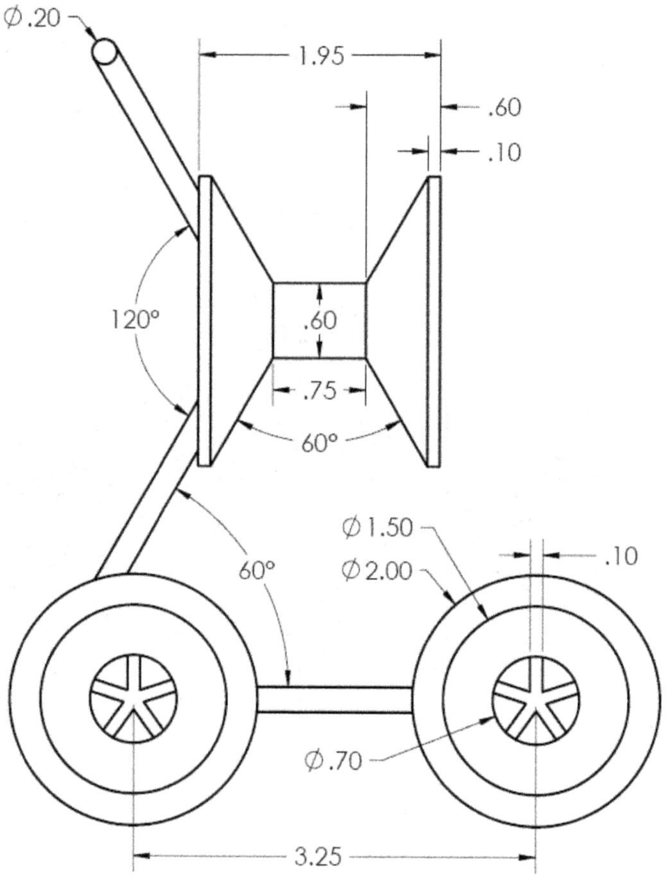

Problem 4—Reel Cart

Project: Landscape

Complete a plan layout for a multi-sport facility that will be placed at the given location. This facility will be used to host national competitions for various club sports. These sports include softball, soccer, flag football, basketball, and tennis. An image of the location, with estimated dimensions for the facility footprint is included.

This facility must include sidewalks, bleachers, appropriate fencing, parking, restrooms, and sports field layouts. Use the space efficiently. Detailed estimated cost of construction must be included. You can exclude from cost estimation any permanent buildings and foundations, but you need to consider costs of sinks, toilets, sidewalks, paving, sports equipment (basketball goals, football goal posts, soccer goals, tennis nets, etc). Although the permanent buildings are not part of your cost estimation, you must have the locations of these entities drawn on your layout. Assume all earthwork, utilities, electrical, etc. have been completed prior to your design.

Provide all major dimensions. In addition to the overall site plan, you must submit detailed layouts of each sports field/court. The overall layout must be the first sheet in your set.

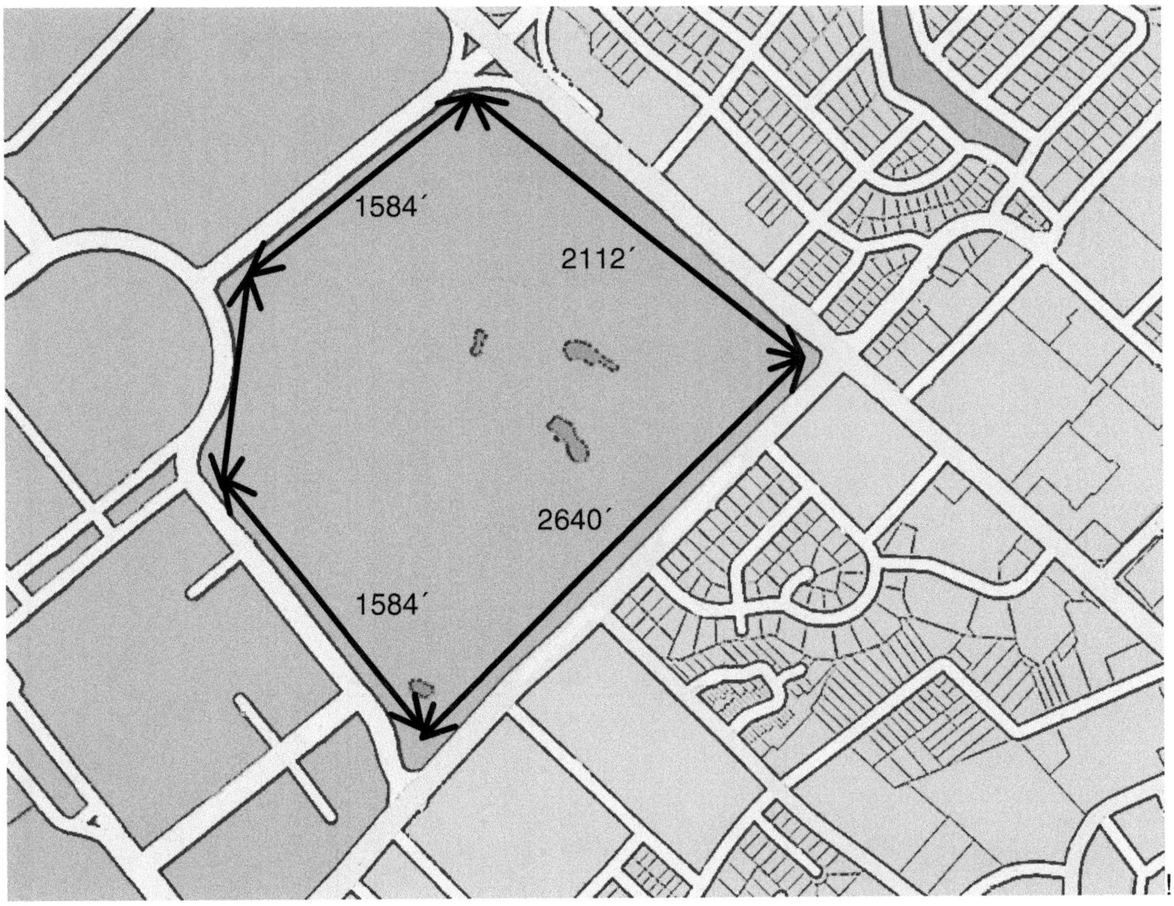

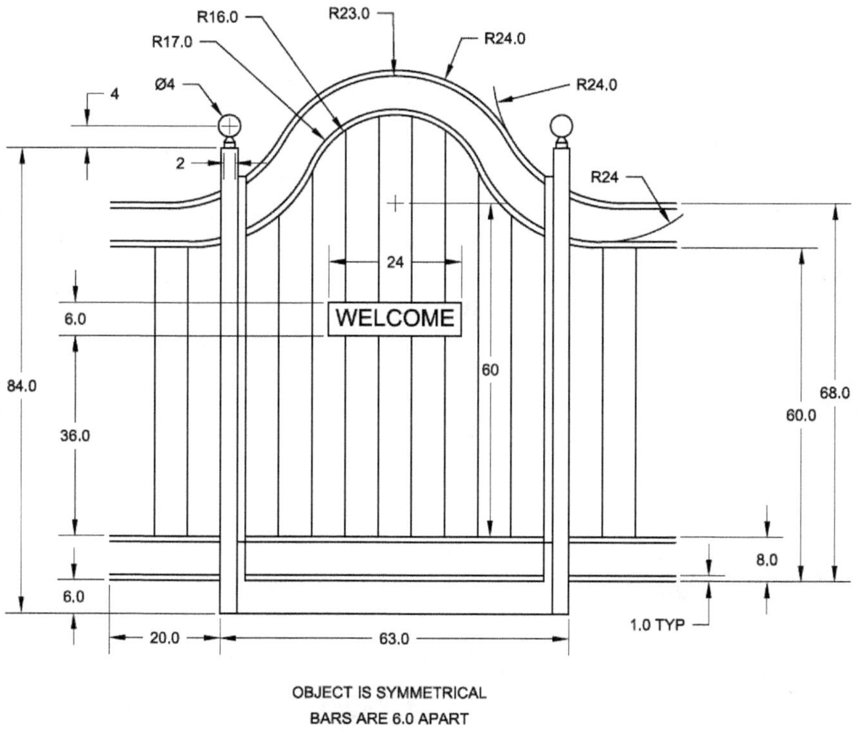

Problem 5—Gate

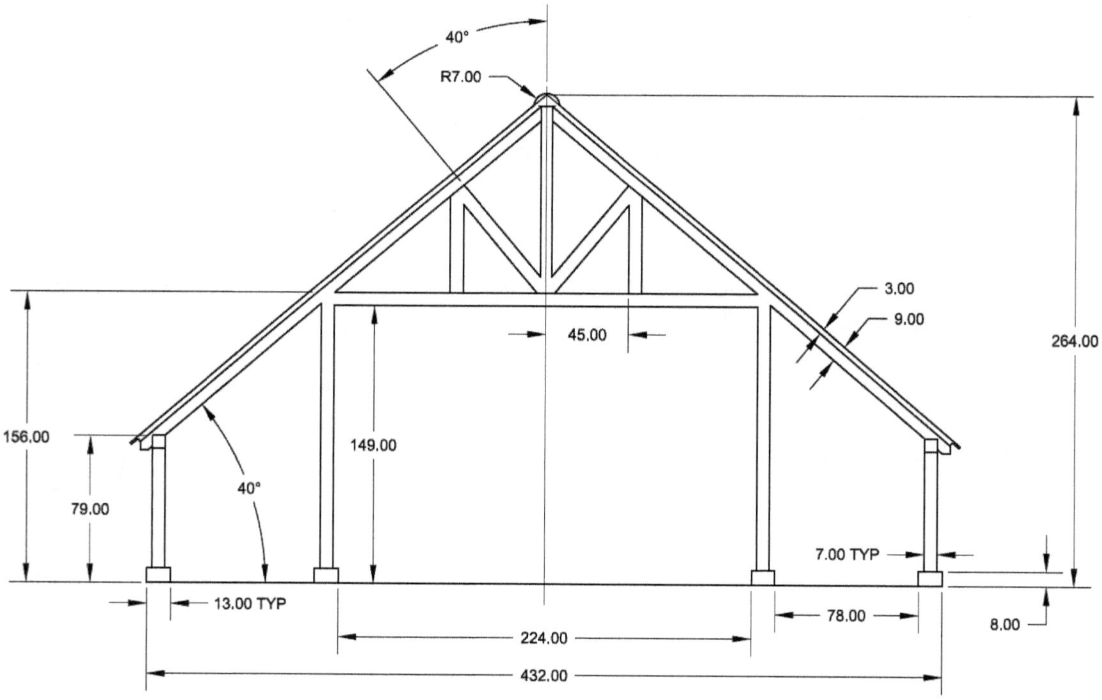

Problem 6—Barn Structure

Parametric Solid Modeling

12
CHAPTER

1. Introduction

Engineers have been increasingly using and taking advantage of computer technology to the point where its role is essential in many engineering fields. With the advent of more powerful computers and software, engineering design has been gradually drifting from pure two-dimensional tools to more powerful three-dimensional (3D) applications. Since objects in the real world are three-dimensional, working in a 3D environment and being able to directly manipulate three-dimensional information offers obvious advantages over traditional 2D tools.

- 3D models are more realistic than 2D drawings, especially when visualizing and manipulating the objects in real time. A 3D model can be rotated and viewed from any angle without having to draw it repeatedly.
- 3D models are more flexible than traditional 2D drawings. 2D drawings require that the object be drawn from multiple viewpoints (and also for sectioned and auxiliary views). Furthermore, when changes are made, each view must be updated.
- Design errors can be identified and remediated more easily in a 3D model. Clearance and interference checks can be performed to ensure accuracy, physical properties can be calculated, and designs can be analyzed and optimized using analysis tools. In fact, 3D models allow engineers to evaluate properties that cannot be defined in 2D drawings.
- 3D models are an excellent tool to communicate with other members of a design team and with non-technical audiences. They provide clearer visualization for people unskilled in technical drawing.
- 3D models are reusable. Multiple design configurations can be generated and evaluated quickly with minor additional effort.

2. 3D Modeling

Before exploring different modeling techniques and practices used in engineering, it is necessary to understand what a model is, particularly a 3D computer model. In general, a model is a representation of a system or a device, and how it works, with the purpose of predicting its behavior under certain conditions. There are different types of models (mathematical models, descriptive models, analytical models, computer models, etc) depending on its purpose. Sketches and orthographic drawings, for example, are descriptive models. They represent how an object or design needs to be manufactured in order to perform correctly.

A 3D computer model is a simulation of a physical entity built inside the computer. This simulation can be very simple or extremely complex, depending on its purpose. Some 3D computer models are used exclusively for visualization purposes. Others are used for marketing and presentations.

In engineering, 3D models are normally required to combine both descriptive and analytical information. A 3D model must be an accurate representation of the geometry of an object and at the same time provide physical and technical information needed for analysis and testing.

There are different types of 3D computer models.

2.1 Wireframe Models

The first 3D computer models were created by orienting basic graphic entities such as lines and arcs in 3D space. In essence, a wireframe model is a collection of XYZ coordinates stored in a computer file. This simple spatial data includes endpoints of lines, centers and diameters of circles and other geometric information (see Figure 12.1). The computer connects those coordinates through edges and displays the model.

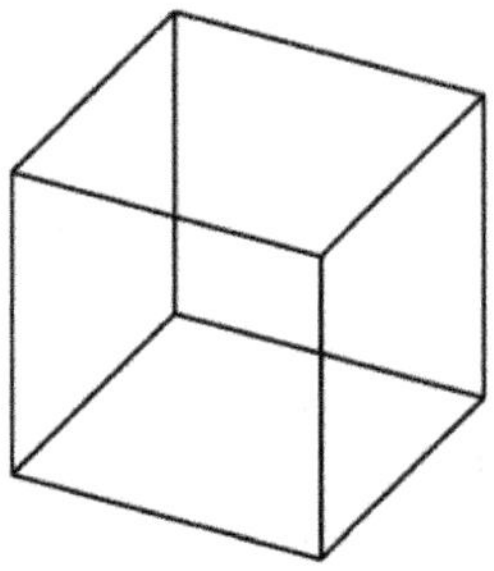

```
v 0.000  0.000  0.000
v 4.000  0.000  0.000
v 0.000  4.000  0.000
v 4.000  4.000  0.000
v 0.000  4.000  -4.000
v 4.000  4.000  -4.000
v 0.000  0.000  -4.000
v 4.000  0.000  -4.000
```

Figure 12.1 Wireframe model and its internal representation in a computer file

Wireframe models can be generated and manipulated very quickly within the computer, but they are extremely simple representations of actual objects. Wireframe models do not keep information about surfaces, volumes, or physical properties, so their applications are very limited. Moreover, the lack of shading or even hidden surfaces makes wireframe models unrealistic and often ambiguous (see Figure 12.2), so to a certain extent the interpretation of the model lies on the person viewing it.

Wireframe models can be generated using very simple tools and software, but as the complexity of the object being modeled increases, the wireframe model becomes very confusing (see Figure 12.3).

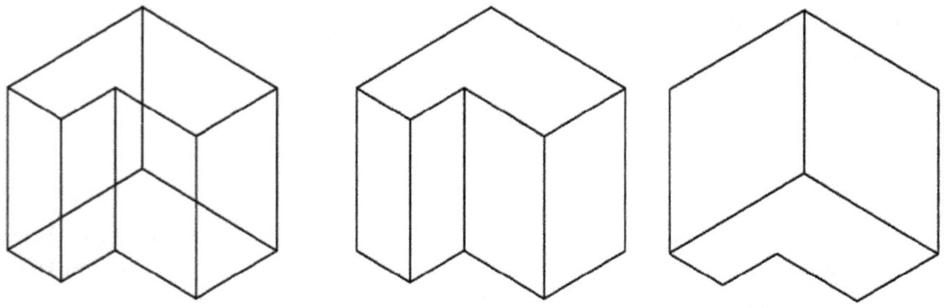

Figure 12.2 Ambiguous wireframe model

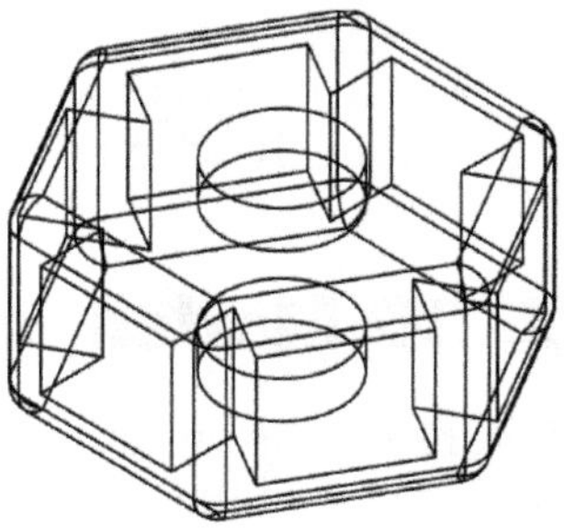

Figure 12.3 Wireframe models become confusing as complexity increases.

2.2 Surface Models

Surface models are the natural evolution of classic wireframe models. Surface models store geometrical data about the object and mathematically describe information related to the surfaces and faces. This information includes textures, reflectivity, and transparency. A surface model can be understood as an empty shell, although the surfaces that form the shell do not have actual thickness. With the powerful computers and software available today, very realistic surface models can be created.

Surface models can be defined as a matrix of vertices, known as polygonal meshes, or as a set of curves that mathematically define a smooth surface, known as a NURBS (Non-Uniform Rational B-Spline) surface. A polygonal mesh is a collection of XYZ coordinates of the vertices that form the mesh, much like a wireframe model, plus information and properties of the faces (normal vectors, color, transparency, etc). Polygonal models are very popular in the videogame and movie industries. 3D meshes may reach thousands or even millions of polygons of complexity (see Figure 12.4).

Figure 12.4 Polygonal model (courtesy of Donivan Potter)

A NURBS surface is the result of applying special mathematical functions to a set of curves. Although the mathematics behind these functions are rather complex, the idea is to define surfaces that blend and connect curves with other curves (see Figure 12.5). NURBS surfaces can be controlled and edited through some of the properties of the originating curves and the resultant surfaces can produce exceptionally smooth 3D models. The popularity of NURBS is increasing and its use is being expanded to many visual arenas. NURBS models are tremendously useful in applications that involve realistic images, such as product design or architectural and scientific visualizations.

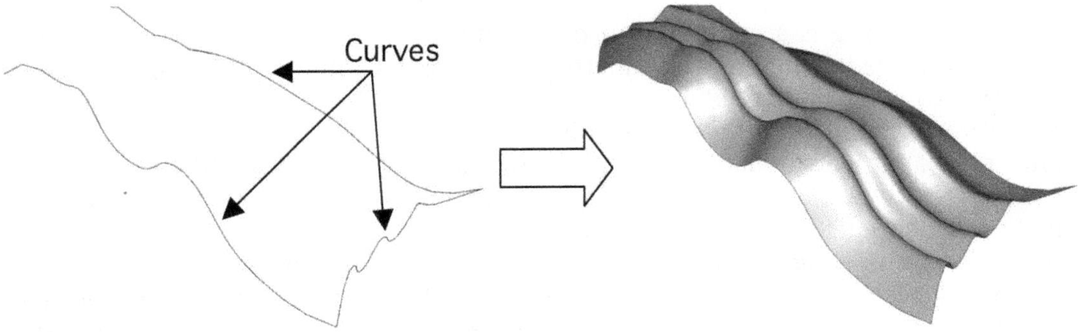

Figure 12.5 Creation of a NURBS surface from three curves

Although surface models can be extremely realistic, they do not retain data about the physical properties of the object. Similar to polygonal meshes, surface models resemble hollow shells. Therefore, it is not possible to perform analysis with them. For this reason, surface models are used almost exclusively in situations where only visualization is required.

2.3 Solid Models

A solid model is a valid, unambiguous, and fairly complete representation of an object. Solid models include accurate geometrical information about an object as well as physical data, such as volume, mass, center of gravity, moments of inertia, etc. Therefore, they can be used for analysis and testing. Solid models are the most useful computer models in engineering (see Figure 12.6).

Material: *ABS Plastic*
Density *= 0.04 pounds per cubic inch*
Mass *= 0.29 pounds*
Volume *= 8.00 cubic inches*
Surface area *= 24.00 inches^2*
Center of mass: *(inches): X = 1.00   Y = 1.00   Z = –1.00*
Moments of inertia: (pounds*square in)

Lxx = 0.20	*Lxy = 0.00*	*Lxz = 0.00*
Lyx = 0.00	*Lyy = 0.20*	*Lyz = 0.00*
Lzx = 0.00	*Lzy = 0.00*	*Lzz = 0.20*

Figure 12.6 Solid models store physical properties.

Historically, solid models were very simple because of the limitations of early computers. They were created using a technique known as constructive solid geometry (CSG). CSG consists of combining a set of simple, predefined geometric primitives, such as cubes, cylinders, spheres, or cones into a more complex solid. These combinations are made through Boolean operations, which will be discussed later in the chapter. Geometric primitives are easy to define, and the properties of the final model can be calculated using the information from the individual primitives that were used to model it. However, better tools and more useful techniques were needed to accurately model more complex objects.

A technique called feature-based modeling was developed in the mid 1990's to provide more intuitive and powerful tools to produce 3D solid models. A feature is a general term that refers to any 3D aspect, component, or characteristic of a model. Examples of features include fillets, chamfers, holes, patterns, ribs, etc. Feature-based modeling applications provide tools that allow the user to define and combine these features to create solid models.

3. Parametric Solid Modeling

Parametric solid modeling is a combination of two different modeling techniques: feature-based solid modeling, discussed earlier, and parametric modeling. Parametric modeling is the ability of a software package to use parameters, properties, and relationships between parameters to define a given model. Parametric modeling has many advantages in engineering design. The most important is the ability to change and update models quickly and with relatively little effort, compared to other modeling techniques. If a certain part of a parametric model needs to be altered, all the other parts related to that specific one will also change.

Parametric solid modeling has become very popular in recent years. In fact, it is the most common and powerful technique implemented in modern 3D CAD packages. In addition, parametric modelers provide a full range of functionality that extends beyond creating single 3D parts. Examples include relating and linking different 3D parts together in an assembly, providing comprehensive 2D working drawings for documentation and manufacturing, performing a variety of tests and analysis, and creating animations and presentations.

The general steps of the design process using a parametric solid modeling package are shown in Figure 12.7. The rest of this chapter will be devoted to exploring the techniques and capabilities of parametric solid modeling.

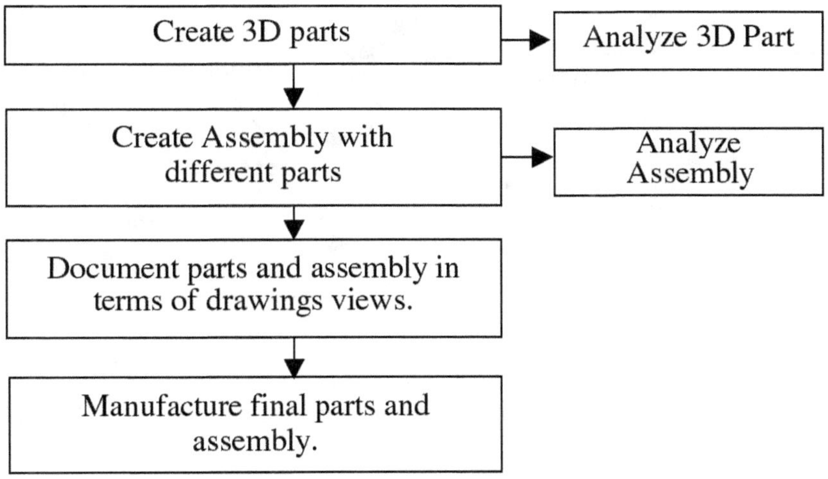

Figure 12.7 General design process using solid modeling packages

3.1 The Parametric Modeling Process

Most parametric solid modeling packages use similar user interfaces and tools, but it is important to become familiar with the particular application that is being used. When a new model is created, most packages present a 3D world with an indicator of the origin (0,0,0) and the three primary planes (XY, XZ, and YZ). A set of 3D manipulation tools (zoom, pan, and rotate) to move around the virtual world are also provided, as well as some common built-in cameras, such as different types of axonometric and orthographic views. Most solid modeling packages can be fully customized, including interface colors, toolbars, workspace, and menus. Important options that should always be checked are the unit system (English, metric) and dimensioning standards (ANSI, ISO, etc).

The required steps to create parametric solid models are also similar across packages. First, 3D solid models usually start with a 2D sketch. A sketch is not drawn to actual size. It is literally a rough, approximate sketch. Second, constraints are applied to the previous sketch. As these constraints are applied, the sketch will change its size and shape, automatically adjusting itself to the new values. This finished constrained sketch, commonly known as a profile, will be the basis of the 3D model. Finally, a 3D operation is applied to give volume to the 2D sketch or profile. Each step will be discussed in detail in following sections. The general steps of the parametric modeling process are shown in Figure 12.8.

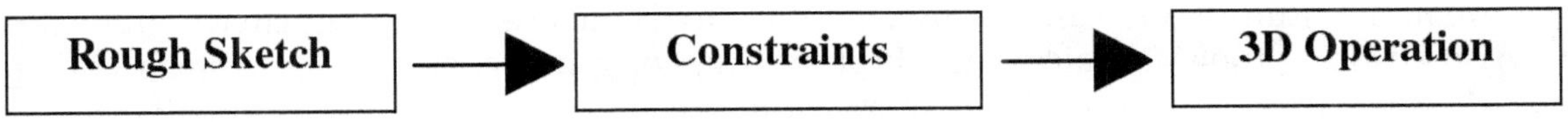

Figure 12.8 General steps of the parametric modeling process

3.2 Sketches

As mentioned earlier, 2D sketches do not have to be perfectly sized, shaped, or even include features such as fillets and chamfers. They are just simple, unscaled sketches (see Figure 12.9). Obviously, the closer the rough sketch is to the final shape, the easier it will be to apply constraints later. When a sketch is significantly larger or smaller than the actual object, it will contort when the parameters are applied, causing confusion when adding missing dimensions or constraints.

Sketches are always created on a plane. When a new model is first created, the first sketch is usually constructed on one of the primary planes (XY, XZ, and YZ). However, 2D sketches can be created on any plane or surface, as long as that surface is flat. See Figure 12.10 for an example. In general, it is good practice to start the first sketch of a solid model at the origin; otherwise the sketch must be dimensioned to the origin so that the sketch is properly located in 3D space.

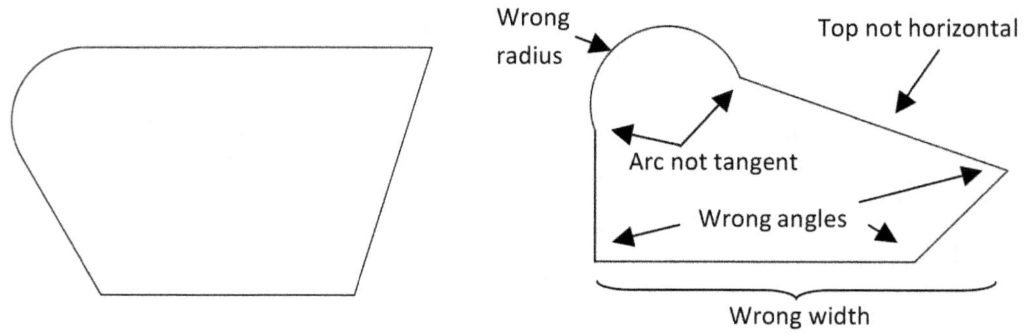

Accurate Sketch — Rough Sketch

Figure 12.9 Example of rough 2D sketch which will serve as the basis for a 3D solid object

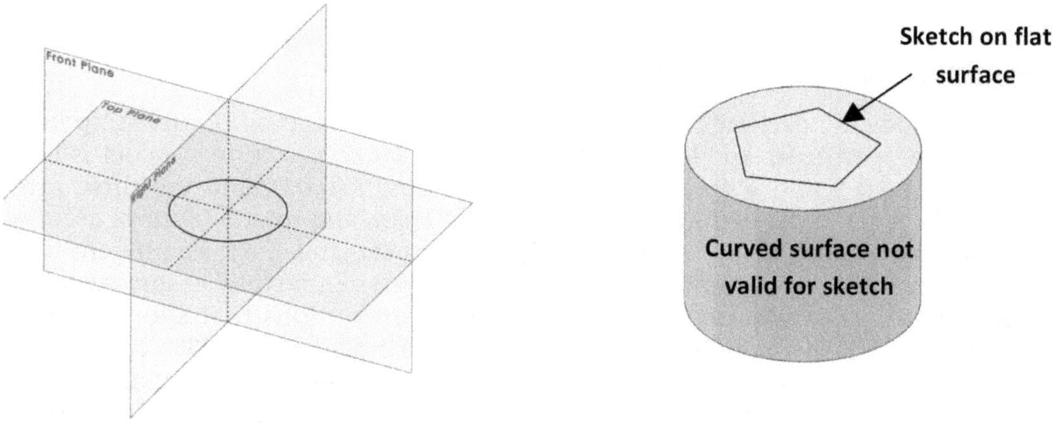

Sketch created on the primary top plane — Sketch created on a flat surface of an existing object

Figure 12.10 Examples of sketches created on a primary plane and on a flat surface of an object

Solid modeling packages provide tools to create and edit sketches. These tools are very similar to those provided by traditional 2D CAD applications. However, many of them have been optimized and they are more intuitive and easier to use. Creation tools include basic lines, circles, arcs, ellipses, and polygons. Editing tools contain trim, offset, mirror, copy, move, and pattern (linear and circular) tools. It is essential to become familiar with the specific tools of the package being used and how those tools work.

Certain conditions need to be satisfied in order to create a valid sketch for solid modeling. Generally, a valid profile must be a closed area or region without lines intersecting or gaps between line segments. See Figure 12.11 for examples of invalid sketches.

Intersecting lines — Gap

Figure 12.11 Examples of invalid sketches

Sketches with multiple objects or entities are valid as long as their boundaries do not intersect and the objects are not nested multiple times. Although not a good practice, some packages allow sketches with overlapping boundaries. In such a case, the user needs to select the portion of the sketch in which to apply the 3D operation. Also, the same sketch can be shared between multiple features. Examples illustrating this concept are shown in Figures 12.12 and 12.13.

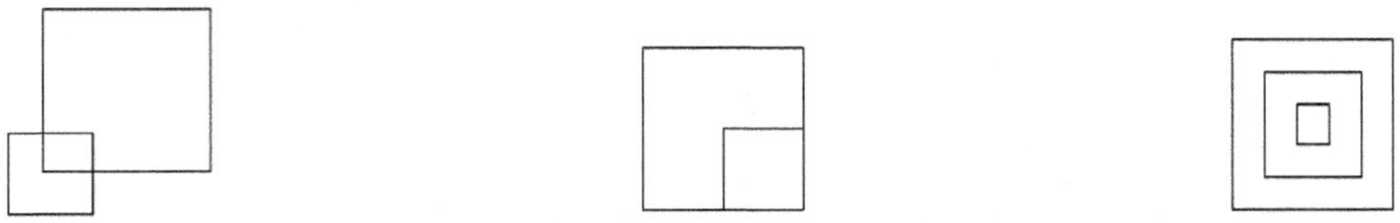

Figure 12.12 Examples of invalid sketches with multiple entities

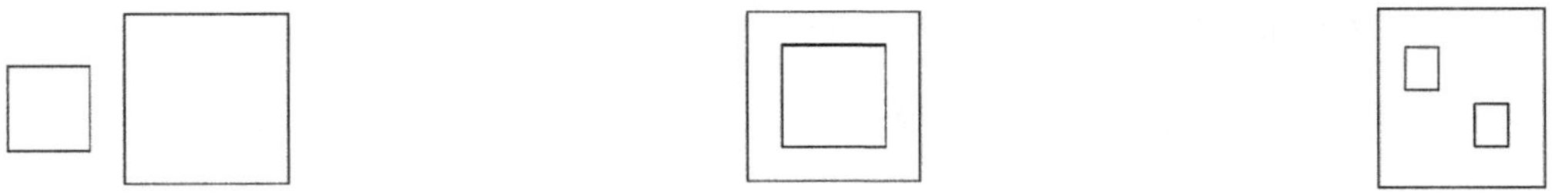

Figure 12.13 Examples of valid sketches with multiple entities

3.3 Constraints

The application of constraints is a crucial step in the parametric modeling process. A constraint is a restriction over a certain portion of a sketch. This restriction can be a dimension (forcing a line to be a certain length, or a circle to have a certain diameter), a geometric relationship (making two lines parallel, or a line tangent to an arc), or an algebraic equation (forcing a line to be twice as long as another line, or making a circle three times as big as the distance between two lines). When a constraint is applied to a sketch, the sketch will adjust and modify itself automatically, changing its shape, to become consistent with the restriction. A discussion of the three types of constraints in further detail follows.

Dimensional Constraints Dimensional constraints can be understood as regular dimensions similar to those used in general 2D CAD software. Dimensional constraints are values that represent magnitudes of entities, such as the length of a line, the diameter of a circle, or the angle between two lines. However, the effect of a dimensional constraint differs from the effect of a regular dimension in non-parametric CAD software. A dimensional constraint forces a line to have certain length or an angle to have certain value. In parametric sketches, the dimensions drive the sketch. Examples of dimensional constraints are shown in Figure 12.14.

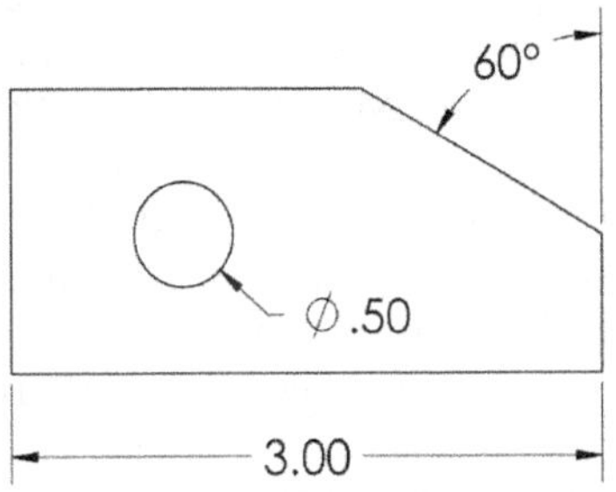

Figure 12.14 Examples of dimensional constraints

Geometric Constraints Geometric constraints are relationships between entities, or certain geometric properties of a specific entity. For example, a line can be forced to be horizontal or vertical by applying a constraint, or the user may force two circles to be concentric. Geometric constraints vary among different parametric modeling packages, so the user should become familiar with what constraints are available in the software, what icons or symbols are used to represent them, and more importantly, how those constraints are applied and affect the sketch.

Applying geometric constraints is fairly simple. Most parametric modeling packages allow the user to select the type of constraint to be applied and the entities in the sketch that need to be constrained. After those selections are made, the sketch is adjusted automatically and the results are displayed immediately. A list of the most common geometric constraints is shown in Figure 12.15.

Most software packages are configured so that some assumptions are made as a sketch is created. Some geometric constraints are automatically applied and its effects are visible immediately as the sketch is being built. For example, if a line is drawn within 3° of horizontal, most software packages

Constraint	Applies to	Description
Horizontal / Vertical	A line	Forces the selected line to be horizontal or vertical
Parallel	Two or more lines	Forces the selected lines to be parallel
Perpendicular	Two lines	Forces the selected lines to be perpendicular
Collinear	Two or more lines	Forces the selected lines to lay along the same straight line
Equal	Two or more lines / Two or more circles / Two or more arcs	Forces the selection to be equal in size
Tangent	A line and an arc / A line and a circle / Two arcs / Two circles	Forces the selection to be tangent
Concentric	Two or more circles or arcs	Forces the selection to have the same center
Coincident	Two or more points	Forces the selected points to coincide

Figure 12.15 Examples of typical geometric constraints

assume the designer's intent to draw a horizontal line, and therefore will automatically apply a horizontal constraint. When a line is created at the endpoint of an arc, and the line is almost tangent to the arc, most programs will apply a tangent constraint automatically.

Automatic constraints are a helpful feature that provides assistance to the designer. However, they can also lead to unexpected problems if the user is not aware of how they are being applied. Automatic constraints can be turned off if necessary, removed or modified at a later time, or configured so that only some automatic constraints are applied. Specific parameters such as threshold values can also be utilized. The 3° value for automatic horizontal constraints, for example, could be changed to any other value defined by the user.

Algebraic Constraints Algebraic constraints, also called equations, are used to relate a dimensional constraint to another through an algebraic expression. This expression can be as simple or as complex as the user desires. Expressions are usually arithmetical, but trigonometric functions or even comparisons can be used. Algebraic constraints allow the user to define multiple dimensions in terms of other dimensions. For example, the user can dimension the width of a rectangle as three times its height, so every time the height of the rectangle is modified; its width is changed automatically.

To create an algebraic constraint, driving and driven dimensions need to be defined. A driving dimension is a regular dimension with a unique identifier (basically, a name) associated to it. A driven dimension uses that unique identifier as a variable in the algebraic equation. As a result, if a driving dimension changes, the associated driven dimension will automatically update according to the equation. See Figure 12.16 for an example.

Status of a Sketch Constraints are the information that parametric software uses to calculate equations needed to define the shape of a sketch. Sketches need to be properly defined (constrained) so that there are no errors or any missing information. Depending on the number of constraints and how they have been applied, sketches can be under-defined, over-defined, or fully-defined.

An under-defined sketch has incomplete information or missing constraints. The parametric modeling application does not have sufficient information to determine the location, position, and shape of the sketch. Although sketches should always be completely defined, an under-defined sketch is acceptable as long as the user is aware of the reason why it is under-defined and the consequences it may bring to the final design. Under-defined sketches allow more flexibility when the user is evaluating different ideas, but final designs should always be fully defined. Parts comprised of under-defined sketches will encounter mathematical inconsistencies, possibly causing early termination of specific calculations, if finite element simulations are performed.

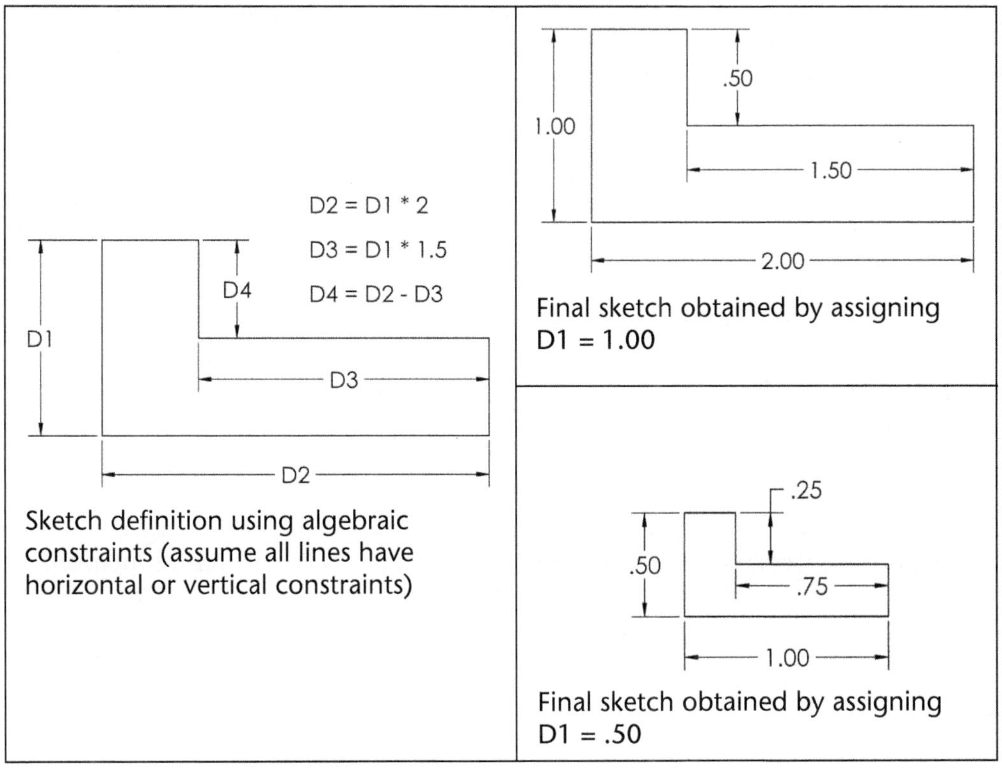

Figure 12.16 Example of algebraic constraints

An over-defined sketch has duplicate constraints or extra information, which causes a conflict for the software when trying to solve and determine the shape of the sketch. Over-defined sketches are never acceptable and cause many problems when creating 3D features. They should always be remedied by deleting the extra constraints that create the conflict.

A fully-defined sketch is fully constrained. All the constraints applied to a fully-defined sketch are necessary to determine, without conflict, the position and location of every line, arc, or circle in the sketch. Fully-defined sketches ensure predictability when a part is modified or updated.

Most parametric modeling applications provide visual aids (such as different colors) so it is easier to determine if a sketch is under-defined, over-defined, or fully-defined. Examples of the three different statuses of a sketch based on its constraints are shown in Figure 12.17.

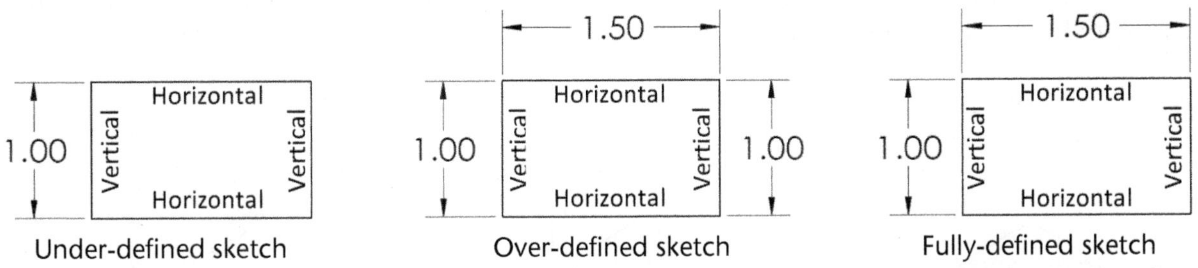

Figure 12.17 Examples of the three different statuses of a sketch. "Horizontal" and "Vertical" indicate geometric constraints.

3.4 Converting to 3D

The final step in the parametric modeling process is to apply a 3D operation to a fully defined profile. Sophisticated methods exist to create 3D objects from 2D sketches, but most are combinations of four basic techniques: extrude, revolve, sweep, and loft.

Extrude An extruded feature or extrusion is created by stretching a 2D sketch along a straight path. Some of the parameters that can be defined are the thickness of the extrusion or the taper angle. See Figure 12.18 for examples of extrusions.

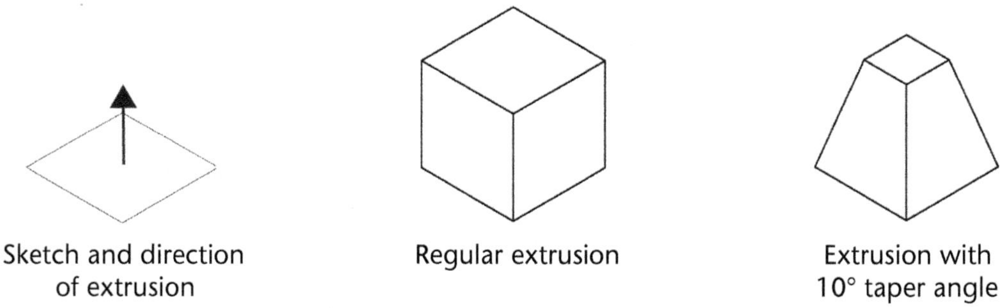

Figure 12.18 Examples of extrusion

In most cases, the extruded feature will be perpendicular to the sketching plane. In fact, this setting is the default in most parametric modeling packages. Certain situations may require the extrusion angle to be modified (see Figure 12.19).

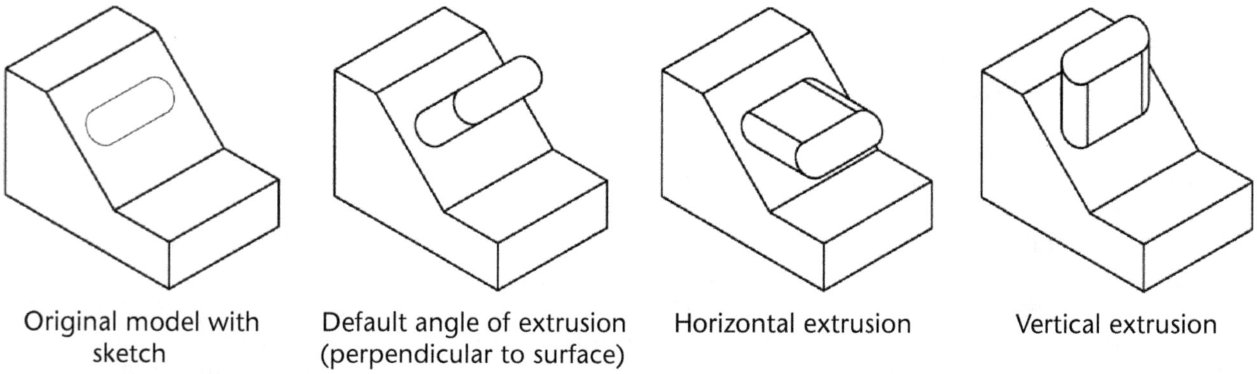

Figure 12.19 Different angles of extrusion

The length of an extruded feature can also be parametric. For example, instead of explicitly indicating a numeric value for the extrusion thickness, a sketch can be extruded up to an existing surface. The options available depend on the specific software package. The most common options include: extrusion to the next surface, extrusion to a selected surface, and extrude through all (see Figure 12.20).

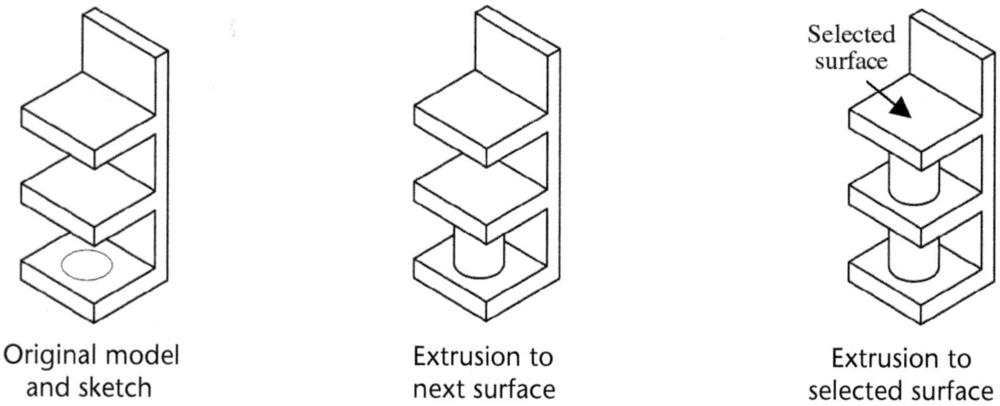

Figure 12.20 Extruding to a surface

Revolve A revolved feature is created by rotating a sketch around an axis. Revolving a sketch requires the user to indicate the axis of revolution and the number of degrees that the sketch will revolve. The axis of revolution can be a separate line outside the sketch, or an edge that is part of the sketch to be revolved. See Figures 12.21 and 12.22 for examples of revolved features. If a hole is created using the Revolve command, it is useful to rotate around a center line, as when the part is brought into a drawing file, the corresponding dimension will appear with the appropriate diameter symbol. There are different methods to revolve a sketch, such as clockwise, counterclockwise, or both directions.

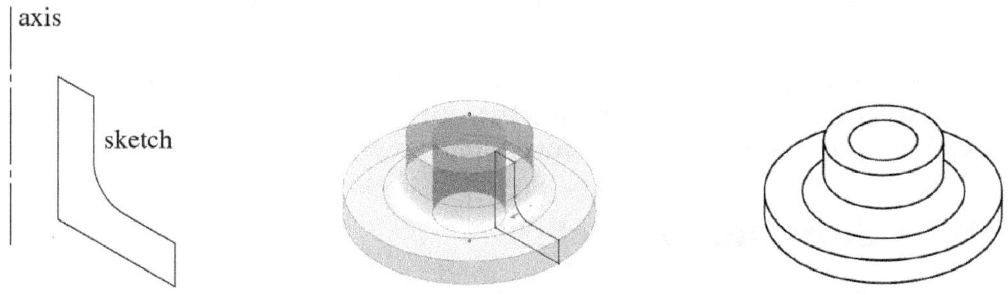

Sketch and axis of revolution — Revolve process — 360° revolved solid object

Figure 12.21 Example of full revolve (360°) using an external axis

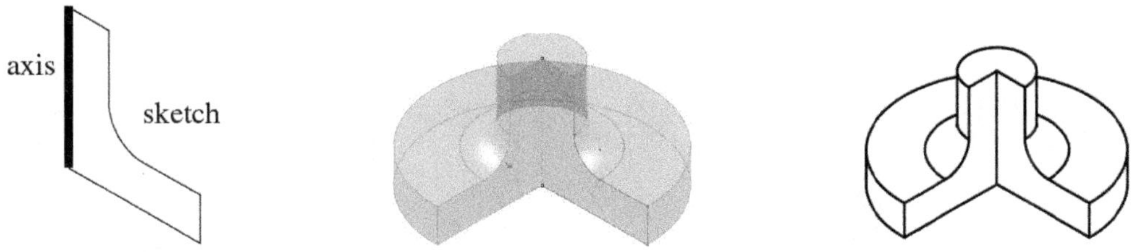

Sketch and axis of revolution — Revolve process — 270° revolved solid object

Figure 12.22 Example of 270° revolve using an edge of the sketch as axis of revolution

Sweep A sweep is created when a 2D sketch is extended along a path. Sweeps are very similar to extrusions. In fact, an extrusion can be considered a special type of sweep, where the path is a straight line. What's more, the revolve operation is also a special type of sweep, where the path is completely circular. Two sketches are required to create a swept feature: the profile to be swept, and the path. The profile to be swept is sketched on the normal plane at one of the endpoints of the path. See Figure 12.23 for an example of sweep.

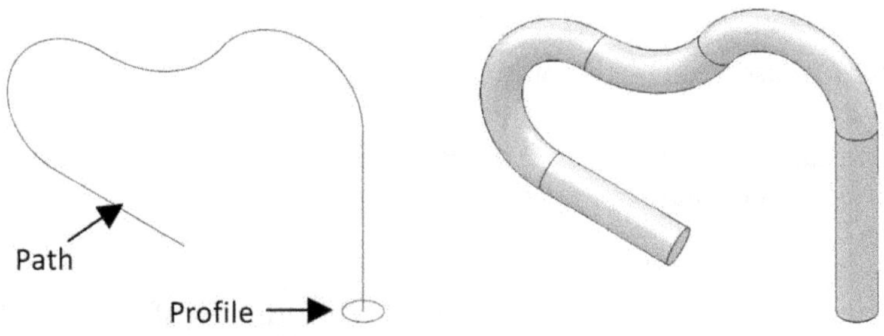

Figure 12.23 Example of swept feature

Loft Loft transforms one sketch into another through a series of varying sketches. At least two sketches are required to create a lofted feature. See Figure 12.24 for an example of a loft.

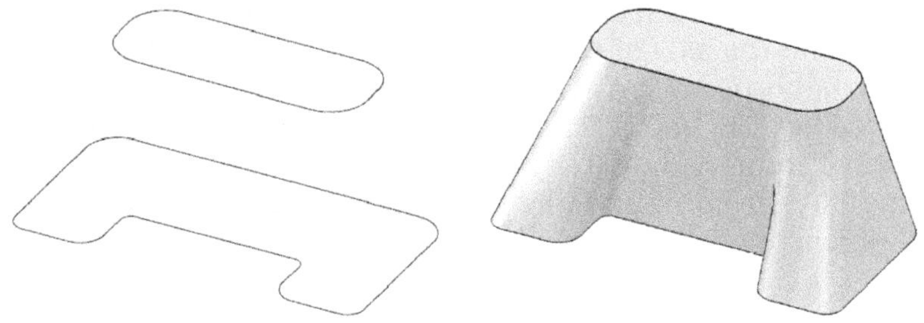

Figure 12.24 Example of lofted feature with two sketches

The previous examples illustrated situations where volume is added to an existing object. However, there are also corresponding commands that remove volume from a model (extrude cut, revolved cut, etc). This discussion is omitted for simplicity.

4. Modeling Complex Objects

The four methods used to create 3D objects from 2D sketches are a good starting point to build simple parts, but not every object can be created with a single sketch and a basic extrude, revolve, sweep, or loft operation. Multiple sketches and solids need to be combined to create more complex objects.

The first step required to model a complex object is to recognize the basic form that best represents the overall shape of the object. This basic shape is called the base feature and it is created using a single extrude or revolve. See Figure 12.25 for example of base features.

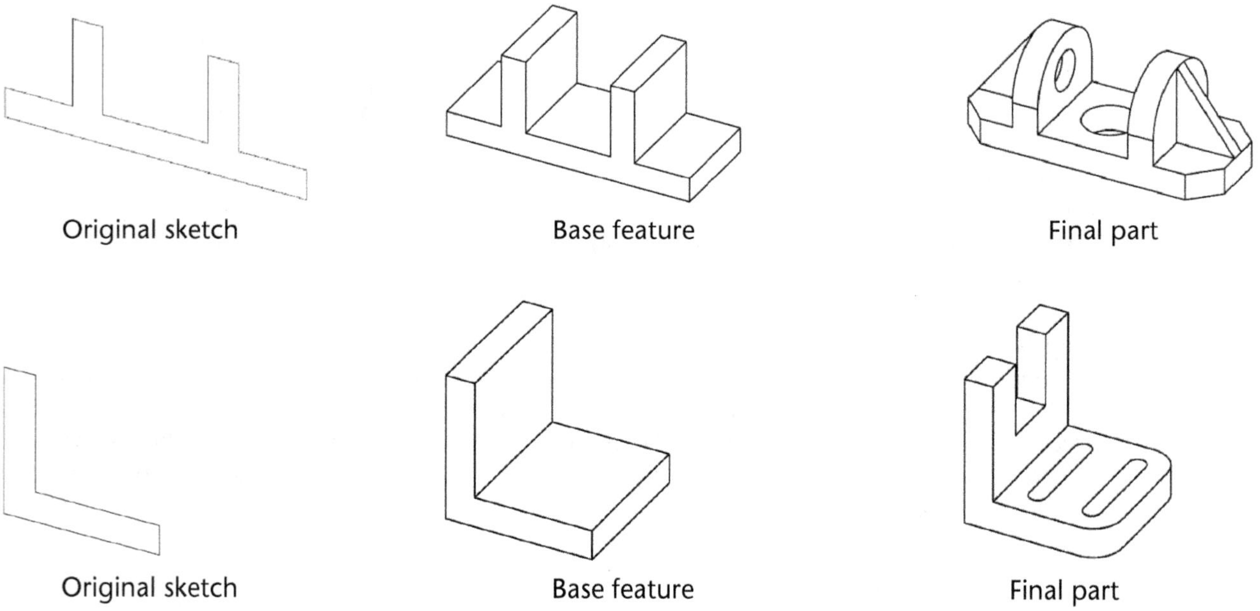

Figure 12.25 Base features used to create complex solids

Subsequent features are combined with the base feature to create the final object. The way solids are combined is defined as Boolean operations. There are three types of Boolean operations: union, difference, and intersection.

- Union ($\cup$): the union operation is essentially an addition of two overlapping solids, counting only the overlapping volume once.
- Difference (-): it removes the overlapping volume between two solids from the one listed first in the operation. Most parametric modelers refer to this operation as "cut."
- Intersection ($\cap$): the result of the intersection between two solid parts is a solid that keeps only the volume of overlap (counted once) between the two parts.

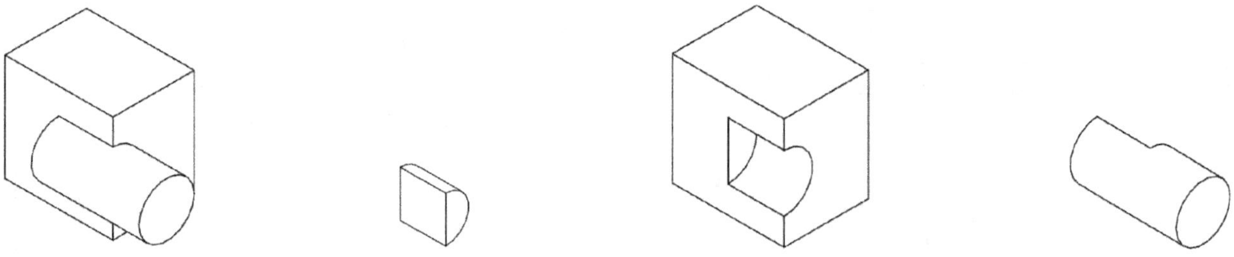

Cube ∪ Cylinder   Cube ∩ Cylinder   Cube - Cylinder   Cylinder - Cube

Figure 12.26 Boolean operations

See Figure 12.26 for examples of each Boolean operation using a regular cube and a cylinder.

Examples on how basic extrude and revolve operations can be combined witing a solid through a Boolean operation are shown in Figure 12.27.

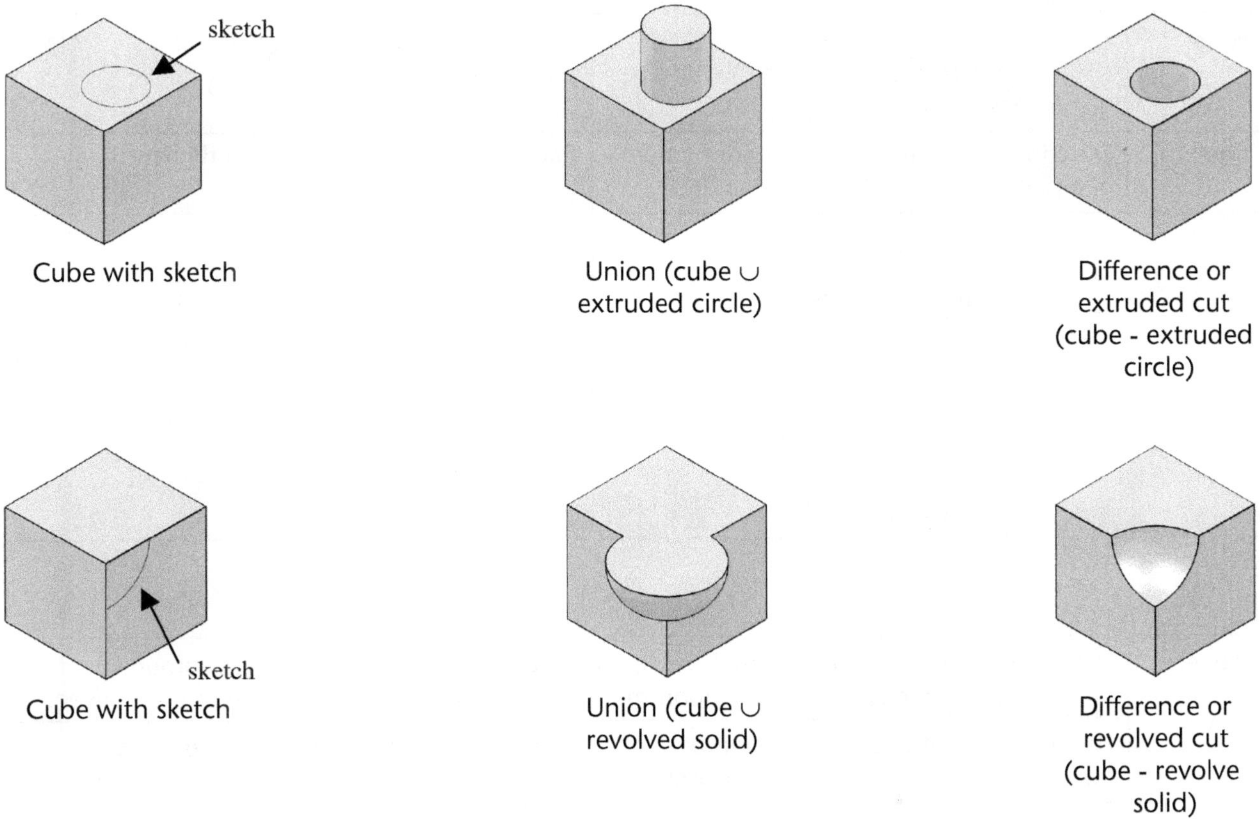

Figure 12.27 Extrude and revolve unions and subtractions

There are special cases, as when two solids do not overlap, or when one solid completely encompasses the other, where Boolean operations may result in objects with no volume (null objects) or objects that equal one of the solids used in the operation. Such cases can be mathematically described, but their practical applications in engineering design are very limited.

5. Common 3D Tools

Any 3D feature can be created by combining different Boolean operations. However, modern parametric modeling applications provide tools to facilitate and simplify the creation of some of the most popular features found in engineering modeling. Although these tools are consistent across different software packages, familiarity with how the tools available in any specific application are used is required. See Figure 12.28 for examples of common 3D tools found in parametric modeling packages. Notice that some of the parameters may differ from application to application.

Fillet	Description: It rounds a corner.	Parameters: Edges to fillet, fillet radius
Chamfer	Description: It creates a beveled edge at a corner.	Parameters: Edges to chamfer, angle, chamfer dimension

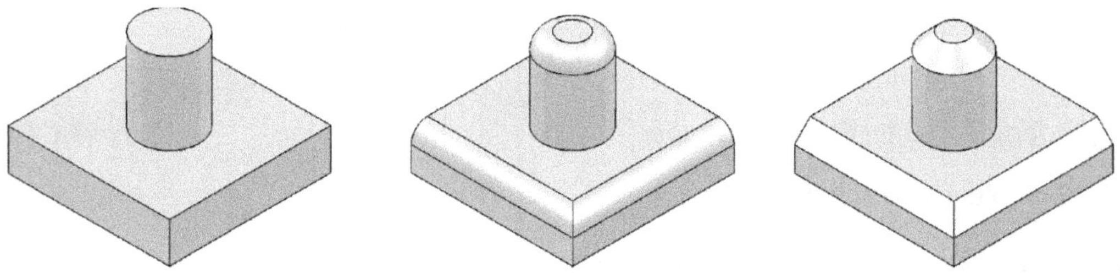

Base object Fillets Chamfers

Hole	Description: It creates a machined hole (counterbored, countersink, etc) with the given parameters.	Parameters: Hole location, type, specific hole parameters
Shell	Description: It removes the interior volume of a part.	Parameters: Base surface, wall thickness

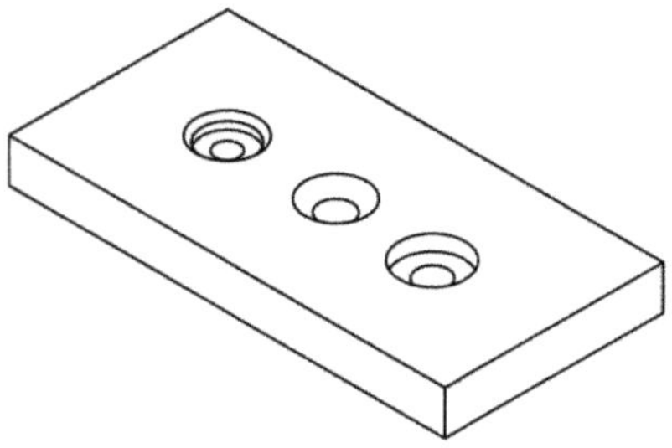

Machined holes

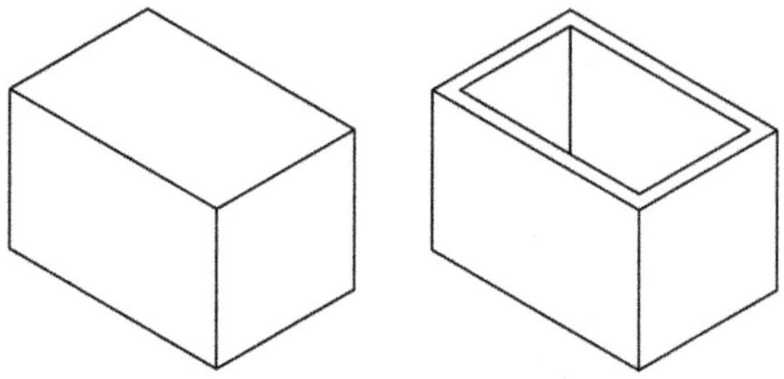

Shell

Pattern	Description: It creates multiple copies of features along one or two directions, or around a reference circle. Some applications support more complex patterns, such as path or table driven.
	Parameters: Features to copy, number of copies, copies to skip. For linear patterns: distance between copies, and directions. For circular patterns: reference circle, degrees to cover.

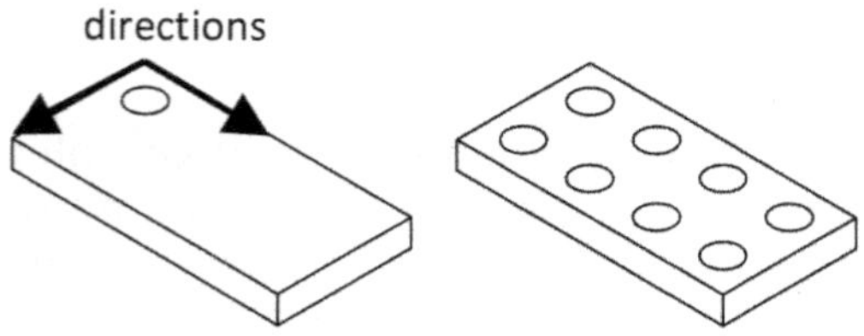

Linear pattern

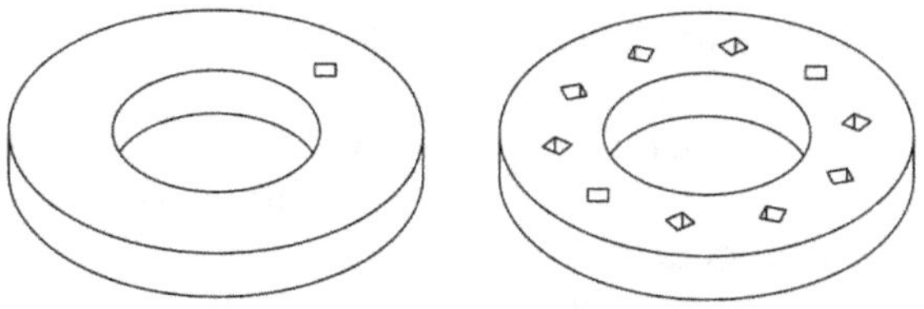

Circular pattern

Mirror	Description: It creates the symmetrical pattern of the given features.
	Parameters: Features to mirror, and mirror plane

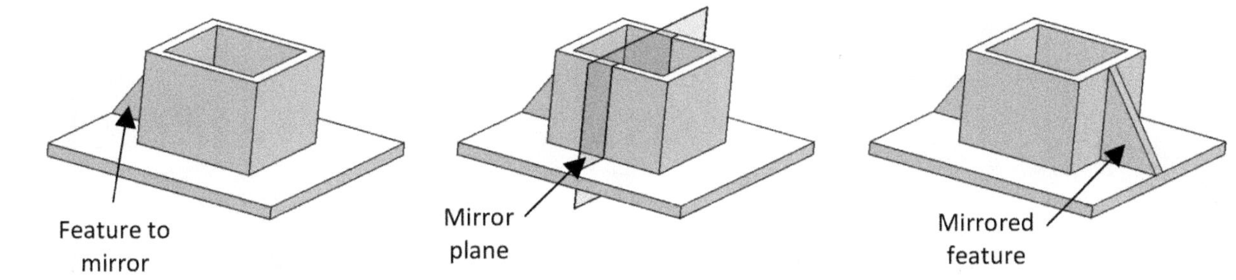

Rib	Description: It creates a rib from a reference line.
	Parameters: Reference line, rib thickness, type (open/closed)

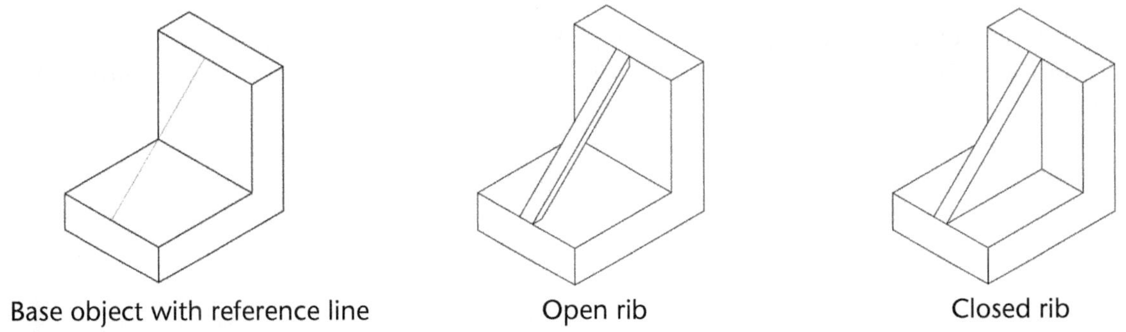

Figure 12.28 Examples of common 3D tools

6. Aids to Create Complex Geometry

Combining features through Boolean operations is a powerful way to create complex models. However, all sketching and creation of new features need to be constructed on one of the standard planes, or on an existing surface of the solid. Modern parametric modeling packages provide reference tools to create even more complex geometry. These tools create auxiliary geometries, often called datums, used to build and add features to a model (see Figure 12.29 for an example). Datum geometries are not solid objects, so they have no mass or physical properties, but actual features and sketches can be dimensioned and constrained relative to them.

There are three types of reference geometry or datums in parametric modeling applications: reference points, reference lines or axes, and reference planes.

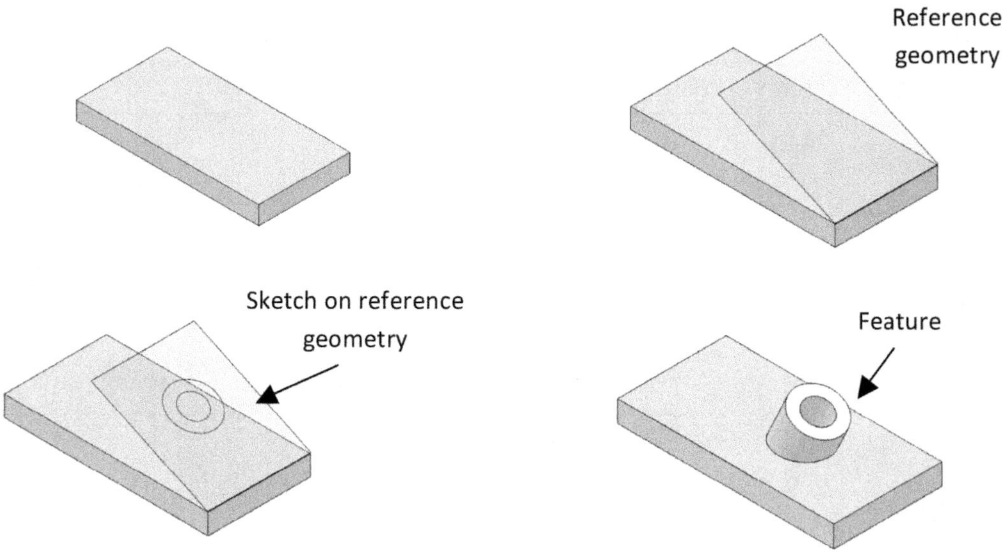

Figure 12.29 Reference geometry used to create a feature

Reference Points

Reference points are used when a specific point needs to be located to create a new feature; for example, to indicate the location of a machined hole on a surface or to locate the midpoint of a certain edge. See Figure 12.30 for an example.

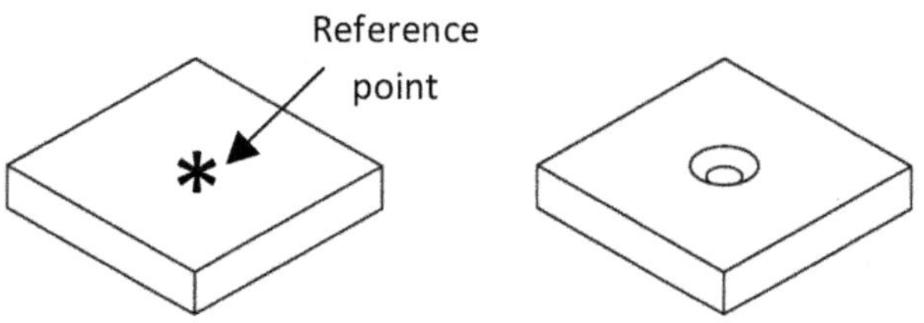

Figure 12.30 Example of reference point to create a countersunk hole

Reference Lines

Reference lines are used as construction lines in a sketch or to define the axis of revolution when creating a revolved feature. There are several ways to define reference lines, including through two points, at the intersection of two planar surfaces, or parallel to another line through a point. See Figure 12.31 for an example of reference line. Depending on the software used, reference lines can be shown as center lines, which are useful (especially for revolved features) when creating drawing files, as the package infers a diameter symbol is needed for proper dimensioning.

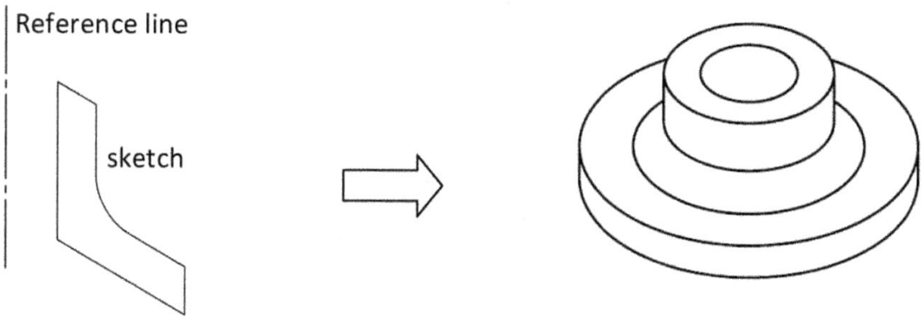

Figure 12.31 Example of reference line used to create a revolved solid

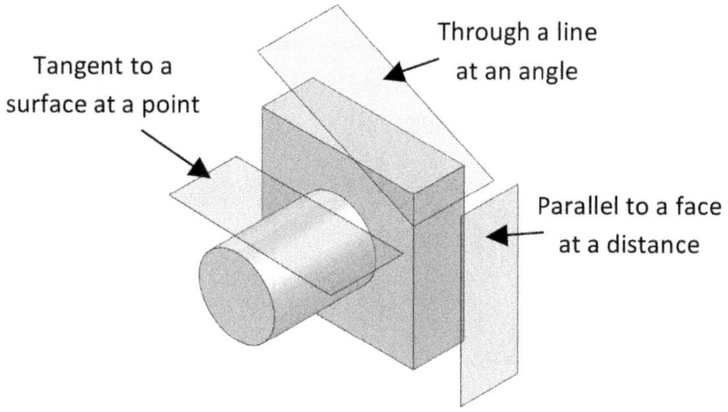

Figure 12.32 Examples of reference planes

Reference Planes

Reference planes are commonly used when a part's planar surface is not available as a sketch plane or when an intermediate position is required to define other sketches (for example, at an angle to a face at an offset distance).

When creating a reference plane, specific information is required: orientation and location in 3D space. With that in mind, there are several ways to create reference planes. Some examples include through three non-collinear points, through two intersecting lines, parallel to a flat surface through a point, tangent to a surface at a point, through a line at a certain angle from a surface, normal to a curve at a point, etc. See Figure 12.32 for examples of reference planes.

Parametric modeling packages provide distinctive methods in which to create reference planes. The steps required to define a reference plane a certain way also differs from application to application. The tools and techniques available within specific applications require familiarity with that program.

7. Design Intent

There are multiple ways to create a 3D solid model and there are several approaches you can take to define and constrain a sketch, but some of them are more useful than others, depending on the applications. The method chosen to create an efficient 3D solid model depends on its design intent. Design intent is the plan of how the model should behave when designed. It is determined by the function of the part to be modeled and most importantly, how the model will react if dimension values or geometric constraints need to be changed in the future.

Capturing design intent to build effective 3D models is not easily achieved. It is the result of knowledge, experience, skills, and most importantly, practice. Because of the intuitive interfaces and tools provided by modern parametric modeling applications, it is common for inexperienced users to begin modeling objects without planning or consideration of design intent.

Design intent is what separates a good model from a bad one. It is a good habit to think about possible modifications or updates that may be needed in the future. The most effective way to create a parametric model is the one that allows the user to update the model with the least amount of effort and time. Design intent can be implemented in all stages of the parametric modeling process, but the way in which constraints are applied plays a very important role in the behavior of the model. In fact, the way a sketch is constrained determines how a feature will relate to other features of the object and how the object will update.

There are several factors that can be used to determine design intent. Knowledge about the design should be used to decide how new features are added: What features are most likely to change in the future? What condition is the design required to satisfy at all times (equally spaced holes, proportions, etc)?

Inexperienced parametric modelers are most familiar with the dimensioning tools available in 2D drafting packages, which are similar to dimensional constraints in parametric modeling applications. While these types of constraints are useful, in order to fully take advantage of the power of parametric modeling and better capture design intent, geometric and algebraic constraints should

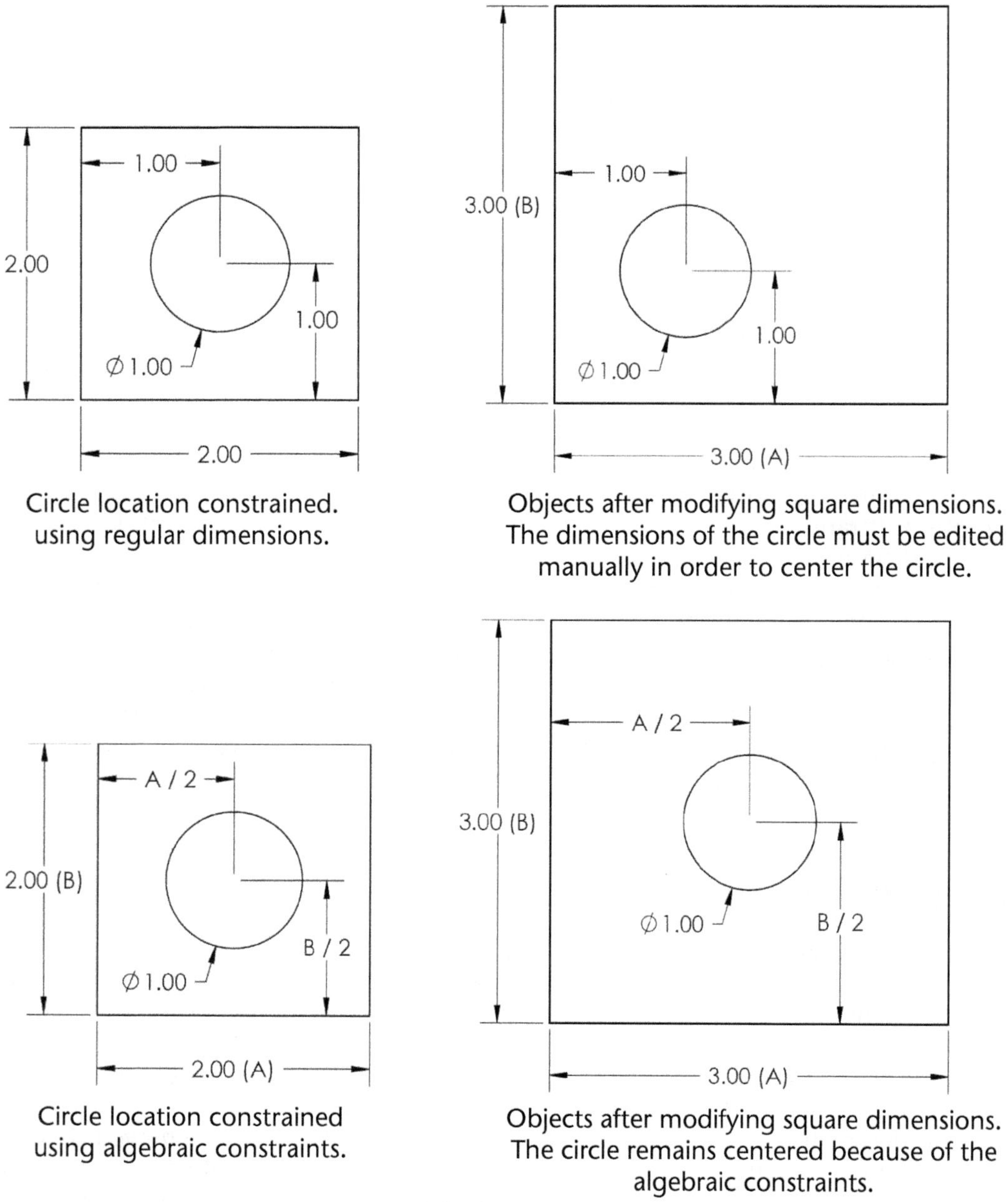

Figure 12.33 Example of using algebraic constraints to better capture design intent

be explored. When used intelligently, these types of constraints can drastically reduce design time and increase the flexibility of the model.

A square with a circular hole in the middle is illustrated in Figure 12.33. If the size of the square is likely to change, but the circular hole is always to remain centered inside the square, it is more efficient to use algebraic constraints to dimension the location of the circle than it is to use regular dimensions.

The example shown in Figure 12.34 illustrates a situation where a sketch needs to be extruded 3.00 inches in order to create a rectangular block connected to both ends of the part. This extrusion can be performed in two different ways: explicitly indicating its thickness (3.00 inches) or up to an existing surface. However, if the overall length of the part is likely to change, extruding up to an existing surface will make the rectangular block adjust automatically as the overall length of the part increases, eliminating the need for manual adjustments, and better capturing design intent.

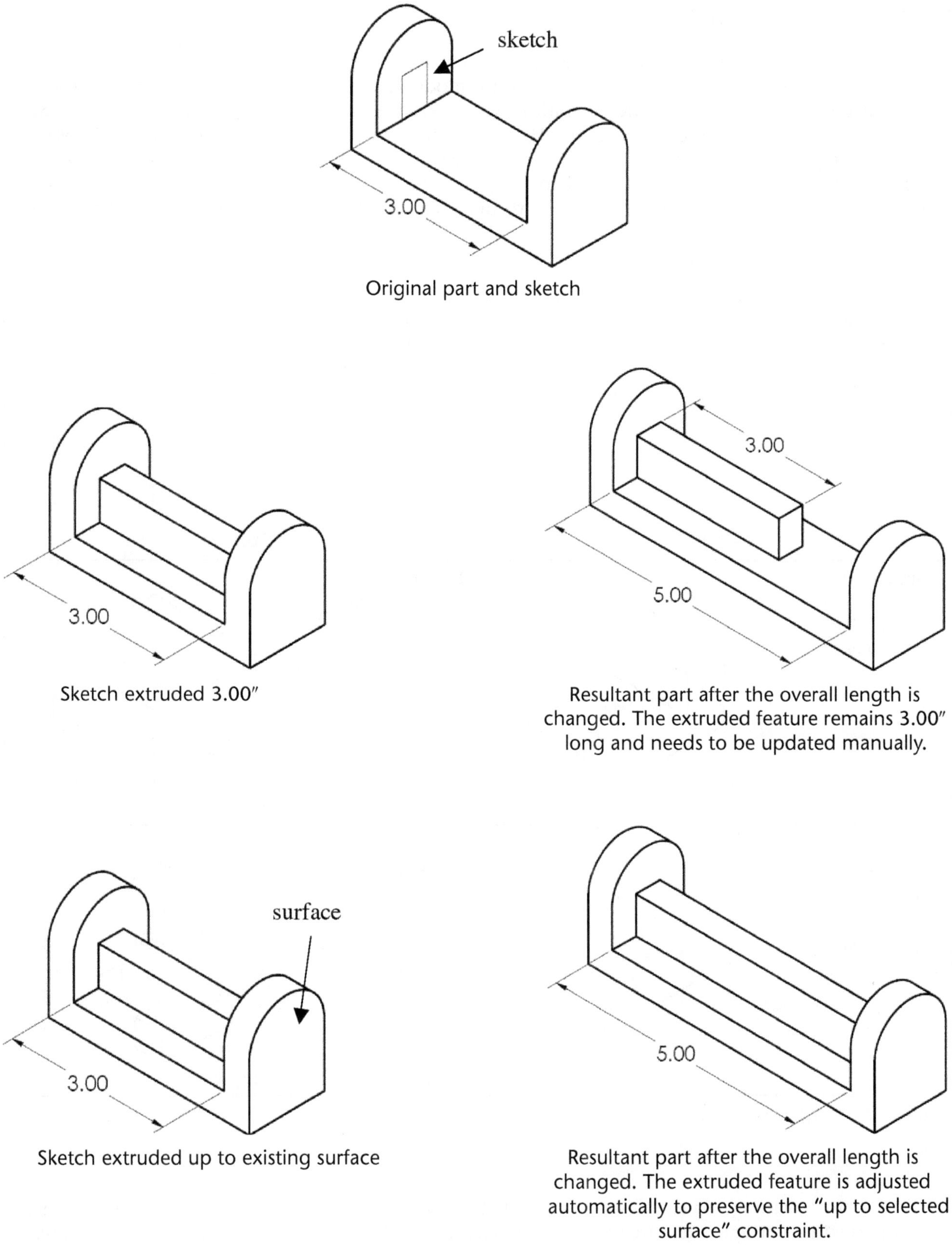

Figure 12.34 Situation where extruding up to a surface better captures design intent

8. The Design Tree

The design tree, sometimes referred to as model tree or history tree, is without a doubt the most important editing element available in parametric modeling packages. The design tree is not a tool itself, but a list of all the steps and operations that were executed to model a specific object. Such steps are presented in a tree-like structure in the same sequence they were performed. New commands are sequentially inserted at the bottom of the tree. In other words, the design tree provides a history of the order of commands used to design an object. The example shown in Figure 12.35 illustrates how new items are added at the bottom of the design tree as the model is being created.

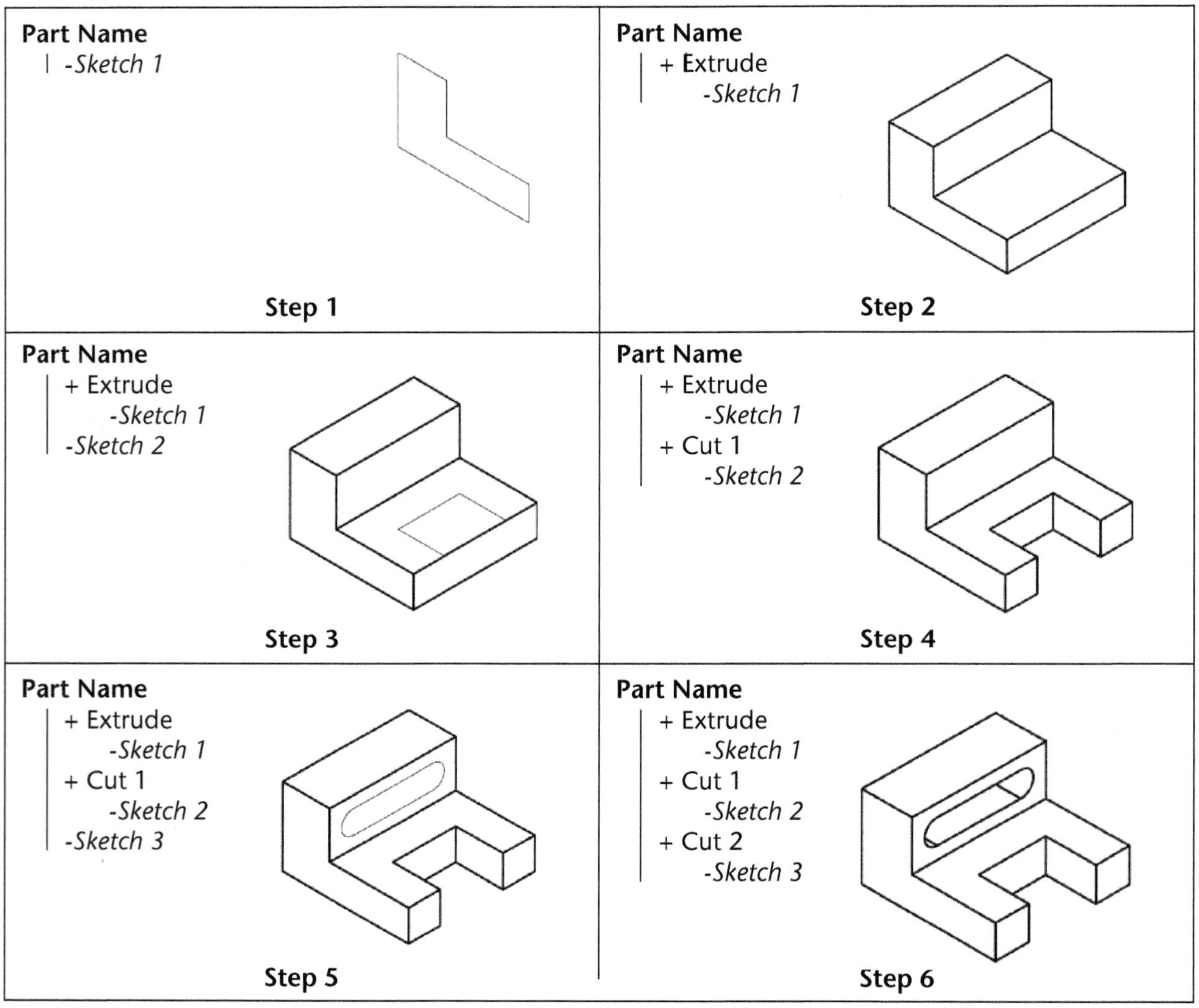

Figure 12.35 The design tree

The main advantage of the design tree is that the user can go back to any specific point in the design and edit a particular feature or sketch. In such cases, the appearance of the model will adjust to the original appearance the model had at that specific point in time. See Figure 12.36 for an example.

Deleting features that were used to define subsequent features will have an unwanted effect on the model. A situation where a feature (*Cut 1*) is dependent upon an existing cylindrical feature (*Extrude 2*) is illustrated in Figure 12.37. If *Extrude 2* is deleted, the dimensional constraint (*20*) will have a conflict. The example shown in Figure 12.38 illustrates a better way to constrain the same object if the cylindrical feature is likely to change in future updates of the part.

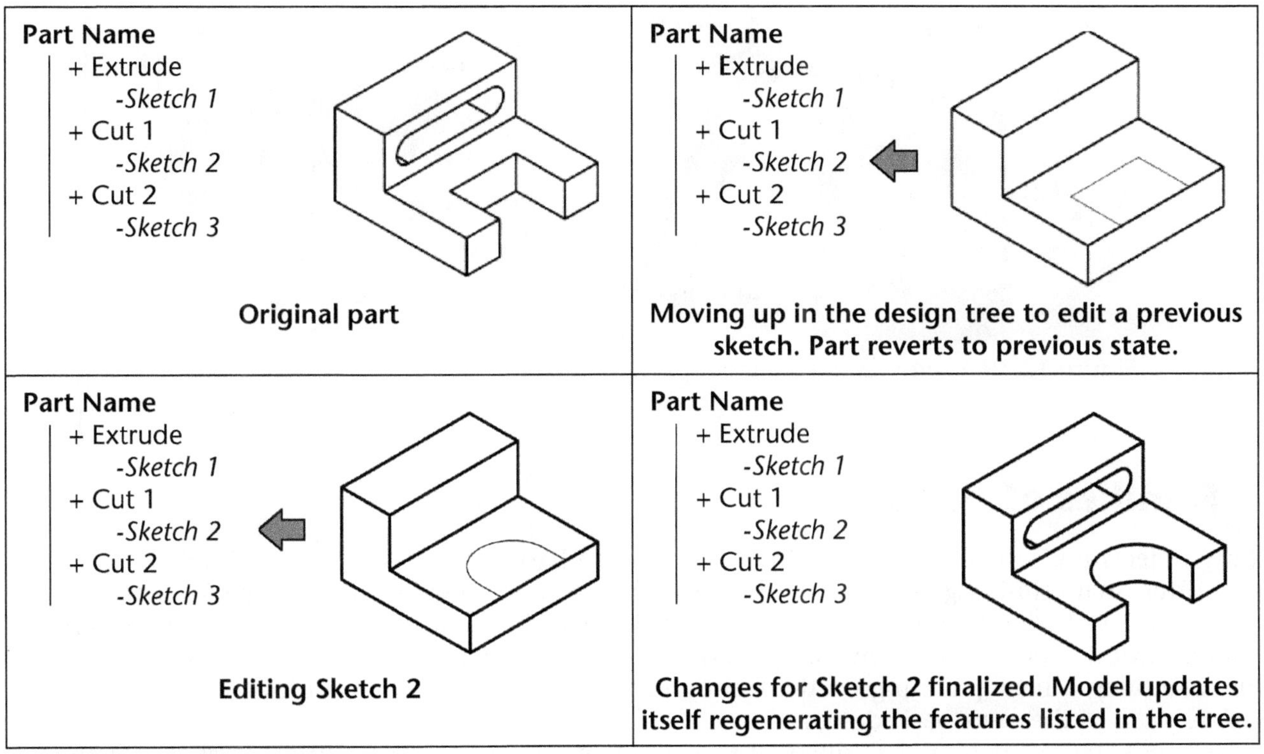

Figure 12.36 Editing an existing feature by moving up in the design tree

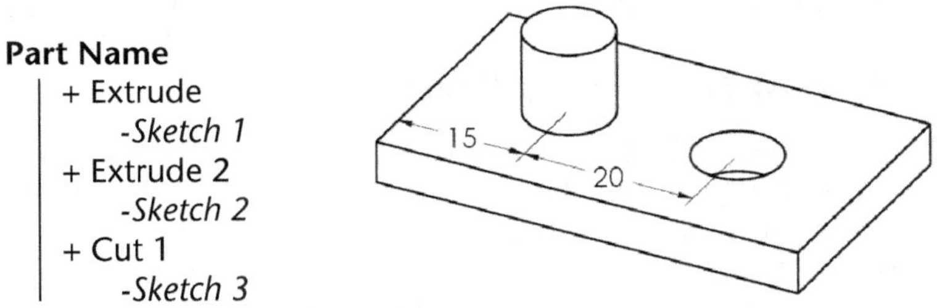

Figure 12.37 Feature (hole) depending on other feature (cylin-

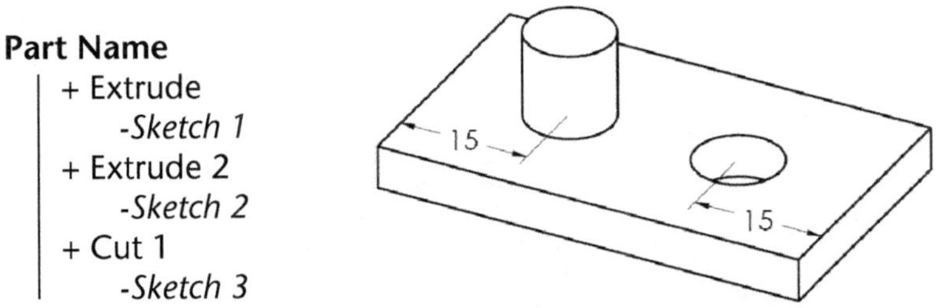

Figure 12.38 Feature (hole) constrained independently from the cylindrical feature

Some software packages will display a warning message to the designer about the effect of deleting or editing certain features. Other packages may delete all the dependent features without notice.

In order to make a model easy to update, these kinds of dependencies should be avoided, or at least minimized. As a general rule, avoid making a feature that is likely to remain unchanged dependent upon another feature that is likely to change in future updates of the model. This procedure is, again, capturing design intent.

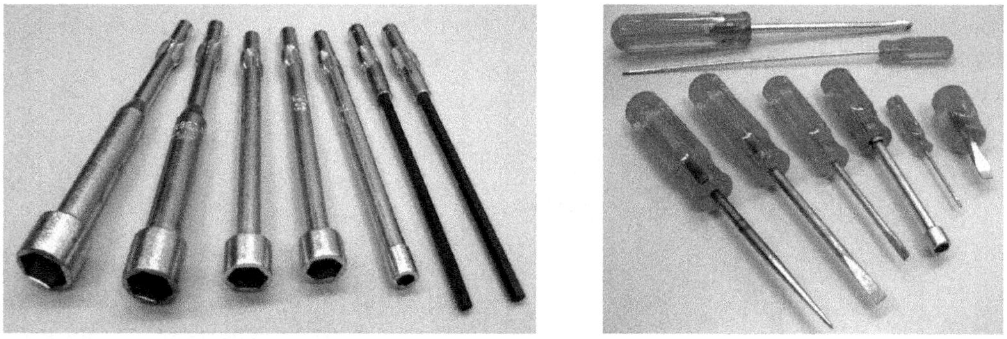

Figure 12.39 Examples of families of parts

9. Families of Parts

Many objects are just different variations of the same basic design (see Figure 12.39). Families of parts often share similar geometry and dimensions and only a few features change from version to version. When a company is making a product that will have variable features, it is necessary to build a flexible model where new design variations can be quickly implemented. In a world where time is money, a little planning on the front end will allow a designer to quickly and easily create new versions of the same basic design.

It is possible to build each individual version of the design separately from scratch, but this is certainly not an efficient approach. Instead, most parametric modelers allow the creation of families of parts or assemblies within a single document. They provide convenient ways to develop and manage families of models with different dimensions, components, or other parameters very efficiently. There are two methods in which to create families of parts: manually and automatically.

9.1 Manual Creation of Families of Parts

Creating families of parts manually only requires the use of a base model. The process is very straight forward. For every configuration the user desires to create, the model is modified manually by changing dimensions and/or activating or deactivating features. The object shown in Figure 12.40 will be used to illustrate the creation of families of parts manually.

For this example, it is assumed that two variations of a part are needed. The first variation, which will be called "regular - no fillets", will have the same dimensions of the generic part, but no filleted corners. The second variation, called "large – no fillets" will have larger cylindrical features (bottom cylinder Ø3.00, top cylinder Ø2.00, and hole Ø1.00), and no filleted corners.

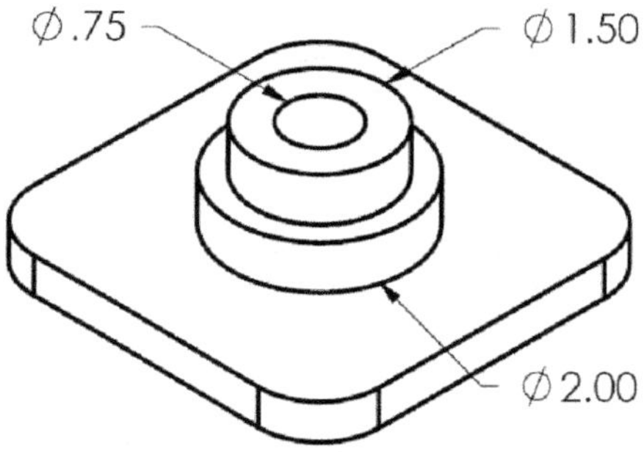

Figure 12.40 Generic model used to create configurations manually

Although different software may take different approaches, most parametric modeling packages provide a menu to manage the different variations of the design. This menu normally displays a list with all the existing versions of the design and provides a set of tools to control this list, such as adding, deleting, or editing an existing version.

The first variation is simply created by adding a version to this menu and suppressing the filleted features in the original part. The activation or suppression of features is normally done through the design tree.

See Figure 12.41 for an illustration of the steps used to create the first variation.

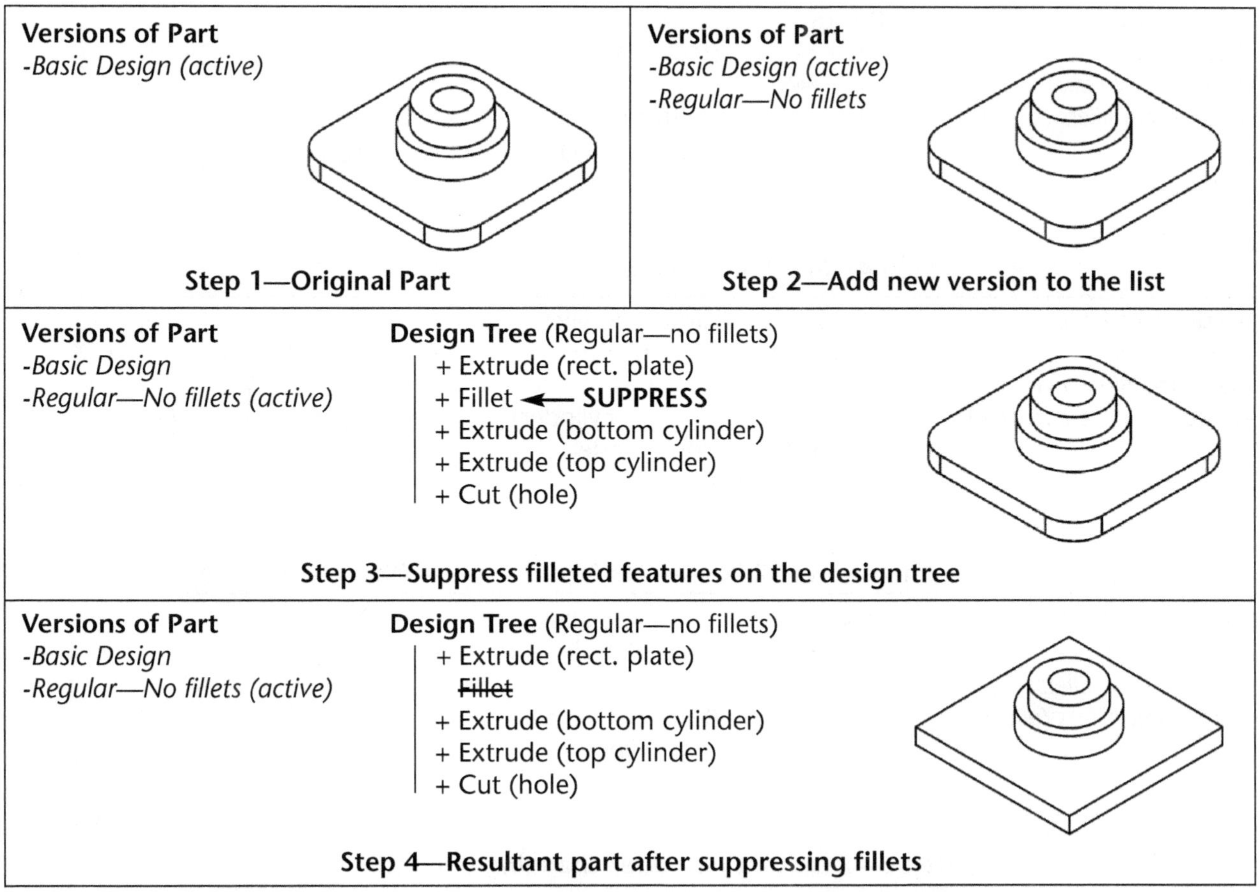

Figure 12.41 General steps required to create the first version of the basic design

The second variation, "Large – no fillets", requires additional steps. First, a new version needs to be added to the versions manager menu. The filleted features are suppressed using the design tree, and three dimensions need to be changed. See Figure 12.42 for an illustration of the steps used to create the second variation.

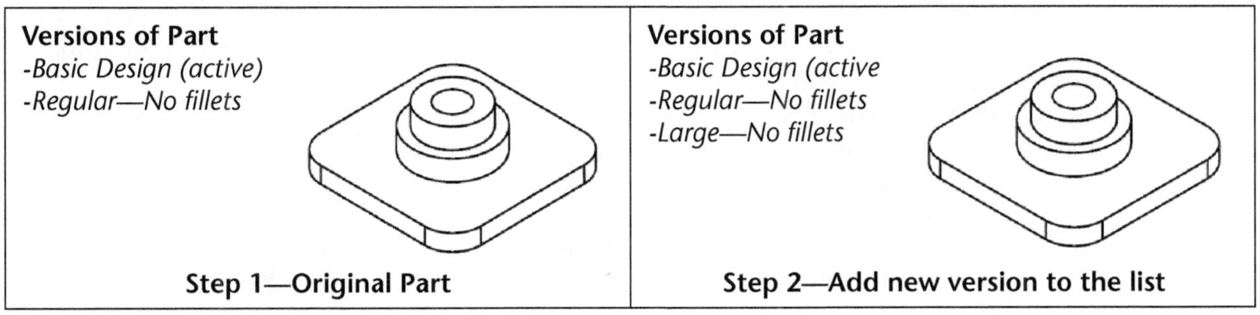

Versions of Part
-Basic Design
-Regular—No fillets
-Large—No fillets (active)

Design Tree (Large—no fillets)
+ Extrude (rect. plate)
+ Fillet ⟵ **SUPPRESS**
+ Extrude (bottom cylinder)
+ Extrude (top cylinder)
+ Cut (hole)

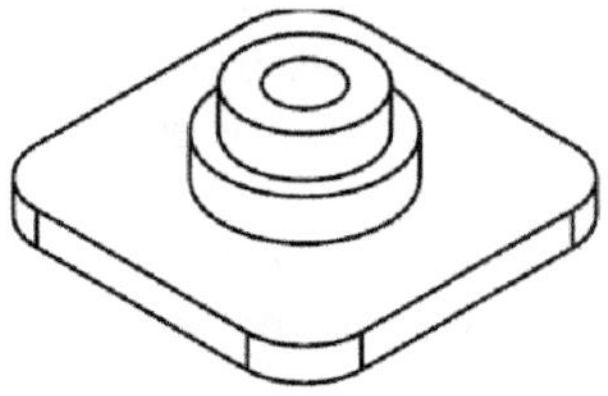

Step 3—Suppress filleted features on the design tree

Versions of Part
-Basic Design
-Regular—No fillets
-Large—No fillets (active)

Design Tree (Large—no fillets)
+ Extrude (rect. plate)
~~Fillet~~
+ Extrude (bottom cylinder)
+ Extrude (top cylinder)
+ Cut (hole)

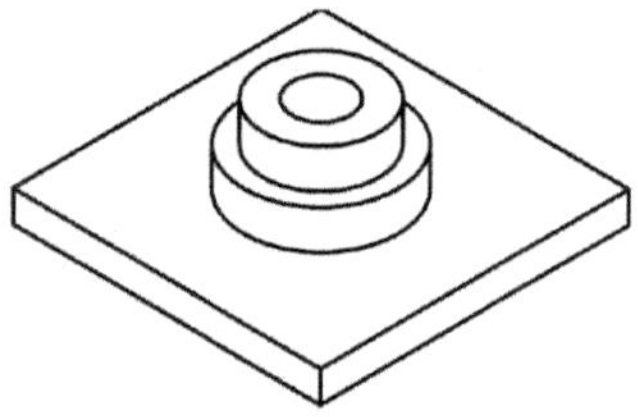

Step 4—Resultant part after suppressing fillets

Versions of Part
-Basic Design
-Regular—No fillets
-Large—No fillets (active)

Design Tree (Large—no fillets)
+ Extrude (rect. plate)
~~Fillet~~
+ Extrude (bottom cylinder)
+ Extrude (top cylinder)
+ Cut (hole)

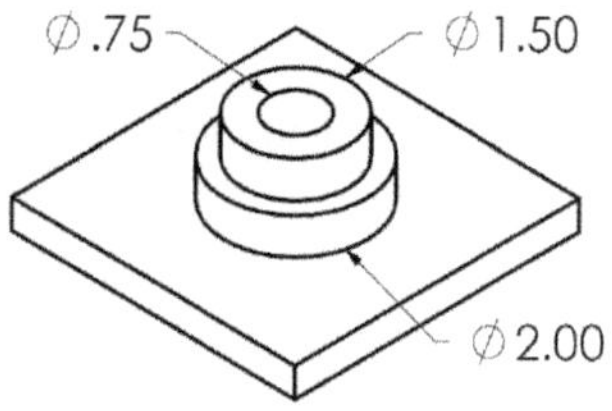

Step 5—Modify Dimensions (Ø2.00 ⟶ Ø3.00, Ø1.50 ⟶ Ø2.00, Ø.75 → Ø1.00)

Versions of Part
-Basic Design
-Regular—No fillets
-Large—No fillets (active)

Design Tree (Large—no fillets)
+ Extrude (rect. plate)
~~Fillet~~
+ Extrude (bottom cylinder)
+ Extrude (top cylinder)
+ Cut (hole)

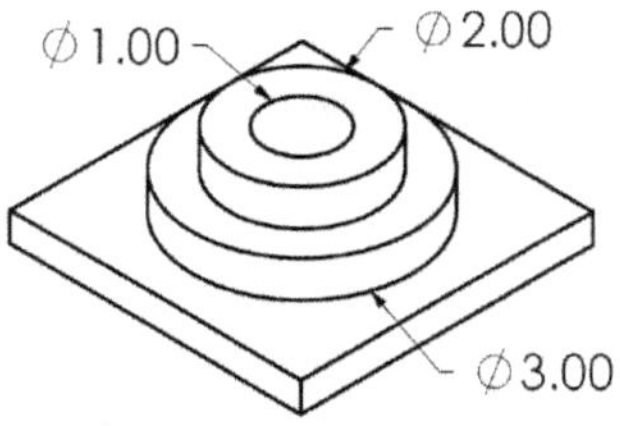

Step 6—Resultant part after changing dimensions

Figure 12.42 General steps required to create the second version of the basic design

9.2 Automatic Creation of Families of Parts

Creating families of parts automatically requires the use of a generic model and a design table. The generic model works essentially as a template that is used to define all the features of the part. The design table stores information about each version of the design and how each specific version differs from the generic model. The attributes that change from version to version are usually dimensions and the suppression state of features.

The first step required to create a family of parts is to design the generic model and indicate the features and dimensions that need to be included in the design table. The object and attributes shown in Figure 12.43 will be used in following examples to illustrate the creation of configurations.

The next step involves the creation of the design table. The design table is a regular table or spreadsheet where each row represents a different version of the design and each column represents an attribute that will change from version to version. When the design table is first created, only one row with the values of the generic model is present (See Figure 12.44). This spreadsheet is electronically linked to the model and drives the geometry.

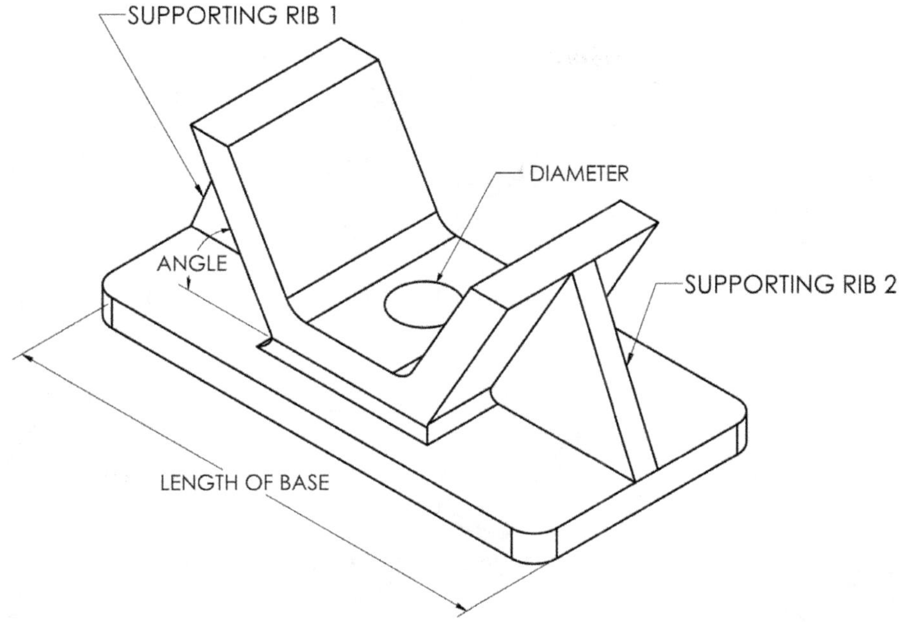

Figure 12.43 Generic model and parameters to be included in the design table

Config name\Parameters	Angle	Length of Base	Diameter	Supporting Rib 1	Supporting Rib 2
Default	60	6	.75	Unsuppressed	Unsuppressed

Figure 12.44 Example of initial design table

For every desired new version of the generic model, a new row must be inserted in the design table. The values for each cell are either numerical values, if they represent a dimension, or state values (suppressed or unsuppressed), to indicate the presence or absence of a particular feature. A design table with five different versions of the generic model is shown in Figure 12.45.

Config name\Parameters	Angle	Length of Base	Diameter	Supporting Rib 1	Supporting Rib 2
Default	60	6	.75	Unsuppressed	Unsuppressed
Version 1	60	8	.75	Unsuppressed	Unsuppressed
Version 2	60	8	.75	Suppressed	Suppressed
Version 3	35	6	.25	Unsuppressed	Unsuppressed
Version 4	35	8	.25	Unsuppressed	Unsuppressed
Version 5	35	8	.25	Suppressed	Suppressed

Figure 12.45 Final design table with definitions of a generic model and five different configurations

The example shown in Figure 12.46 illustrates the different 3D models generated automatically by a parametric modeling package using the generic model and design table created in the previous example.

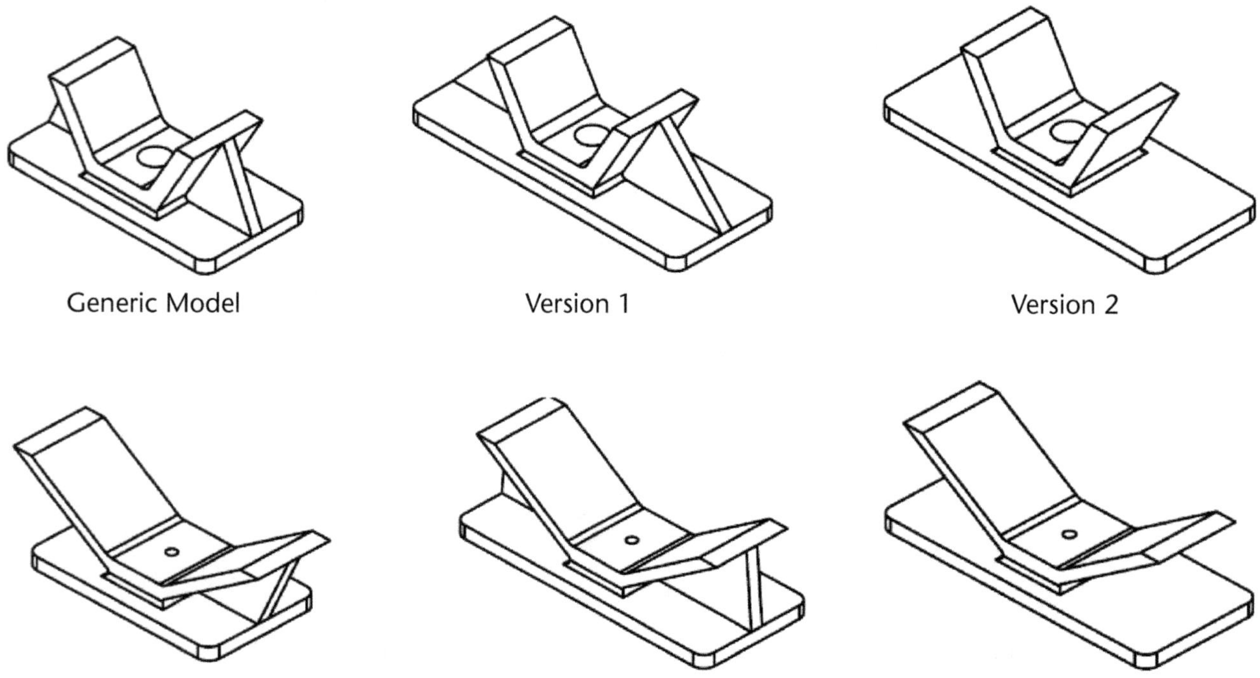

Figure 12.46 Models created automatically using the definitions from the previous design table

10. Obtaining Mass Properties

The parts modeled with a solid modeling package are accurate simulations of real world objects. Mass properties are the static physical properties associated to a solid model. Common physical properties include volume, surface area, density, mass, center of gravity, and moments of inertia.

Acquiring the mass properties of a model is essential for engineering design and analysis. The volume of an object may determine how much material is needed to manufacture it. The surface area of a part can determine how a part dissipates heat or verify if the air flow traveling through the part gets disrupted.

Material: Titanium	
Elastic Modulus	1.59541e+007 lb/in^2
Poissons Ratio	0.3
Shear Modulus	6.23662e+006 lb/in^2
Thermal Expansion Coefficient	8.8e-006
Density	0.166185 lb/in^3
Thermal Conductivity	22 W/m K
Specific Heat	460 J/kg K
Tensile Strength	34083.9 lb/in^2
Yield Strength	20305.3 lb/in^2

Figure 12.47 Titanium material properties stored in a CAD system

The mass properties of an object are related to one another, but calculating them by hand can be complex and time-consuming. Most CAD packages provide commands to acquire the mass properties of solid models and allow the user to export this information for analysis and reports.

Before calculating mass properties, it is necessary to assign material properties to the part. In most CAD packages, the properties of standard and commonly used materials are internally stored in a library, although custom material properties can be defined and assigned to parts. See Figure 12.47 for an example of material properties stored in a CAD package.

Some materials may have other properties defined: hardening factor, compressive strength, structural damping coefficient, etc.

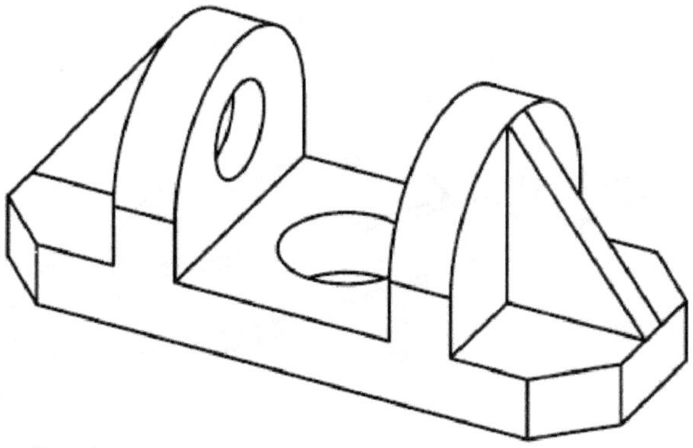

Material: Cast Carbon Steel
Density = 0.01 grams per cubic millimeter
Mass = 444.85 grams
Volume = 57032.61 cubic millimeters
Surface area = 16807.06 millimeters^2
Center of mass: (millimeters) X = 50.00 Y = 11.85 Z = 20.00
Principal axes of inertia and principal moments of inertia: (grams*square millimeters)

Ix = (1.00, 0.00, 0.00)	Px = 101074.45
Iy = (0.00, 0.00, -1.00)	Py = 381846.13
Iz = (0.00, 1.00, 0.00)	Pz = 389872.02

Moments of inertia: (grams*square millimeters)

Lxx = 101074.45	Lxy = 0.00	Lxz = -0.00
Lyx = 0.00	Lyy = 389872.02	Lyz = 0.00
Lzx = -0.00	Lzy = 0.00	Lzz = 381846.13

Moments of inertia: (grams*square millimeters)

Ixx = 341441.30	Ixy = 263486.63	Ixz = 444854.39
Iyx = 263486.63	Iyy = 1679949.74	Iyz = 105394.65
Izx = 444854.39	Izy = 105394.65	Izz = 1556407.20

Figure 12.48 Acquiring mass properties of an object

In some cases, appearance properties such as color, reflectivity, or even hatch patterns are stored along with physical properties.

The mass properties of a solid object are calculated automatically by the CAD system using the material properties and the geometry of the part (see Figure 12.48).

11. Assembly Modeling

In engineering, very rarely does a design consist of one single part. Instead, different components are combined and linked together to work as a unique system known as an assembly. Modern parametric modeling packages provide tools to establish geometric relationships between different 3D models and examine the interfaces between them. Designers can study how the components will react when assembled and perform analysis, interference checking, and other physical tests. This process lowers design cost, reduces design time to market, and increases profit.

The techniques and methods required to work with assemblies are often referred to as assembly modeling. Parametric modelers provide separate environments, tools, and file formats for this purpose.

Assemblies contain two or more parts. Each individual part is referred to as a component in assembly terminology (see Figures 12.49 and 12.50). With large assemblies, subgroups of components may be created if necessary. Each subgroup is called a subassembly. A subassembly is a group of parts that fit together and work as a functional unit within a larger assembly.

The relationship between the parts and the assembly is associative. Any changes made in parts are automatically reflected in the associated assembly, and vice versa.

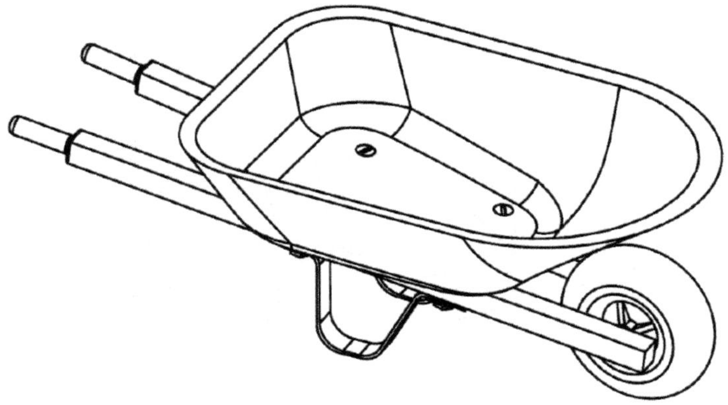

Figure 12.49 Assembly drawing of a wheelbarrow

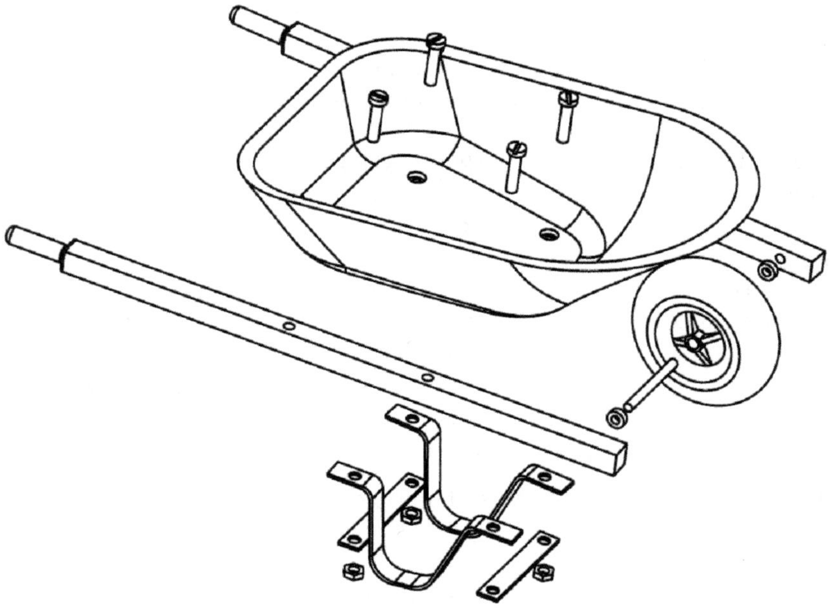

Figure 12.50 Exploded assembly drawing of a wheelbarrow showing its components

There are two different approaches to assembly modeling: Top-down and Bottom-up.

Top-down assembly modeling requires an initial definition of the assembly, specifying but not detailing its individual parts. Each part is refined later in greater detail. The definition of the assembly is normally done using generic structures provided by the software that hold information on the overall layout of the assembly. Individual parts are later modeled using these generic structures. In top-down assembly modeling, the overall layout of the assembly is built before each individual part is designed. The general steps of the top-down assembly modeling process are illustrated in Figure 12.51.

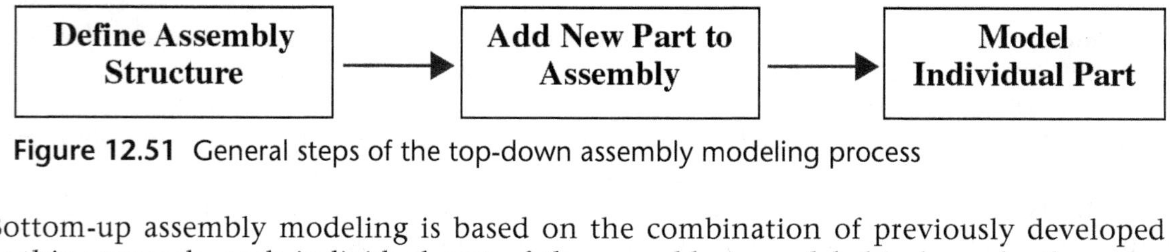

Figure 12.51 General steps of the top-down assembly modeling process

Bottom-up assembly modeling is based on the combination of previously developed parts. Using this approach, each individual part of the assembly is modeled independently and once all the parts are created, they are inserted into the assembly and connected. In bottom-up assembly

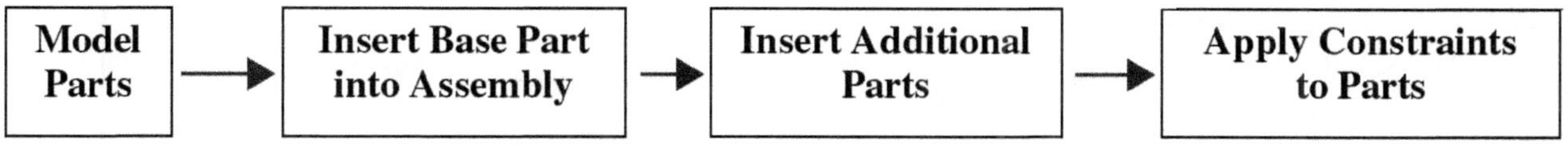

Figure 12.52 General steps of the bottom-up assembly modeling process

modeling, each individual part is built before the overall assembly is designed. The general steps of the bottom-up assembly modeling process are illustrated in Figure 12.52.

Different factors determine how an assembly should be created. Some time should be spent planning for the best way to put the components together before the assembly process begins. Capturing design intent should always be considered. Decisions should be made whether subassemblies are needed and how the parts will interact with each other.

The first part inserted in an assembly is called the base component. The base component is fixed in space, thus it will not move. All other components will eventually relate to the base component. Whenever a design is assembled, a decision needs to be made as to which part is most suited to be the base part. In most assemblies, it is fairly easy to identify the base component, but occasionally different possibilities exist.

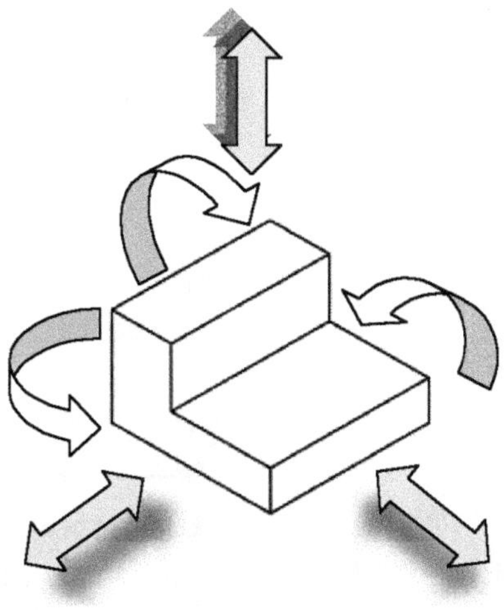

Figure 12.53 Six degrees of freedom for parts without any restrictions

Once the base component is established, the next step is to insert additional components. These components are not fixed in space, so they can be freely moved and rotated. If multiple copies of the same part are required in the assembly, there is no need to model each one individually. Multiple instances of the same component can be brought into the assembly with relatively little effort.

The number of different ways a part can be moved in 3D space is defined by its degrees of freedom. Without any restrictions, a part can move with six degrees of freedom: translations along X, Y, and Z axis, and rotations around X, Y, and Z axis (See Figure 12.53). If a part is fully constrained or fixed in space and it is not allowed to move, the part is said to have zero degrees of freedom.

The most important step in the assembly modeling process is to locate and orient the components relative to one another and define how each component is allowed to move. This step determines how many degrees of freedom will be available for each component. The number of degrees of freedom for all the components in an assembly must be minimized, only allowing the rotations or translations required for analysis or simulations.

The reduction of degrees of freedom is accomplished though the use of assembly constraints. An assembly constraint is a geometric relationship between two components. Similar to regular geometric constraints for individual parts, assembly constraints are applied to the vertices, edges, or surfaces of two different components. For example, a planar surface of one component can be made parallel to a planar surface of another component, or a cylindrical surface of one part concentric to a circular hole of a different part, or a straight edge of one component coincident with a straight edge of another component.

Different types of assembly constraints can be used. Normally, software packages will display a list of possible constraints that can be applied to components based on the objects that are currently selected. Although there is a fairly consistent group of constraints available in almost all parametric modelers (coincident, parallel, perpendicular, concentric, tangent, angle, and distance), some applications may provide specific and advanced constraints, such as symmetric, paths, screws, or gear constraints that may not be available in other packages.

Different approaches can be taken to constrain an assembly. Some approaches require the use of more constraints while others are more flexible. In general, the same guidelines used in the individual part modeling process apply to assemblies. Assembly constraints should resemble the physical constraints of the design. The selection and application of constraints to a particular assembly should be the result of a prior planning where design intent is always the major consideration.

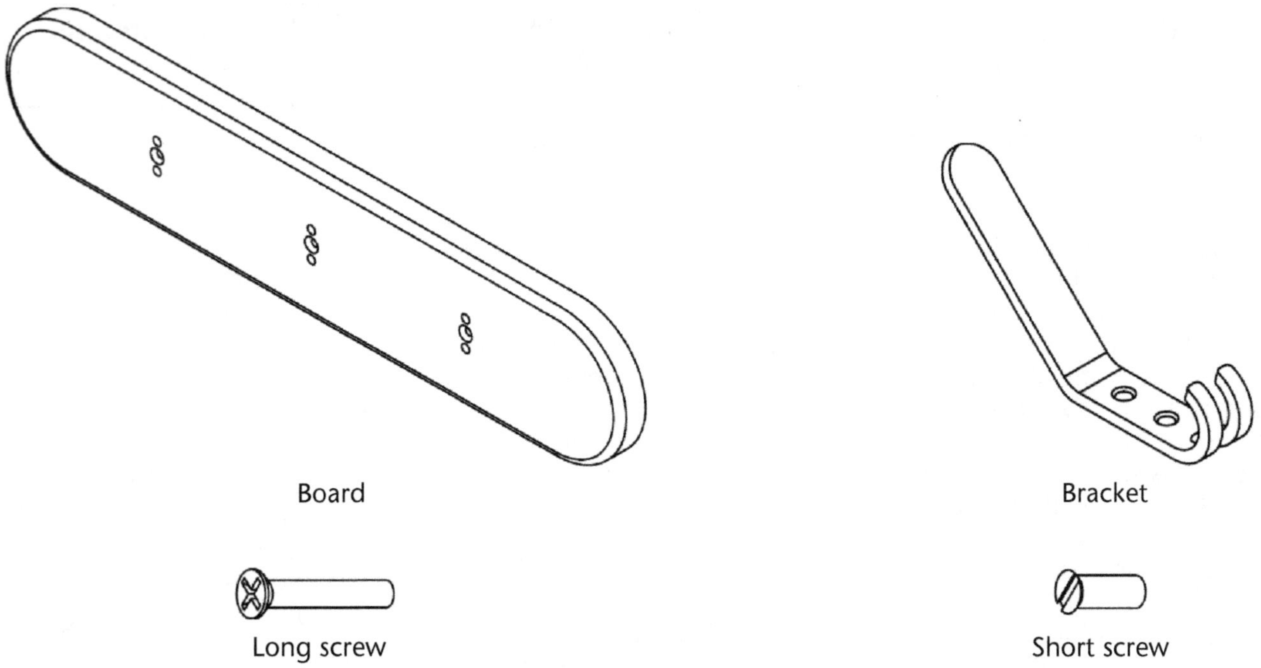

Figure 12.54 Parts required for the coat hanger assembly

Consider a simple assembly, such as a coat rack, to illustrate the assembly modeling process and to understand how constraints are applied. There are multiple ways to constrain the components and create the assembly. This example illustrates several techniques, so different types of constraints are used. The parts required to build this assembly are shown in Figure 12.54.

Only the board and the bracket were modeled for this example. Most CAD packages provide built-in libraries of standard parts that can be used in assemblies. Common objects like screws, nuts, bolts, washers, etc. do not need to be modeled from scratch every time, but can be imported from the library into the assembly.

For this simple assembly, it appears that the board is the natural base component. Three instances of "Long Screw" are added as additional components. These three screws will fix the board to the wall. Additional components are initially inserted into the assembly without any constraints, with six degrees of freedom. Thus, these components can be moved and rotated freely using the manipulation tools provided by the parametric modeling software. It is easier if additional components are located and oriented so that they do not overlap with any of the existing components and the surfaces are easy to select when creating constraints (See Figure 12.55).

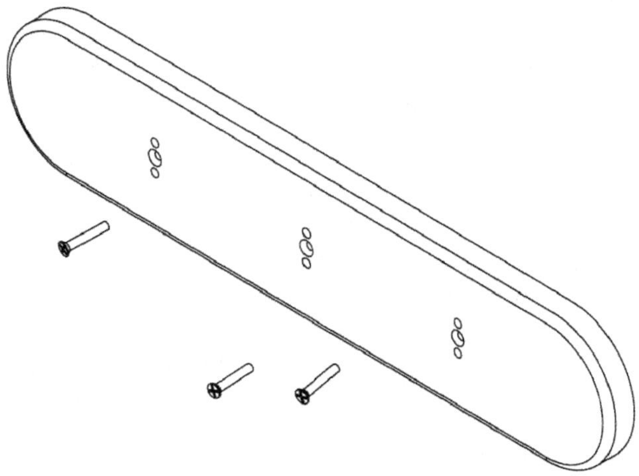

Figure 12.55 Base component "Board" and three instances of component "Long screw"

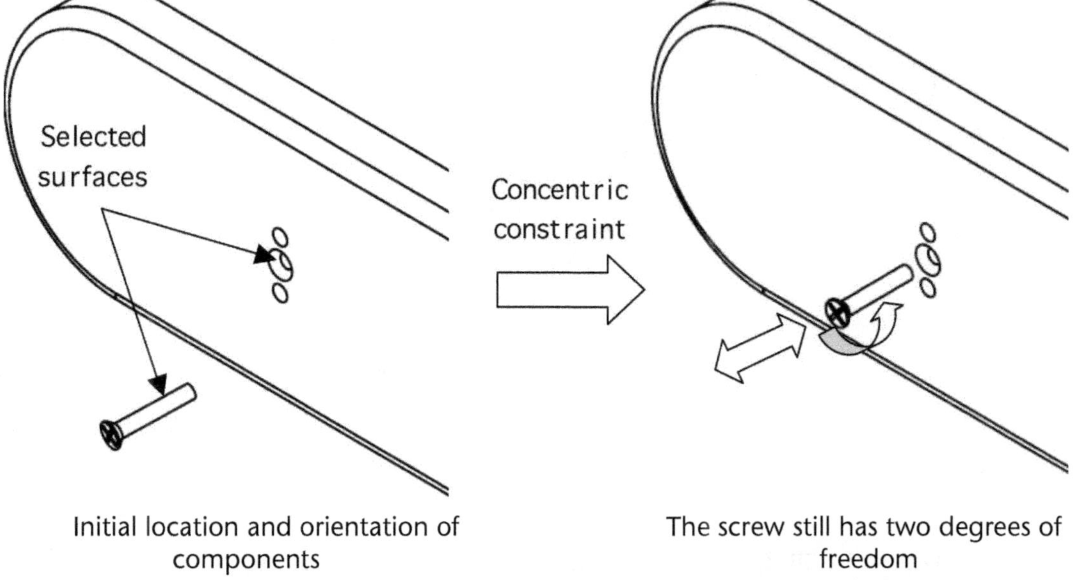

Figure 12.56 Result of applying a concentric constraint to the cylindrical surfaces of the screw and the hole

The next assembly step is to create a relationship between the first screw and the first hole on the board. The screw and the hole must align so the screw can fit inside the hole. A concentric constraint between one of the cylinders of the screw and the interior surface of the hole will provide the desired result. The concentric constraint eliminates four degrees of freedom on the screw, but the screw can still move horizontally (in and out) and rotate around its longitudinal axis (See Figure 12.56).

Next, another constraint is applied in order to eliminate the horizontal translation of the screw so that it is fully inside the hole. There are multiple options that can be considered in order to obtain this result. In this case, a coincident constraint is created between the top surfaces of the screw and the board (See Figure 12.57).

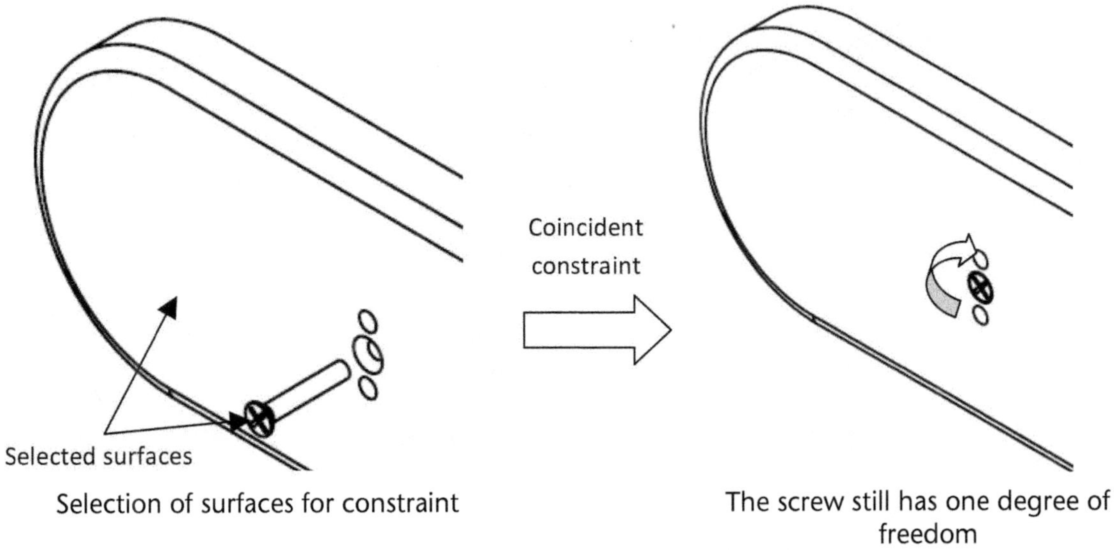

Figure 12.57 Result of applying a coincident constraint to the top surface of the screw and the board

Removal of the last degree of freedom of the screw is not required since this state reflects how the component is allowed to move in reality. The same set of constraints (concentric-coincident) can be created to relate the remaining two long screws to the board (see Figure 12.58).

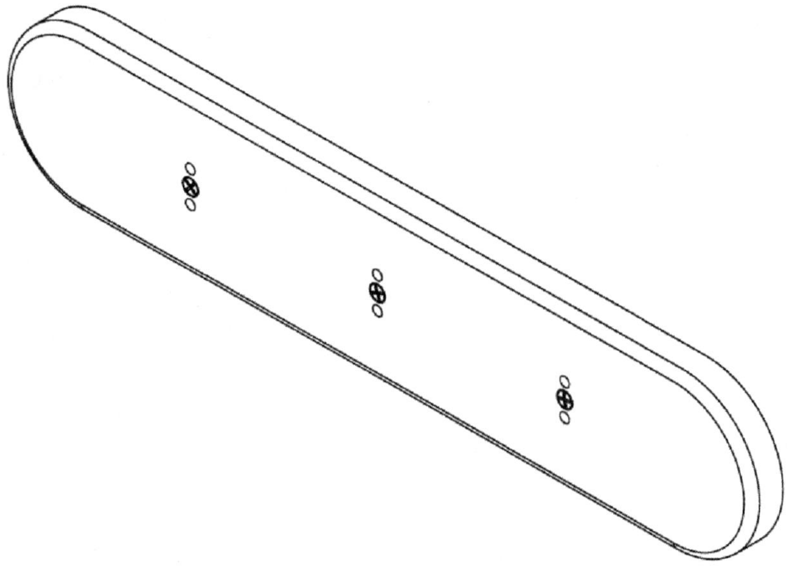

Figure 12.58 Board and screws after applying constraints

Continued assembly requires that three instances of the bracket be inserted into the assembly. Again, the instances are located and oriented so that they do not overlap with any of the existing components (see Figure 12.59).

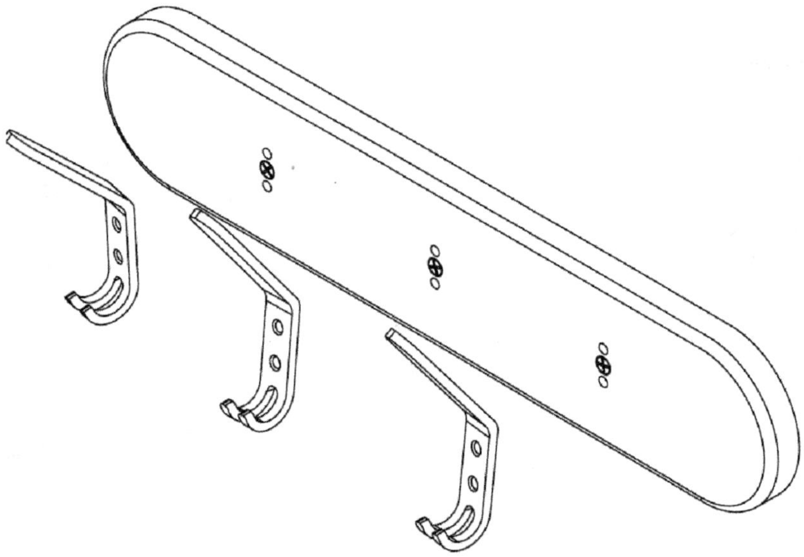

Figure 12.59 Bracket instances inserted into the assembly

To relate the first bracket to the board, two concentric constraints can be created between the holes on the bracket and the holes on the board (see Figure 12.60).

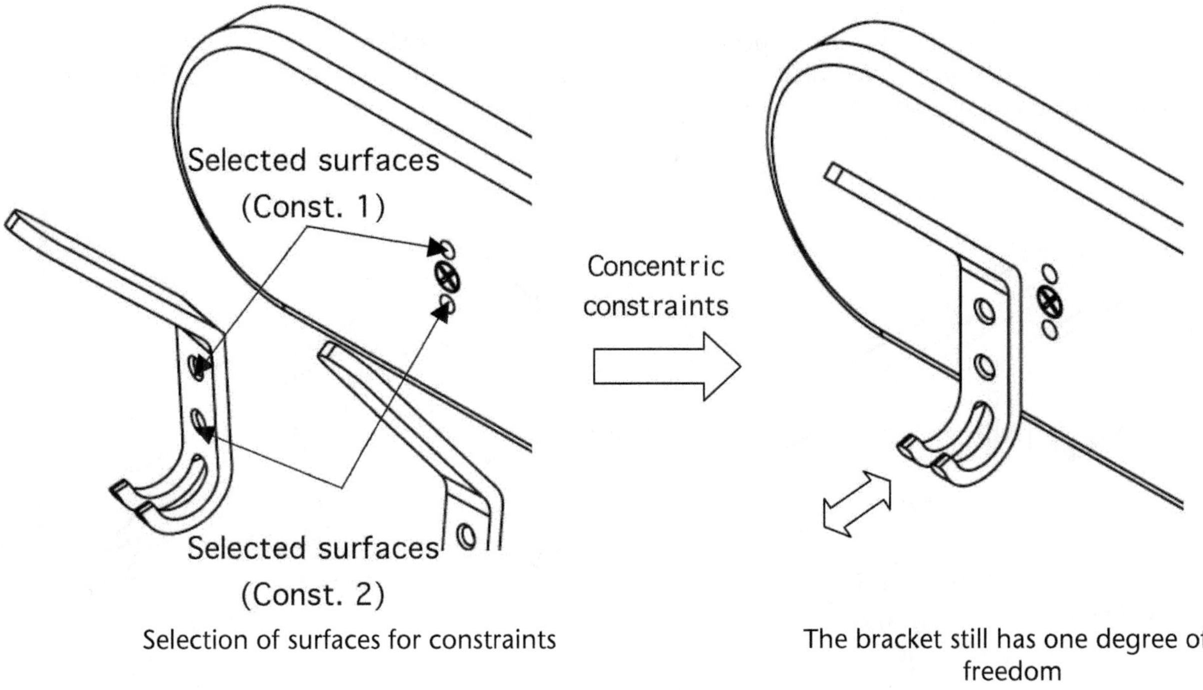

Figure 12.60 Concentric constraints between the holes on the bracket and the holes on the board

To fully constrain the bracket to the board and eliminate the last degree of freedom, a coincident constraint needs to be created between the back surface of the bracket and the front surface of the board (see Figure 12.61).

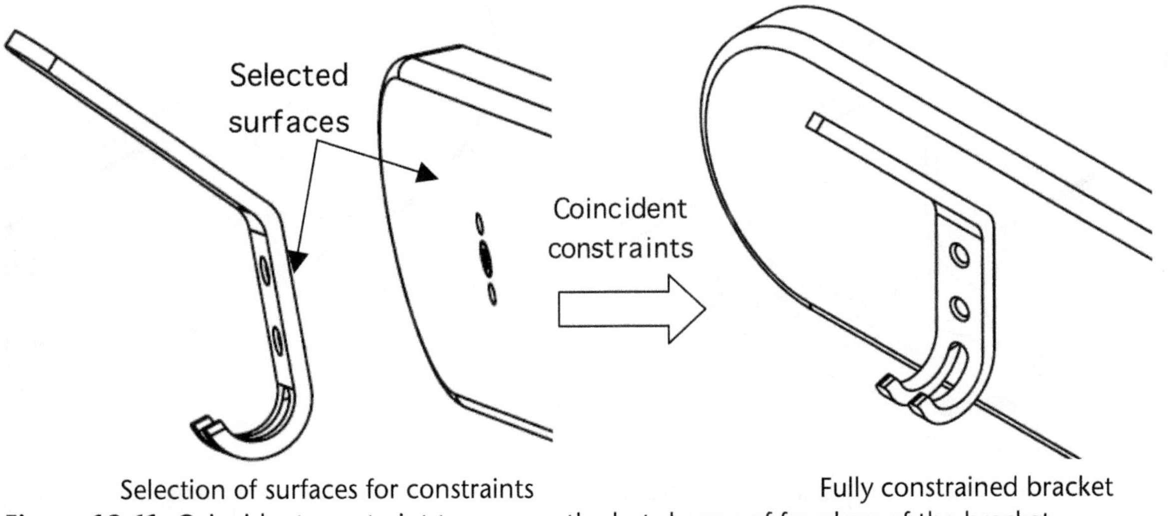

Figure 12.61 Coincident constraint to remove the last degree of freedom of the bracket

The remaining two brackets can be constrained to the board using the same approach that was used for the first bracket (two concentric constraints and a coincident constraint). A coincident constraint and two distance constraints can be used (see Figures 12.62, 12.63, and 12.64). Advanced modelers can create a linear pattern of the bracket component to save steps.

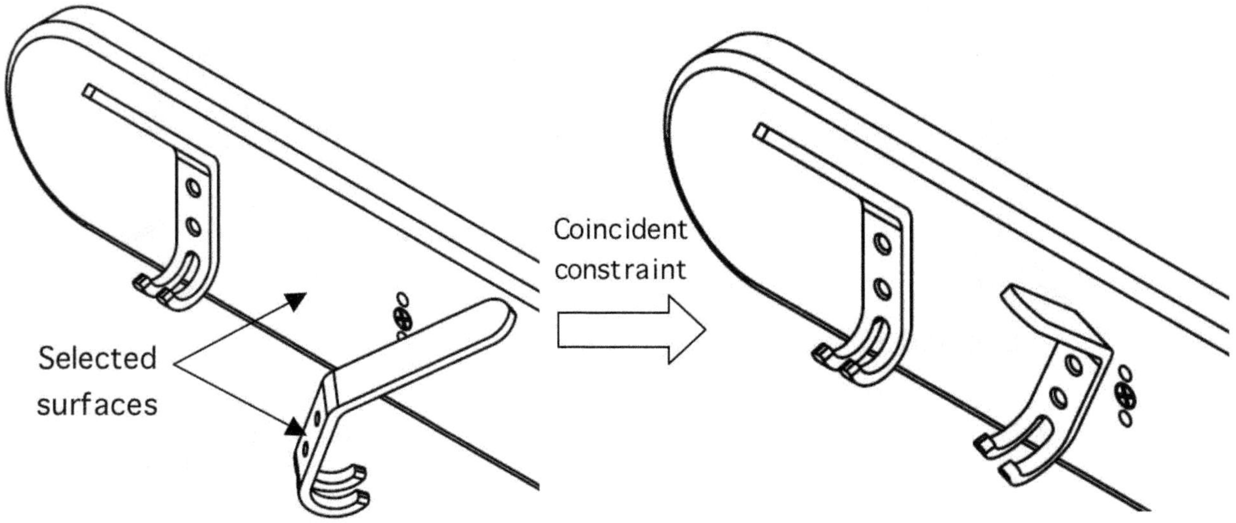

Selection of surfaces for constraints

Bracket after being constrained

Figure 12.62 Coincident constraint for second bracket

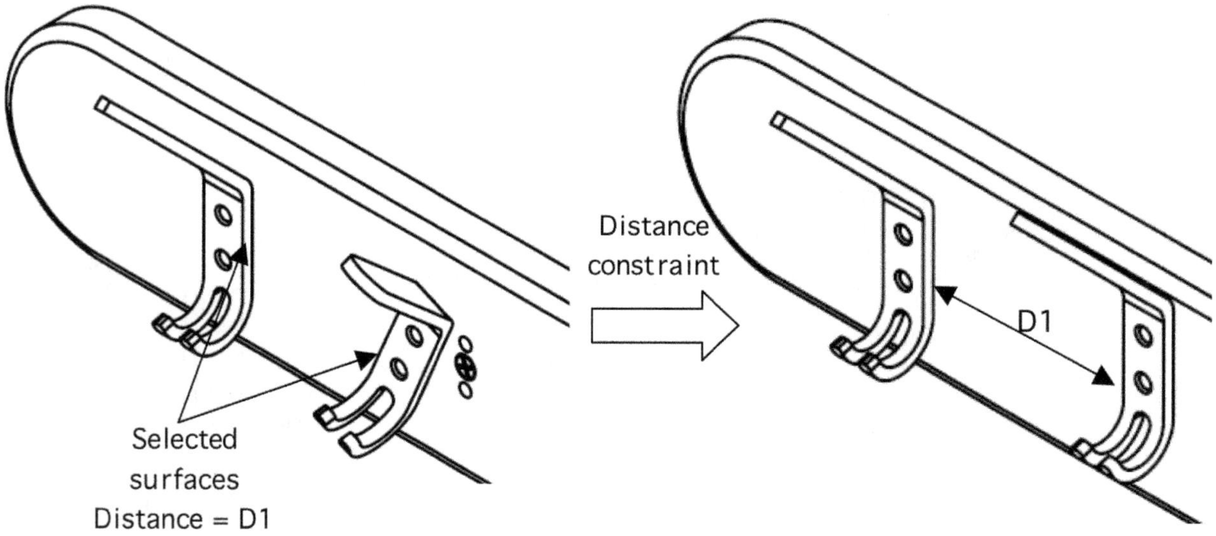

Selection of surfaces for constraint indicating distance

Bracket after being constrained

Figure 12.63 First distance constraint for second bracket

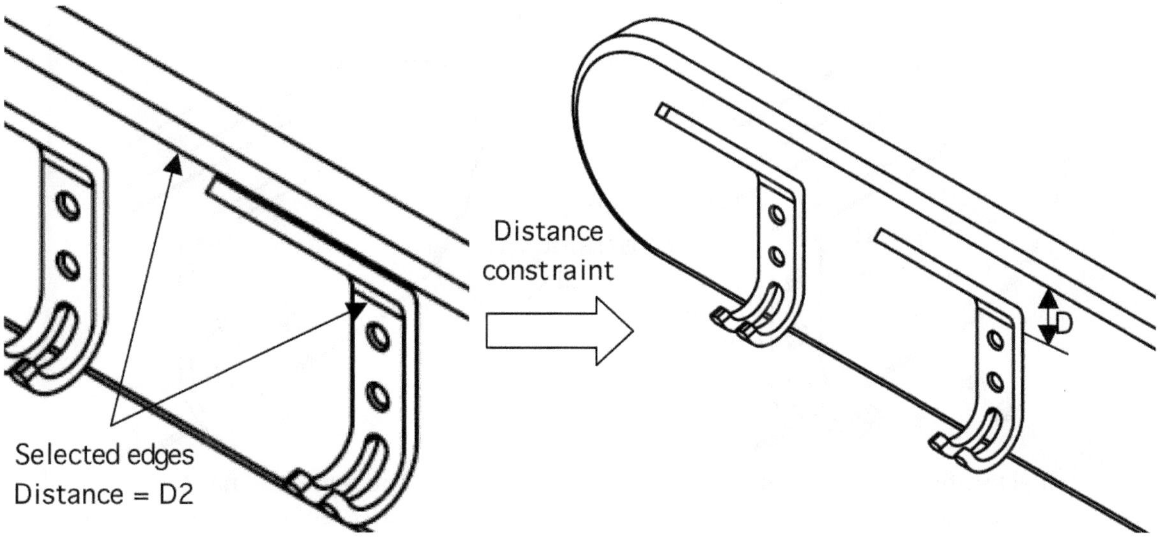

Selection of edges for constraint Bracket after being constrained

Figure 12.64 Second distance constraint for second bracket

Finally, six instances of "short screw" must be inserted to complete the assembly. The instances are originally located and oriented close to its final position so that they are easy to select when applying constraints (see Figure 12.65).

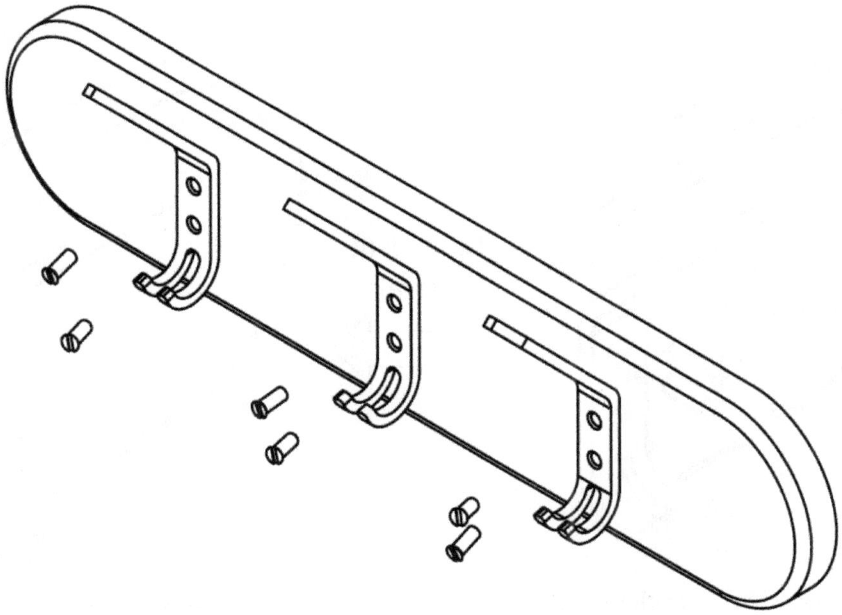

Figure 12.65 Instances of "short screw" inserted in the assembly

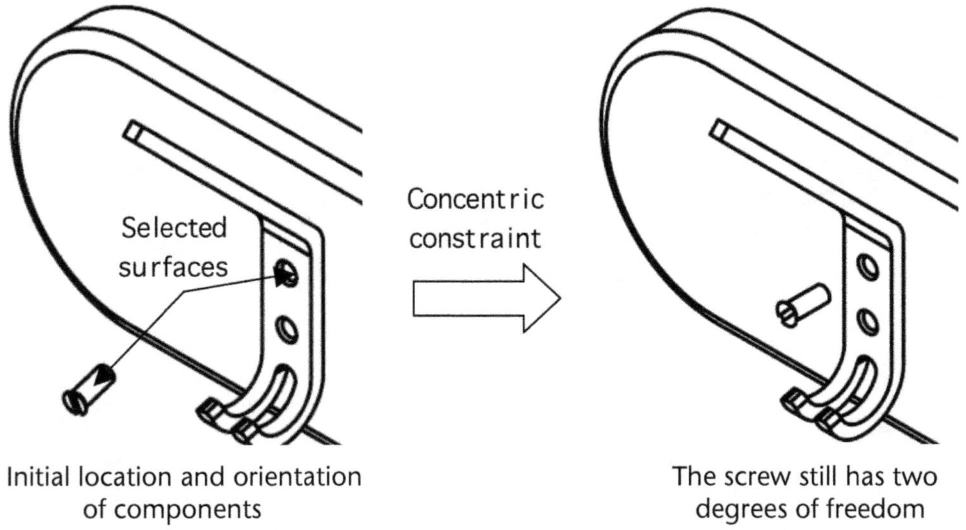

Figure 12.66 Result of applying a concentric constraint to the cylindrical surfaces of the screw and hole

Each of these new instances needs to be constrained to its corresponding bracket. Concentric and coincident constraints can be used again in this case (see Figure 12.66).

A coincident constraint between the planar surface of the screw and the planar surface of the bracket will eliminate the translational degree of freedom (see Figure 12.67). Again, there is no need to remove the last degree of freedom of the screw.

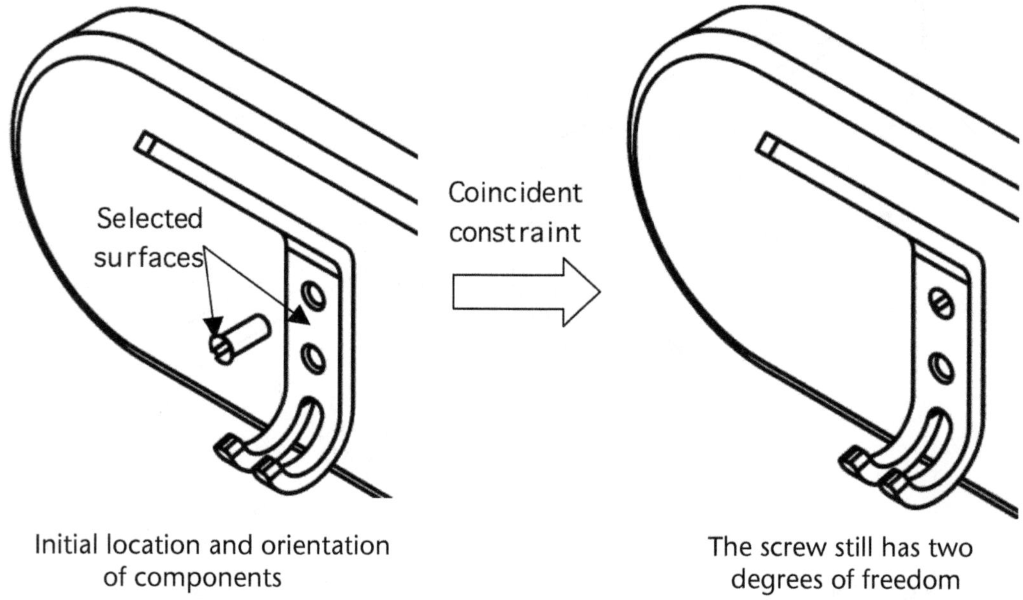

Figure 12.67 Result of applying a coincident constraint to the planar surface of the screw and the bracket

The finished assembly, after the components are inserted and the required constraints are defined, is shown in Figure 12.68.

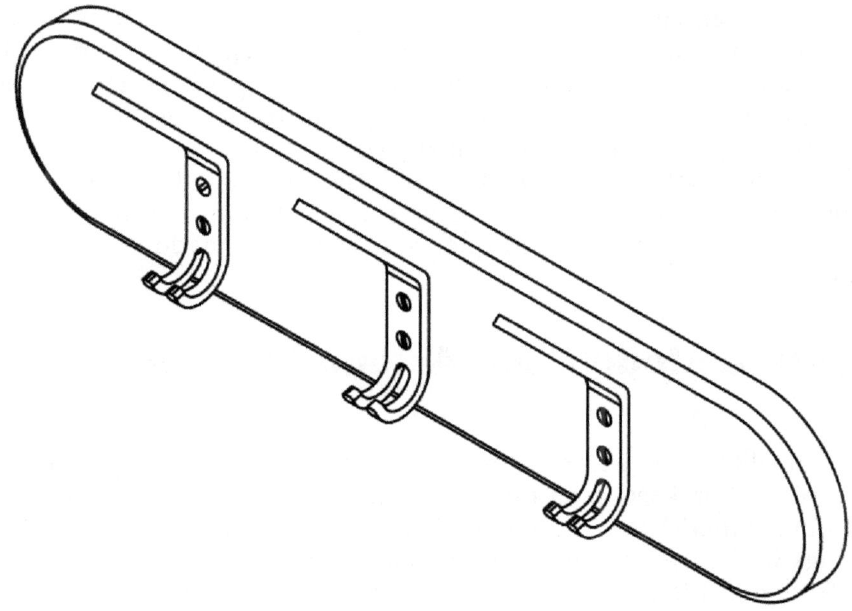

Figure 12.68 Finished coat hanger assembly

For assemblies with moving parts, such as a door hinge or a table vice, numerical limits can be set for components to establish the distance or the degrees a part is allowed to move or rotate with relation to another part. Also, most parametric modeling applications provide collision detection capabilities that can be used to determine if two parts come in contact with each other when they are moved or rotated or if two components interfere with each other. Checking for interference is especially useful when working with meshing components, such as gear trains.

12. Families of Assemblies and Exploded Views

The tools to manage families of parts provide a convenient way to develop variations of models with different dimensions, features, or other parameters. These tools also allow the creation of multiple variations of an assembly model. Simplified versions of the design can be easily created by suppressing or hiding components. Families of assemblies with different modifications of the components, different parameters for assembly features, different dimensions, or configuration-specific custom properties can be generated with little effort.

Exploded assemblies are certainly the most common type of assembly representations. The purpose of an exploded assembly is to show how the components are placed together. The components are shown slightly separated by distance revealing the order of assembly (see Figure 12.69).

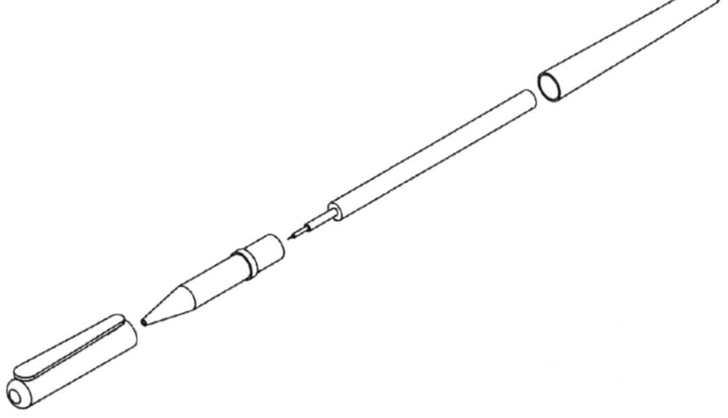

Figure 12.69 Exploded configuration of a pen showing the order of assembly

Generating exploded configurations is usually a very straight forward task in most parametric modelers. Once the assembly model is put together, each individual component can be selected and moved along the X, Y, and Z directions to the desired position. For clarity, each part should be moved far enough from the assembly so that it does not overlap with the rest of the components when the assembly is shown. For complex assemblies with many components, some applications provide automatic explosion tools so the user is not required to move the components one at a time.

Exploded configurations of assemblies can be used to create drawings and presentations. Modern packages even provide tools to generate exploded assembly animations that can be exported to popular video formats.

13. Documenting Models and Assemblies. Generation of Drawings

The creation of individual three-dimensional parts is the main purpose of parametric modeling applications. However, nearly all packages offer other tools that provide extra functionality. An important feature of most parametric modelers is the automatic generation of drawings from 3D models.

Creating drawings with a parametric modeler usually requires the use of a different module of the program and a different file type. Most packages provide a separate interface just to work with 2D drawings. The tools available for this purpose are explicitly related to page layout and getting a drawing ready for printing. A two-dimensional environment resembling a sheet of paper is displayed, as opposed to the 3D world used when modeling.

There are several steps that need to be completed to generate professional quality drawings from 3D models. Although some steps do not necessarily have to be performed in the same sequence they are discussed, the user should be aware of how their specific application handles each step. It is a good idea to check the general settings of the page before a drawing is created. Selecting the correct paper size, orientation, and creating a title block beforehand will save time and effort.

The generation of drawings from 3D models is fairly automatic in most parametric modeling software and only a few parameters need to be set. First, the type of drawing to generate needs to be indicated. Most packages provide orthographic views, isometrics, auxiliary views, and various types of sections. See Figure 12.70 for an example.

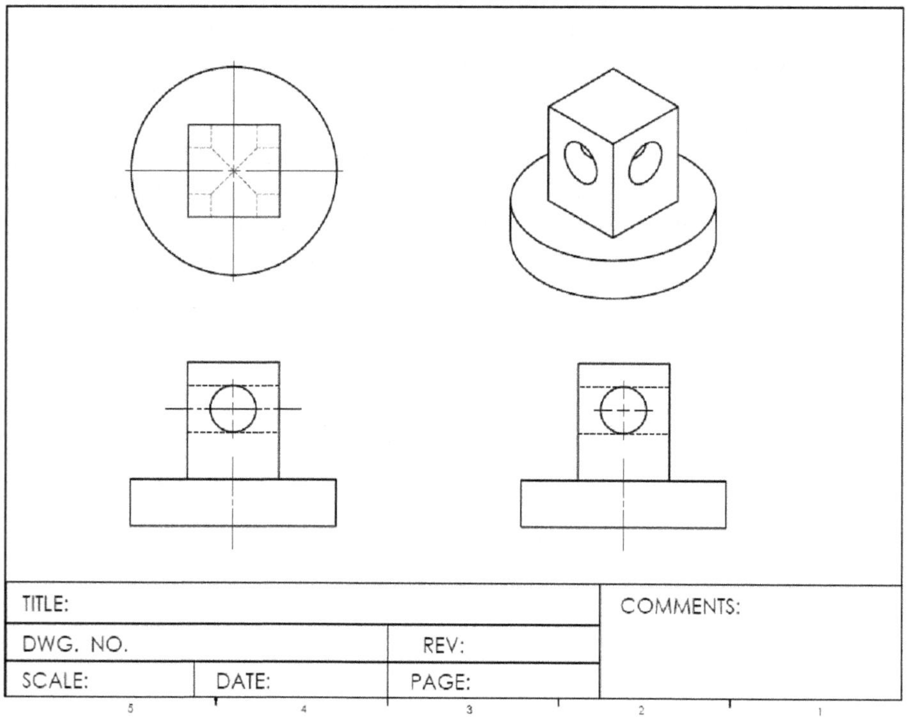

Figure 12.70 Sample page with custom title block, orthographic and pictorial views

When creating orthographic drawings, the projection method to use (First angle / Third angle) needs to be specified. First, a base view is created. A base view is used as a reference to generate other views, known as projected views. In orthographic drawings, the base view is usually the front view (which is commonly the most descriptive), and is used to generate the top, side, and pictorial views. If the 3D object was modeled with the correct orientation in 3D space, the process of finding the desired front view becomes easier, although the views can be easily redefined in the part file. In addition, the software ensures orthographic alignment, even when one of the views is moved.

The creation of special types of views is usually very intuitive. Special views are always projected from another view. For example, the steps required to create an auxiliary view are usually the selection of the base view to project from, and the line that represents the sloped surface. The creation of sectional views involves the definition of the cutting plane line on a regular orthographic view first. Once the cutting plane line is drawn, the section view is projected.

Each view of the drawing can be customized separately. Properties such as visibility of hidden lines, centerlines, shading mode (shaded, wireframe, etc), tangent edges, line thickness, colors, etc. are available in most parametric modelers (see Figure 12.71).

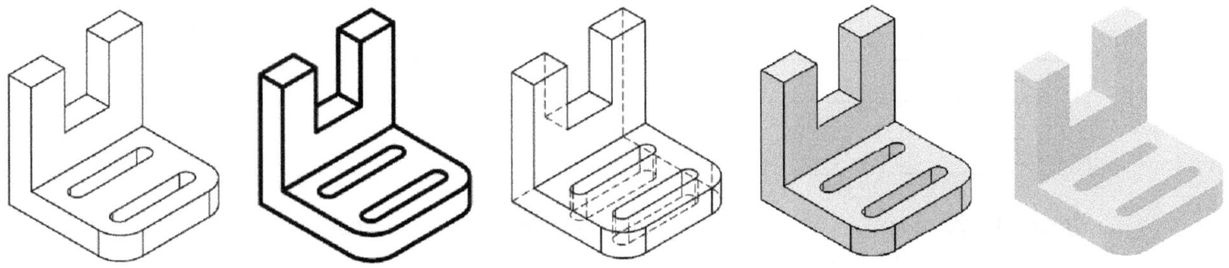

Figure 12.71 Customization of views in a drawing

13.1 Including Dimensions in Drawings

The views inserted in a drawing are directly connected to the 3D model. All the information located in the 3D model, such as dimensional constraints, can be brought into the drawing. In addition, editing and changing the 3D model will have a direct effect on the 2D drawing, and vice versa.

Retrieving dimensions from a 3D model to include them in a 2D drawing is usually a very straight forward operation. However, there is an important point that must be remembered. Dimensions are placed in the 2D drawing in accordance with how the sketches were constrained when the object was modeled. The software does not pay attention to aesthetics or dimensioning standards. In fact, the software does not even check if the drawing is entirely dimensioned or if there are any dimensions missing or duplicated. Missing dimensions are fairly common, especially when geometric constraints are used to define features in the model.

There are two types of dimensions that can be inserted in a drawing from a 3D model: driving dimensions and driven dimensions. Driving dimensions are those that are imported directly from the 3D model. These dimensions are directly linked to the 3D model, which means that editing a driving dimension in drawing mode will update the corresponding dimension in the 3D part. Driven dimensions are those that are applied to further annotate the drawing. Driven dimensions will conform to the part and they cannot be edited to change the 3D model. The example shown in Figure 12.72 illustrates a drawing after dimensions are automatically retrieved from the 3D model, the drawing is cleaned up following proper dimensioning standards and additional dimensions and notes are added.

13.2 Assembly Drawings

Assembly Drawings, especially exploded representations, are very common in engineering. These drawings contain a pictorial view of the exploded assembly (usually isometric), balloons, and a parts list, also called a Bill of Materials. Balloons are numbers pointing toward components of the assembly in the exploded view. These numbers correspond to the parts as described in the parts list. The parts list is a table that lists all assembly components and the quantity required. Other information such as material, reference number, or supplier can also be included (see Figure 12.73).

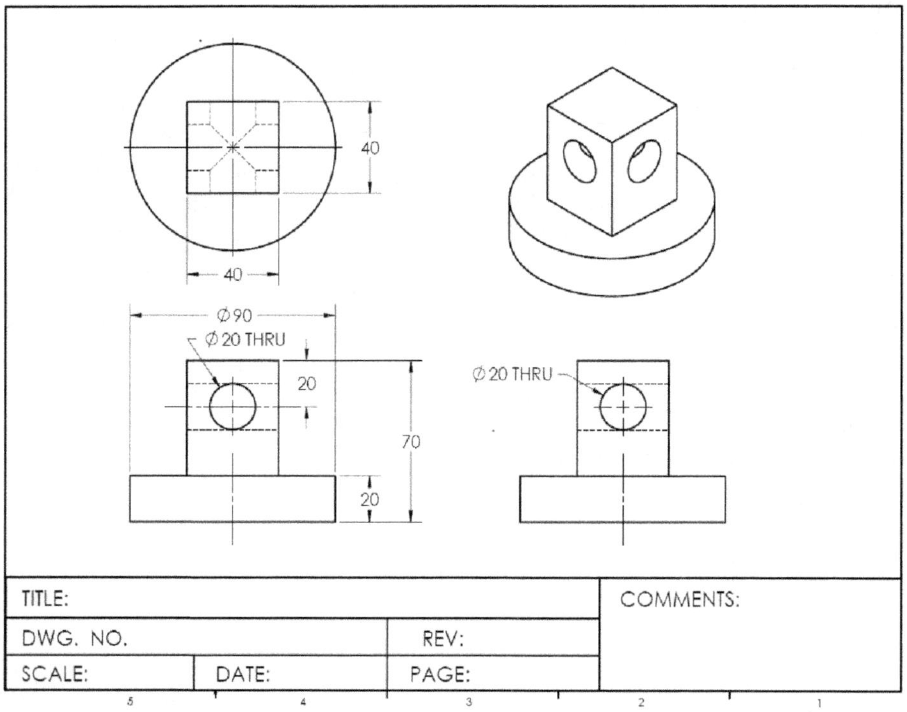

Figure 12.72 Sample page with custom title block, and dimensioned orthographic and pictorial views

ITEM NO.	PART	QTY.
1	BUCKET	1
2	HANDLE	2
3	LEG	2
4	WHEEL	1
5	AXLE	1
6	LONG SUPPORT	1
7	BOLT	4
8	NUT	4
9	SHORT SUPPORT	1
10	SPACER	2

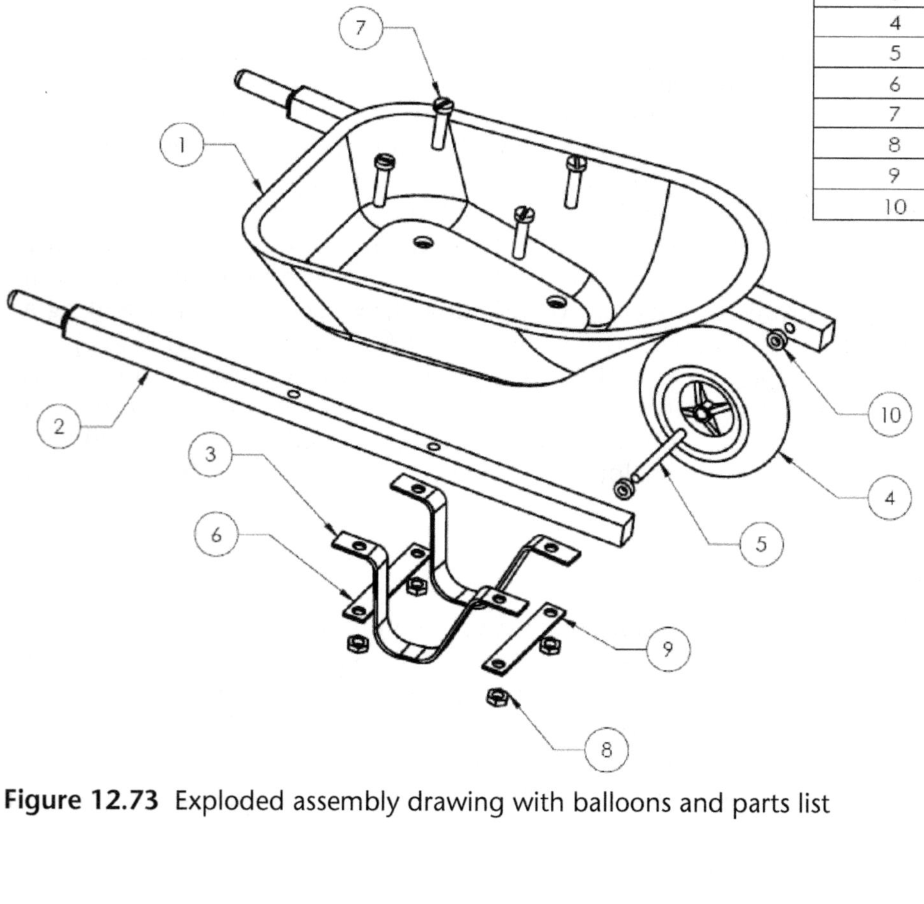

Figure 12.73 Exploded assembly drawing with balloons and parts list

Practice Test

1. **Which of the following is not a type of 3D computer model?**
   A) Wireframe B) Surface C) Solid D) CSG E) None of the above

2. **Which is the only type of 3D model that stores information about physical properties?**
   A) Wireframe B) Surface C) Solid D) CSG E) None of the above

3. **The ability of software to use specific parameters or properties to define a given model is called:**
   A) Parametric Modeling
   B) Assembly Modeling
   C) Constrained Modeling
   D) Sketch-based Modeling
   E) None of the above

4. **What is the definition of Design Intent?**
   A) Term used to describe the procedure to model a solid object
   B) Term used to describe how the model should perform once it is altered
   C) Term applied after the sketch is completed
   D) Term used to relate dimensions to each other algebraically
   E) None of the above

5. **A sketch that has duplicated dimensions and/or geometric constraints is _____**
   A) Over-defined
   B) Dimensionally dependent
   C) Fully defined
   D) Dimensionally independent
   E) Under-defined

6. **Examples of Geometric Constraints include:**
   A) Parallel B) Coincident C) Vertical D) Tangent E) All of the above

7. **In Parametric Modeling applications, all solids must first start out with a/an _____**
   A) Feature B) Extrusion C) Sketch D) Assembly E) None of the above

8. **Which Boolean operation is sensitive to the order of the operands?**
   A) Union B) Difference C) Intersection D) All of the above E) None of the above

9. **Which of the following is not a common method to create solid objects?**
   A) Extrude B) Revolve C) Sweep D) Lathe E) None of the above

10. **Revolve takes a sketch and rotates it about a/an _____**
    A) Axis B) Point C) Plane D) B and C are correct E) A, B, and C are correct

11. **In Parametric Solid Modeling applications, assembly constraints allow to align and fit together parts in an assembly. These constraints can be established between _____**
    A) Two edges B) A face and a vertex C) Two faces
    D) A and C are correct E) None of the above

12. The overall height of the figure shown below will change if dimension(s) ____ are increased

A) 4.00 B) 25.00° C) 1.50 D) B and C are correct E) A, B, and C are correct

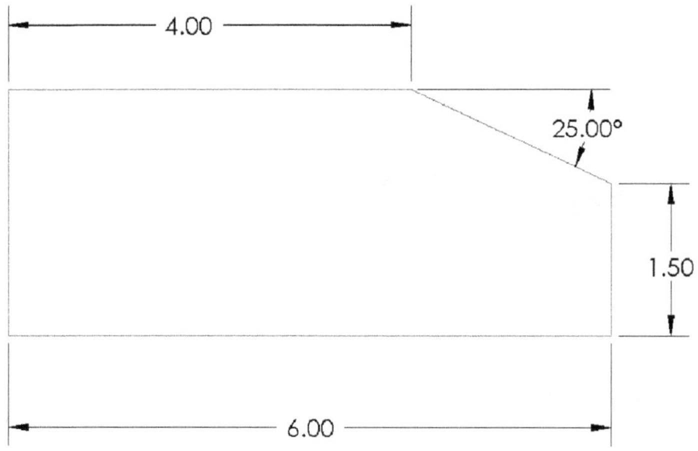

13. Assuming the sketch shown to the right has horizontal, vertical, and tangent constraints applied, how many dimensions are required to fully define the sketch?

A) 2 B) 3 C) 4 D) 5 E) More than 5

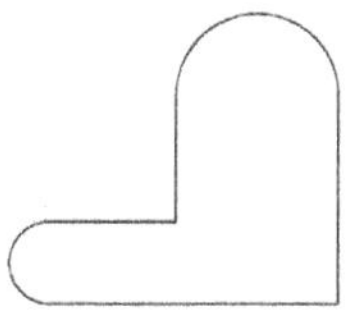

14. How many degrees of freedom does a part with zero constraints have?

A) 0 B) 2 C) 6 D) 8 E) It depends on the part

15. Most parametric modeling software packages allow the user _______

A) To create 3D solid part models.
B) To relate different parts in an assembly model.
C) To provide 2D working drawings from a finalized design.
D) A, and B are correct
E) A, B, and C are correct

Problems—Part Modeling

Create 3D models of the following objects using your parametric solid modeling application.

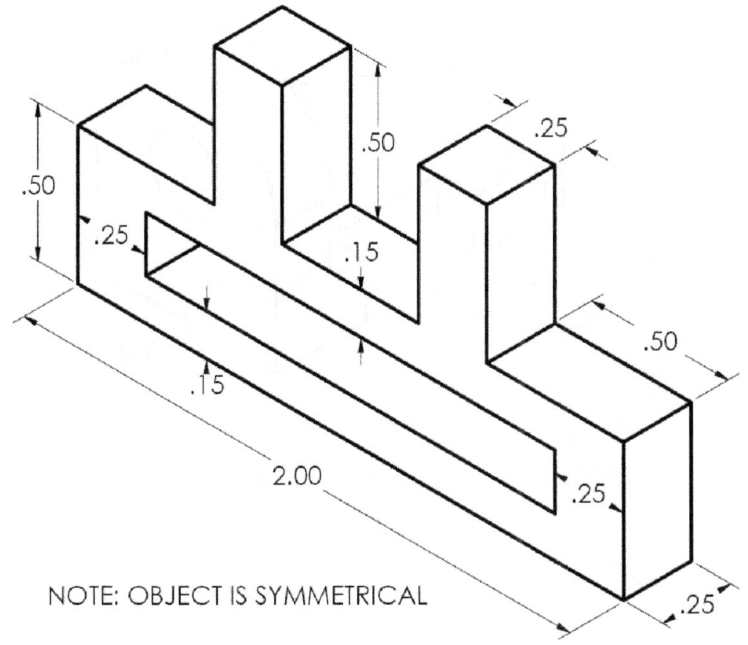

Problem 1

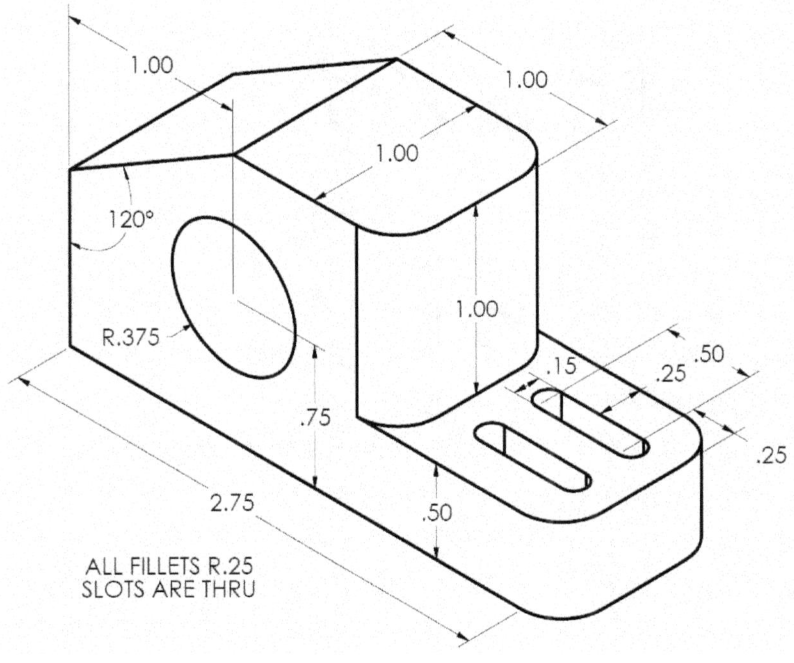

Problem 2

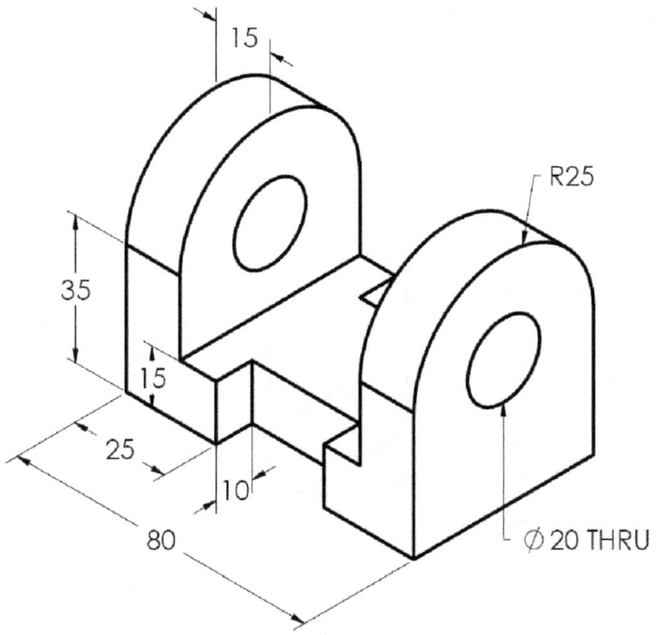

NOTE: OBJECT IS SYMMETRICAL

Problem 3

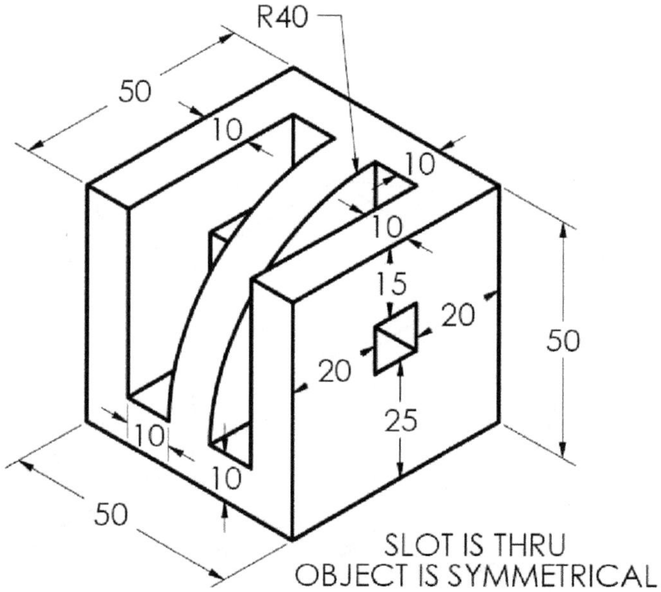

SLOT IS THRU
OBJECT IS SYMMETRICAL

Problem 4

Problems—Assembly Modeling

Create the components and the assembly of the following objects using your parametric solid modeling application.

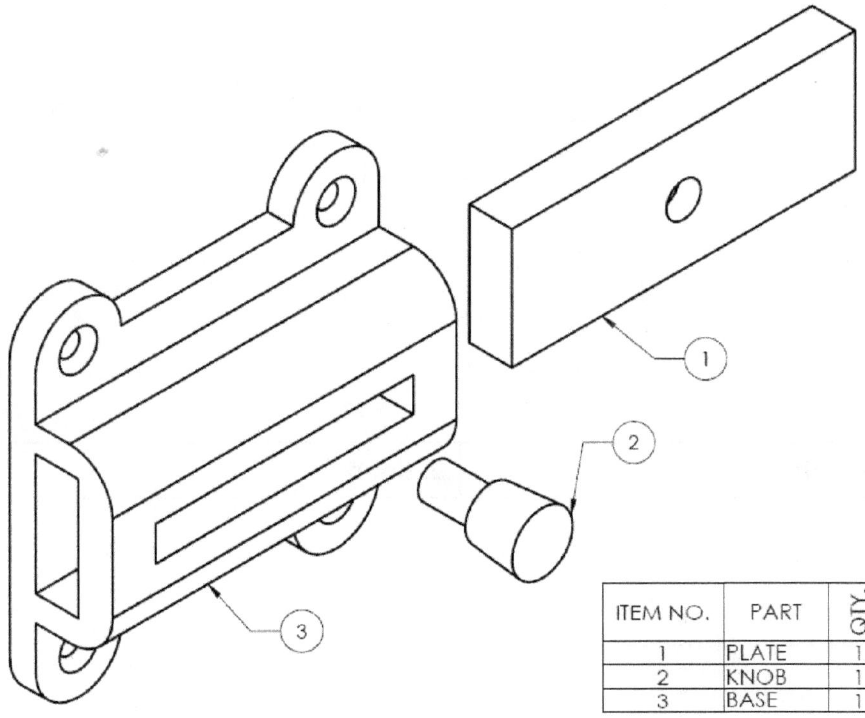

ITEM NO.	PART	QTY.
1	PLATE	1
2	KNOB	1
3	BASE	1

Problem 1: Barrel Bolt | **Exploded View**

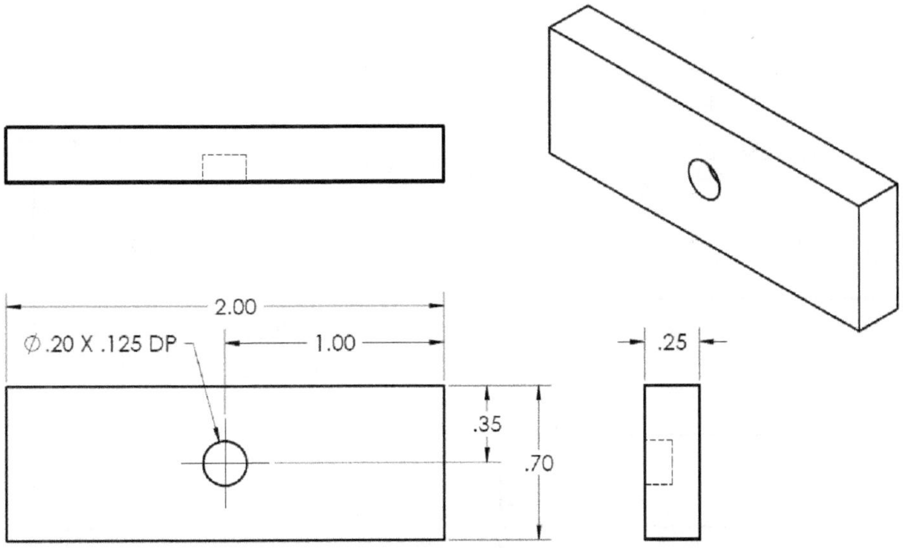

Problem 1: Barrel Bolt | **1. Plate**

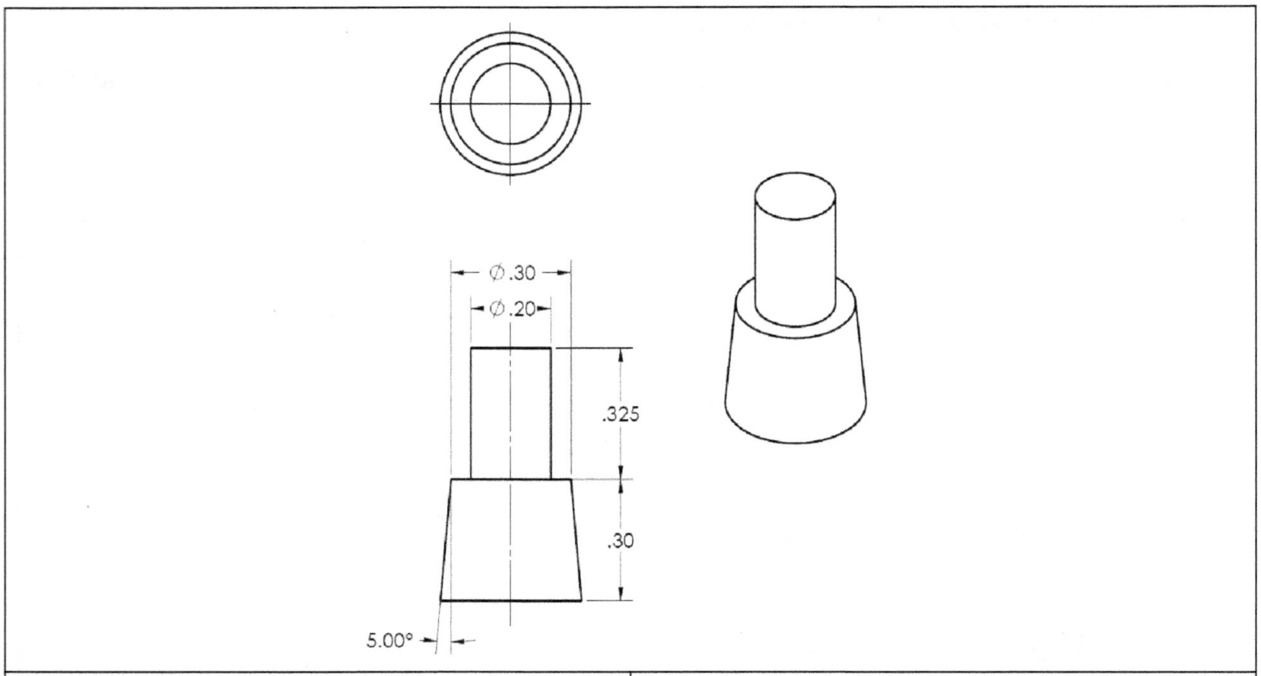

Problem 1: Barrel Bolt | **2. Knob**

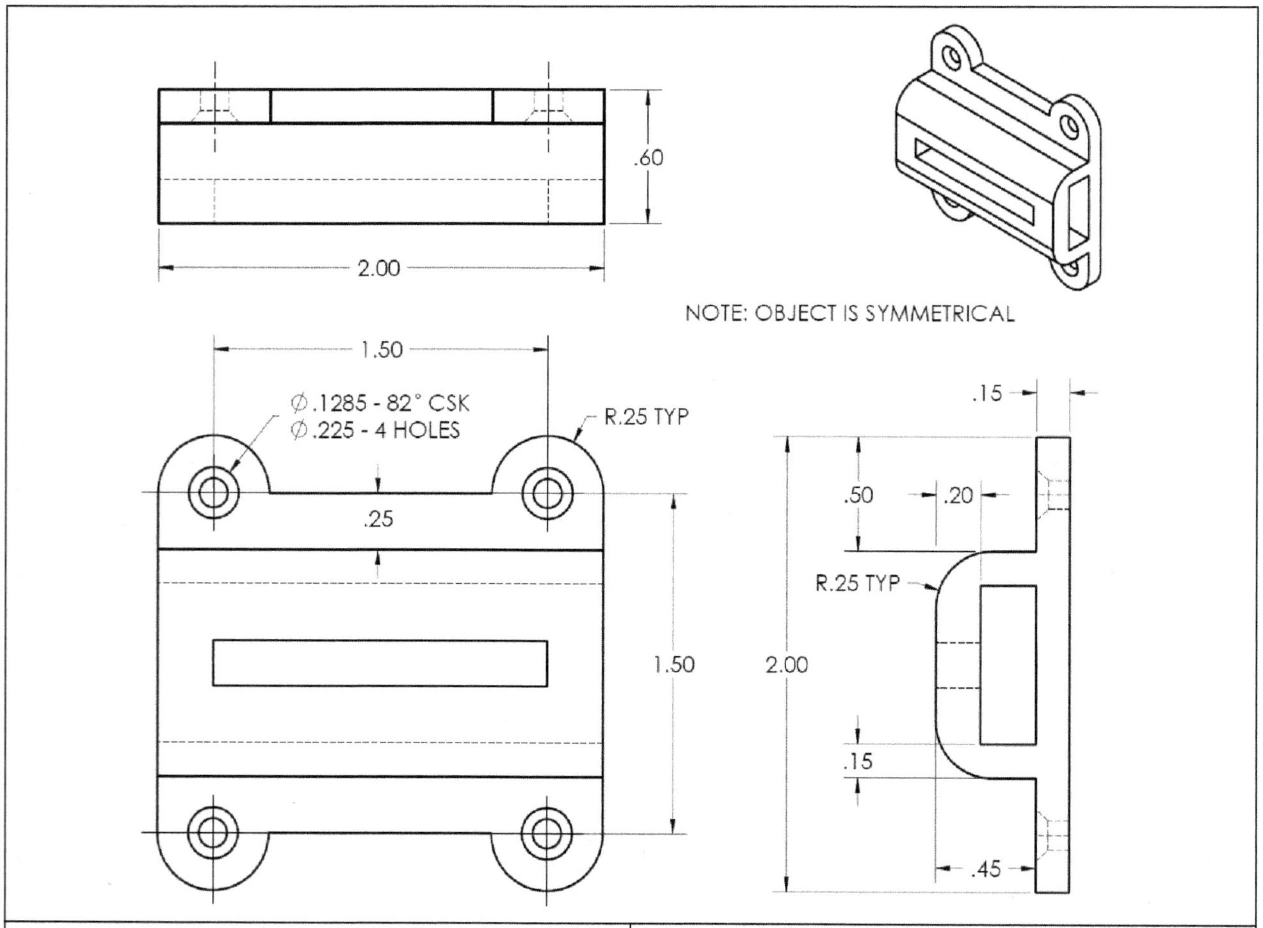

Problem 1: Barrel Bolt | **3. Base**

ITEM NO.	PART	QTY.
1	HEAD	1
2	COVER	1
3	INSET	1
4	SPRING	1
5	COUPLER	1
6	TUBE	1
7	BOTTLE	1

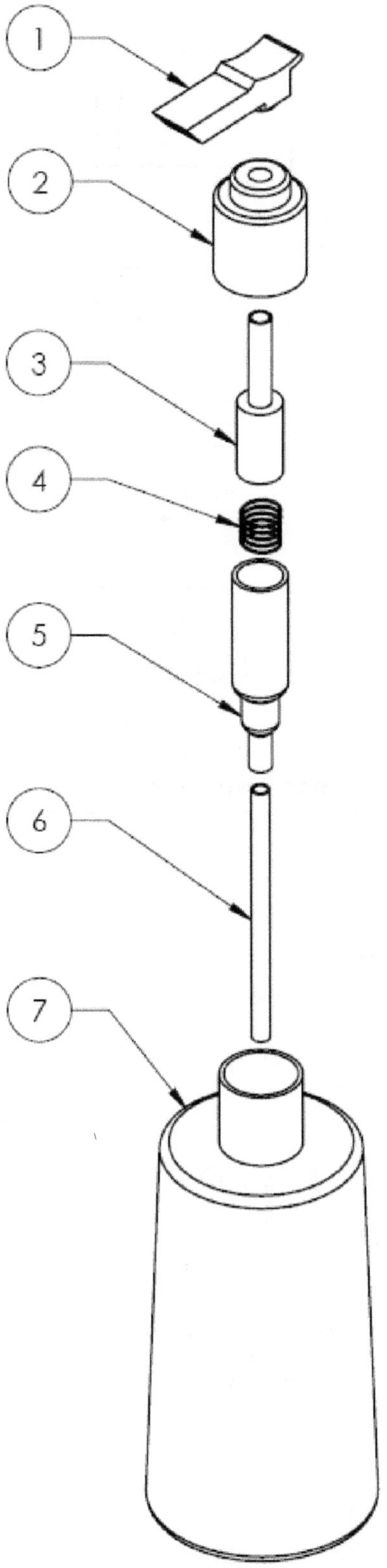

Problem 2: Soap Dispenser	**Exploded View**

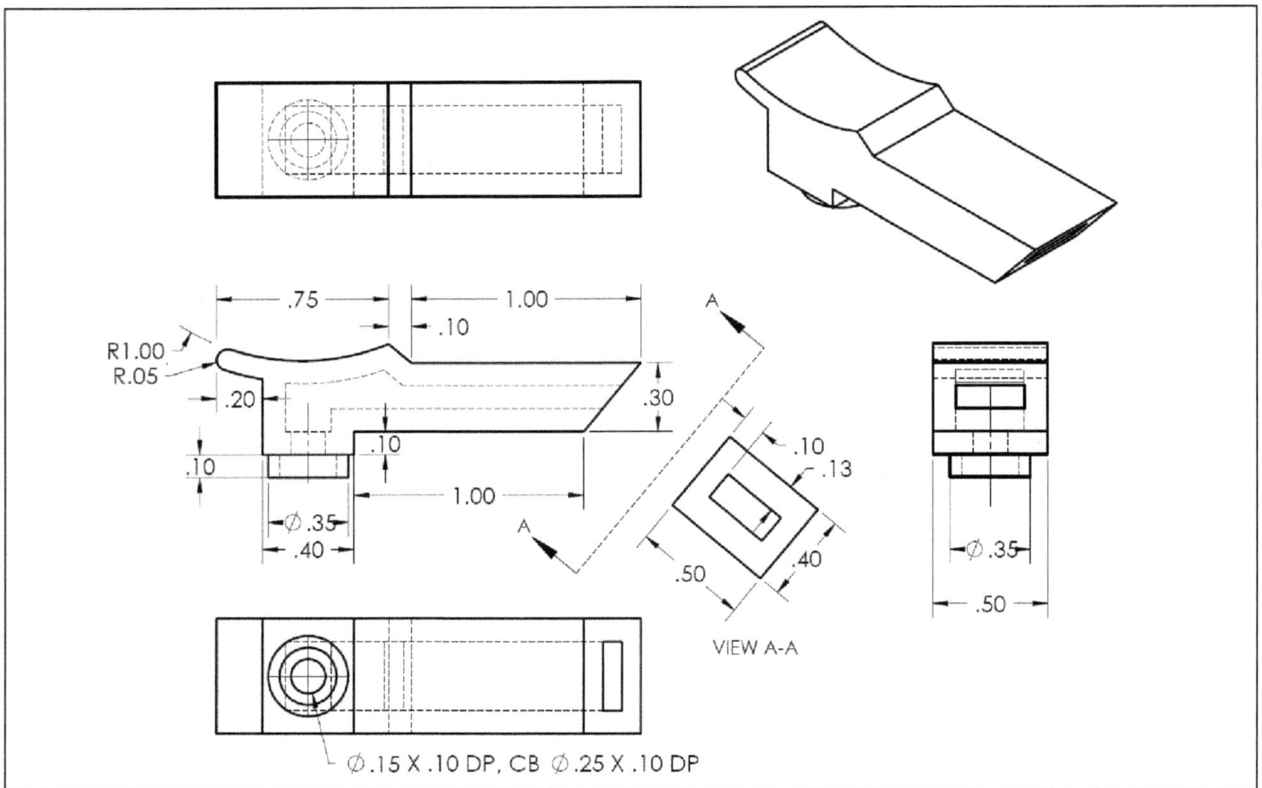

Problem 2: Soap Dispenser | **1. Head**

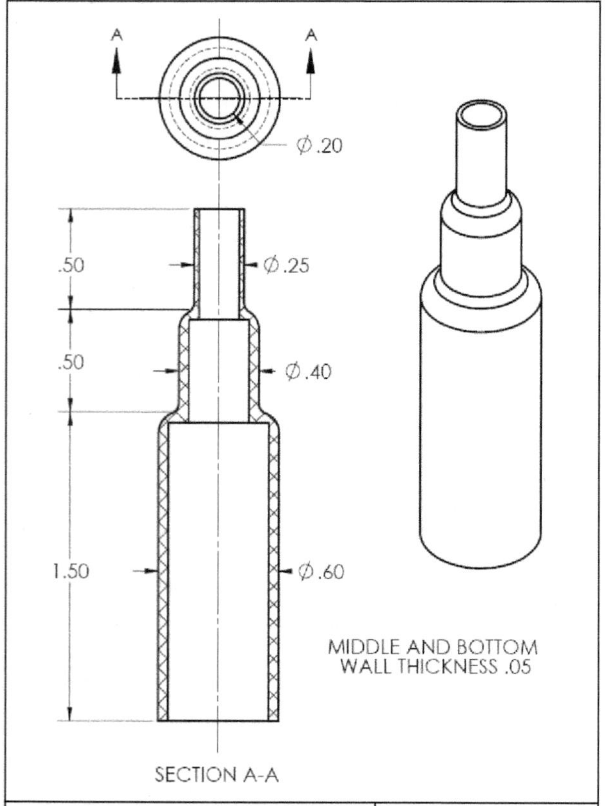

Problem 2: Soap Dispenser | **2. Cover**

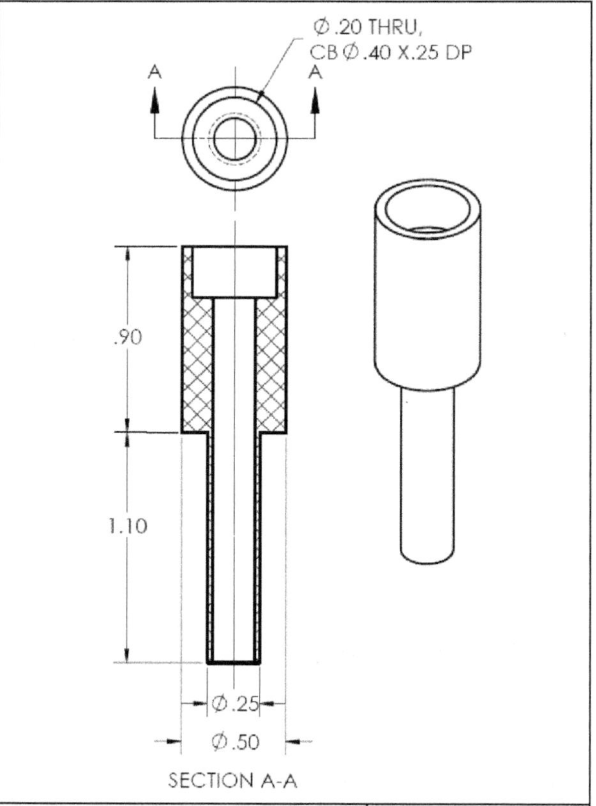

Problem 2: Soap Dispenser | **3. Inset**

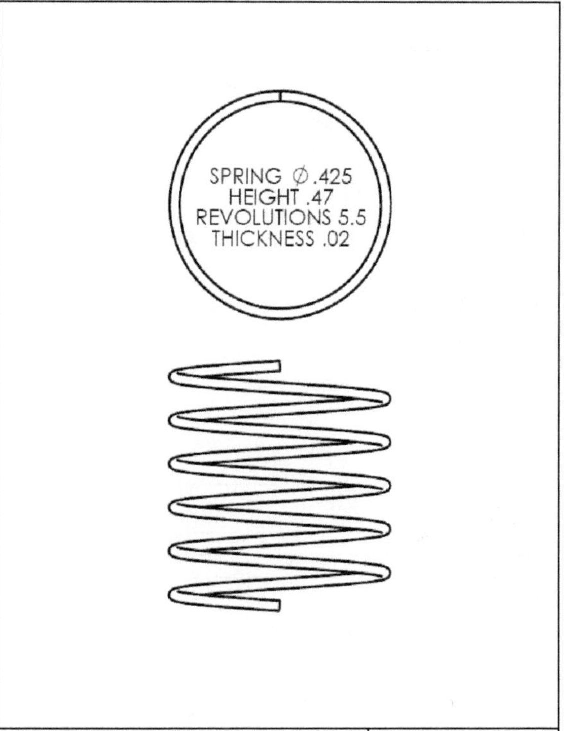

Problem 2: Soap Dispenser	4. Spring

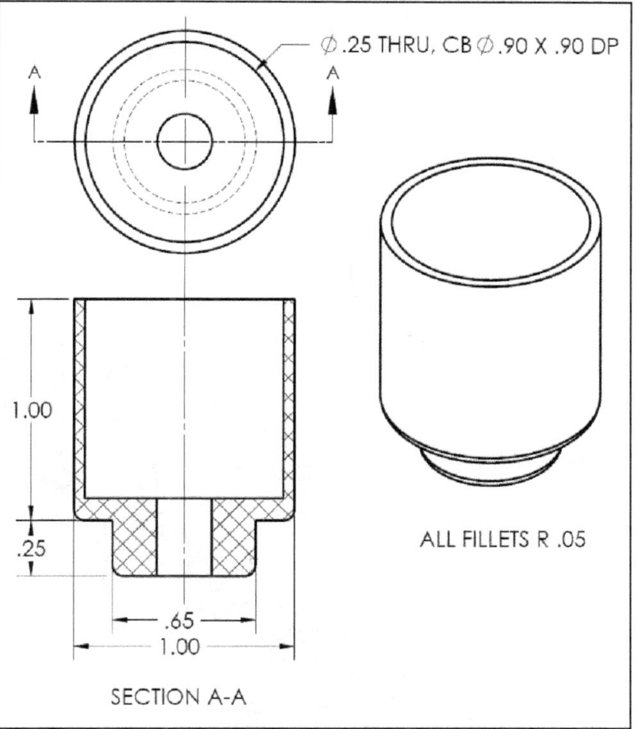

Problem 2: Soap Dispenser	5. Coupler

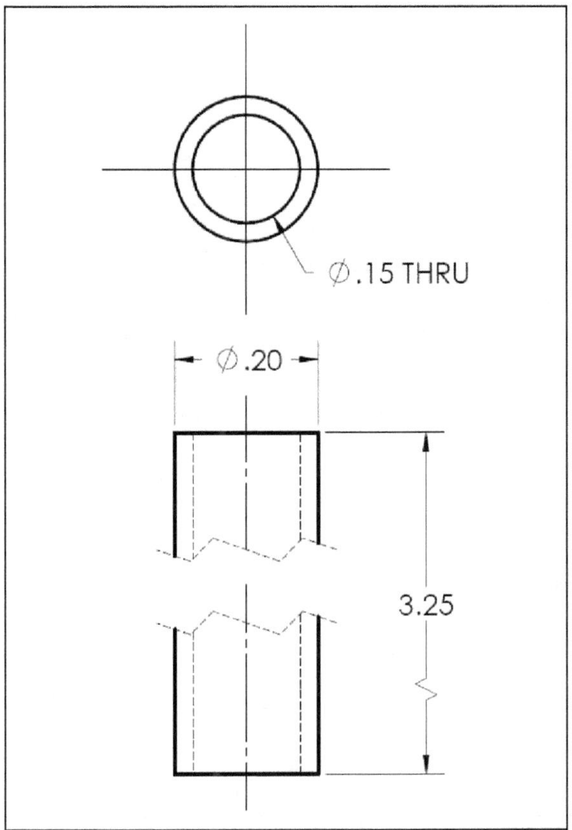

Problem 2: Soap Dispenser	6. Tube

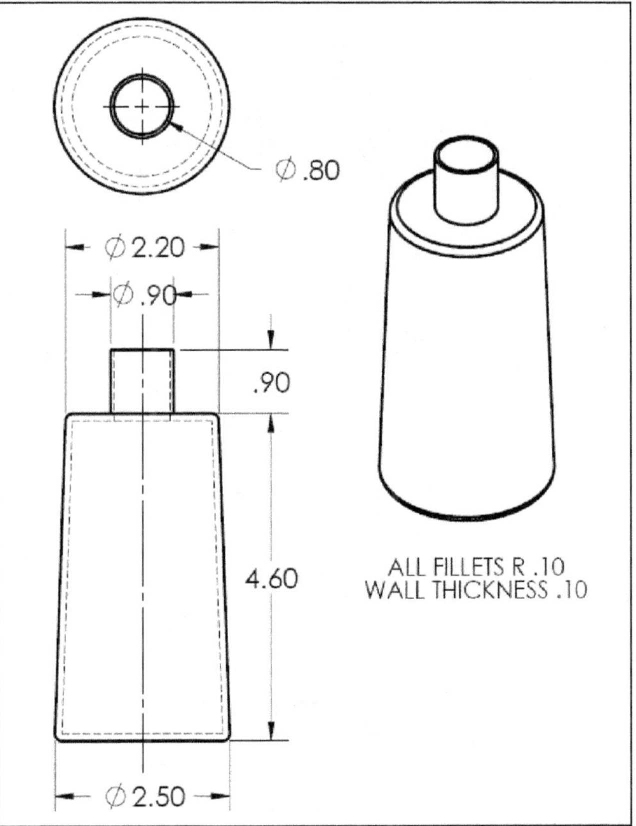

Problem 2: Soap Dispenser	7. Bottle

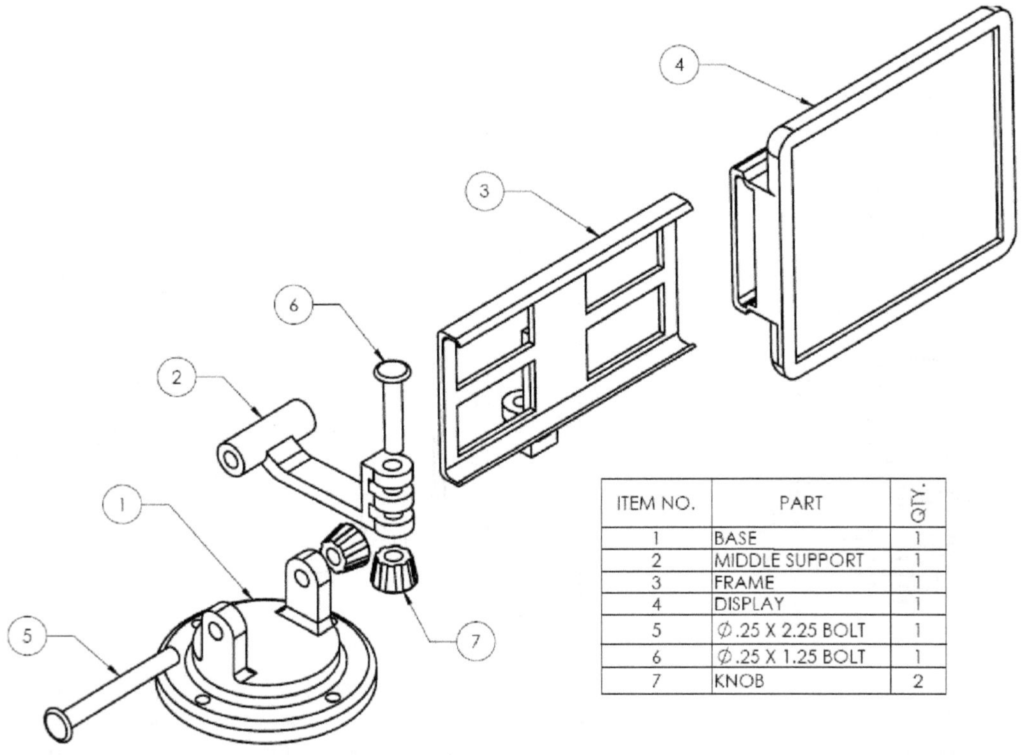

ITEM NO.	PART	QTY.
1	BASE	1
2	MIDDLE SUPPORT	1
3	FRAME	1
4	DISPLAY	1
5	Ø.25 X 2.25 BOLT	1
6	Ø.25 X 1.25 BOLT	1
7	KNOB	2

Problem 3: Display Mount	Exploded View

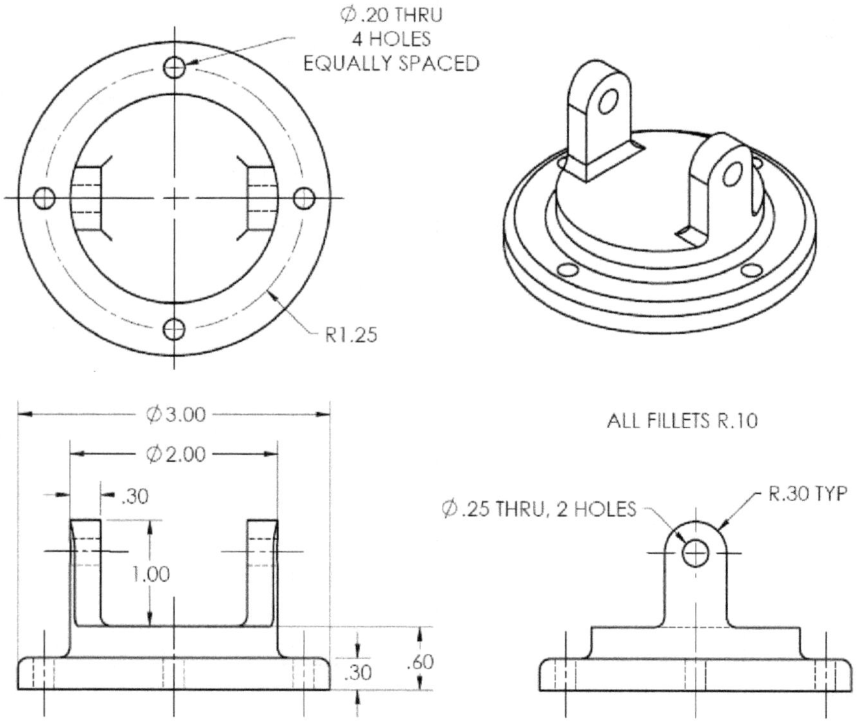

Problem 3: Display Mount	1. Base

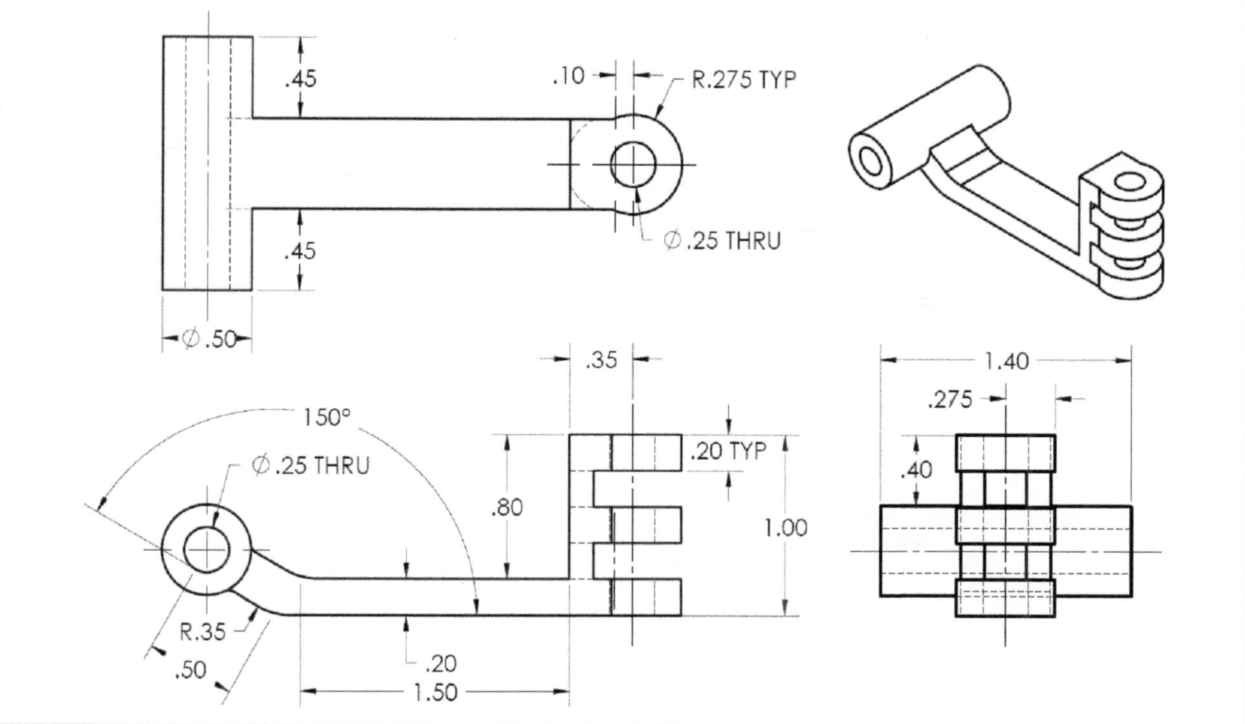

Problem 3: Display Mount	2. Middle Support

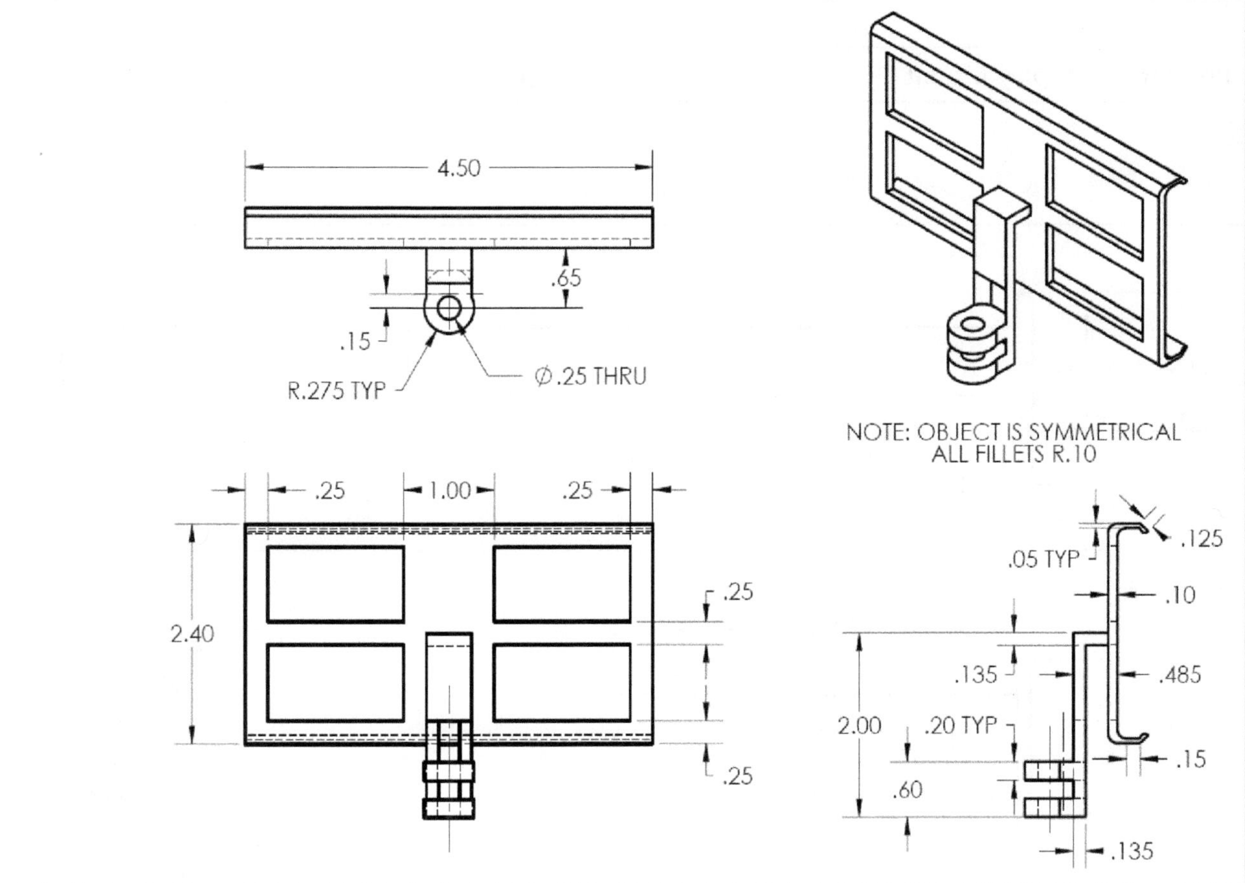

Problem 3: Display Mount	3. Frame

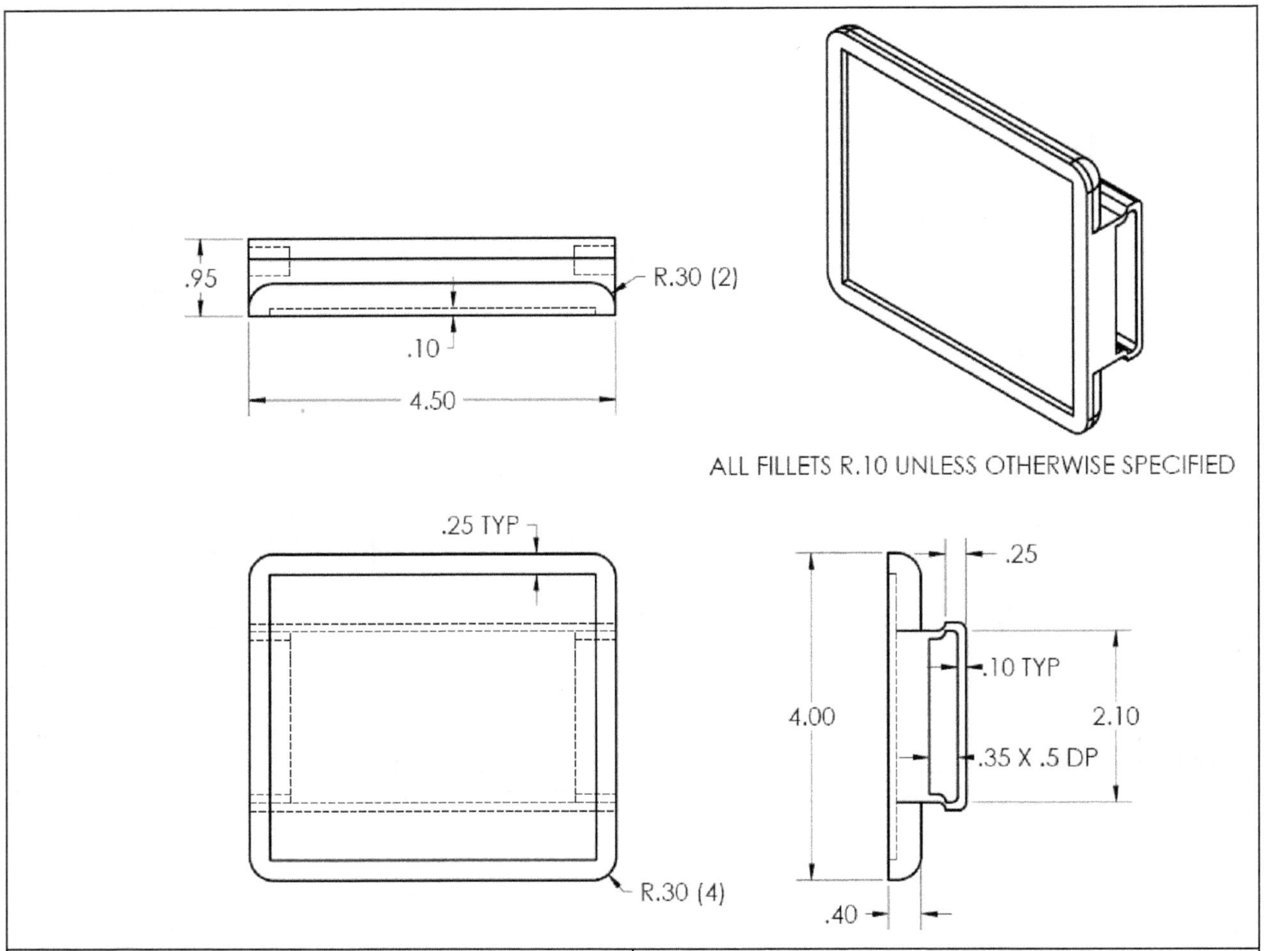

Problem 3: Display Mount	**4. Display**

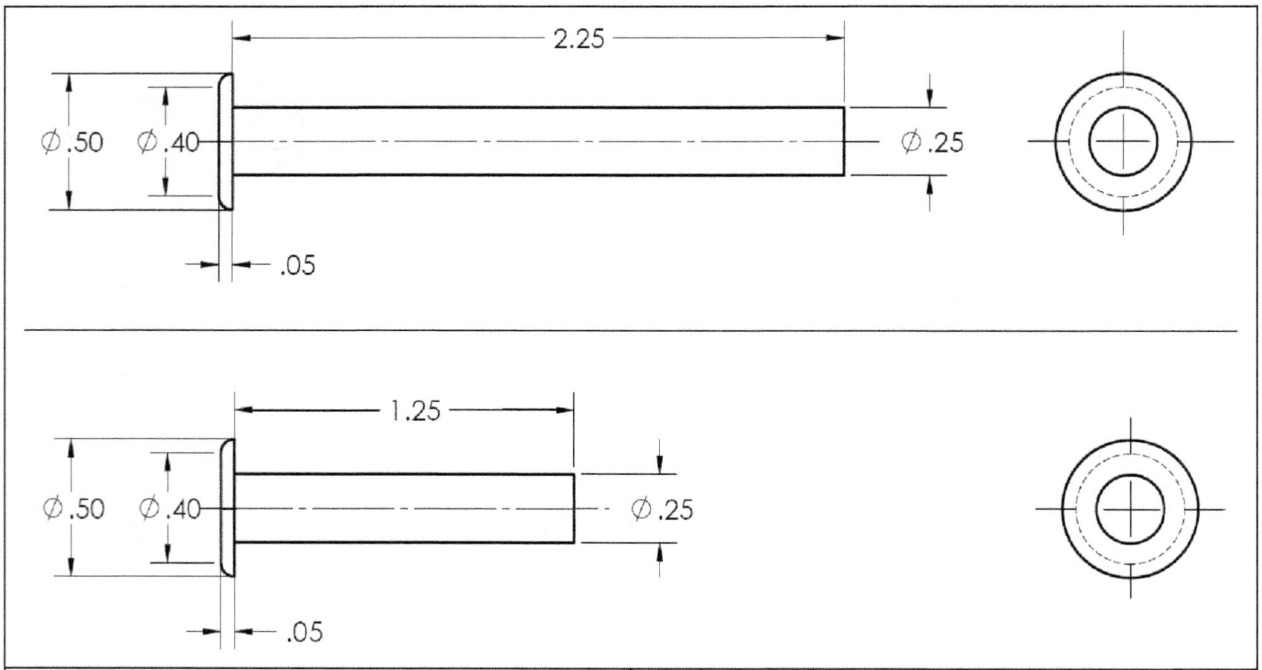

Problem 3: Display Mount	**5 & 6. Bolts**

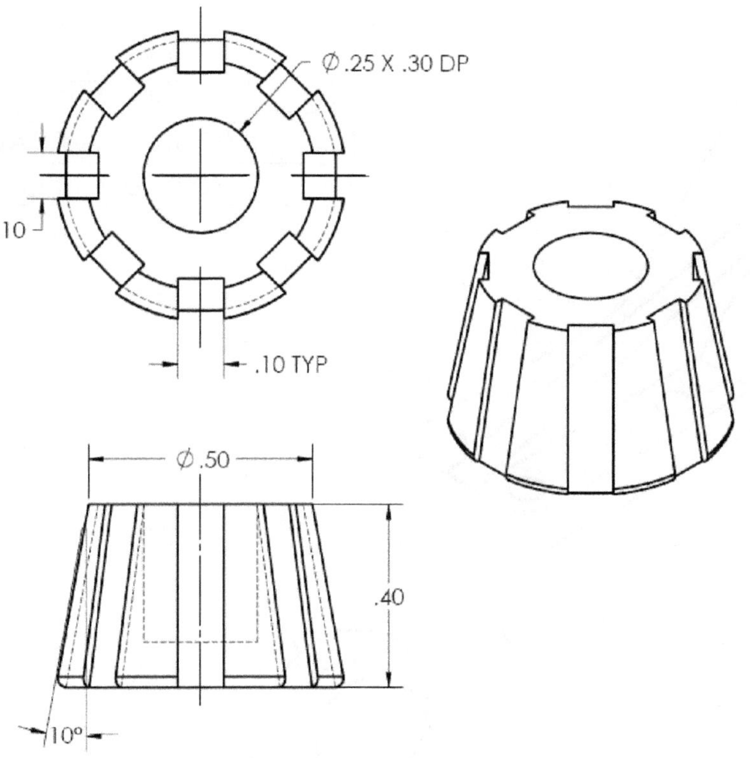

Problem 3: Display Mount	7. Knob

TENSION CABLE HOLDER

ITEM NO.	DESCRIPTION	QTY.
1	MOUNTING BRACKET	1
2	CABLE LOCK COLLAR	2

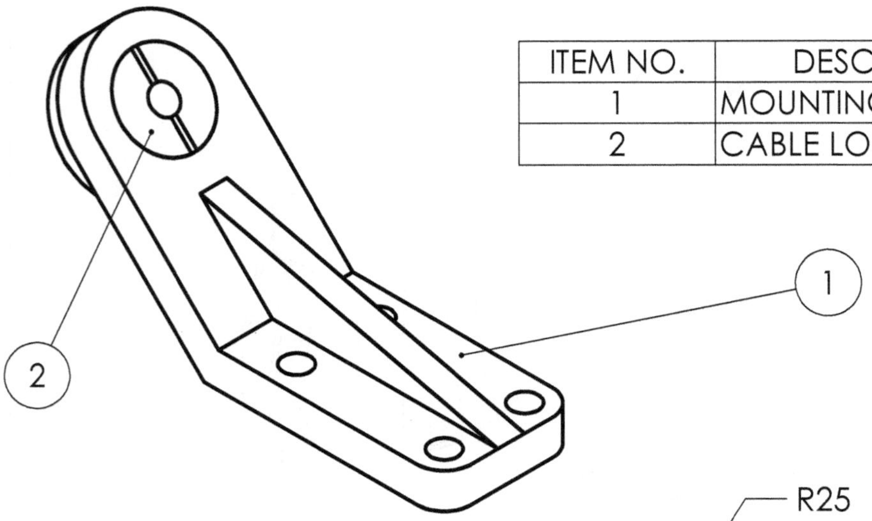

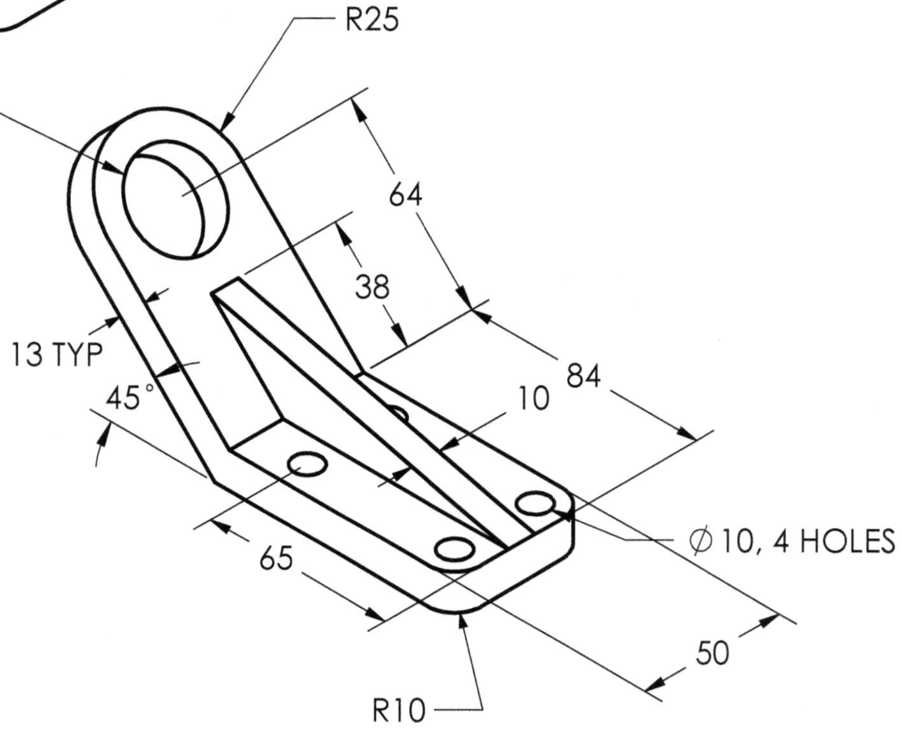

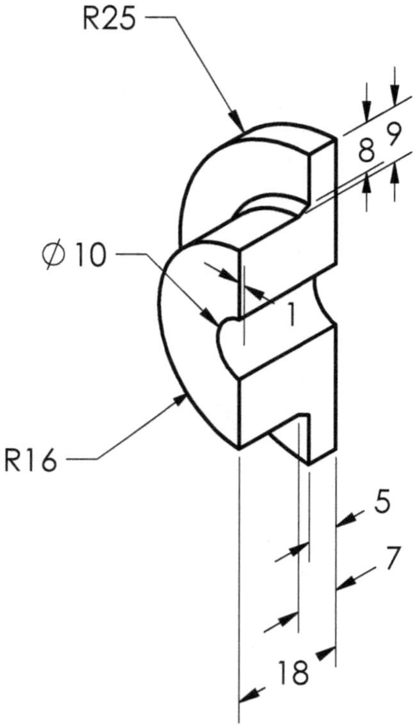

(1) MOUNTING BRACKET
1020 STEEL

(2) CABLE LOCK COLLAR
1020 STEEL
2 REQ.

IDLER YOKE

ITEM NO.	DESCRIPTION	QTY.
1	BASE	1
2	YOKE	1
3	AXLE	1
4	PIVOT PIN	2
5	PULLEY	2
6	ϕ.19 SET SCREW, .375 LG	2

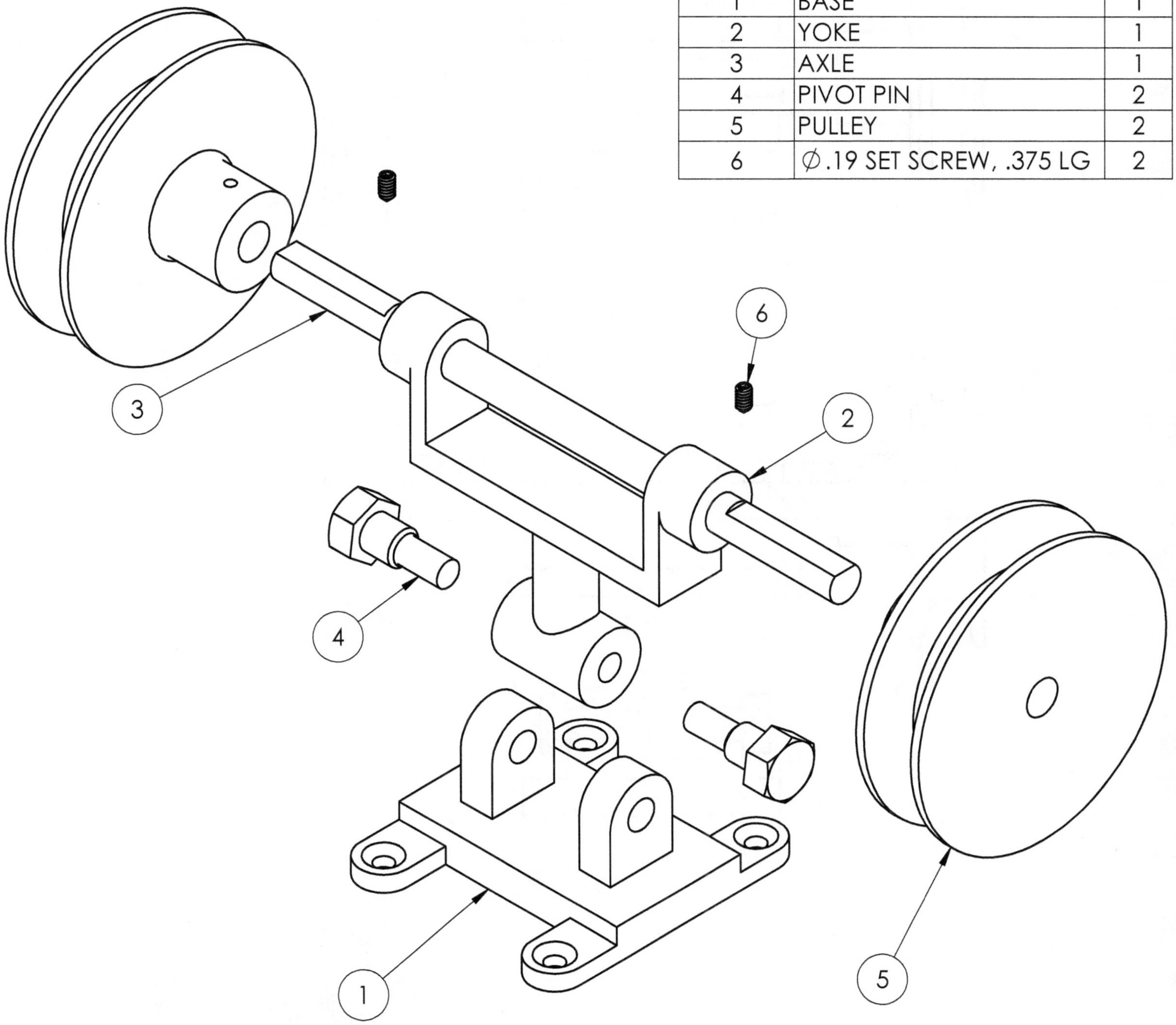

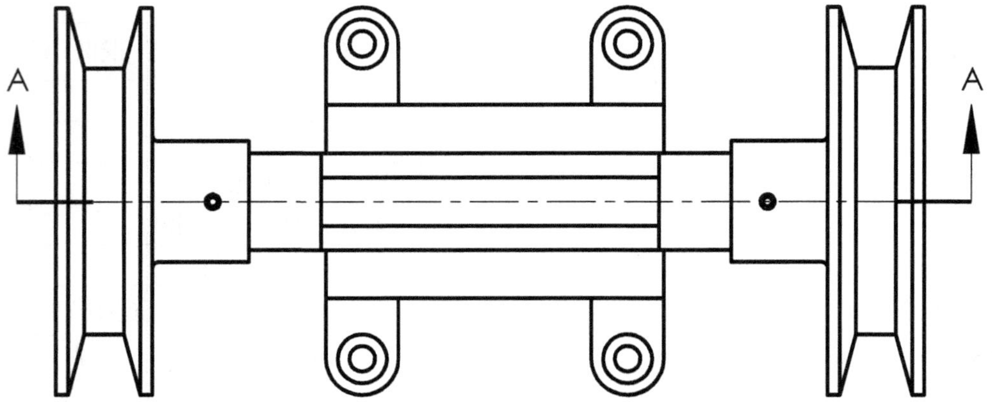

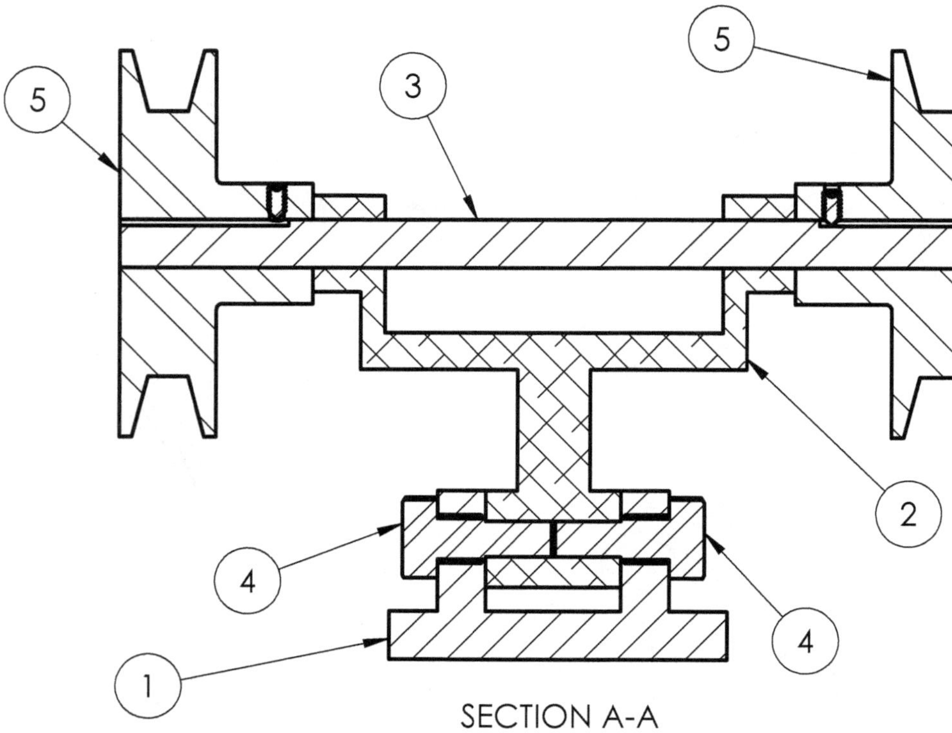

SECTION A-A

ITEM NO.	DESCRIPTION	MATERIAL	QTY.
1	BASE	GREY CAST IRON	1
2	YOKE	ALUMINUM	1
3	AXLE	PLAIN CARBON STEEL	1
4	PIVOT PIN	PLAIN CARBON STEEL	2
5	PULLEY	STAINLESS STEEL	2
6	∅.19 SET SCREW, .375 LG	STEEL	2

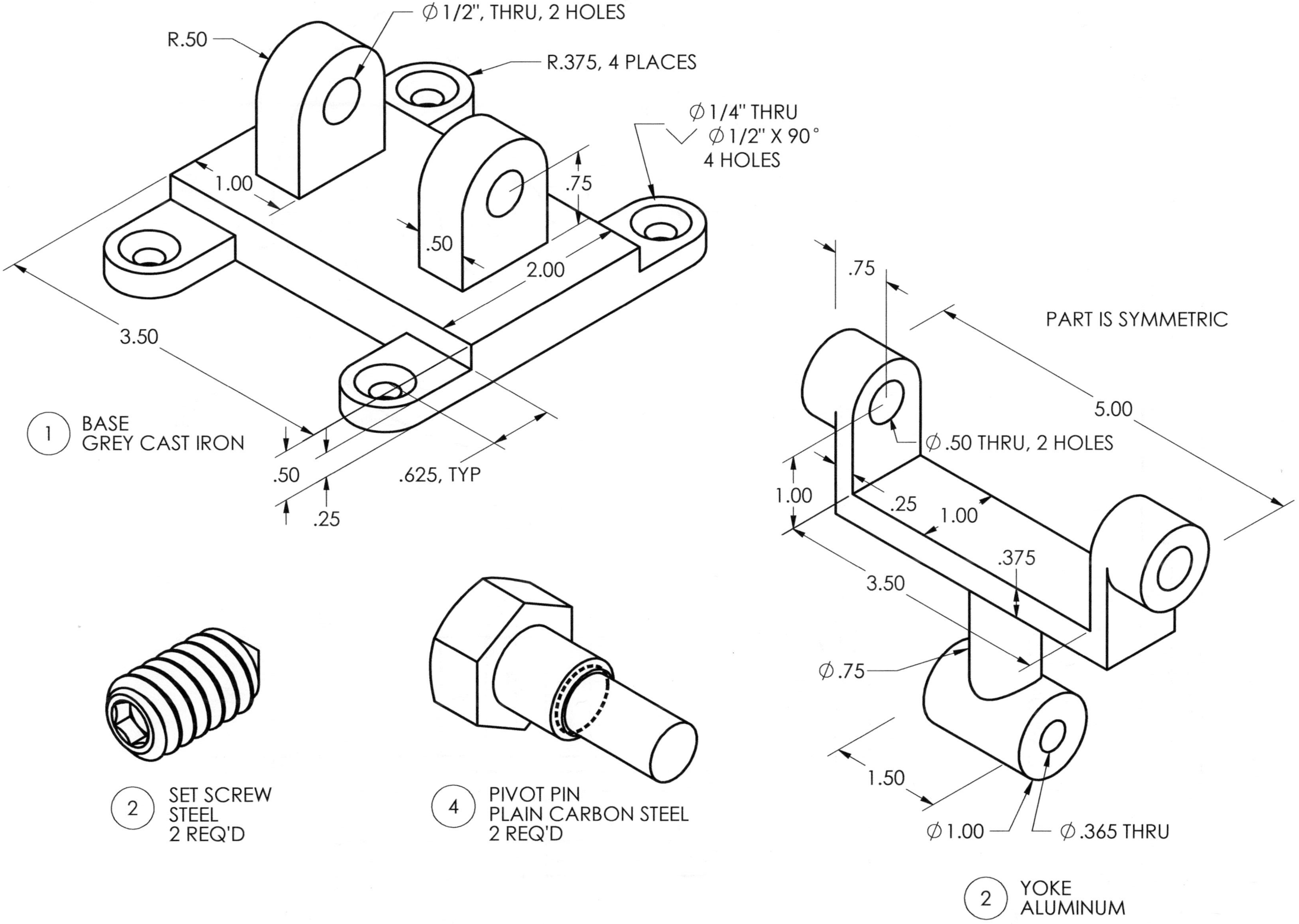

⌀1/2", THRU, 2 HOLES
R.50
R.375, 4 PLACES
⌀1/4" THRU
⌀1/2" X 90°
4 HOLES
1.00
.75
.50
2.00
3.50
.50
.625, TYP
.25
1 BASE
GREY CAST IRON
.75
PART IS SYMMETRIC
5.00
⌀.50 THRU, 2 HOLES
1.00
.25
1.00
.375
3.50
⌀.75
1.50
⌀1.00
⌀.365 THRU
2 YOKE
ALUMINUM
2 SET SCREW
STEEL
2 REQ'D
4 PIVOT PIN
PLAIN CARBON STEEL
2 REQ'D

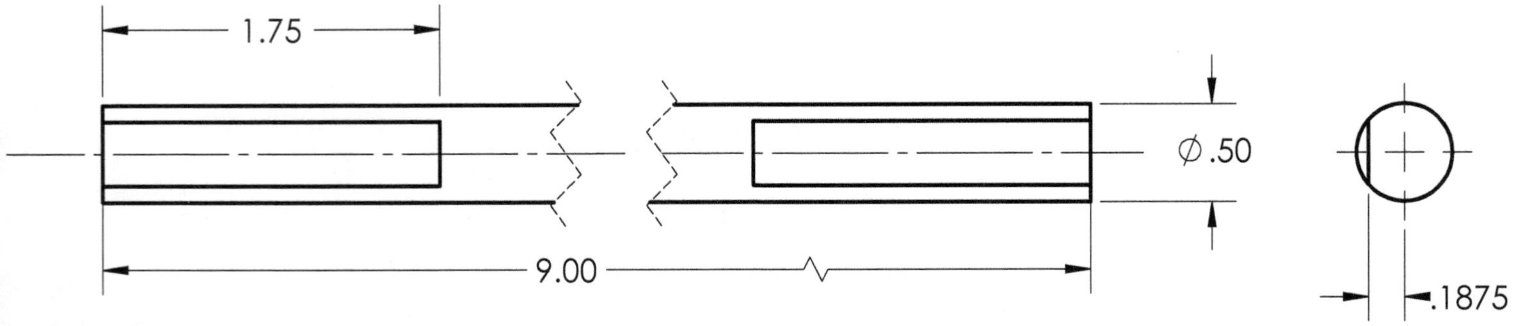

3 AXLE
PLAIN CARBON STEEL

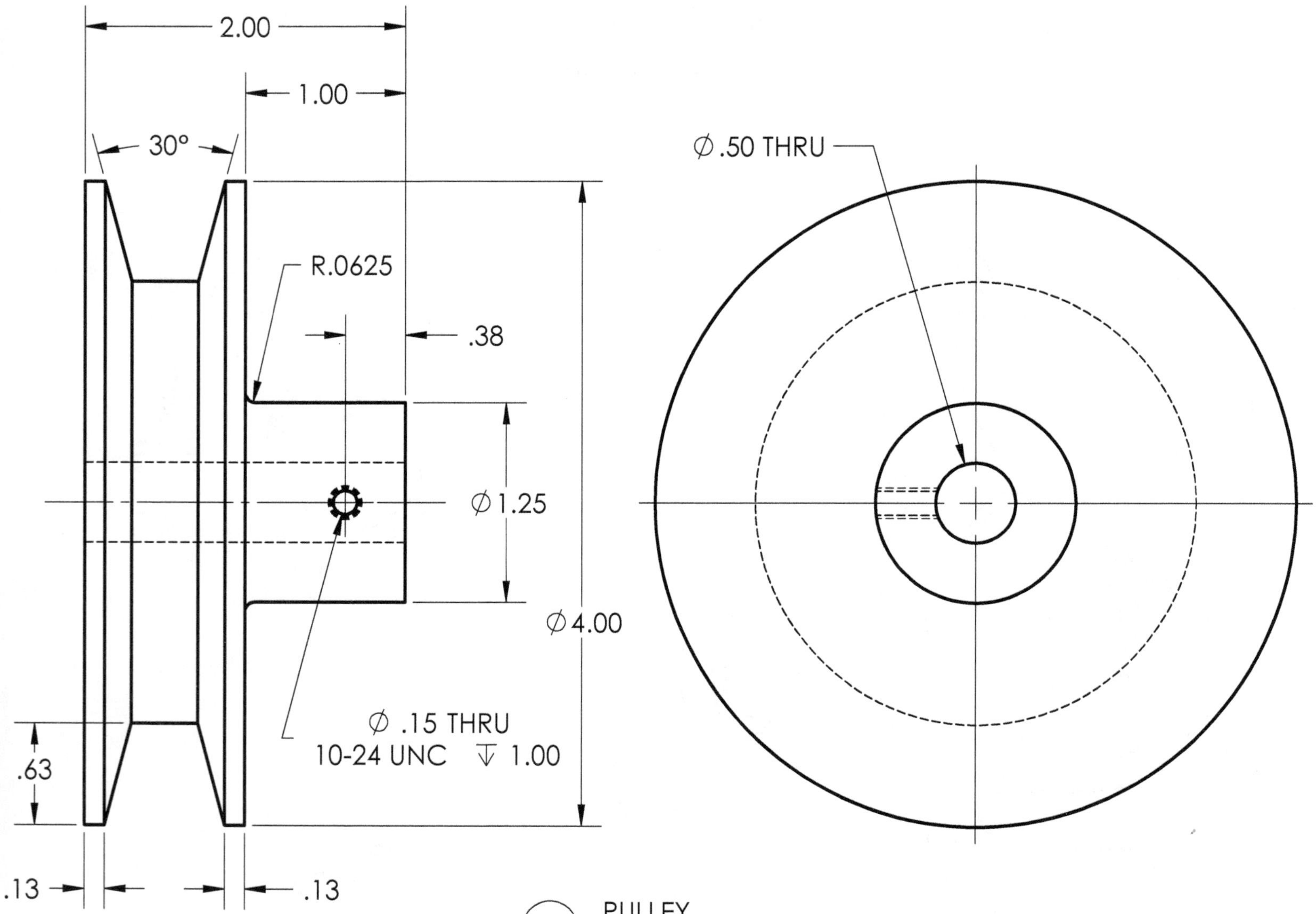

5 PULLEY
STAINLESS STEEL
2 REQ'D

ITEM NO.	DESCRIPTION	MATERIAL	QTY.
1	BASE	ALUMINUM	1
2	CW SPRING	SPRING STEEL	1
3	POST	CARBON FIBER RESIN	2
4	LEFT JAW	CARBON FIBER RESIN	1
5	RIGHT JAW	CARBON FIBER RESIN	1
6	CCW SPRING	SPRING STEEL	1
7	BALL BEARING	STEEL	22
8	ROPE GUIDE	STEEL	1
9	FLAT HD SCREW	STEEL	2

JAM CLEAT

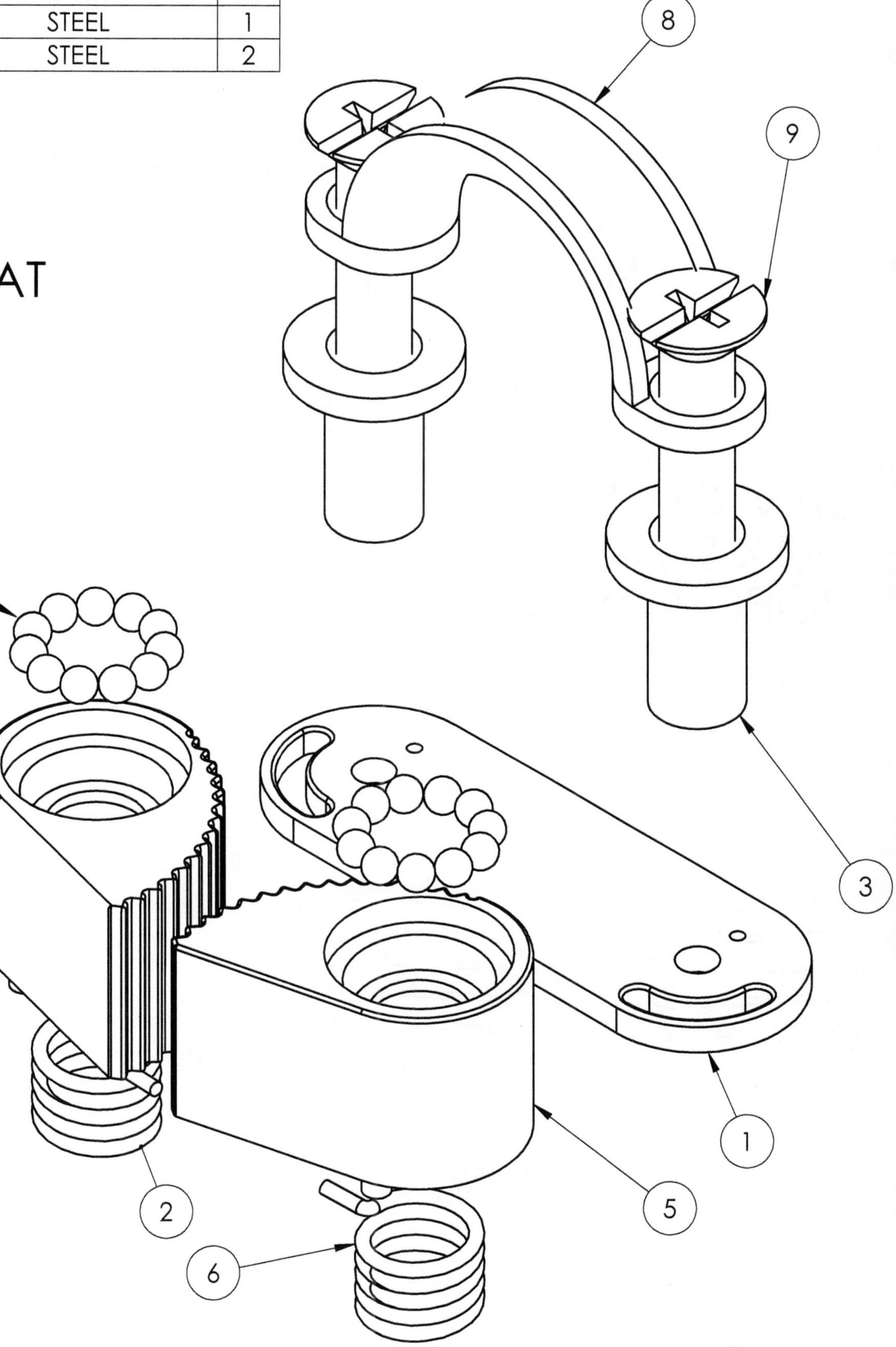

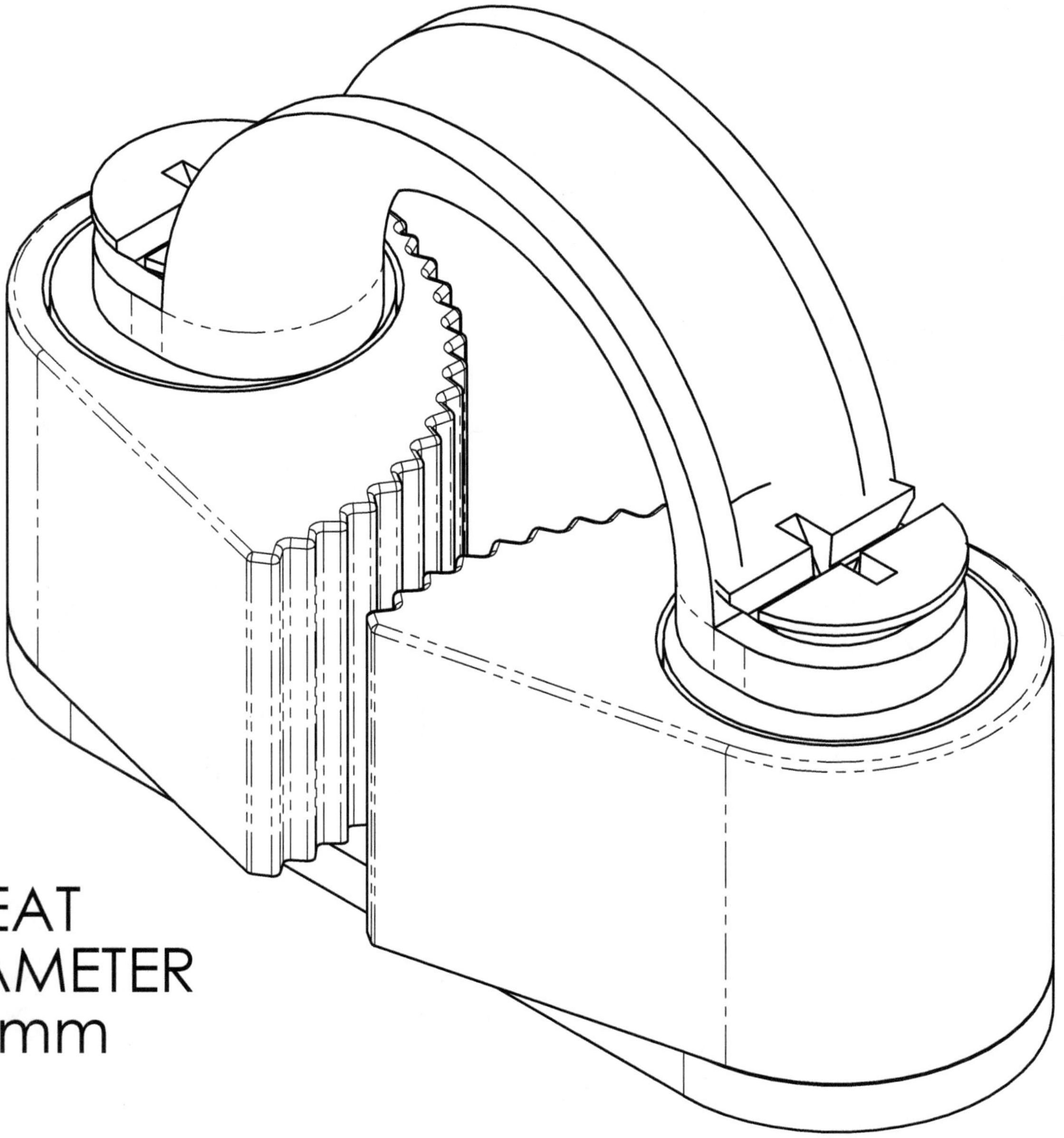

JAM CLEAT
MAX DIAMETER
ROPE 24mm

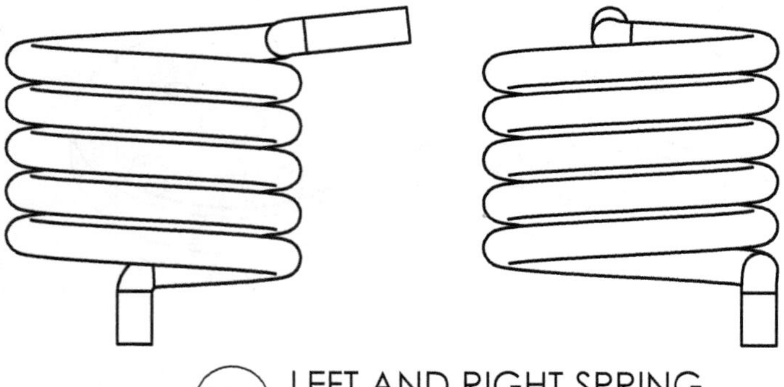

(2&6) LEFT AND RIGHT SPRING
15 ID, ∅2 WIRE
MAKE MIRROR VERSIONS

(7) ∅5.0 BALL BEARING
STEEL
22 REQ'D

(9) M6 X 1 X 40 LG FLAT HEAD SCREW
STEEL
2 REQ'D

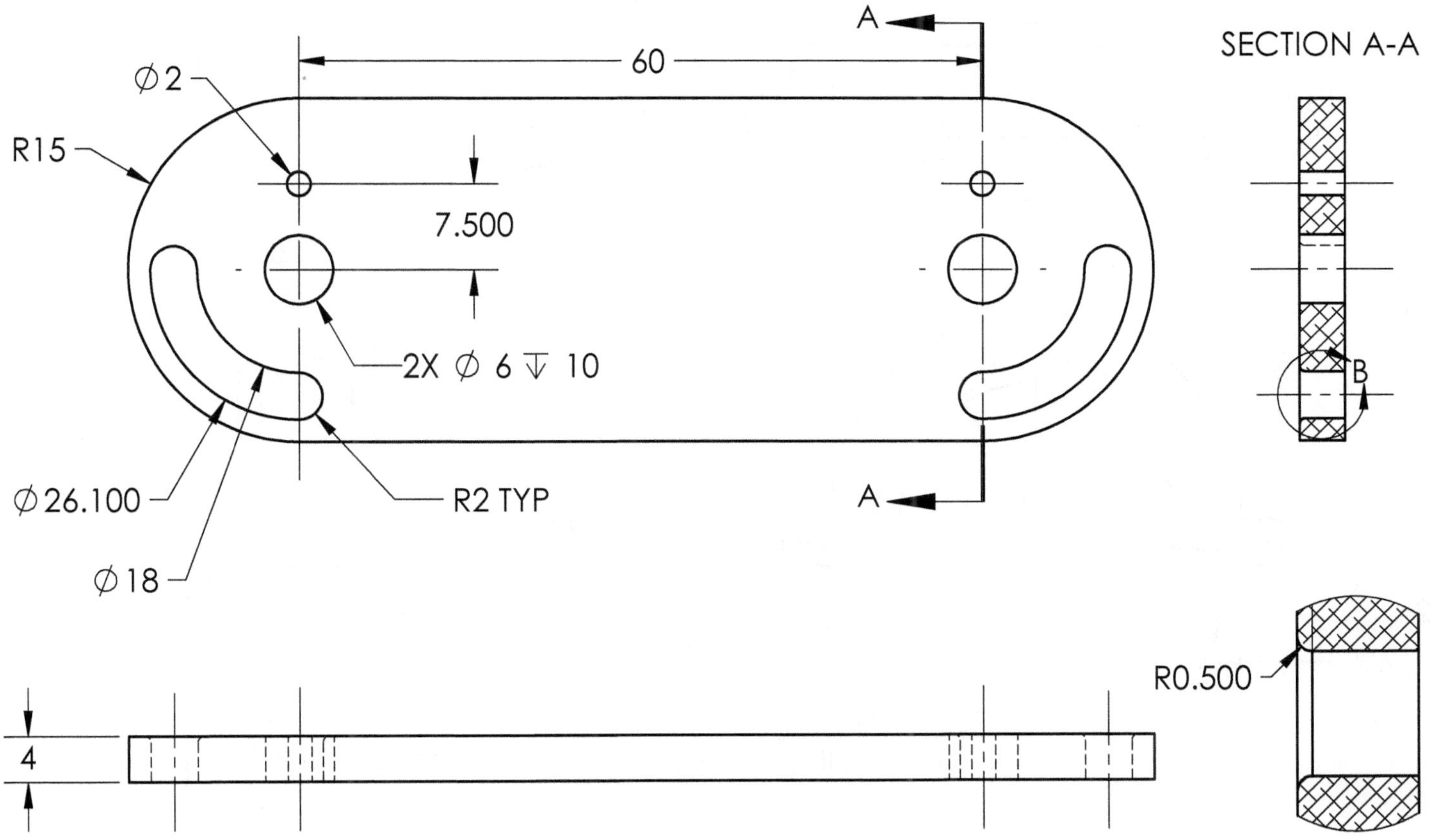

1 BASE
ALUMINUM

DETAIL B

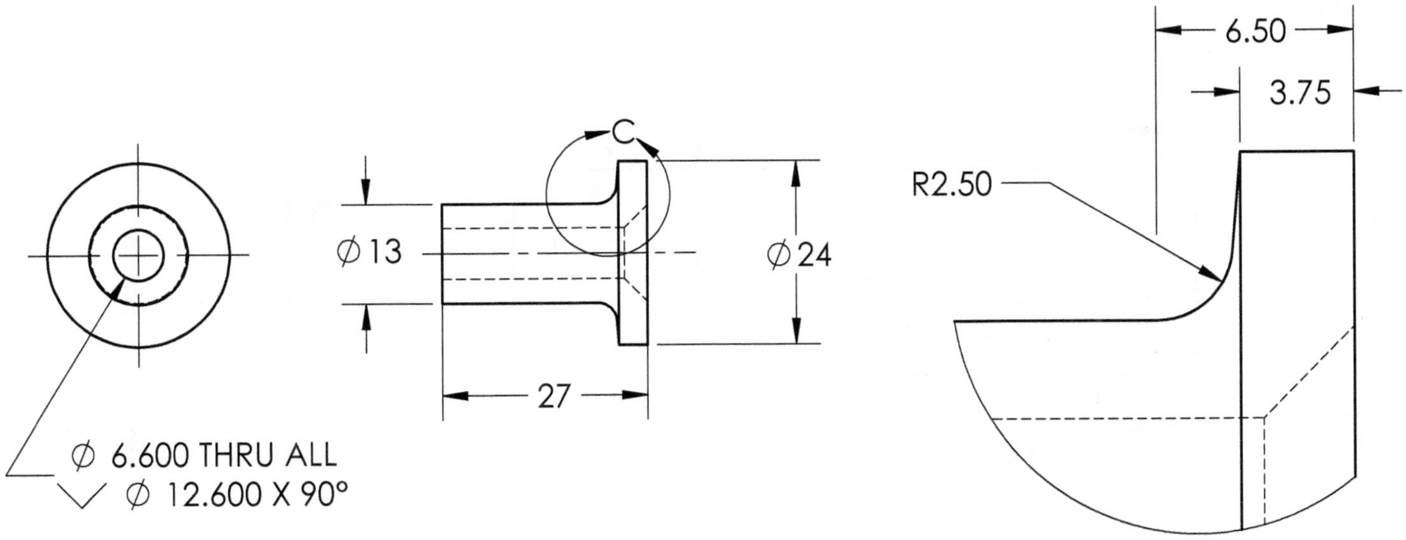

3 POST
STAINLESS STEEL

DETAIL C
SCALE 4 : 1

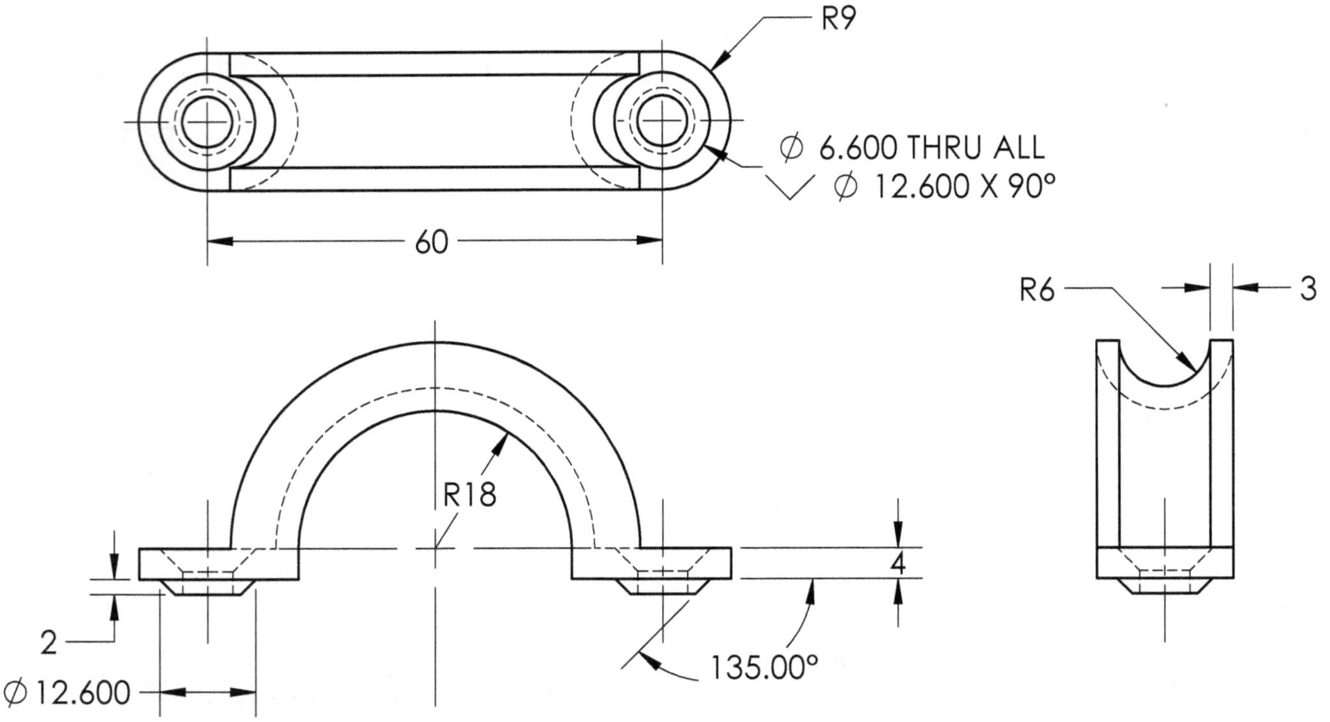

8 ROPE GUIDE
STAINLESS STEEL

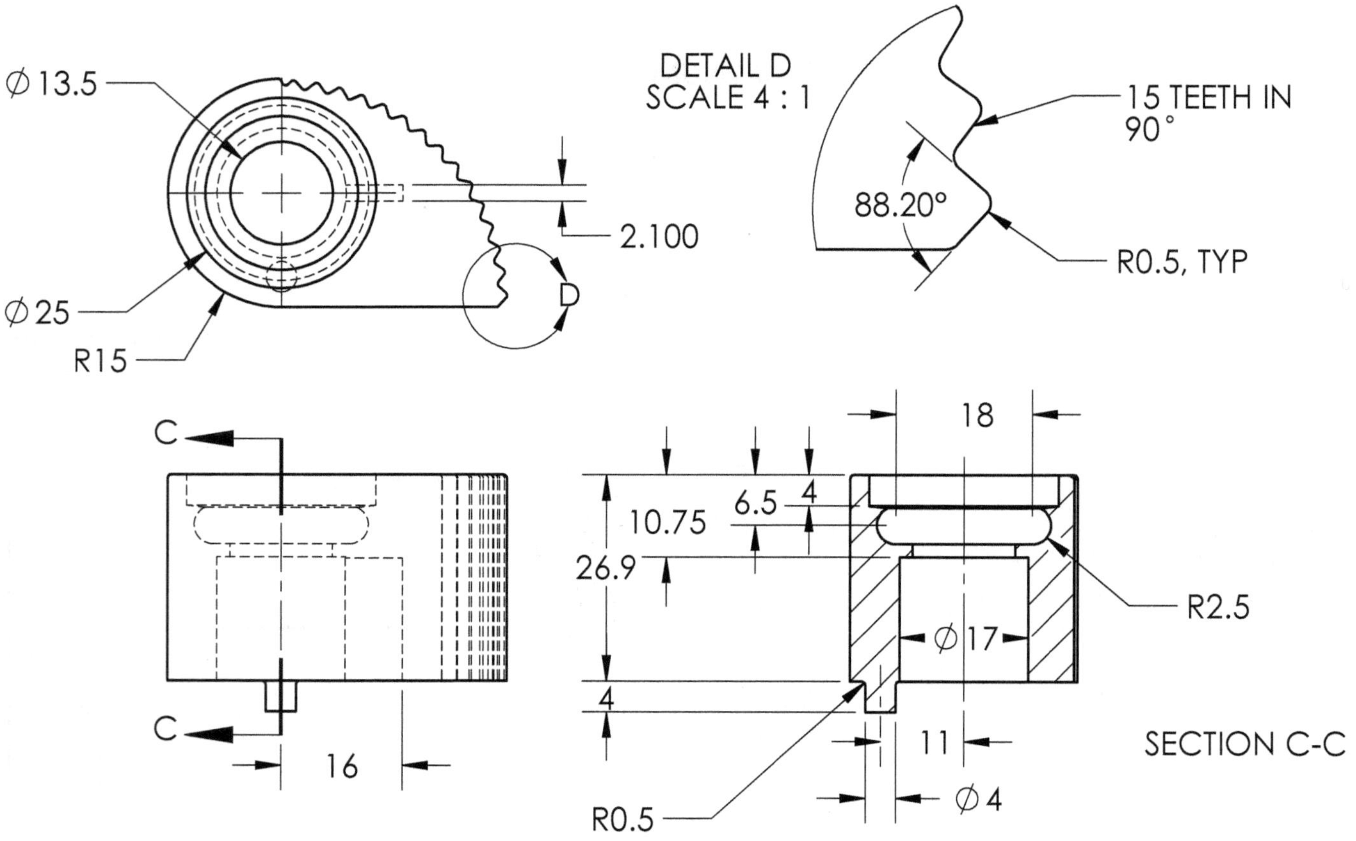

4&5 LEFT AND RIGHT JAW
CARBON FIBER RESIN
MAKE MIRROR IMAGE PARTS

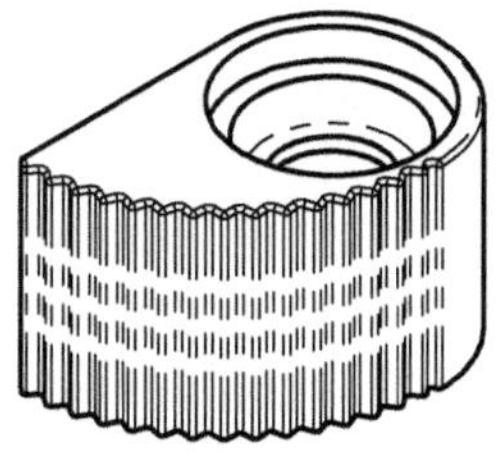

ITEM NO.	DESCRIPTION	MATERIAL	QTY.
1	RIGHT BODY	PE HIGH DENSITY PLASTIC	1
2	LEFT BODY	PE HIGH DENSITY PLASTIC	1
3	TRIGGER	PE HIGH DENSITY PLASTIC	1
4	BASKET	PE HIGH DENSITY PLASTIC	1
5	FLAP	PE HIGH DENSITY PLASTIC	1

"POP 'N CATCH"
LAUNCHER ASSEMBLY

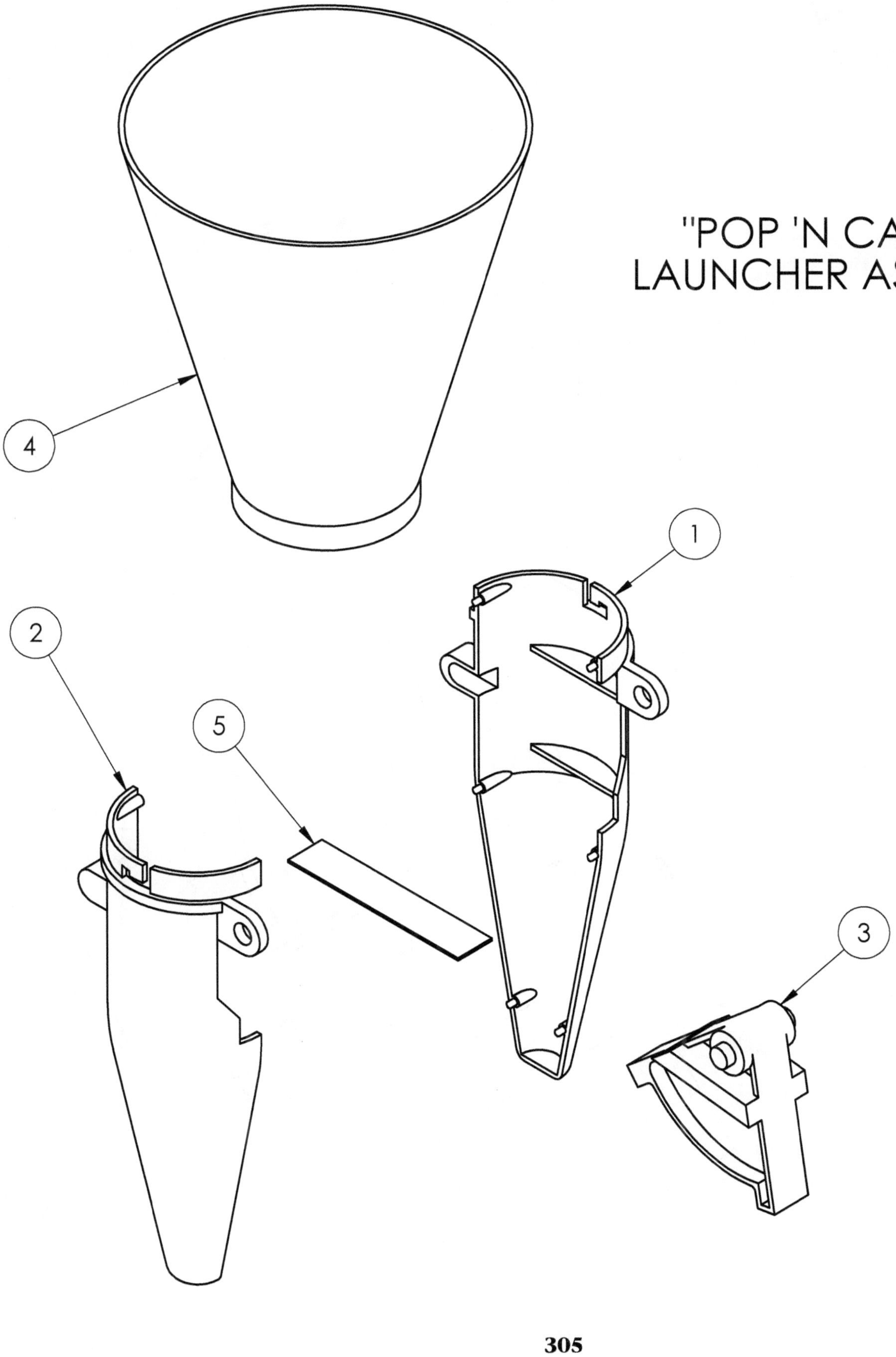

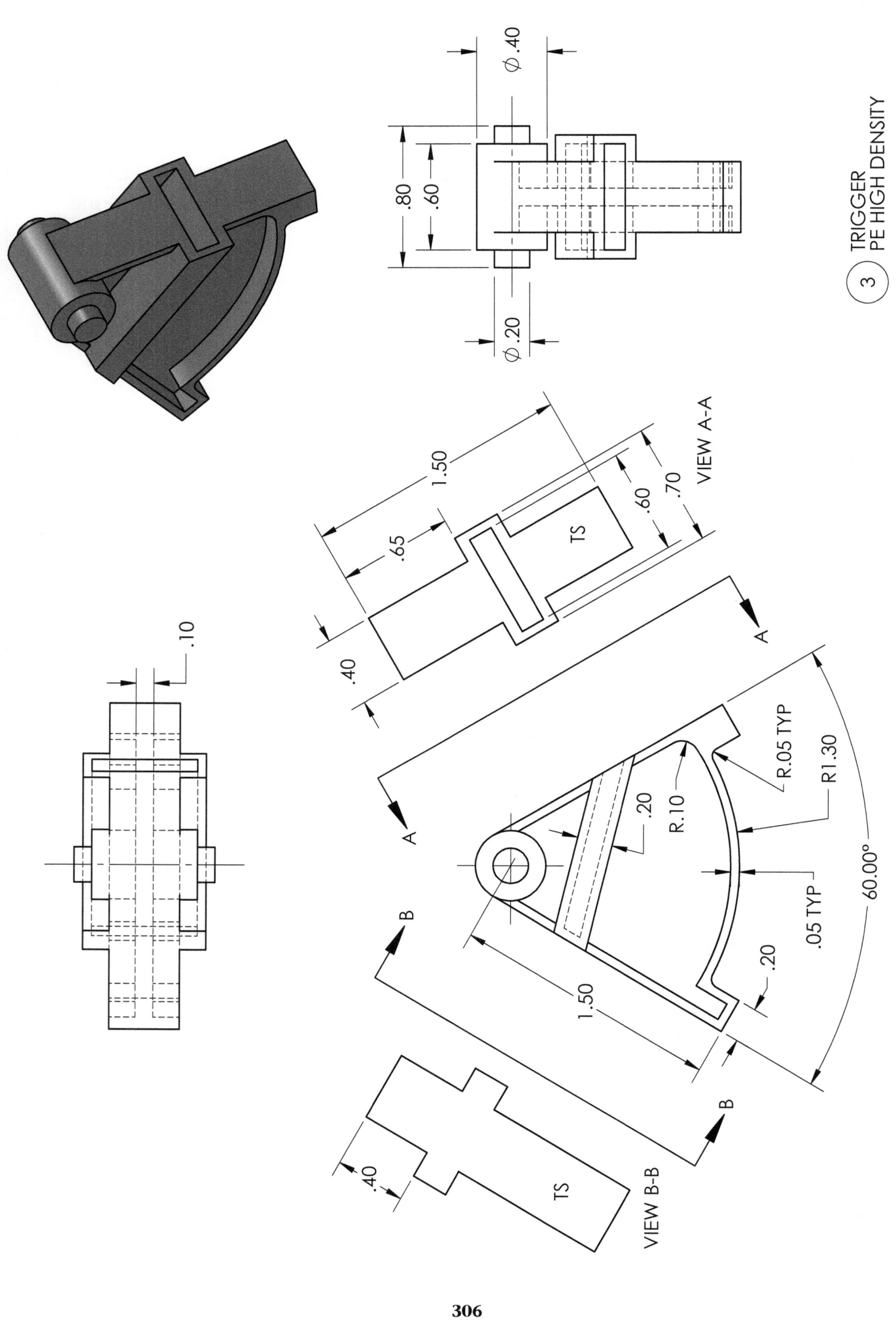

3 TRIGGER
PE HIGH DENSITY
VIEW A-A
VIEW B-B
TS
TS
A
A
B
B
.80
.60
⌀.40
⌀.20
.10
1.50
.65
.60
.70
.40
.40
R.05 TYP
R1.30
R.10
.20
.20
.05 TYP
60.00°
1.50

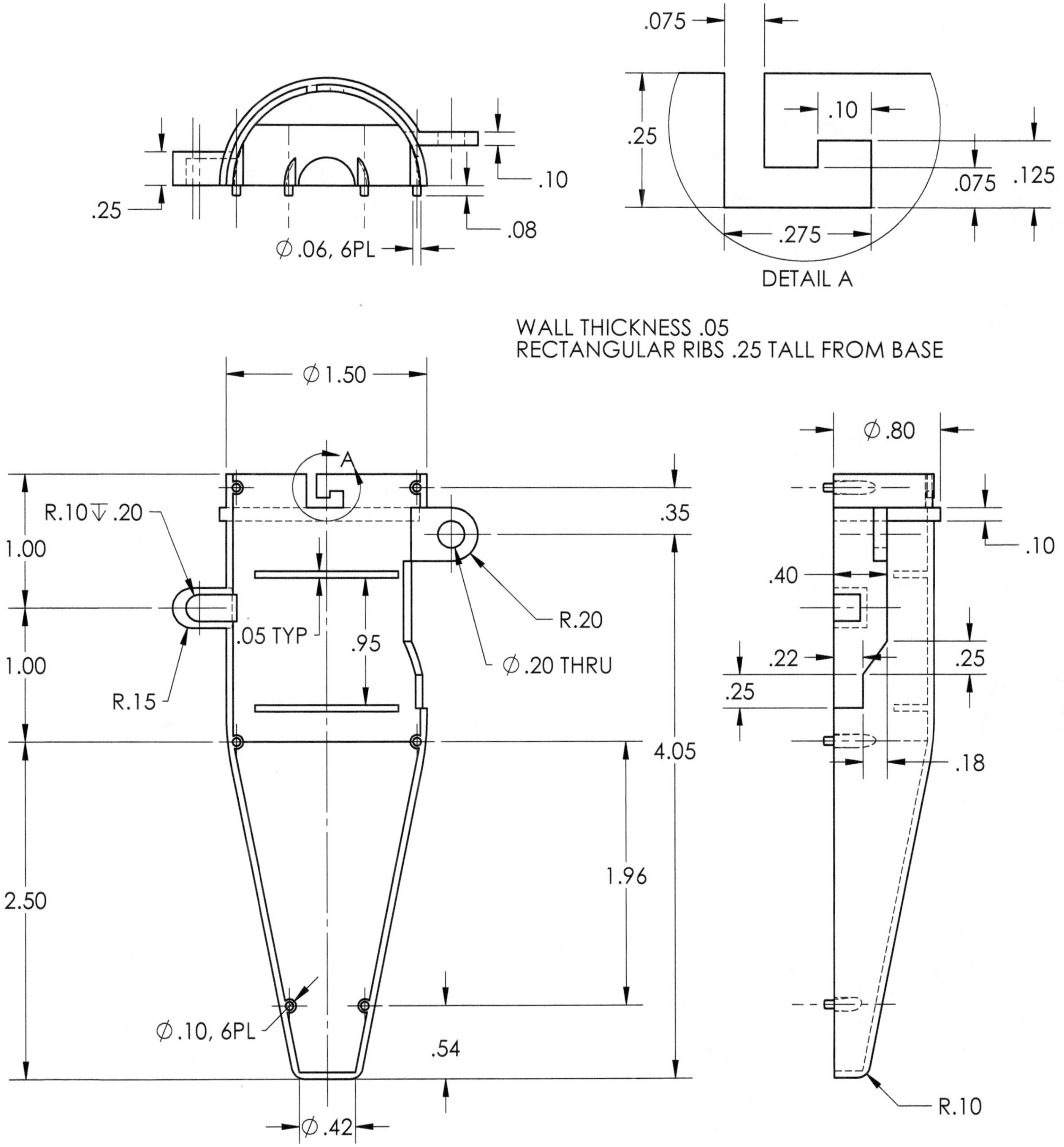

(1&2) LEFT AND RIGHT BODY
PE HIGH DENSITY PLASTIC
MAKE MIRROR IMAGE PARTS

ON MIRROR VERSION:
REVERSE DETAIL A
REPLACE 6 STUDS WITH 6 HOLES (∅.10↧.08) TO CONNECT BODIES

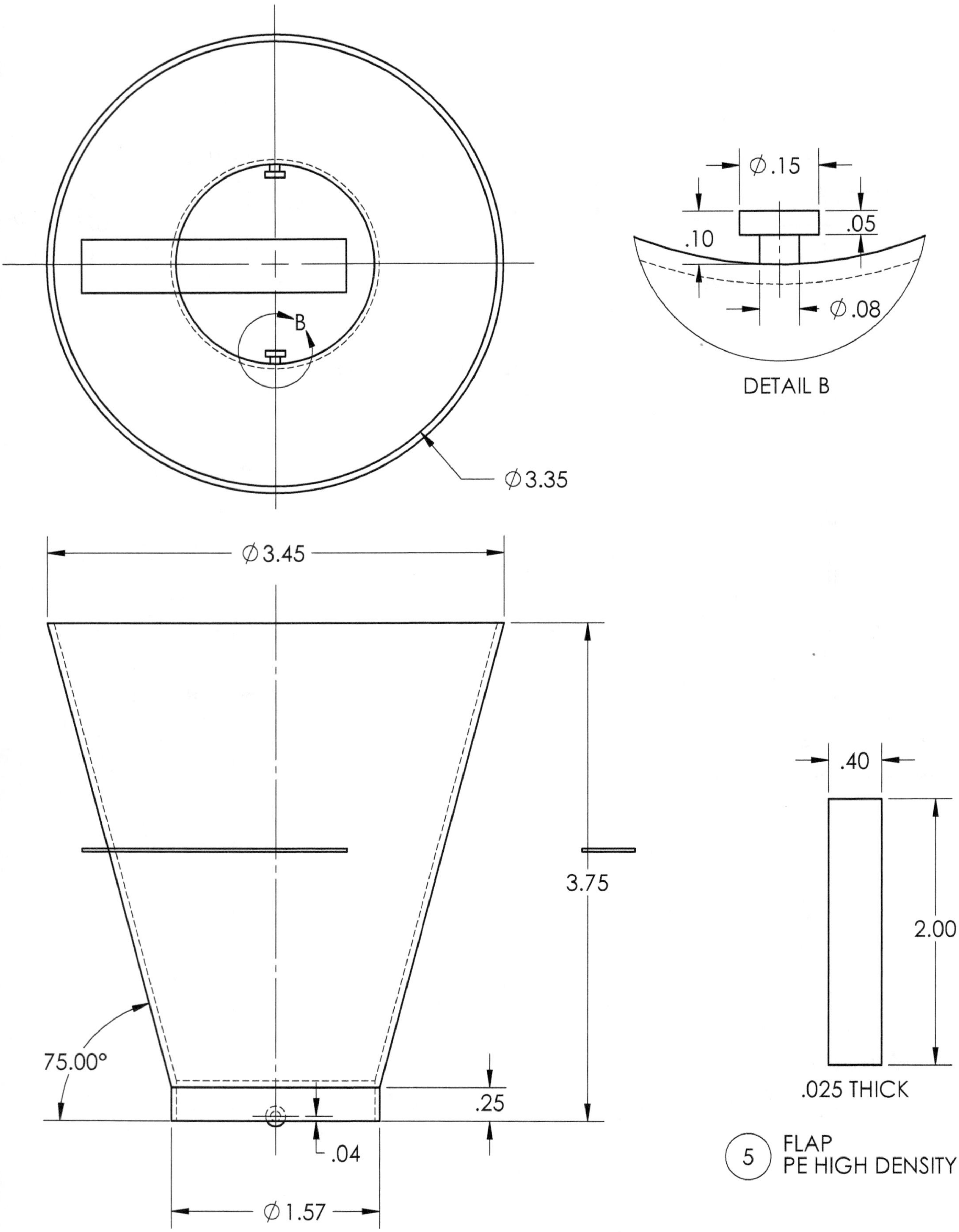

5 FLAP
PE HIGH DENSITY PLASTIC

4 BASKET
PE HIGH DENSITY PLASTIC

13

CHAPTER

Rapid Prototyping

1. Introduction

In engineering, there is often a certain level of uncertainty as to whether a new design will perform as designed, even when analysis has been done. At this point in the design process, prototypes play an important role.

In general, a prototype is a draft version or an approximation of a final product. There are several reasons why prototypes are developed: to identify possible problems that were not identified in previous stages of the design process, to confirm the suitability of a design prior to starting mass production, to conduct tests and verify performance, or simply for visualization purposes. Some prototypes are also used as market research and promotional tools to validate consumer interest in a specific product. But most importantly, it is cheaper to manufacture, test and make changes to a prototype than it is to a final product.

When a functional prototype is developed, different tests and analyses are performed. Using results obtained from these tests, the original design is modified, refined, and a new prototype is created. The process repeats until satisfactory levels of performance are reached. At this point, the design advances to its production stages.

A prototype can be very simple or extremely complex, depending on its purpose. As an example, a prototype developed to study the ergonomics or visual aspects of a design does not require the accuracy and strength as a functional prototype created for testing.

Prototyping used to be, and in some sense still is, a costly and time consuming process, but the combination of modern computer aided manufacturing (CAM) technologies and computer numerical controlled (CNC) machine tools, such as rapid prototyping and rapid manufacturing processes, have drastically transformed the development of prototypes.

2. Rapid Prototyping

Rapid prototyping, also known as solid free-form manufacturing and layered manufacturing, is a broad term that comprises many different technologies used to quickly fabricate a physical model directly from computer data. The first rapid prototyping method, called stereo lithography, was developed in the late 1980s, but more sophisticated techniques are available today.

The term "rapid" is relative. Some prototypes may take hours or even days to build, which is still much faster than the weeks that may be required for a technician to machine a design out of metal. Rapid prototyping systems are additive manufacturing processes that work on the basic principle of producing a 3D part by building and stacking multiple 2D layers together. The most common types of rapid prototyping systems include SLA (Stereo Lithography Application), SLS (Selective Laser Sintering), LOM (Laminate Object Manufacturing), and FDM (Fused Deposition Modeling). Different technologies use different materials to produce the parts.

2.1 Stereo Lithography Apparatus (SLA)

Stereo Lithography uses a liquid photosensitive resin as building material and a low power laser to build the part one layer at a time. The 3D part is produced on a flat platform that is gradually submerged on a pool filled with photosensitive liquid resin.

For each layer, a laser beam traces out the corresponding cross-section pattern of the part on the surface of the liquid container. The pattern is then solidified and added to the layer below. The platform descends one layer thickness (the layer thickness depends on the precision of the machine) and the process repeats. When the process finishes, the part is immersed in a chemical bath for cleaning and removing excess material. The system is illustrated in Figure 13.1.

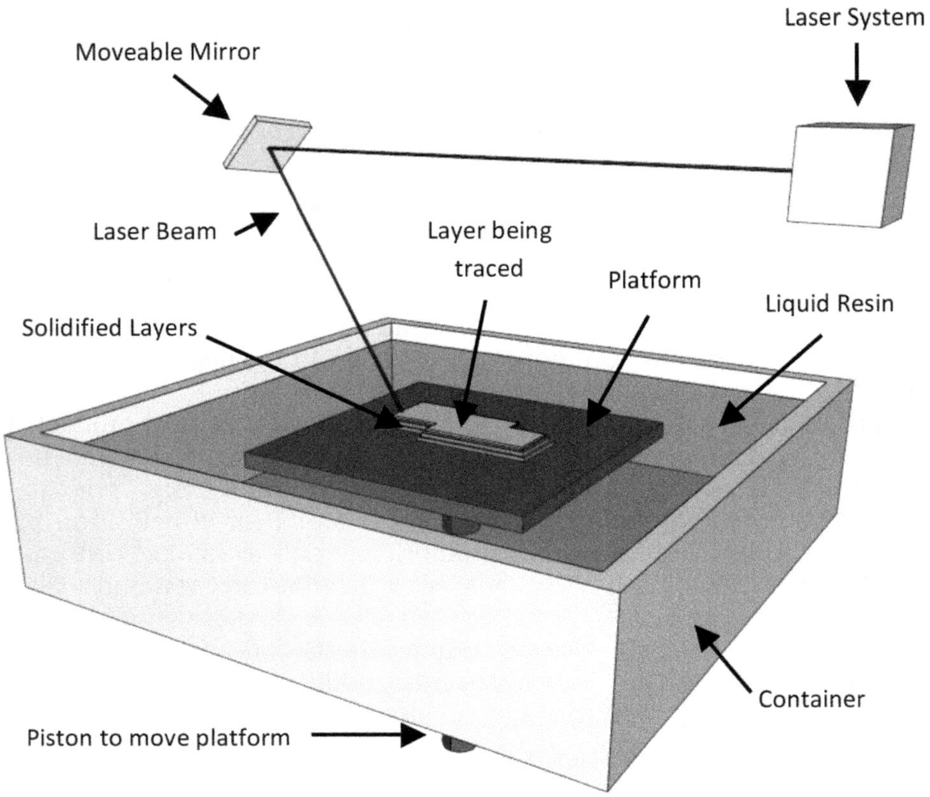

Figure 13.1 Stereo Lithography Apparatus (SLA)

SLA prototyping normally results in visually appealing parts with very smooth surfaces (see Figure 13.2). However, they are not appropriate for most functional tests, due to the physical properties of the materials used.

Figure 13.2 SLA prototype

2.2 Selective Laser Sintering (SLS)

Selective Laser Sintering is a technology that uses a high power laser, normally a hot CO_2 laser, and powdered materials to build a prototype. A wide variety of materials can be used, ranging from thermoplastic polymers, such as nylon and polystyrene, to some metals. 3D parts are produced by fusing a thin slice of the powdered material onto the layers below it.

Because of its powdered material base, the surfaces of SLS prototypes are not as smooth as those produced by SLA processes. However, SLS parts are sufficiently strong and resistant for many functional tests. See Figure 13.3 for examples of SLS prototypes made of corn starch.

The powdered material is kept on a delivery platform and supplied to the building area by a roller. For each layer, a laser traces the corresponding shape of the part on the surface of the building area, by heating the powder until it melts, fusing it with the layer below it. The platform containing the part lowers one layer thickness and the platform supplying the material elevates, providing more material to the system. The roller moves the new material to the building platform, leveling the surface, and the process repeats. The system is illustrated in Figure 13.4. Some SLS prototype machines use two delivery platforms, one on each side of the building platform, for efficiency, so the roller can supply material to the building platform in both directions.

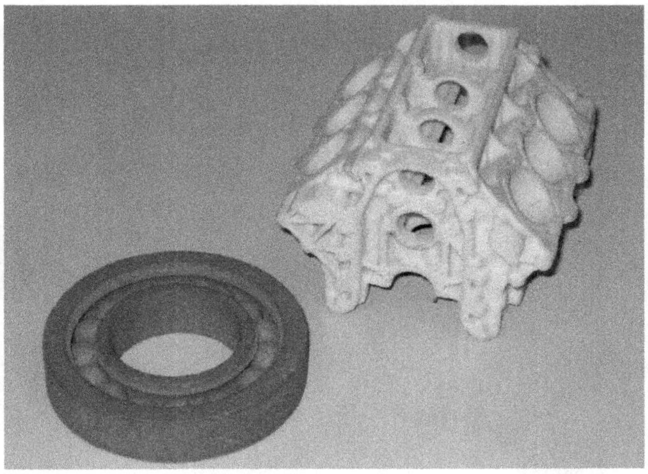

Figure 13.3 SLS prototypes

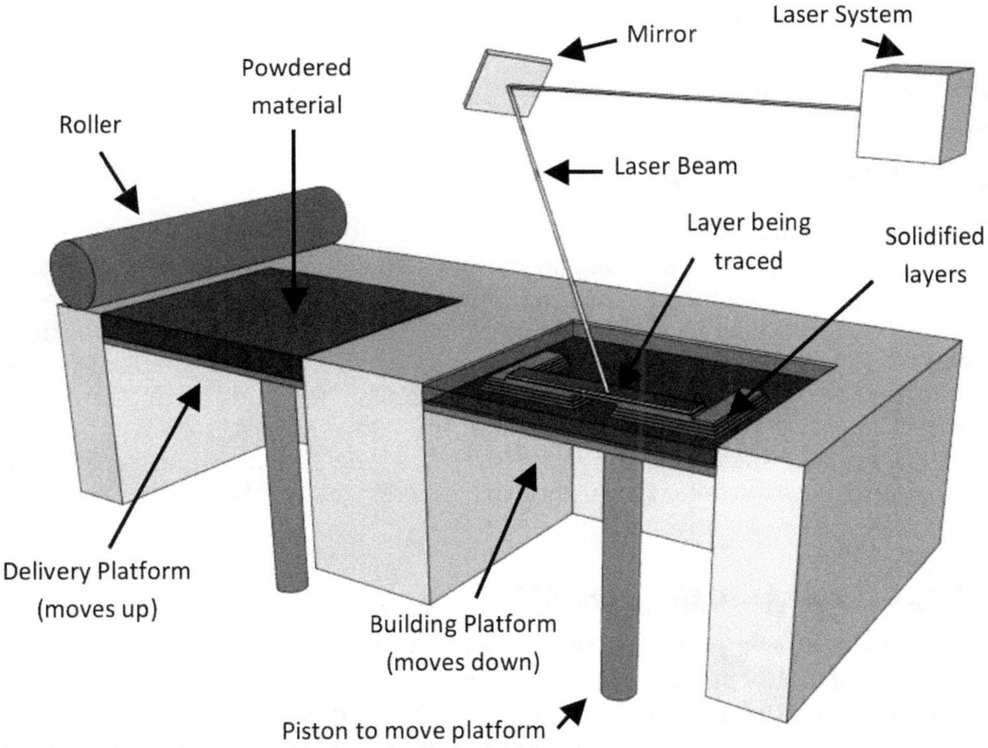

Figure 13.4 Selective Laser Sintering (SLS) machine (sectional view)

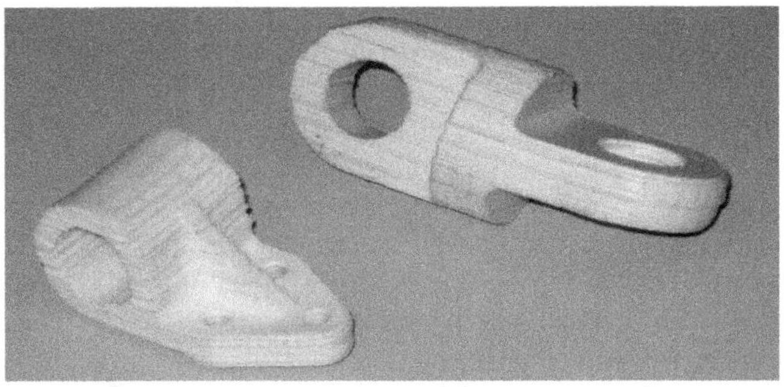

Figure 13.5 LOM prototypes

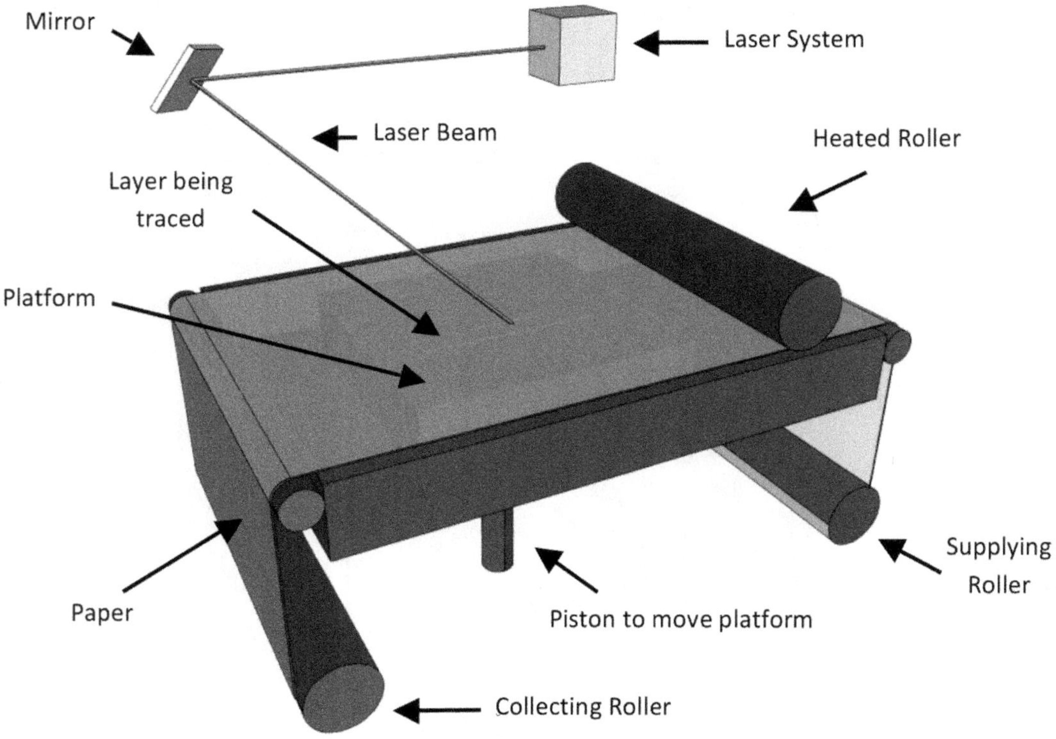

Figure 13.6 Laminated Object Manufacturing (LOM)

2.3 Laminated Object Manufacturing (LOM)

Laminated Object Manufacturing is a relatively low cost rapid prototyping technology where thin slices of material (usually paper or wood) are successively glued together to form a 3D shape. Examples of LOM prototypes are shown in Figure 13.5. The process uses two rollers to control the supply of paper with heat-activated glue to a building platform. When new paper is in position, it is flattened and added to the previously created layers using a heated roller. The shape of the new layer is traced and cut by a blade or a laser. When the layer is complete, the building platform descends and new paper is supplied. When the paper is in position, the platform moves back up so the new layer can be glued to the existing stack, and the process repeats. The system is illustrated in Figure 13.6. LOM manufacturing fell out of favor because of the inherent risk of laser generated fires with the paper material.

2.4 Fused Deposition Modeling (FDM)

Fused Deposition Modeling is a rapid prototyping technique that involves melting a thermoplastic polymer (usually polyester, ABS plastic, or casting wax) and squeezing thin filaments out of a nozzle, layer by layer, on a building platform). As the material descends to the platform, it is immediately hardened. A second nozzle supplies soluble support material, if necessary, to the building platform to

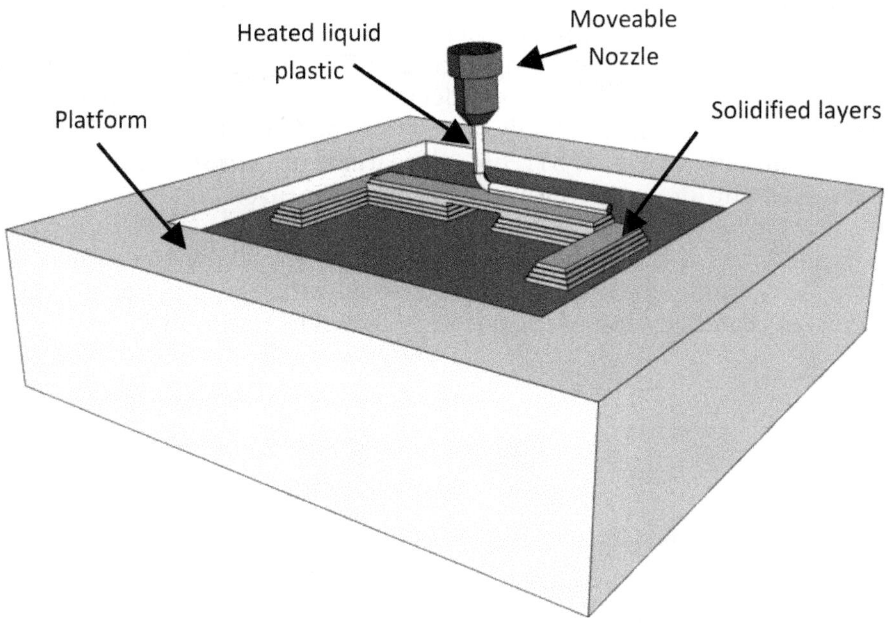

Figure 13.7 Fused Deposition Modeling (FDM)

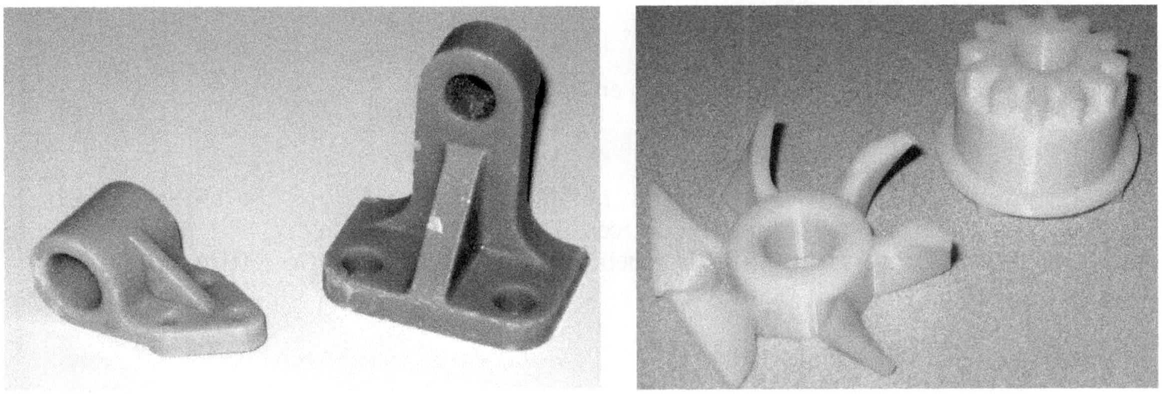

Wax-based

Plastic-based

Figure 13.8 FDM prototypes

prevent certain geometry from deflecting because of gravity. When the process finishes, the FDM part is immersed in an alkali chemical bath to remove the support material. The system is illustrated in Figure 13.7. Depending on the machine, the nozzle can move across a stationary platform or the nozzle may remain stationary while the platform moves. Generally, a combination of both is found, where the nozzle moves front and back, and left and right, and the platform moves up and down.

Wax based prototyping machines are being gradually replaced by more efficient plastic-based equipment. Although a slight layering texture is noticeable on the surface of the parts, plastic-based FDM machines produce very strong and durable prototypes that are ideal for functional testing.

3. Stereo Lithography Files

The stereo lithography file format, known as STL (Standard Tessellation Language), is the current industry standard data interface for rapid prototyping and manufacturing. Before a 3D model is sent to a rapid prototype machine, it must be converted to an STL file. Most modeling packages support this format, and the conversion process is automatic in most cases. From a user standpoint, the process typically requires only exporting or saving the model as an STL file. Some software packages, however, allow the user to define some specific parameters about how the conversion will be performed.

The STL file format defines the geometry of a model as a single mesh of triangles. Information about color, textures, materials, and other properties of the object are ignored in the STL file. Each triangular face of the mesh is stored along with a unit vector (known as facet normal) that is normal to the face pointing outwards from the object. This facet normal is used to determine the outside boundaries of the model that will ultimately define the surface of the part to be built. When a solid model is converted into an STL file, all features are consolidated into one geometric figure. The resulting STL file does not allow individual features created with the parametric modeling application to be edited.

The process of approximating the actual surfaces of the object with a closed mesh of triangles is known as tessellation. An example of a solid modeling created with a parametric modeling application and the corresponding tessellated version of the model is shown in Figure 13.9.

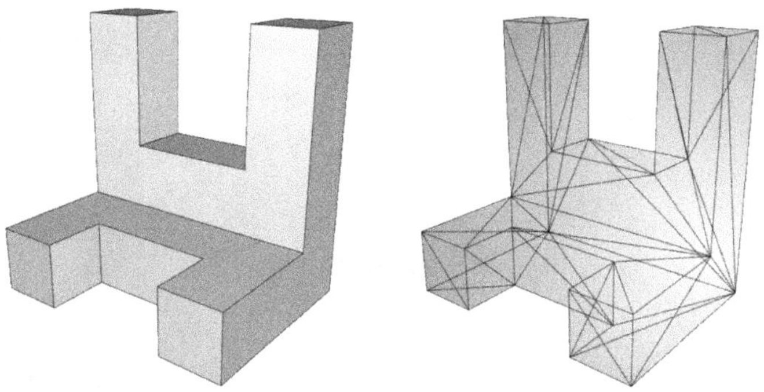

Solid model   Model after tessellation

Figure 13.9 Solid model and STL model

When the tessellated STL file is sent to the rapid prototype machine, the model is sliced into multiple horizontal layers that are later reproduced physically by the device. The thickness of these slices depends on the prototyping machine being used and the specific settings defined by the controller. See Figure 13.10 for an example.

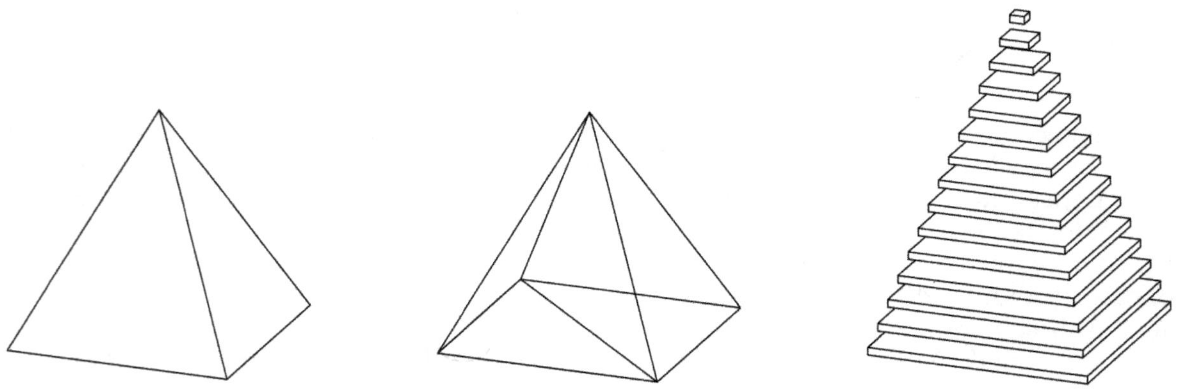

Solid model   STL model   Sliced STL ready for prototyping

Figure 13.10 Model conversion

Practice Test

1. **Which of the following is not a common use of prototypes?**
   A) Visualization
   B) Sharing ideas in early design stages
   C) Marketing and promotion
   D) Testing
   E) All of the above are common uses of prototypes

2. **Which of the following is not a rapid prototyping technology?**
   A) FDM B) SLS C) LOM D) STL E) SLA

3. **In rapid prototyping, what does the abbreviation LOM stand for?**
   A) Laser Oriented Mechanism
   B) Low Operating Machine
   C) Linear Output Modeling
   D) Laminated Object Manufacturing
   E) None of the above

4. **What file type is the current industry standard for rapid prototyping?**
   A) RPT B) TXT C) STL D) PDF E) DXF

5. **In rapid prototyping, what does the abbreviation SLS stand for?**
   A) Single Laminated Section
   B) Selective Laser Sintering
   C) Synchronized Laser System
   D) Stable Light Segmentation
   E) None of the above

6. **What rapid prototyping technology uses paper as building material?**
   A) LOM B) SLS C) FDM D) SLA E) None of the above

7. **The process of approximating the surfaces of a 3D object with a mesh of triangles is known as _______**
   A) Prototyping
   B) Reduction
   C) Tessellation
   D) Normalization
   E) None of the above

8. **In general, plastic-based FDM parts are appropriate for functional testing.**
   A) True B) False

9. **In general, SLA parts have very smooth surfaces and are ideal for functional testing.**
   A) True B) False

10. **An STL file needs to be manually sliced before a rapid prototype machine can build it.**
    A) True B) False

GIVEN THE FRONT AND TOP VIEWS, SELECT THE CORRECT RIGHT SIDE VIEW FROM THE CHOICES PROVIDED

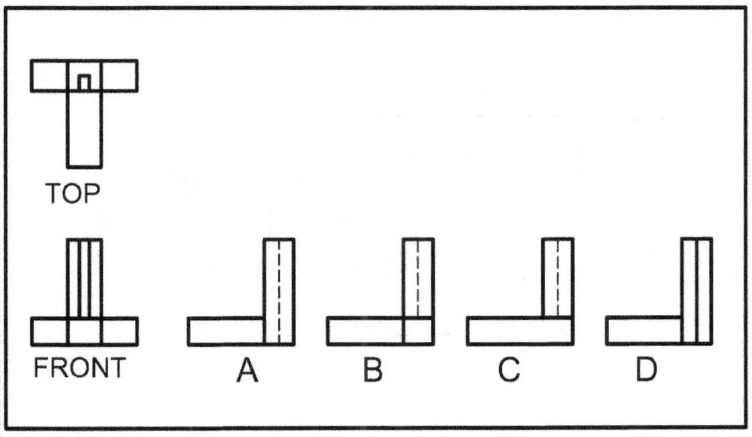

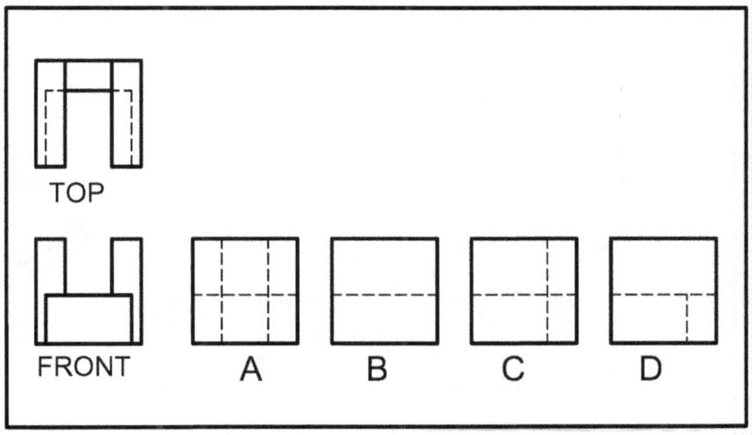

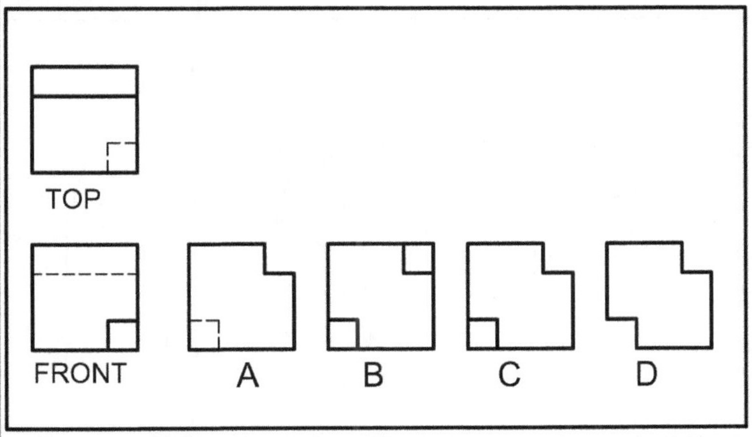

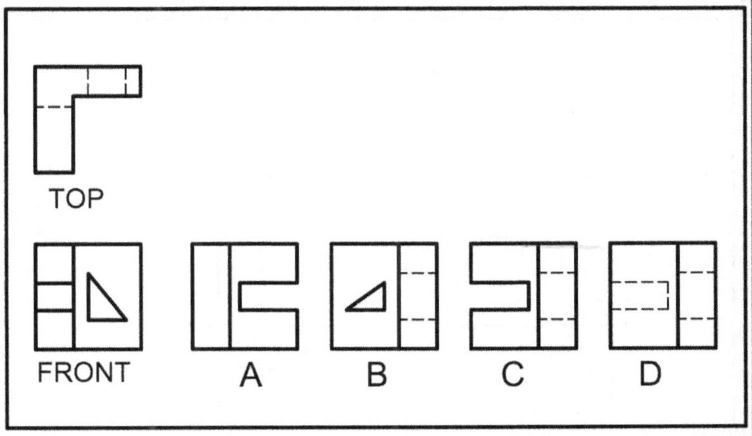

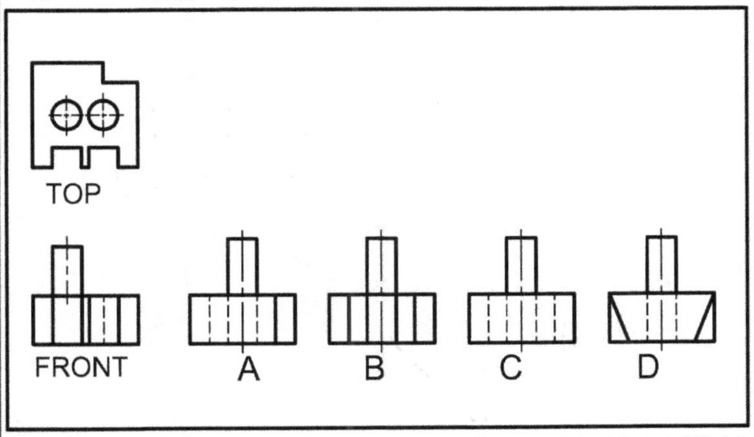

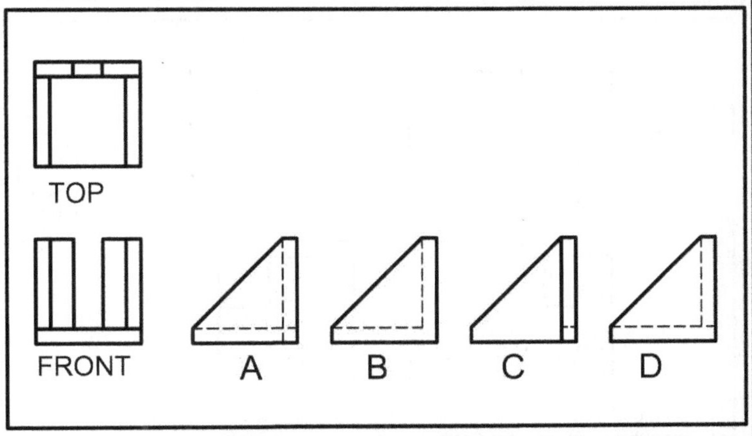

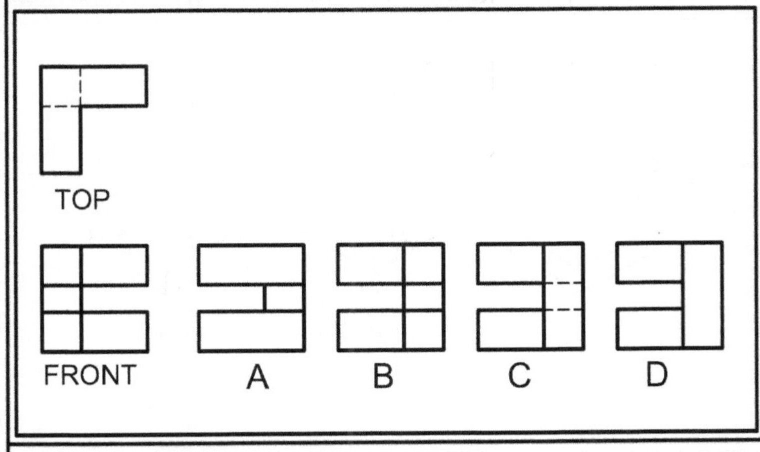

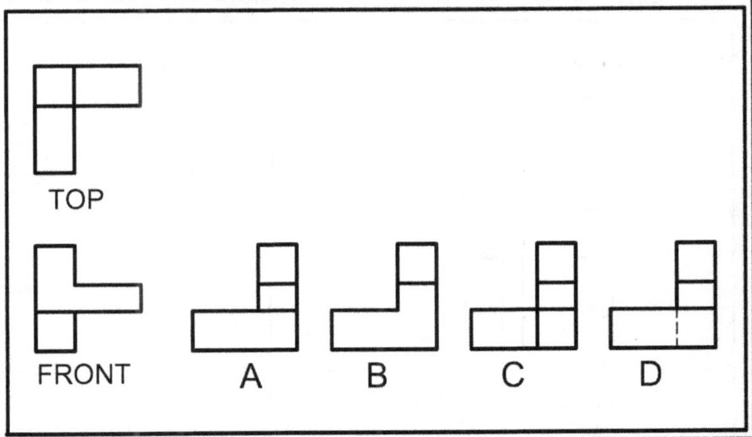

AUTHOR:	SCALE:
TEAM: SECTION:	DATE:

GIVEN THE ORTHOGRAPHIC VIEWS, SELECT THE CORRECT ISOMETRIC FROM THE CHOICES PROVIDED

1.

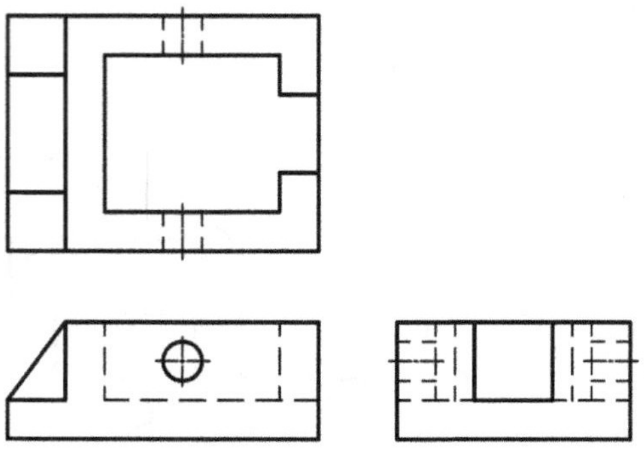

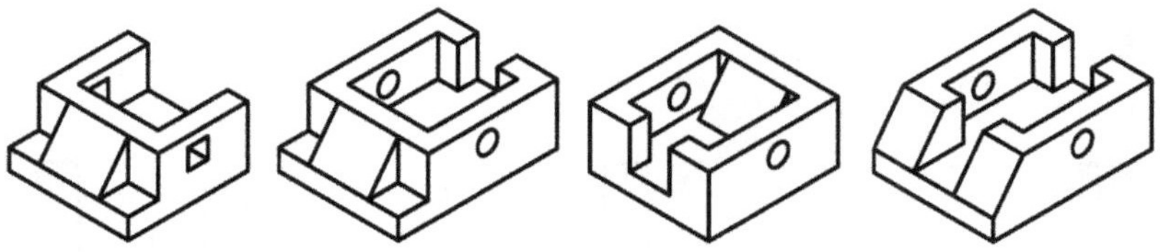

A B C D

2.

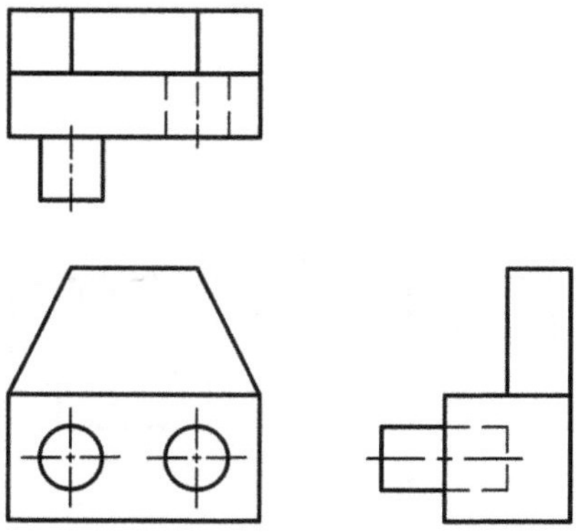

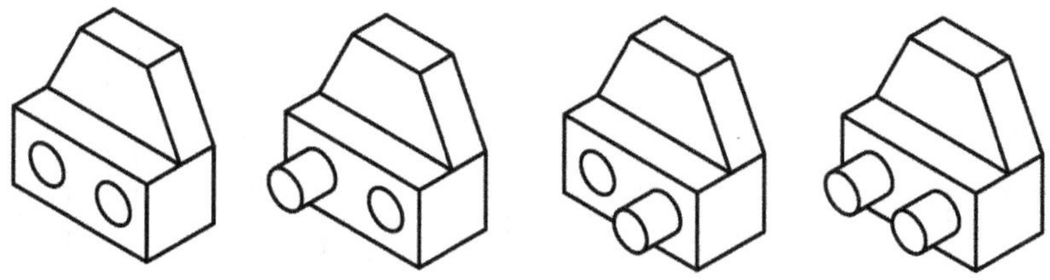

A B C D

3.

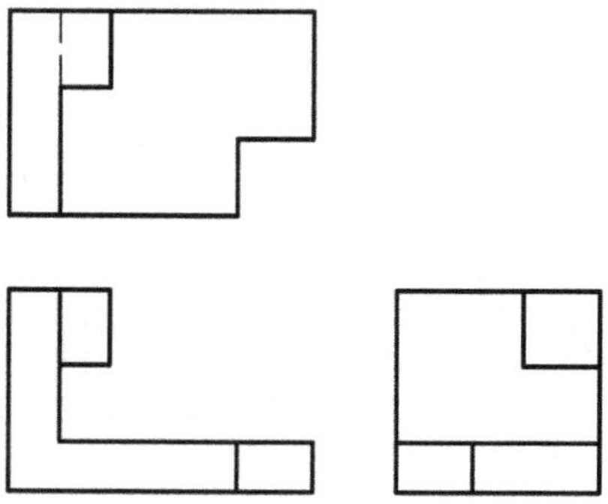

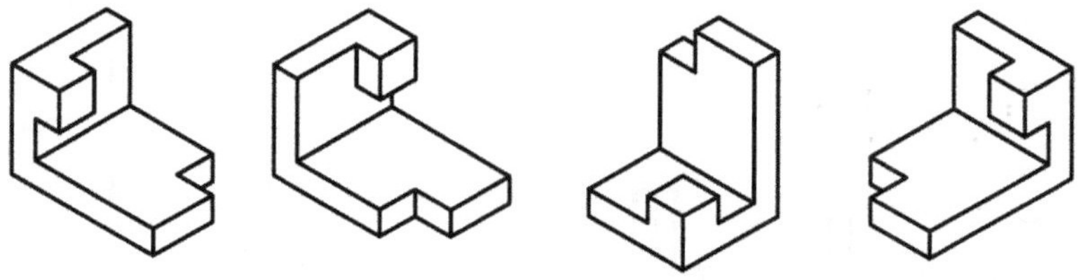

A B C D

4.

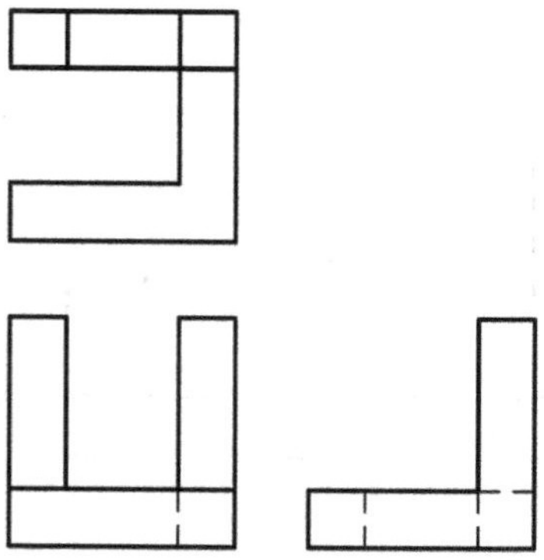

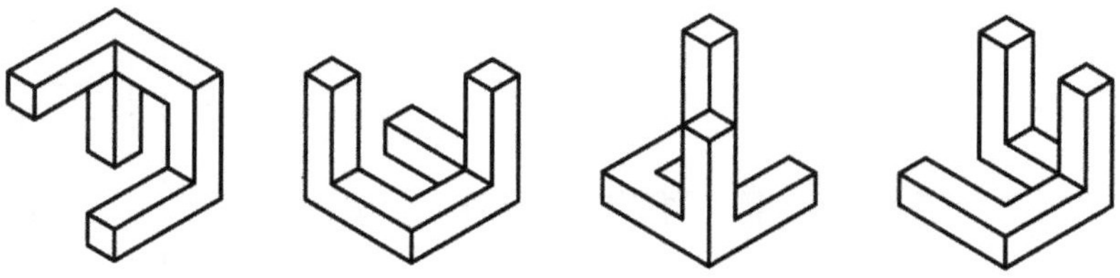

A B C D

AUTHOR:	SCALE: ....
TEAM: .... SECTION:	DATE: ....

DRAW ALL MISSING LINES (VISIBLE, HIDDEN, AND CENTER). LINES MAY BE MISSING FROM ANY VIEW.

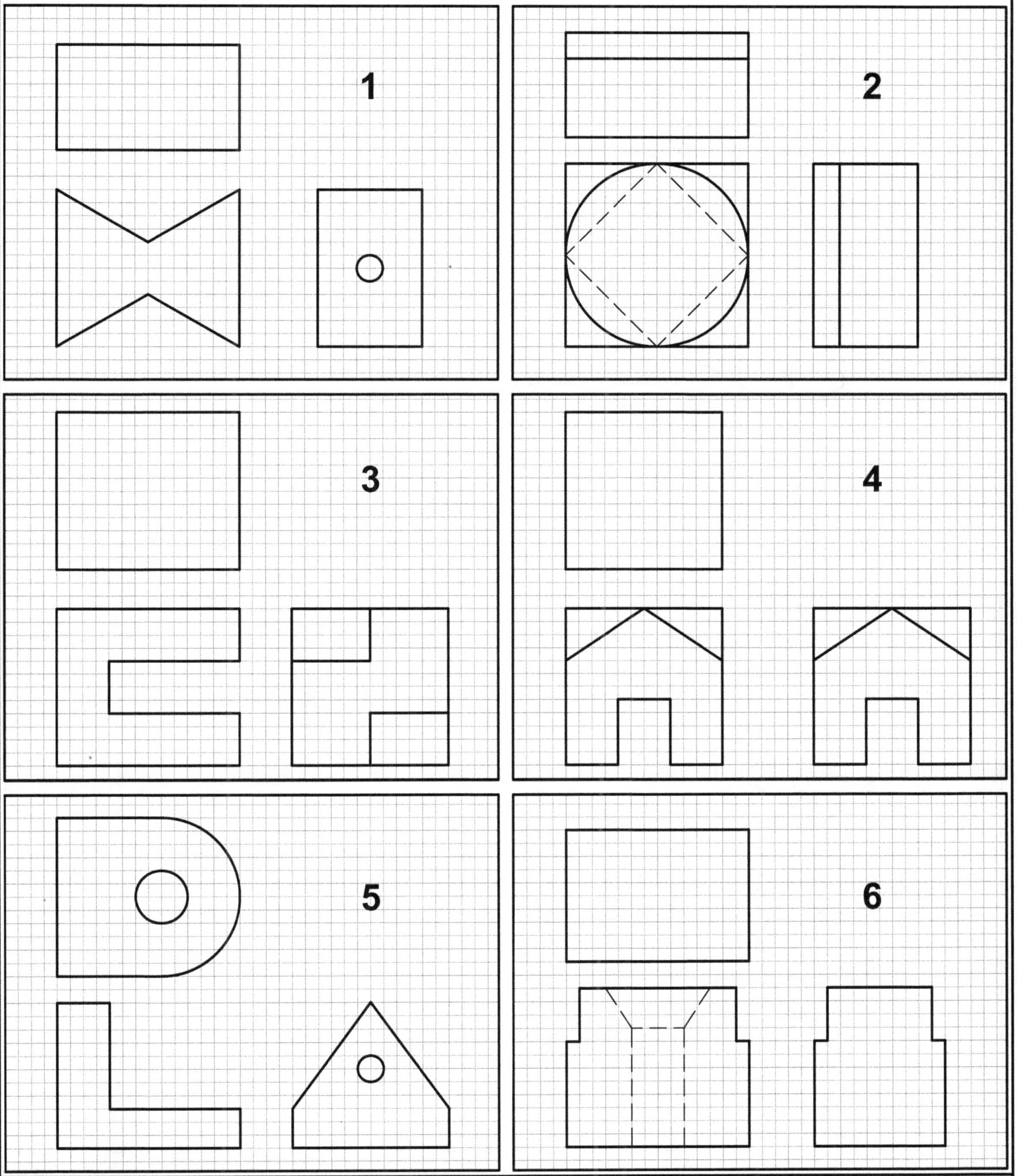

AUTHOR: ....

SCALE: ....

TEAM: .... SECTION: ....

DATE: ....

DRAW THE OBJECTS PROVIDED AFTER THEY ARE ROTATED AROUND THE GIVEN AXIS THE INDICATED AMOUNT.

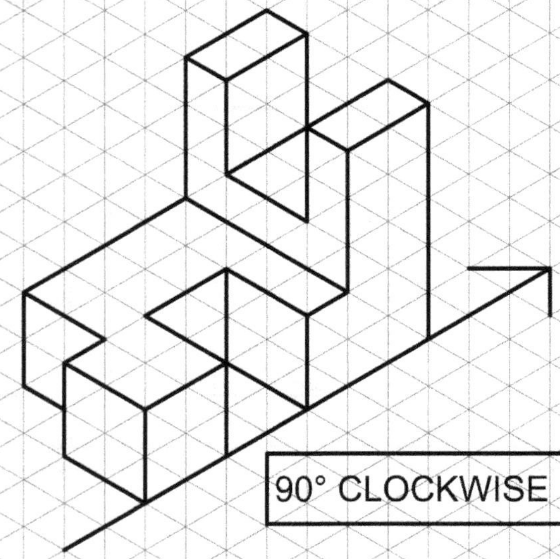

90° CLOCKWISE

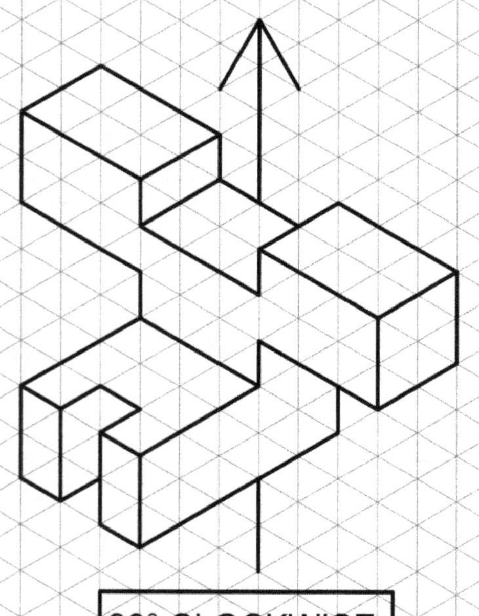

90° CLOCKWISE

AUTHOR:		SCALE:
TEAM:	SECTION:	DATE:

DRAW THE OBJECTS PROVIDED AFTER THEY ARE ROTATED AROUND THE GIVEN AXIS THE INDICATED AMOUNT.

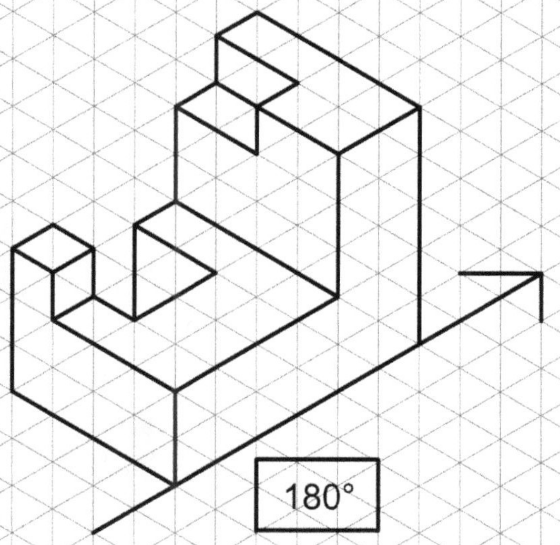

180°

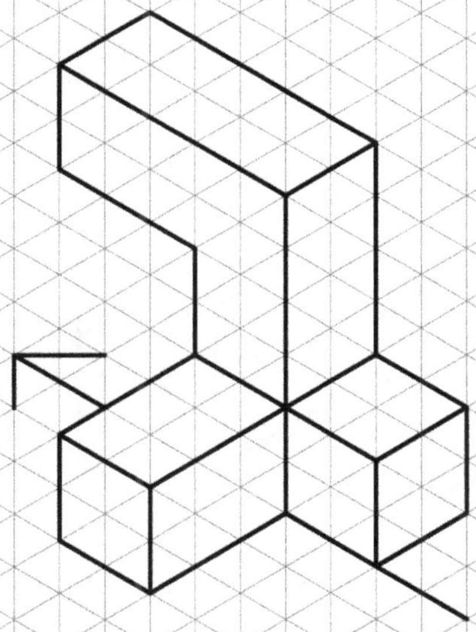

90° COUNTERCLOCKWISE

AUTHOR:		SCALE:
TEAM:	SECTION:	DATE:

DRAW THE ISOMETRIC THAT IS SYMMETRICAL TO THE OBJECT ACROSS THE INDICATED PLANE.

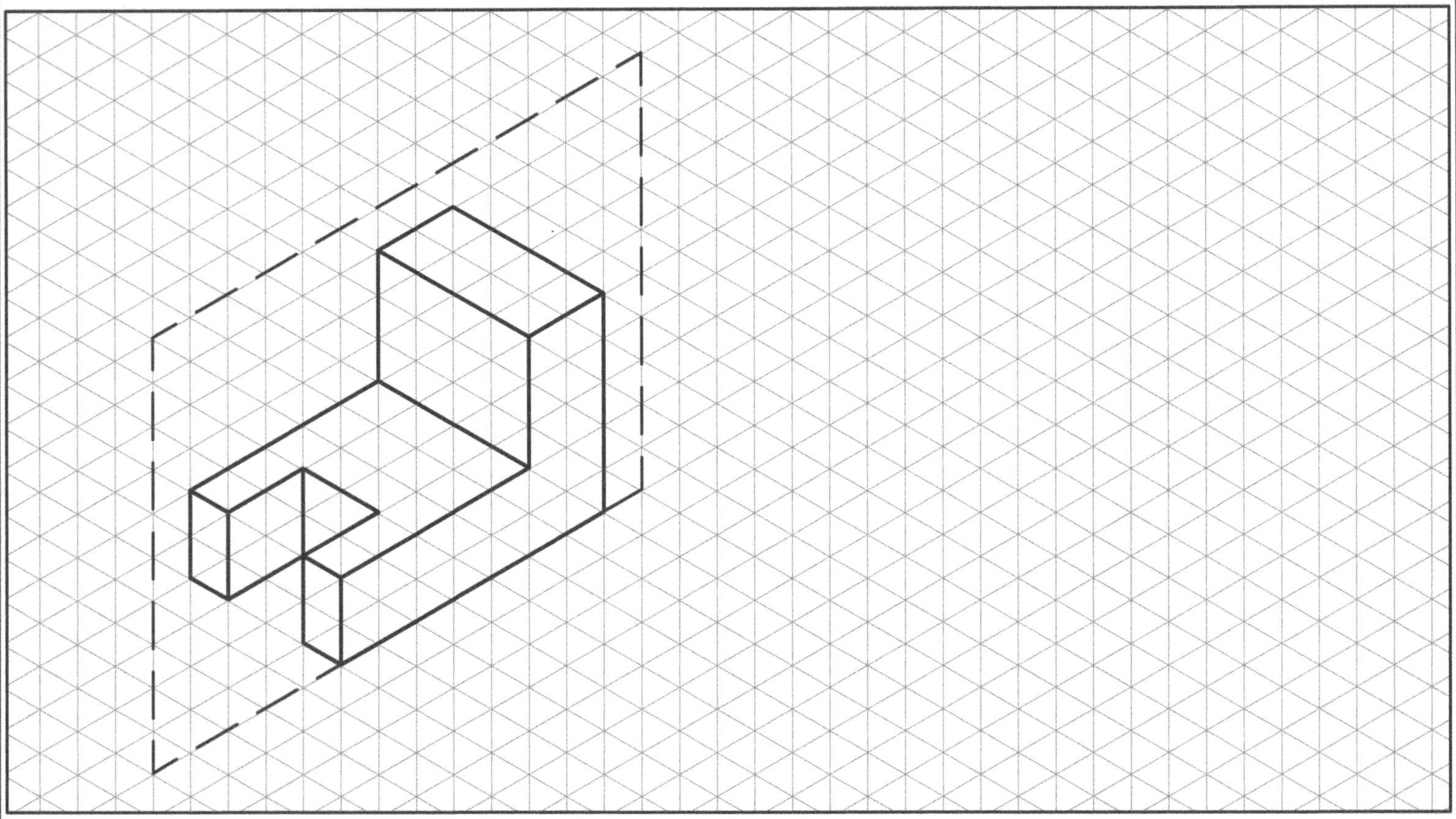

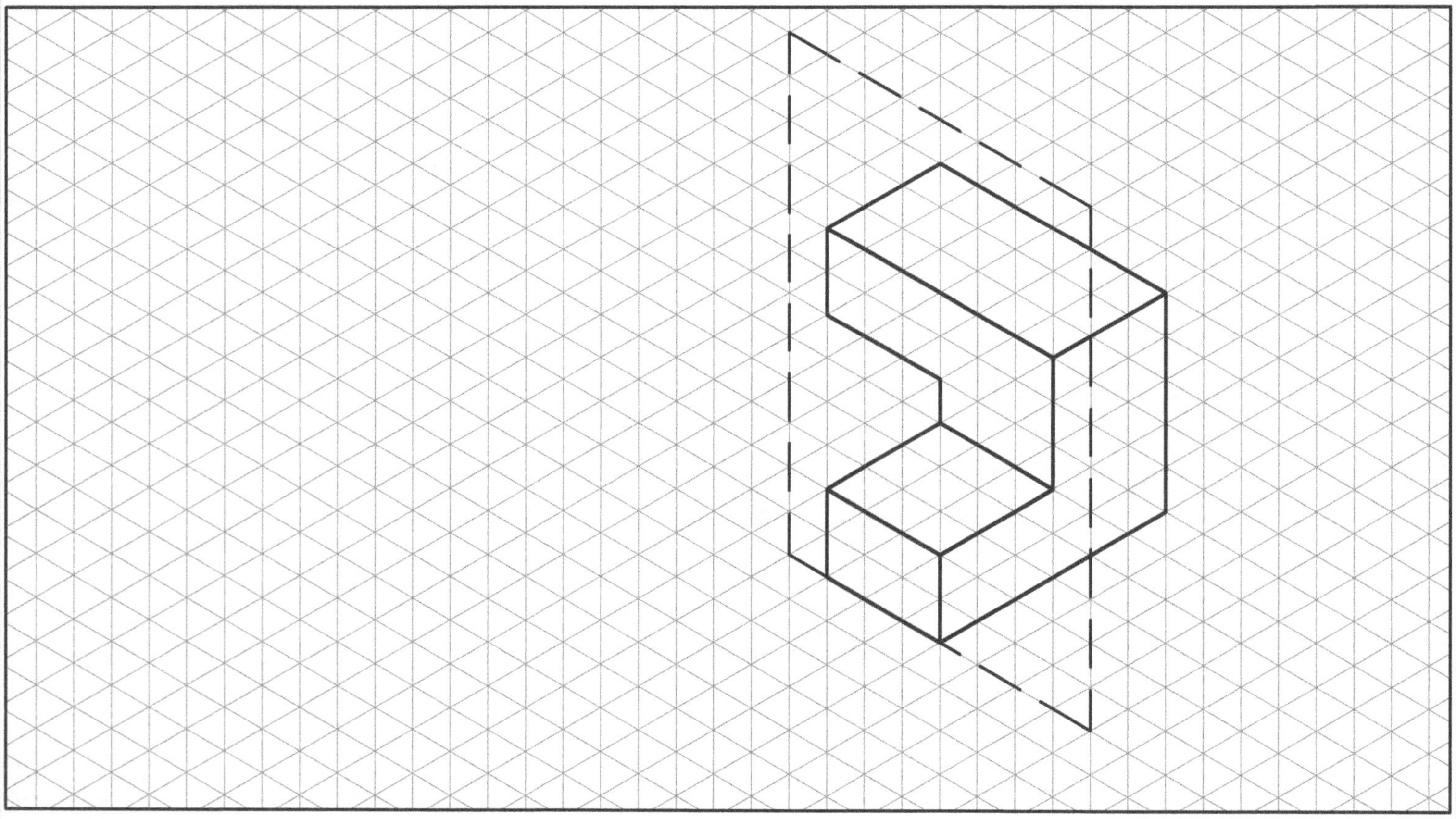

AUTHOR:	SCALE:
TEAM: SECTION:	DATE:

DRAW THE ISOMETRIC THAT IS SYMMETRICAL TO THE OBJECT ACROSS THE INDICATED PLANE.

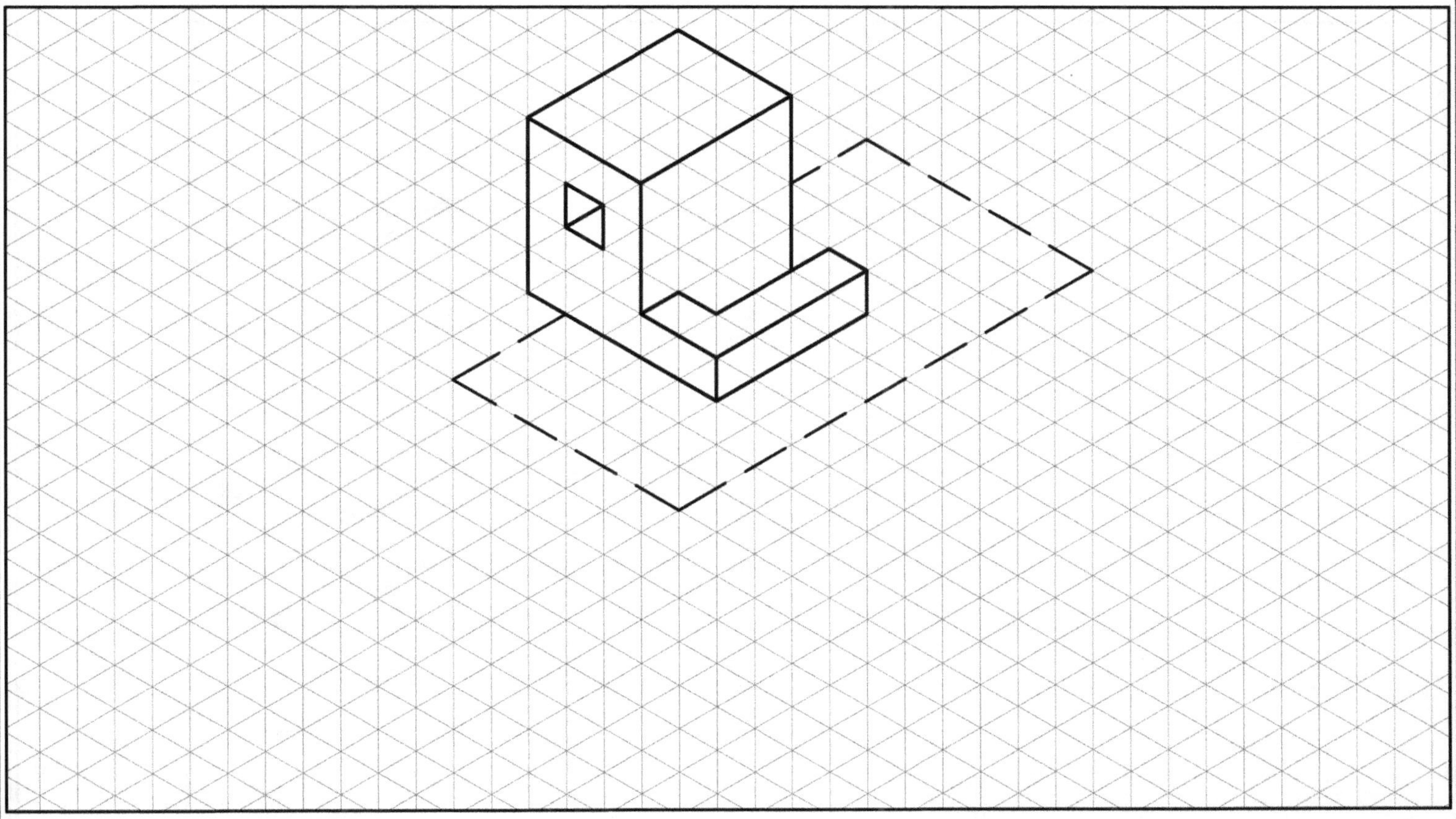

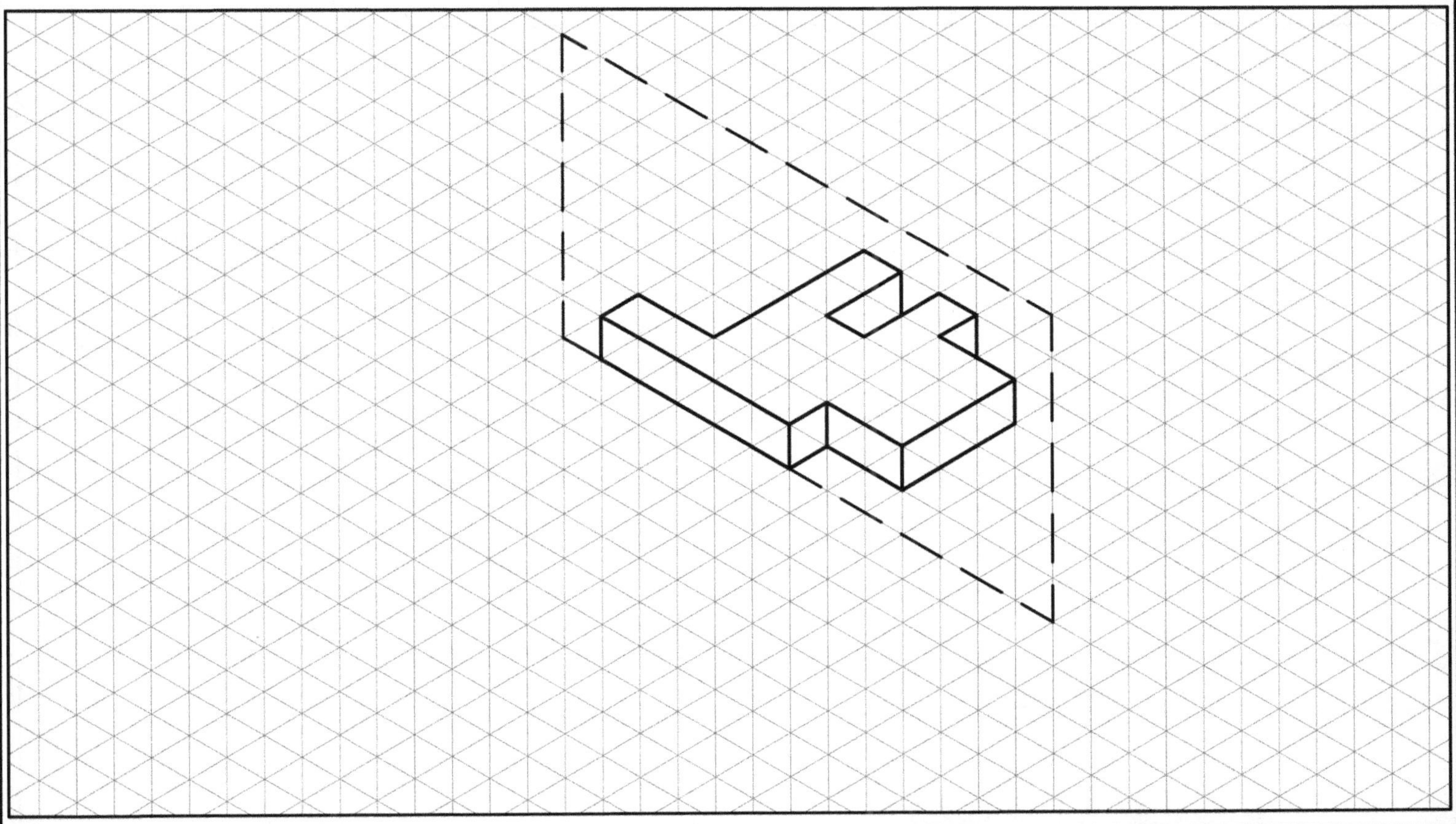

AUTHOR: . . . . SCALE: . . . .

TEAM: . . . . SECTION: . . . . DATE: . . . .

DRAW THE ISOMETRIC VIEW OF THE MATING PART TO FIT ON TOP OF THE OBJECT PROVIDED.

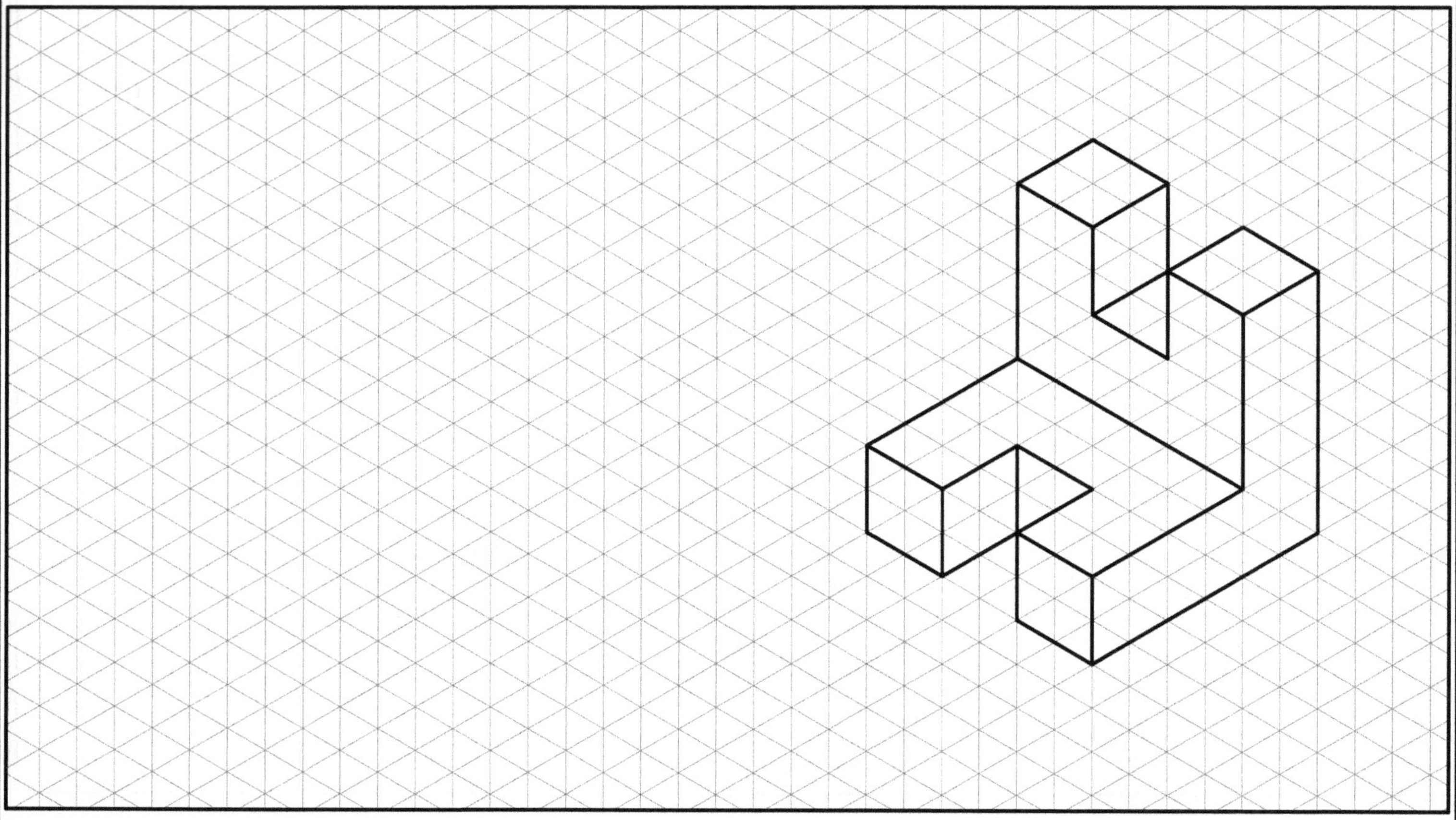

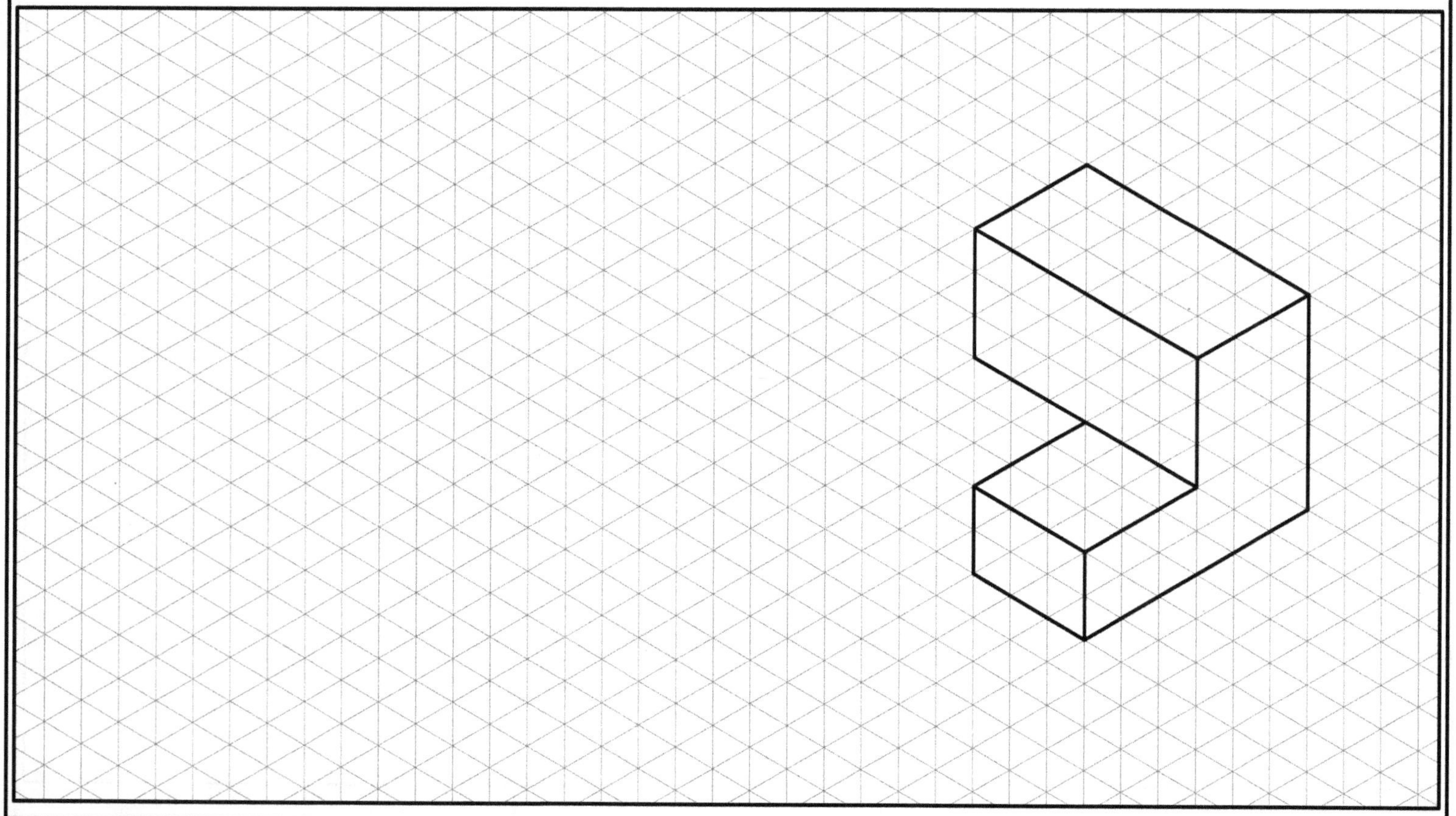

AUTHOR: ....

SCALE: ....

TEAM: .... SECTION: ....

DATE: ....

DRAW THE ISOMETRIC VIEW OF THE MATING PART TO FIT ON TOP OF THE OBJECT PROVIDED.

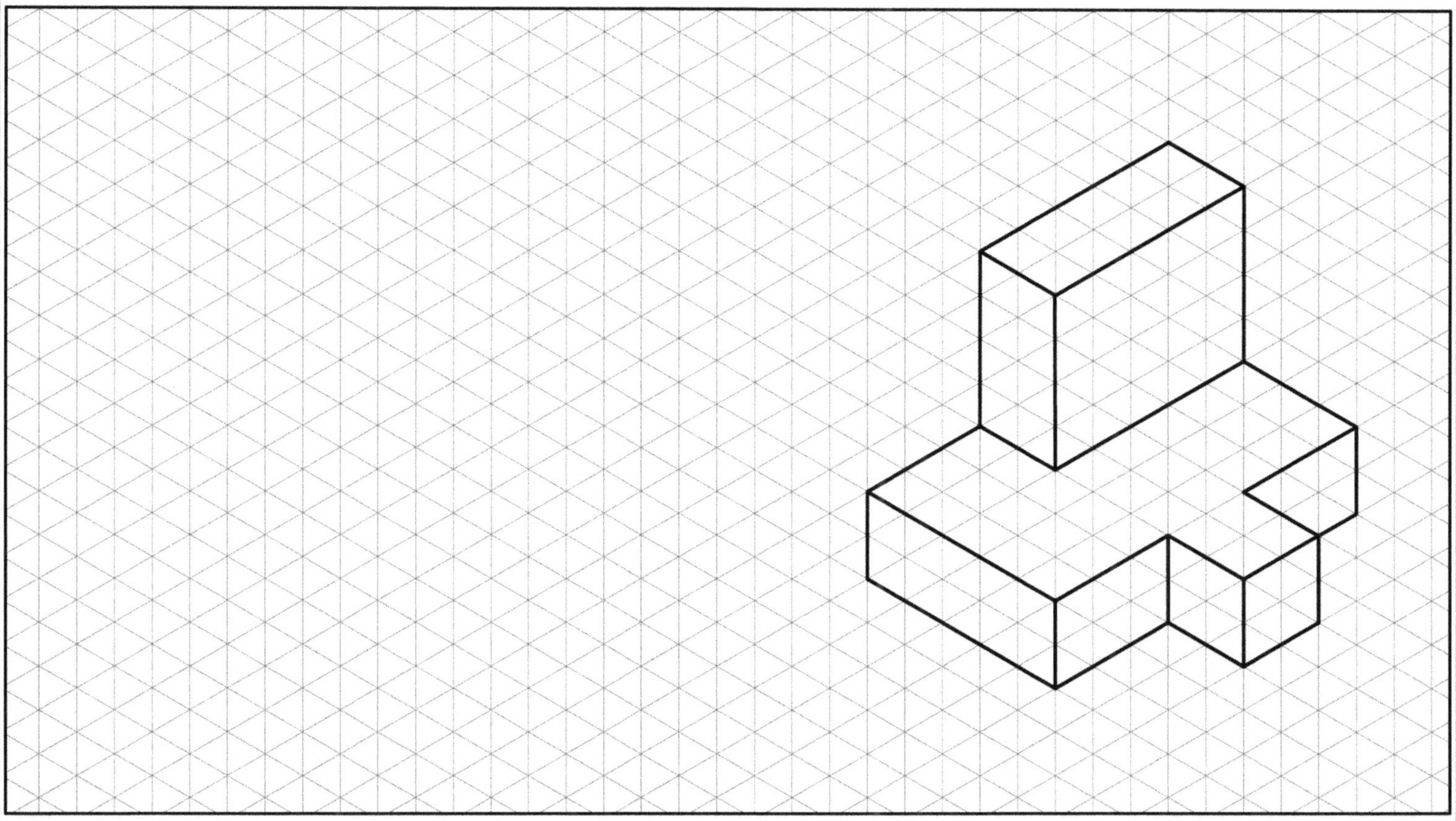

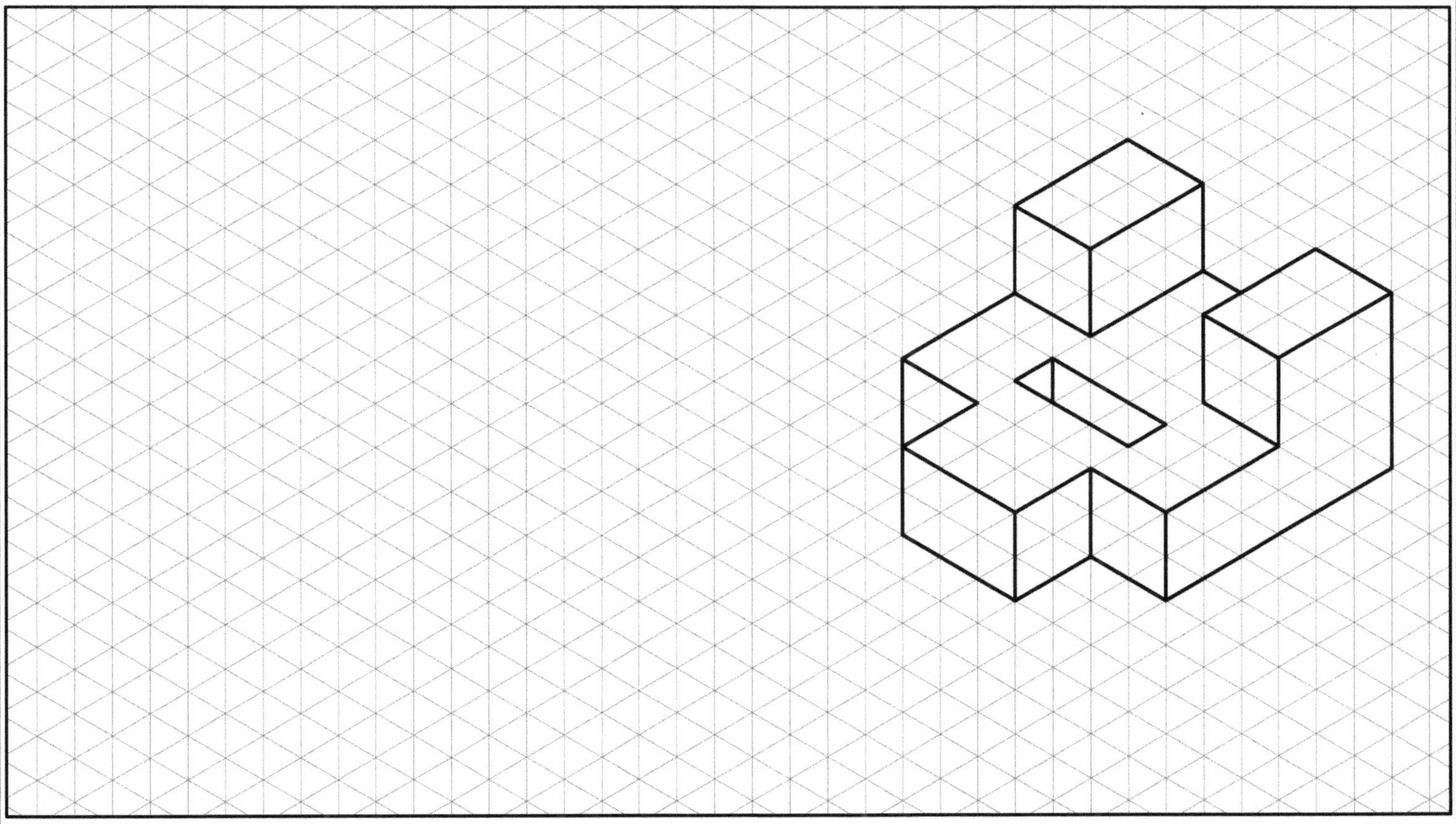

AUTHOR:	SCALE:
TEAM: SECTION:	DATE:

Appendix

English Threads

Sizes	Basic Major Diameter	THREADS PER INCH										
		Series with graded pitches			Series with constant pitches							
		Coarse UNC	Fine UNF	Extra fine UNEF	4UN	6UN	8UN	12UN	16UN	20UN	28UN	32UN
0	0.060	-	80	-	-	-	-	-	-	-	-	-
1	0.073	64	72	-	-	-	-	-	-	-	-	-
2	0.086	56	64	-	-	-	-	-	-	-	-	-
3	0.099	48	56	-	-	-	-	-	-	-	-	-
4	0.112	40	48	-	-	-	-	-	-	-	-	-
5	0.125	40	44	-	-	-	-	-	-	-	-	-
6	0.138	32	40	-	-	-	-	-	-	-	-	UNC
8	0.164	32	36	-	-	-	-	-	-	-	-	UNC
10	0.190	24	32	-	-	-	-	-	-	-	-	UNF
12	0.216	24	28	32	-	-	-	-	-	-	UNF	UNEF
1/4	0.250	20	28	32	-	-	-	-	-	UNC	UNF	UNEF
5/16	0.3125	18	24	32	-	-	-	-	-	20	28	UNEF
3/8	0.375	16	24	32	-	-	-	-	UNC	20	28	UNEF
7/16	0.4375	14	20	28	-	-	-	-	16	UNF	UNEF	32
1/2	0.500	13	20	28	-	-	-	-	16	UNF	UNEF	32
9/16	0.5625	12	18	24	-	-	-	UNC	16	20	28	32
5/8	0.625	11	18	24	-	-	-	12	16	20	28	32
11/16	0.6875	-	-	24	-	-	-	12	16	20	28	32
3/4	0.750	10	16	20	-	-	-	12	UNF	UNEF	28	32
13/16	0.8125	-	-	20	-	-	-	12	16	UNEF	28	32
7/8	0.875	9	14	20	-	-	-	12	16	UNEF	28	32
15/16	0.9375	-	-	20	-	-	-	12	16	UNEF	28	32
1	1.000	8	12	20	-	-	UNC	UNF	16	UNEF	28	32
1-1/16	1.063	-	-	18	-	-	8	12	16	20	28	-
1-1/8	1.125	7	12	18	-	-	8	UNF	16	20	28	-
1-3/16	1.188	-	-	18	-	-	8	12	16	20	28	-
1-1/4	1.250	7	12	18	-	-	8	UNF	16	20	28	-
1-5/16	1.313	-	-	18	-	-	8	12	16	20	28	-
1-3/8	1.375	6	12	18	-	UNC	8	UNF	16	20	28	-
1-7/16	1.438	-	-	18	-	6	8	12	16	20	28	-
1-1/2	1.500	6	12	18	-	UNC	8	UNF	16	20	28	-
1-9/16	1.563	-	-	18	-	6	8	12	16	20	-	-
1-5/8	1.625	-	-	18	-	6	8	12	16	20	-	-
1-11/16	1.688	-	-	18	-	6	8	12	16	20	-	-

English Threads

		THREADS PER INCH										
		Series with graded pitches			Series with constant pitches							
Sizes	Basic Major Diameter	Coarse UNC	Fine UNF	Extra fine UNEF	4UN	6UN	8UN	12UN	16UN	20UN	28UN	32UN
1-3/4	1.750	5	-	-	-	6	8	12	16	20	-	-
1-13/16	1.8125	-	-	-	-	6	8	12	16	20	-	-
1-7/8	1.875	-	-	-	-	6	8	12	16	20	-	-
1-15/16	1.9375	-	-	-	-	6	8	12	16	20	-	-
2	2.000	4.5	-	-	-	6	8	12	16	20	-	-
2-1/8	2.125	-	-	-	-	6	8	12	16	20	-	-
2-1/4	2.225	4.5	-	-	-	6	8	12	16	20	-	-
2-3/8	2.375	-	-	-	-	6	8	12	16	20	-	-
2-1/2	2.500	4	-	-	UNC	6	8	12	16	20	-	-
2-5/8	2.625	-	-	-	4	6	8	12	16	20	-	-
2-3/4	2.750	4	-	-	UNC	6	8	12	16	20	-	-
2-7/8	2.875	-	-	-	4	6	8	12	16	20	-	-
3	3.000	4	-	-	UNC	6	8	12	16	20	-	-
3-1/8	3.125	-	-	-	4	6	8	12	16	-	-	-
3-1/4	3.250	4	-	-	UNC	6	8	12	16	-	-	-
3-3/8	3.375	-	-	-	4	6	8	12	16	-	-	-
3-1/2	3.500	4	-	-	UNC	6	8	12	16	-	-	-
3-5/8	3.625	-	-	-	4	6	8	12	16	-	-	-
3-3/4	3.750	4	-	-	UNC	6	8	12	16	-	-	-
3-7/8	3.875	-	-	-	4	6	8	12	16	-	-	-
4	4.000	4	-	-	UNC	6	8	12	16	-	-	-
4-1/8	4.125	-	-	-	4	6	8	12	16	-	-	-
4-1/4	4.250	-	-	-	4	6	8	12	16	-	-	-
4-3/8	4.375	-	-	-	4	6	8	12	16	-	-	-
4-1/2	4.500	-	-	-	4	6	8	12	16	-	-	-
4-5/8	4.625	-	-	-	4	6	8	12	16	-	-	-
4-3/4	4.750	-	-	-	4	6	8	12	16	-	-	-
4-7/8	4.875	-	-	-	4	6	8	12	16	-	-	-
5	5.000	-	-	-	4	6	8	12	16	-	-	-
5-1/8	5.125	-	-	-	4	6	8	12	16	-	-	-
5-1/4	5.250	-	-	-	4	6	8	12	16	-	-	-
5-3/8	5.375	-	-	-	4	6	8	12	16	-	-	-
5-1/2	5.500	-	-	-	4	6	8	12	16	-	-	-
5-5/8	5.625	-	-	-	4	6	8	12	16	-	-	-
5-3/4	5.750	-	-	-	4	6	8	12	16	-	-	-
5-7/8	5.875	-	-	-	4	6	8	12	16	-	-	-
6	6.000	-	-	-	4	6	8	12	16	-	-	-

Metric Threads

Major Diameter	Tap Drill	Pitch	
		Coarse	Fine
M1.6	1.25	0.35	-
M1.8	1.45	0.35	-
M2	1.60	0.4	-
M2.2	1.75	0.45	-
M2.5	2.05	0.45	-
M3	2.50	0.5	-
M3.5	2.90	0.6	-
M4	3.30	0.7	-
M4.5	3.75	0.75	-
M5	4.20	0.8	-
M6	5.00	1	-
M7	6.00	1	-
M8	6.80	1.25	1
M9	7.75	1.25	-
M10	8.50	1.5	1.25
M12	10.30	1.75	1.25
M14	12.00	2	1.5
M16	14.00	2	1.5
M18	15.50	2.5	1.5
M20	17.50	2.5	1.5
M22	19.50	2.5	1.5
M24	21.00	3	2
M27	24.00	3	2
M30	26.50	3.5	2
M33	29.50	3.5	2
M36	32.00	4	3
M39	35.00	4	3
M42	37.50	4.5	3
M45	40.50	4.5	3
M48	43.50	5	3
M56	50.50	5.5	4
M64	58.00	6	4
M72	66.00	6	4
M80	74.00	6	4
M90	84.00	6	4
M100	94.00	6	4

Running and Sliding Fits (English)

Nominal Size Range		Class RC1			Class RC2			Class RC3			Class RC4			Class RC5		
		Clearance	Standard Limits		Clearance	Standard Limits		Clearance	Standard Limits		Clearance	Standard Limits		Clearance	Standard Limits	
Over	To		Hole H5	Shaft g4		Hole H6	Shaft g5		Hole H7	Shaft f6		Hole H8	Shaft f7		Hole H8	Shaft e7
0	0.12	0.1 0.45	+0.2 0	-0.1 -0.25	0.1 0.55	+0.25 0	-0.1 -0.3	0.3 0.95	+0.4 0	-0.3 -0.55	0.3 1.3	+0.6 0	-0.3 -0.7	0.6 1.6	+0.6 -0	-0.6 -1.0
0.12	0.24	0.15 0.5	+0.2 0	-0.15 -0.3	0.15 0.65	+0.3 0	-0.15 -0.35	0.4 1.12	+0.5 0	-0.4 -0.7	0.4 1.6	+0.7 0	-0.4 -0.9	0.8 2.0	+0.7 -0	-0.8 -1.3
0.24	0.4	0.2 0.6	0.25 0	-0.2 -0.35	0.2 0.85	+0.4 0	-0.2 -0.45	0.5 1.5	+0.6 0	-0.5 -0.9	0.5 2	+0.9 0	-0.5 -1.1	1.0 2.5	+0.9 -0	-1.0 -1.6
0.4	0.71	0.25 0.75	+0.3 0	-0.25 -0.45	0.25 0.95	+0.4 0	-0.25 -0.55	0.6 1.7	+0.7 0	-0.6 -1.0	0.6 2.3	+1.0 0	-0.6 -1.3	1.2 2.9	+1.0 -0	-1.2 -1.9
0.71	1.19	0.3 0.95	+0.4 0	-0.3 -0.55	0.3 1.2	+0.5 0	-0.3 -0.7	0.8 2.1	+0.8 0	-0.8 -1.3	0.8 2.8	+1.2 0	-0.8 -1.6	1.6 3.6	+1.2 -0	-1.6 -2.4
1.19	1.97	0.4 1.1	+0.4 0	-0.4 -0.7	0.4 1.4	+0.6 0	-0.4 -0.8	1 2.6	+1.0 0	-1.0 -1.6	1.0 3.6	+1.6 0	-1.0 -2.0	2.0 4.6	+1.6 -0	-2.0 -3.0
1.97	3.15	0.4 1.2	+0.5 0	-0.4 -0.7	0.4 1.6	+0.7 0	-0.4 -0.9	1.2 3.1	+1.2 0	-1.2 -1.9	1.2 4.2	+1.8 0	-1.2 -2.4	2.5 5.5	+1.8 -0	-2.5 -3.7
3.15	4.73	0.5 1.5	+0.6 0	-0.5 -0.9	0.5 2.0	+0.9 0	-0.5 -1.1	1.4 3.7	+1.4 0	-1.4 -2.3	1.4 5	+2.2 0	-1.4 -2.8	3.0 6.6	+2.2 -0	-3.0 -4.4
4.73	7.09	0.6 1.8	+0.7 0	-0.6 -1.1	0.6 2.3	+1.0 0	-0.6 -1.3	1.6 4.2	+1.6 0	-1.6 -2.6	1.6 5.7	+2.5 0	-1.6 -3.2	3.5 7.6	+2.5 -0	-3.0 -5.1
7.09	9.85	0.6 2	+0.8 0	-0.6 -1.2	0.6 2.6	+1.2 0	-0.6 -1.4	2 5	+1.8 0	-2.0 -3.2	2 6.6	+2.8 0	-2.0 -3.8	4.0 8.6	+2.8 -0	-4.0 -5.8
9.85	12.41	0.8 2.3	+0.9 0	-0.8 -1.4	0.8 2.9	+1.2 0	-0.8 -1.7	2.5 5.7	+2.0 0	-2.5 -3.7	2.5 7.5	+3.0 0	-2.5 -4.5	5.0 10.0	+3.0 0	-5.0 -7.0
12.41	15.75	1 2.7	+1.0 0	-1.0 -1.7	1 3.4	+1.4 0	-1.0 -2.0	3 6.6	+2.2 0	-3.0 -4.4	3 8.7	+3.5 0	-3.0 -5.2	6.0 11.7	+3.5 0	-6.0 -8.2
15.75	19.69	1.2 3	+1.0 0	-1.2 -2.0	1.2 3.8	+1.6 0	-1.2 -2.2	4 8.1	+1.6 0	-4.0 -5.6	4 10.5	+4.0 0	-4.0 -6.5	8.0 14.5	+4.0 0	-8.0 -10.5
19.69	30.09	1.6 3.7	+1.2 0	-1.6 -2.5	1.6 4.8	+2.0 0	-1.6 -2.8	5 10	+3.0 0	-5.0 -7.0	5 13	+5.0 0	-5.0 -8.0	10.0 18.0	+5.0 0	-10.0 -13.0
30.09	41.49	2 4.6	+1.6 0	-2.0 -3.0	2 6.1	+2.5 0	-2.0 -3.6	6 12.5	+4.0 0	-6.0 -8.5	6 16	+6.0 0	-6.0 -10.0	12.0 22.0	+6.0 0	-12.0 -16.0
41.49	56.19	2.5 5.7	+2.0 0	-2.5 -3.7	2.5 7.5	+3.0 0	-2.5 -4.5	8 16	+5.0 0	-8.0 -11.0	8 21	+8.0 0	-8.0 -13.0	16.0 29.0	+8.0 0	-16.0 -21.0
56.19	76.39	3 7.1	+2.5 0	-3.0 -4.6	3 9.5	+4.0 0	-3.0 -5.5	10 20	+6.0 0	-10.0 -14.0	10 26	+10.0 0	-10.0 -16.0	20.0 36.0	+10.0 0	-20.0 -26.0
76.39	100.9	4 9	+3.0 0	-4.0 -6.0	4 12	+5.0 0	-4.0 -7.0	12 25	+8.0 0	-12.0 -17.0	12 32	+12.0 0	-12.0 -20.0	25.0 45.0	+12.0 0	-25.0 -33.0

Running and Sliding Fits (English)

Nominal Size Range		Class RC6			Class RC7			Class RC8			Class RC9		
		Clearance	Standard Limits		Clearance	Standard Limits		Clearance	Standard Limits		Clearance	Standard Limits	
Over	To		Hole H9	Shaft e8		Hole H9	Shaft d8		Hole H10	Shaft c9		Hole H11	Shaft
0	0.12	0.6 2.2	+1.0 -0	-0.6 -1.2	1.0 2.6	+1.0 0	-1.0 -1.6	2.5 5.1	+1.6 0	-2.5 -3.5	4.0 8.1	+2.5 0	-4.0 -5.6
0.12	0.24	0.8 2.7	+1.2 -0	-0.8 -1.5	1.2 3.1	+1.2 0	-1.2 -1.9	2.8 5.8	+1.8 0	-2.8 -4.0	4.5 9.0	+3.0 0	-4.5 -6.0
0.24	0.4	1.0 3.3	+1.4 -0	-1.0 -1.9	1.6 3.9	+1.4 0	-1.6 -2.5	3.0 6.6	+2.2 0	-3.0 -4.4	5.0 10.7	+3.5 0	-5.0 -7.2
0.4	0.71	1.2 3.8	+1.6 -0	-1.2 -2.2	2.0 4.6	+1.6 0	-2.0 -3.0	3.5 7.9	+2.8 0	-3.5 -5.1	6.0 12.8	+4.0 -0	-6.0 -8.8
0.71	1.19	1.6 4.8	+2.0 -0	-1.6 -2.8	2.5 5.7	+2.0 0	-2.5 -3.7	4.5 10.0	+3.5 0	-4.5 -6.5	7.0 15.5	+5.0 0	-7.0 -10.5
1.19	1.97	2.0 6.1	+2.5 -0	-2.0 -3.6	3.0 7.1	+2.5 0	-3.0 -4.6	5.0 11.5	+4.0 0	-5.0 -7.5	8.0 18.0	+6.0 0	-8.0 -12.0
1.97	3.15	2.5 7.3	+3.0 -0	-2.5 -4.3	4.0 8.8	+3.0 0	-4.0 -5.8	6.0 13.5	+4.5 0	-6.0 -9.0	9.0 20.5	+7.0 0	-9.0 -13.5
3.15	4.73	3.0 8.7	+3.5 -0	-3.0 -5.2	5.0 10.7	+3.5 0	-5.0 -7.2	7.0 15.5	+5.0 0	-7.0 -10.5	10.0 24.0	+9.0 0	-10.0 -15.0
4.73	7.09	3.5 10.0	+4.0 -0	-3.5 -6.0	6.0 12.5	+4.0 0	-6.0 -8.5	8.0 18.0	+6.0 0	-8.0 -12.0	12.0 28.0	+10.0 0	-12.0 -18.0
7.09	9.85	4.0 11.3	+4.5 0	-4.0 -6.8	7.0 14.3	+4.5 0	-7.0 -9.8	10.0 21.5	+7.0 0	-10.0 -14.5	15.0 34.0	+12.0 0	-15.0 -22.0
9.85	12.41	5.0 13.0	+5.0 0	-5.0 -8.0	8.0 16.0	+5.0 0	-8.0 -11.0	12.0 25.0	+8.0 0	-12.0 -17.0	18.0 38.0	+12.0 0	-18.0 -26.0
12.41	15.75	6.0 15.5	+6.0 0	-6.0 -9.5	10.0 19.5	+6.0 0	-10.0 -13.5	14.0 29.0	+9.0 0	-14.0 -20.0	22.0 45.0	+14.0 0	-22.0 -31.0
15.75	19.69	8.0 18.0	+6.0 0	-8.0 -12.0	12.0 22.0	+6.0 0	-12.0 -16.0	16.0 32.0	+10.0 0	-16.0 -22.0	25.0 51.0	+16.0 0	-25.0 -35.0
19.69	30.09	10.0 23.0	+8.0 0	-10.0 -15.0	16.0 29.0	+8.0 0	-16.0 -21.0	20.0 40.0	+12.0 0	-20.0 -28.0	30.0 62.0	+20.0 0	-30.0 -42.
30.09	41.49	12.0 28.0	+10.0 0	-12.0 -18.0	20.0 36.0	+10.0 0	-20.0 -26.0	25.0 51.0	+16.0 0	-25.0 -35.0	40.0 81.0	+25.0 0	-40.0 -56.
41.49	56.19	16.0 36.0	+12.0 0	-16.0 -24.0	25.0 45.0	+12.0 0	-25.0 -33.0	30.0 62.0	+20.0 0	-30.0 -42.0	50.0 100	+30.0 0	-50.0 -70.0
56.19	76.39	20.0 46.0	+16.0 0	-20.0 -30.0	30.0 56.0	+16.0 0	-30.0 -40.0	40.0 81.0	+25.0 0	-40.0 -56.0	60.0 125	+40.0 0	-60.0 -85.0
76.39	100.9	25.0 57.0	+20.0 0	-25.0 -37.0	40.0 72.0	+20.0 0	-40.0 -52.0	50.0 100	+30.0 0	-50.0 -70.0	80.0 160	+50.0 0	-80.0 -110

Clearance Locational Fits (English)

Nominal Size Range		Class LC1			Class LC2			Class LC3			Class LC4			Class LC5		
		Clearance	Standard Limits		Clearance	Standard Limits		Clearance	Standard Limits		Clearance	Standard Limits		Clearance	Standard Limits	
Over	To		Hole H6	Shaft h5		Hole H7	Shaft h6		Hole H8	Shaft h7		Hole H10	Shaft h9		Hole H7	Shaft g6
0	0.12	0 0.45	0.25 -0	+0 -0.2	0 0.65	+0.4 -0	+0 -0.25	0 1	+0.6 -0	+0 -0.4	0 2.6	+1.6 -0	+0 -1.0	0.1 0.75	+0.4 -0	-0.1 -0.35
0.12	0.24	0 0.5	+0.3 -0	+0 -0.2	0 0.8	+0.5 -0	+0 -0.3	0 1.2	+0.7 -0	+0 -0.5	0 3.0	+1.8 -0	+0 -1.2	0.15 0.95	+0.5 -0	-0.15 -0.45
0.24	0.4	0 0.65	+0.4 -0	+0 -0.25	0 1.0	+0.6 -0	+0 -0.4	0 1.5	+0.9 -0	+0 -0.6	0 3.6	+2.2 -0	+0 -1.4	0.2 1.2	+0.6 -0	-0.2 -0.6
0.4	0.71	0 0.7	+0.4 -0	+0 -0.3	0 1.1	+0.7 -0	+0 -0.4	0 1.7	+1.0 -0	+0 -0.7	0 4.4	+2.8 -0	+0 -1.6	0.25 1.35	+0.7 -0	-0.25 -0.65
0.71	1.19	0 0.9	+0.5 -0	+0 -0.4	0 1.3	+0.8 -0	+0 -0.5	0 2	+1.2 -0	+0 -0.8	0 5.5	+3.5 -0	+0 -2.0	0.3 1.6	+0.8 -0	-0.3 -0.8
1.19	1.97	0 1	+0.6 -0	+0 -0.4	0 1.6	+1.0 -0	+0 -0.6	0 2.6	+1.6 -0	+0 -1	0 6.5	+4.0 -0	+0 -2.5	0.4 2.0	+1.0 -0	-0.4 -1.0
1.97	3.15	0 1.2	+0.7 -0	+0 -0.5	0 1.9	+1.2 -0	+0 -0.7	0 3	+1.8 -0	+0 -1.2	0 7.5	+4.5 -0	+0 -3	0.4 2.3	+1.2 -0	-0.4 -1.1
3.15	4.73	0 1.5	+0.9 -0	+0 -0.6	0 2.3	+1.4 -0	+0 -0.9	0 3.6	+2.2 -0	+0 -1.4	0 8.5	+5.0 -0	+0 -3.5	0.5 2.8	+1.4 -0	-0.5 -1.4
4.73	7.09	0 1.7	+1.0 -0	+0 -0.7	0 2.6	+1.6 -0	+0 -1.0	0 4.1	+2.5 -0	+0 -1.6	0 10	+6.0 -0	+0 -4	0.6 3.2	+1.6 -0	-0.6 -1.6
7.09	9.85	0 2	+1.2 -0	+0 -0.8	0 3.0	+1.8 -0	+0 -1.2	0 4.6	+2.8 -0	+0 -1.8	0 11.5	+7.0 -0	+0 -4.5	0.6 3.6	+1.8 -0	-0.6 -1.8
9.85	12.41	0 2.1	+1.2 -0	+0 -0.9	0 3.2	+2.0 -0	+0 -1.2	0 5	+3.0 -0	+0 -2.0	0 13	+8.0 -0	+0 -5	0.7 3.9	+2.0 -0	-0.7 -1.9
12.41	15.75	0 2.4	+1.4 -0	+0 -1.0	0 3.6	+2.2 -0	+0 -1.4	0 5.7	+3.5 -0	+0 -2.2	0 15	+9.0 -0	+0 -6	0.7 4.3	+2.2 -0	-0.7 -2.1
15.75	19.69	0 2.6	+1.6 -0	+0 -1.0	0 4.1	+2.5 -0	+0 -1.6	0 6.5	+4 -0	+0 -2.5	0 16	+10.0 -0	+0 -6	0.8 4.9	+2.5 -0	-0.8 -2.4
19.69	30.09	0 3.2	+2.0 -0	+0 -1.2	0 5.0	+3 -0	+0 -2	0 8	+5 -0	+0 -3	0 20	+12.0 -0	+0 -8	0.9 5.9	+3.0 -0	-0.9 -2.9
30.09	41.49	0 4.1	+2.5 -0	+0 -1.6	0 6.5	+4 -0	+0 -2.5	0 10	+6 -0	+0 -4	0 26	+16.0 -0	+0 -10	1.0 7.5	+4.0 -0	-1.0 -3.5
41.49	56.19	0 5	+3.0 -0	+0 -2.0	0 8.0	+5 -0	+0 -3	0 13	+8 -0	+0 -5	0 32	+20.0 -0	+0 -12	1.2 9.2	+5.0 -0	-1.2 -4.2
56.19	76.39	0 6.5	+4.0 -0	+0 -2.5	0 10	+6 -0	+0 -4	0 16	+10 -0	+0 -6	0 41	+25.0 -0	+0 -16	1.2 11.2	+6.0 -0	-1.2 -5.2
76.39	100.9	0 8	+5.0 -0	+0 -3.0	0 13	+8 -0	+0 -5	0 20	+12 -0	+0 -8	0 50	+30.0 -0	+0 -20	1.4 14.4	+8.0 -0	-1.4 -6.4
100.9	131.9	0 10	+6.0 -0	+0 -4.0	0 16	+10 -0	+0 -6	0 26	+16 -0	+0 -10	0 65	+40.0 -0	+0 -25	1.6 17.6	+10.0 -0	-1.6 -7.6
131.9	171.9	0 13	+8.0 -0	+0 -5.0	0 20	+12 -0	+0 -8	0 32	+20 -0	+0 -12	0 8	+50.0 -0	+0 -30	1.8 21.8	+12.0 -0	-1.8 -9.8
171.90	200	0 16	+10.0 -0	+0 -6.0	0 26	+16 -0	+0 -10	0 41	+25 -0	+0 -16	0 100	+60.0 -0	+0 -40	1.8 27.8	+16.0 -0	-1.8 -11.8

Clearance Locational Fits (English)

Nominal Size Range		Class LC6			Class LC7			Class LC8			Class LC9			Class LC10			Class LC11		
		Clearance	Standard Limits		Clearance	Standard Limits		Clearance	Standard Limits		Clearance	Standard Limits		Clearance	Standard Limits		Clearance	Standard Limits	
Over	To		Hole H9	Shaft f8		Hole H10	Shaft e9		Hole H10	Shaft c10		Hole H11	Shaft c10		Hole H12	Shaft		Hole H13	Shaft
0	0.12	0.3	+1.0	-0.3	0.6	+1.6	-0.6	1.0	+0.6	-1.0	2.5	+2.5	-2.5	4	+4	-4	5	+6	-5
		1.9	0	-0.9	3.2	0	-1.6	3.6	-0	-2.0	6.6	-0	-4.1	12	-0	-8	17	-0	-11
0.12	0.24	0.4	+1.2	-0.4	0.8	+1.8	-0.8	1.2	+1.8	-1.2	2.8	+3.0	-2.8	4.5	+5	-4.5	6	+7	-6
		2.3	0	-1.1	3.8	0	-2.0	4.2	-0	-2.4	7.6	-0	-4.6	14.5	-0	-9.5	20	-0	-13
0.24	0.4	0.5	+1.4	-0.5	1.0	+2.2	-1.0	1.6	+2.2	-1.6	3.0	+3.5	-3.0	5	+6	-5	7	+9	-7
		2.8	0	-1.4	4.6	0	-2.4	5.2	-0	-3.0	8.7	-0	-5.2	17	-0	-11	25	-0	-16
0.4	0.71	0.6	+1.6	-0.6	1.2	+2.8	-1.2	2.0	+2.8	-2.0	3.5	+4.0	-3.5	6	+7	-6	8	+0	-8
		3.2	0	-1.6	5.6	0	-2.8	6.4	-0	-3.6	10.3	-0	-6.3	20	-0	-13	28	-0	-18
0.71	1.19	0.8	+2.0	-0.8	1.6	+3.5	-1.6	2.5	+3.5	-2.5	4.5	+5.0	-4.5	7	+8	-7	10	+12	-10
		4.0	0	-2.0	7.1	0	-3.6	8.0	-0	-4.5	13.0	-0	-8.0	23	-0	-15	34	-0	-22
1.19	1.97	1.0	+2.5	-1.0	2.0	+4.0	-2.0	3.0	+4.0	-3.0	5	+6	-5	8	+10	-8	12	+16	-12
		5.1	0	-2.6	8.5	0	-4.5	9.5	-0	-5.5	15	-0	-9	28	-0	-18	44	-0	-28
1.97	3.15	1.2	+3.0	-1.2	2.5	+4.5	-2.5	4.0	+4.5	-4.0	6	+7	-6	10	+12	-10	14	+18	-14
		6.0	0	-3.0	10.0	0	-5.5	11.5	-0	-7.0	17.5	-0	-10.5	34	-0	-22	50	-0	-32
3.15	4.73	1.4	+3.5	-1.4	3.0	+5.0	-3.0	5.0	+5.0	-5.0	7	+9	-7	11	+14	-11	16	+22	-16
		7.1	0	-3.6	11.5	0	-6.5	13.5	-0	-8.5	21	-0	-12	39	-0	-25	60	-0	-38
4.73	7.09	1.6	+4.0	-1.6	3.5	+6.0	-3.5	6	+6	-6	8	+10	-8	12	+16	-12	18	+25	-18
		8.1	0	-4.1	13.5	0	-7.5	16	-0	-10	24	-0	-14	44	-0	-28	68	-0	-43
7.09	9.85	2.0	+4.5	-2.0	4.0	+7.0	-4.0	7	+7	-7	10	+12	-10	16	+18	-16	22	+28	-22
		9.3	0	-4.8	15.5	0	-8.5	18.5	-0	-11.5	29	-0	-17	52	-0	-34	78	-0	-50
9.85	12.41	2.2	+5.0	-2.2	4.5	+8.0	-4.5	7	+8	-7	12	+12	-12	20	+20	-20	28	+30	-28
		10.2	0	-5.2	17.5	0	-9.5	20	-0	-12	32	-0	-20	60	-0	-40	88	-0	-58
12.41	15.75	2.5	+6.0	-2.5	5.0	+9.0	-5	8	+9	-8	14	+14	-14	22	+22	-22	30	+35	-30
		12.0	0	-6.0	20.0	0	-11	23	-0	-14	37	-0	-23	66	-0	-44	100	-0	-65
15.75	19.69	2.8	+6.0	-2.8	5.0	+10.0	-5	9	+10	-9	16	+16	-16	25	+25	-25	35	+40	-35
		12.8	0	-6.8	21.0	0	-11	25	-0	-15	42	-0	-26	75	-0	-50	115	-0	-75
19.69	30.09	3.0	+8.0	-3.0	6.0	+12.0	-6	10	+12	-10	18	+20	-18	28	+30	-28	40	+50	-40
		16.0	0	-8.0	26.0	-0	-14	30	-0	-18	50	-0	-30	88	-0	-58	140	-0	-90
30.0 9	41.49	3.5	+10.0	-3.5	7.0	+16.0	-7	12	+16	-12	20	+25	-20	30	+40	-30	45	+60	-45
		19.5	0	-9.5	33.0	-0	-17	38	-0	-22	61	-0	-36	110	-0	-70	165	-0	-105
41.49	56.19	4.0	+12.0	-4.0	8.0	+20.0	-8	14	+20	-14	25	+30	-25	40	+50	-40	60	+80	-60
		24.0	0	-12.0	40.0	-0	-20	46	-0	-26	75	-0	-45	140	-0	-90	220	-0	-140
56.19	76.39	4.5	+16.0	-4.5	9.0	+25.0	-9	16	+25	-16	30	+40	-30	50	+60	-50	70	+100	-70
		30.5	0	-14.5	50.0	-0	-25	57	-0	-32	95	-0	-55	170	-0	-110	270	-0	-170
76.39	100.9	5.0	+20.0	-5	10.0	+30.0	-10	18	+30	-18	35	+50	-35	50	+80	-50	80	+125	-80
		37.0	0	-17	60.0	-0	-30	68	-0	-38	115	-0	-65	210	-0	-130	330	-0	-205
100.9	131.9	6.0	+25.0	-6	12.0	+40.0	-12	20	+40	-20	40	+60	-40	60	+100	-60	90	+160	-90
		47.0	0	-22	67.0	-0	-27	85	-0	-45	140	-0	-80	260	-0	-160	410	-0	-250
131.9	171.9	7.0	+30.0	-7	14.0	+50.0	-14	25	+50	-25	50	+80	-50	80	+125	-80	100	+200	-100
		57.0	0	-27	94.0	-0	-44	105	-0	-55	180	-0	-100	330	-0	-205	500	-0	-300
171.90	200	7.0	+40.0	-7	14.0	+60.0	-14	25	+60	-25	50	+100	-50	90	+160	-90	125	+250	-125
		72.0	0	-32	114.0	-0	-54	125	-0	-65	210	-0	-110	410	-0	-250	625	-0	-375

Transition Locational Fits (English)

Nominal Size Range		Class LT1			Class LT2			Class LT3			Class LT4			Class LT5			Class LT6		
			Standard Limits			Standard Limits			Standard Limits			Standard Limits			Standard Limits			Standard Limits	
Over	To	Fit	Hole H7	Shaft js6	Fit	Hole H8	Shaft js7	Fit	Hole H7	Shaft k6	Fit	Hole H8	Shaft k7	Fit	Hole H7	Shaft n6	Fit	Hole H7	Shaft n7
0	0.12	-0.10 +0.50	+0.4 -0	+0.10 -0.10	-0.2 +0.8	+0.6 -0	+0.2 -0.2							-0.5 +0.15	+0.4 -0	+0.5 +0.25	-0.65 +0.15	+0.4 -0	-0.65 +0.25
0.12	0.24	-0.15 +0.65	+0.5 -0	+0.15 -0.15	-0.25 +0.95	+0.7 -0	+0.25 -0.25							-0.6 +0.2	+0.5 -0	+0.6 +0.3	-0.8 +0.2	+0.5 -0	+0.8 +0.3
0.24	0.4	-0.2 +0.8	+0.6 -0	+0.2 -0.2	-0.3 +1.2	+0.9 -0	+0.3 -0.3	-0.5 +0.5	+0.6 -0	+0.5 +0.1	-0.7 +0.8	+0.9 -0	+0.7 +0.1	-0.8 +0.2	+0.6 -0	+0.8 +0.4	-1.0 +0.2	+0.6 -0	+1.0 +0.4
0.4	0.71	-0.2 +0.9	+0.7 -0	+0.2 -0.2	-0.35 +1.35	+1.0 -0	+0.35 -0.35	-0.5 +0.6	+0.7 -0	+0.5 +0.1	-0.8 +0.9	+1.0 -0	+0.8 +0.1	-0.9 +0.2	+0.7 -0	+0.9 +0.5	-1.2 +0.2	+0.7 -0	+1.2 +0.5
0.71	1.19	-0.25 +1.05	+0.8 -0	+0.25 -0.25	-0.4 +1.6	+1.2 -0	+0.4 -0.4	-0.6 +0.7	+0.8 -0	+0.6 +0.1	-0.9 +1.1	+1.2 -0	+0.9 +0.1	-1.1 +0.2	+0.8 -0	+1.1 +0.6	-1.4 +0.2	+0.8 -0	+1.4 +0.6
1.19	1.97	-0.3 +1.3	+1.0 -0	+0.3 -0.3	-0.5 +2.1	+1.6 -0	+0.5 -0.5	-0.7 +0.9	+1.0 -0	+0.7 +0.1	-1.1 +1.5	+1.6 -0	+1.1 +0.1	-1.3 +0.3	+1.0 -0	+1.3 +0.7	-1.7 +0.3	+1.0 -0	+1.7 +0.7
1.97	3.15	-0.3 +1.5	+1.2 -0	+0.3 -0.3	-0.6 +2.4	+1.8 -0	+0.6 -0.6	-0.8 +1.1	+1.2 -0	+0.8 +0.1	-1.3 +1.7	+1.8 -0	+1.3 +0.1	-1.5 +0.4	+1.2 -0	+1.5 +0.8	-2.0 +0.4	+1.2 -0	+2.0 +0.8
3.15	4.73	-0.4 +1.8	+1.4 -0	+0.4 -0.4	-0.7 +2.9	+2.2 -0	+0.7 -0.7	-1.0 +1.3	+1.4 -0	+1.0 +0.1	-1.5 +2.1	+2.2 -0	+1.5 +0.1	-1.9 +0.4	+1.4 -0	+1.9 +1.0	-2.4 +0.4	+1.4 -0	+2.4 +1.0
4.73	7.09	-0.5 +2.1	+1.6 -0	+0.5 -0.5	-0.8 +3.3	+2.5 -0	+0.8 -0.8	-1.1 +1.5	+1.6 -0	+1.1 +0.1	-1.7 +2.4	+2.5 -0	+1.7 +0.1	-2.2 +0.4	+1.6 -0	+2.2 +1.2	-2.8 +0.4	+1.6 -0	+2.8 +1.2
7.09	9.85	-0.6 +2.4	+1.8 -0	+0.6 -0.6	-0.9 +3.7	+2.8 -0	+0.9 -0.9	-1.4 +1.6	+1.8 -0	+1.4 +0.2	-2.0 +2.6	+2.8 -0	+2.0 +0.2	-2.6 +0.4	+1.8 -0	+2.6 +1.4	-3.2 +0.4	+1.8 -0	+3.2 +1.4
9.85	12.41	-0.6 +2.6	+2.0 -0	+0.6 -0.6	-1.0 +4.0	+3.0 -0	+1.0 -1.0	-1.4 +1.8	+2.0 -0	+1.4 +0.2	-2.2 +2.8	+3.0 -0	+2.2 +0.2	-2.6 +0.6	+2.0 -0	+2.6 +1.4	-3.4 +0.6	+2.0 -0	+3.4 +1.4
12.41	15.75	-0.7 +2.9	+2.2 -0	+0.7 -0.7	-1.0 +4.5	+3.5 -0	+1.0 -1.0	-1.6 +2.0	+2.2 -0	+1.6 +0.2	-2.4 +3.3	+3.5 -0	+2.4 +0.2	-3.0 +0.6	+2.2 -0	+3.0 +1.6	-3.8 +0.6	+2.2 -0	+3.8 +1.6
15.75	19.69	-0.8 +3.3	+2.5 -0	+0.8 -0.8	-1.2 +5.2	+4.0 -0	+1.2 -1.2	-1.8 +2.3	+2.5 -0	+1.8 +0.2	-2.7 +3.8	+4.0 -0	+2.7 +0.2	-3.4 +0.7	+2.5 -0	+3.4 +1.8	-4.3 +0.7	+2.5 -0	+4.3 +1.8

Interference Locational Fits (English)

Nominal Size Range		LN1			LN2			LN3		
		Limits of Interference	Standard Limits		Limits of Interference	Standard Limits		Limits of Interference	Standard Limits	
Over	To		Hole H6	Shaft n5		Hole H7	Shaft p6		Hole H7	Shaft r6
0	0.12	0	+0.25	+0.45	0	+0.4	+0.65	0.1	+0.4	+0.75
		0.45	-0	+0.25	0.65	-0	+0.4	0.75	-0	+0.5
0.12	0.24	0	+0.3	+0.5	0	+0.5	+0.8	0.1	+0.5	+0.9
		0.5	-0	+0.3	0.8	-0	+0.5	0.9	-0	+0.6
0.24	0.4	0	+0.4	+0.65	0	+0.6	+1.0	0.2	+0.6	+1.2
		0.65	-0	+0.4	1.0	-0	+0.6	1.2	-0	+0.8
0.4	0.71	0	+0.4	+-.8	0	+0.7	+1.1	0.3	+0.7	+1.4
		0.8	-0	+-.4	1.1	-0	+0.7	1.4	-0	+1.0
0.71	1.19	0	+0.5	+1.0	0	+0.8	+1.3	0.4	+0.8	+1.7
		1.0	-0	+0.5	1.3	-0	+0.8	1.7	-0	+1.2
1.19	1.97	0	+0.6	+1.1	0	+1.0	+1.6	0.4	+1.0	+2.0
		1.1	-0	+0.6	1.6	-0	+1.0	2.0	-0	+1.4
1.97	3.15	0.1	+0.7	+1.3	0.2	+1.2	+2.1	0.4	+1.2	+2.3
		1.3	-0	+0.7	2.1	-0	+1.4	2.3	-0	+1.6
3.15	4.73	0.1	+0.9	+1.6	0.2	+1.4	+2.5	0.6	+1.4	+2.9
		1.6	-0	+1.0	2.5	-0	+1.6	2.9	-0	+2.0
4.73	7.09	0.2	+1.0	+1.9	0.2	+1.6	+2.8	0.9	+1.6	+3.5
		1.9	-0	+1.2	2.8	-0	+1.8	3.5	-0	+2.5
7.09	9.85	0.2	+1.2	+2.2	0.2	+1.8	+3.2	1.2	+1.8	+4.2
		2.2	-0	+1.4	3.2	-0	+2.0	4.2	-0	+3.0
9.85	12.41	0.2	+1.2	+2.3	0.2	+2.0	+3.4	1.5	+2.0	+4.7
		2.3	-0	+1.4	3.4	-0	+2.2	4.7	-0	+3.5
12.41	15.75	0.2	+1.4	+2.6	0.3	+2.2	+3.9	2.3	+2.2	+5.9
		2.6	-0	+1.6	3.9	-0	+2.5	5.9	-0	+4.5
15.75	19.69	0.2	+1.6	+2.8	0.3	+2.5	+4.4	2.5	+2.5	+6.6
		2.8	-0	+1.8	4.4	-0	+2.8	6.6	-0	+5.0
19.69	30.09		+2.0		0.5	+3	+5.5	4	+3	+9
			-0		5.5	-0	+3.5	9	-0	+7
30.09	41.49		+2.5		0.5	+4	+7.0	5	+4	+11.5
			-0		7.0	-0	+4.5	11.5	-0	+9
41.49	56.19		+3.0		1	+5	+9	7	+5	+15
			-0		9	-0	+6	15	-0	+12
56.19	76.39		+4.0		1	+6	+11	10	+6	+20
			-0		11	-0	+7	20	-0	+16
76.39	100.9		+5.0		1	+8	+14	12	+8	+25
			-0		14	-0	+9	25	-0	+20
100.9	131.9		+6.0		2	+10	+18	15	+10	+31
			-0		18	-0	+12	31	-0	+25
131.9	171.9		+8.0		4	+12	+24	18	+12	+38
			-0		24	-0	+16	38	-0	+30
171.90	200		+10.0		4	+16	+30	24	+16	+50
			-0		30	-0	+20	50	-0	+40

Force and Shrink Fits (English)

Nominal Size Range		Class FN1			Class FN2			Class FN3			Class FN4			Class FN5		
		Limits of Interference	Standard Limits		Limits of Interference	Standard Limits		Limits of Interference	Standard Limits		Limits of Interference	Standard Limits		Limits of Interference	Standard Limits	
Over	To		Hole H6	Shaft		Hole H7	Shaft s6		Hole H7	Shaft t6		Hole H7	Shaft u6		Hole H8	Shaft x7
0	0.12	0.05 0.5	+0.25 -0	+0.5 +0.3	0.2 0.85	+0.4 -0	+0.85 +0.6				0.3 0.95	+0.4 -0	+0.95 +0.7	0.3 1.3	+0.6 -0	+1.3 +0.9
0.12	0.24	0.1 0.6	+0.3 -0	+0.6 +0.4	0.2 1	+0.5 -0	+1.0 +0.7				0.4 1.2	+0.5 -0	+1.2 -0.9	0.5 1.7	+0.7 -0	+1.7 +1.2
0.24	0.4	0.1 0.75	+0.4 -0	+0.75 +0.5	0.4 1.4	+0.6 -0	+1.4 +1.0				0.6 1.6	+0.6 -0	+1.6 +1.2	0.5 2.0	+0.9 -0	+2.0 +1.4
0.4	0.56	0.1 0.8	-0.4 -0	+0.8 +0.5	0.5 1.6	+0.7 -0	+1.6 +1.2				0.7 1.8	+0.7 -0	+1.8 +1.4	0.6 2.3	+1.0 -0	+2.3 +1.6
0.56	0.71	0.2 0.9	+0.4 -0	+0.9 +0.6	0.5 1.6	+0.7 -0	+1.6 +1.2				0.7 1.8	+0.7 -0	+1.8 +1.4	0.8 2.5	+1.0 -0	+2.5 +1.8
0.71	0.95	0.2 1.1	+0.5 -0	+1.1 +0.7	0.6 1.9	+0.8 -0	+1.9 +1.4				0.8 2.1	+0.8 -0	+2.1 +1.6	1.0 3.0	+1.2 -0	+3.0 +2.2
0.95	1.19	0.3 1.2	+0.5 -0	+1.2 +0.8	0.6 1.9	+0.8 -0	+1.9 +1.4	0.8 2.1	+0.8 -0	+2.1 +1.6	1.0 2.3	+0.8 -0	+2.3 +1.8	1.3 3.3	+1.2 -0	+3.3 +2.5
1.19	1.58	0.3 1.3	+0.6 -0	+1.3 +0.9	0.8 2.4	+1.0 -0	+2.4 +1.8	1.0 2.6	+1.0 -0	+2.6 +2.0	1.5 3.1	+1.0 -0	+3.1 +2.5	1.4 4.0	+1.6 -0	+4.0 +3.0
1.58	1.97	0.4 1.4	+0.6 -0	+1.4 +1.0	0.8 2.4	+1.0 -0	+2.4 +1.8	1.2 2.8	+1.0 -0	+2.8 +2.2	1.8 3.4	+1.0 -0	+3.4 +2.8	2.4 5.0	+1.6 -0	+5.0 +4.0
1.97	2.56	0.6 1.8	+0.7 -0	+1.8 +1.3	0.8 2.7	+1.2 -0	+2.7 +2.0	1.3 3.2	+1.2 -0	+3.2 +2.5	2.3 4.2	+1.2 -0	+4.2 +3.5	3.2 6.2	+1.8 -0	+6.2 +5.0
2.56	3.15	0.7 1.9	+0.7 -0	+1.9 +1.4	1.0 2.9	+1.2 -0	+2.9 +2.2	1.8 3.7	+1.2 -0	+3.7 +3.0	2.8 4.7	+1.2 -0	+4.7 +4.0	4.2 7.2	+1.8 -0	+7.2 +6.0
3.15	3.94	0.9 2.4	+0.9 -0	+2.4 +1.8	1.4 3.7	+1.4 -0	+3.7 +2.8	2.1 4.4	+1.4 -0	+4.4 +3.5	3.6 5.9	+1.4 -0	+5.9 +5.0	4.8 8.4	+2.2 -0	+8.4 +7.0
3.94	4.73	1.1 2.6	+0.9 -0	+2.6 +2.0	1.6 3.9	+1.4 -0	+3.9 +3.0	2.6 4.9	+1.4 -0	+4.9 +4.0	4.6 6.9	+1.4 -0	+6.9 +6.0	5.8 9.4	+2.2 -0	+9.4 +8.0
4.73	5.52	1.2 2.9	+1.0 -0	+2.9 +2.2	1.9 4.5	+1.6 -0	+4.5 +3.5	3.4 6.0	+1.6 -0	+6.0 +5.0	5.4 8.0	+1.6 -0	+8.0 +7.0	7.5 11.6	+2.5 -0	+11.6 +10.0
5.52	6.3	1.5 3.2	+1.0 -0	+3.2 +2.5	2.4 5.0	+1.6 -0	+5.0 +4.0	3.4 6.0	+1.6 -0	+6.0 +5.0	5.4 8.0	+1.6 -0	+8.0 +7.0	9.5 13.6	+2.5 -0	+13.6 +12.0
6.3	7.09	1.8 3.5	+1.0 -0	+3.5 +2.8	2.9 5.5	+1.6 -0	+5.5 +4.5	4.4 7.0	+1.6 -0	+7.0 +6.0	6.4 9.0	+1.6 -0	+9.0 +8.0	9.5 13.6	+2.5 -0	+13.6 +12.0
7.09	7.88	1.8 3.8	+1.2 -0	+3.8 +3.0	3.2 6.2	+1.8 -0	+6.2 +5.0	5.2 8.2	+1.8 -0	+8.2 +7.0	7.2 10.2	+1.8 -0	+10.2 +9.0	11.2 15.8	+2.8 -0	+15.8 +14.0
7.88	8.86	2.3 4.3	+1.2 -0	+4.3 +3.5	3.2 6.2	+1.8 -0	+6.2 +5.0	5.2 8.2	+1.8 -0	+8.2 +7.0	8.2 11.2	+1.8 -0	+11.2 +10.0	13.2 17.8	+2.8 -0	+17.8 +16.0
8.86	9.85	2.3 4.3	+1.2 -0	+4.3 +3.5	4.2 7.2	+1.8 -0	+7.2 +6.0	6.2 9.2	+1.8 -0	+9.2 +8.0	10.2 13.2	+1.8 -0	+13.2 +12.0	13.2 17.8	+2.8 -0	+17.8 +16.0
9.85	11.03	2.8 4.9	+1.2 -0	+4.9 +4.0	4.0 7.2	+2.0 -0	+7.2 +6.0	7.0 10.2	+2.0 -0	+10.2 +9.0	10.0 13.2	+2.0 -0	+13.2 +12.00	15.0 20.0	+3.0 -0	+20.0 +18.0
11.03	12.41	2.8 4.9	+1.2 -0	+4.9 +4.0	5.0 8.2	+2.0 -0	+8.2 +7.0	7.0 10.2	+2.0 -0	+10.2 +9.0	12.0 15.2	+2.0 -0	+15.2 +14.0	17.0 22.0	+3.0 -0	+22.0 +20.0
12.41	13.98	3.1 5.5	+1.4 -0	+5.5 +4.5	5.8 9.4	+2.2 -0	+9.4 +8.0	7.8 11.4	+2.2 -0	+11.4 +10.0	13.8 17.4	+2.2 -0	+17.4 +16.0	18.5 24.2	+3.5 -0	+24.2 +22.0
13.98	15.75	3.6 6.1	+1.4 -0	+6.1 +5.0	5.8 9.4	+2.2 -0	+9.4 +8.0	9.8 13.4	+2.2 -0	+13.4 +12.0	15.8 19.4	+2.2 -0	+19.4 +18.0	21.5 27.2	+3.5 -0	+27.2 +25.0
15.75	17.72	4.4 7.0	+1.6 -0	+7.0 +6.0	6.5 10.6	+2.5 -0	+10.6 +9.0	9.5 13.6	+2.5 -0	+13.6 +12.0	17.5 21.6	+2.5 -0	+21.6 +20.0	24.0 30.5	+4.0 -0	+30.5 +28.0
17.72	19.69	4.4 7.0	+1.6 -0	+7.0 +6.0	7.5 11.6	+2.5 -0	+11.6 +10.0	11.5 15.6	+2.5 -0	+15.6 +14.0	19.5 23.6	+2.5 -0	+23.6 +22.0	26.0 32.5	+4.0 -0	+32.5 +30.0

Hole Basis Clearance Fits (Metric)

Basic Size		LOOSE RUNNING			FREE RUNNING			CLOSE RUNNING			SLIDING			LOCATIONAL CLEARANCE		
		Hole H11	Shaft c11	Fit	Hole H9	Shaft d9	Fit	Hole H8	Shaft f7	Fit	Hole H7	Shaft g6	Fit	Hole H7	Shaft h6	Fit
1	Max	1.060	0.940	0.180	1.025	0.980	0.070	1.014	0.994	0.030	1.010	0.998	0.018	1.010	1.000	0.016
	Min	1.000	0.880	0.060	1.000	0.955	0.020	1.000	0.984	0.006	1.000	0.992	0.002	1.000	0.994	0.000
1.2	Max	1.260	1.140	0.180	1.225	1.180	0.070	1.214	1.194	0.030	1.210	1.198	0.018	1.210	1.200	0.016
	Min	1.200	1.080	0.060	1.200	1.155	0.020	1.200	1.184	0.006	1.200	1.192	0.002	1.200	1.194	0.000
1.6	Max	1.660	1.540	0.180	1.625	1.580	0.070	1.614	1.594	0.030	1.610	1.598	0.018	1.610	1.600	0.016
	Min	1.600	1.480	0.060	1.600	1.555	0.020	1.600	1.584	0.006	1.600	1.592	0.002	1.600	1.594	0.000
2	Max	2.060	1.940	0.180	2.025	1.980	0.070	2.014	1.994	0.030	2.010	1.998	0.018	2.010	2.000	0.016
	Min	2.000	1.880	0.060	2.000	1.955	0.020	2.000	1.984	0.006	2.000	1.992	0.002	2.000	1.994	0.000
2.5	Max	2.560	2.440	0.180	2.525	2.480	0.070	2.514	2.494	0.030	2.510	2.498	0.018	2.510	2.500	0.016
	Min	2.500	2.380	0.060	2.500	2.455	0.020	2.500	2.484	0.006	2.500	2.492	0.002	2.500	2.494	0.000
3	Max	3.060	2.940	0.180	3.025	2.980	0.070	3.014	2.994	0.030	3.010	2.998	0.018	3.010	3.000	0.016
	Min	3.000	2.880	0.060	3.000	2.955	0.020	3.000	2.984	0.006	3.000	2.992	0.002	3.000	2.994	0.000
4	Max	4.075	3.930	0.220	4.030	3.970	0.090	4.018	3.990	0.040	4.012	3.996	0.024	4.012	4.000	0.020
	Min	4.000	3.855	0.070	4.000	3.940	0.030	4.000	3.978	0.010	4.000	3.988	0.004	4.000	3.992	0.000
5	Max	5.075	4.930	0.220	5.030	4.970	0.090	5.018	4.990	0.040	5.012	4.996	0.024	5.012	5.000	0.020
	Min	5.000	4.855	0.070	5.000	4.940	0.030	5.000	4.978	0.010	5.000	4.988	0.004	5.000	4.992	0.000
6	Max	6.075	5.930	0.220	6.030	5.970	0.090	6.018	5.990	0.040	6.012	5.996	0.024	6.012	6.000	0.020
	Min	6.000	5.855	0.070	6.000	5.940	0.030	6.000	5.978	0.010	6.000	5.988	0.004	6.000	5.992	0.000
8	Max	8.090	7.920	0.260	8.036	7.960	0.112	8.022	7.987	0.050	8.015	7.995	0.029	8.015	8.000	0.024
	Min	8.000	7.830	0.080	8.000	7.924	0.040	8.000	7.972	0.013	8.000	7.986	0.005	8.000	7.991	0.000
10	Max	10.090	9.920	0.260	10.036	9.960	0.112	10.022	9.987	0.050	10.015	9.995	0.029	10.015	10.000	0.024
	Min	10.000	9.830	0.080	10.000	9.924	0.040	10.000	9.972	0.013	10.000	9.986	0.005	10.000	9.991	0.000
12	Max	12.110	11.905	0.315	12.043	11.950	0.136	12.027	11.984	0.061	12.018	11.994	0.035	12.018	12.000	0.029
	Min	12.000	11.795	0.095	12.000	11.907	0.050	12.000	11.966	0.016	12.000	11.983	0.006	12.000	11.989	0.000
16	Max	16.110	15.905	0.315	16.043	15.950	0.136	16.027	15.984	0.061	16.018	15.994	0.035	16.018	16.000	0.029
	Min	16.000	15.795	0.095	16.000	15.907	0.050	16.000	15.966	0.016	16.000	15.983	0.006	16.000	15.989	0.000
20	Max	20.130	19.890	0.370	20.052	19.935	0.169	20.033	19.980	0.074	20.021	19.993	0.041	20.021	20.000	0.034
	Min	20.000	19.760	0.110	20.000	19.883	0.065	20.000	19.959	0.020	20.000	19.980	0.007	20.000	19.987	0.000
25	Max	25.130	24.890	0.370	25.052	24.935	0.169	25.033	24.980	0.074	25.021	24.993	0.041	25.021	25.000	0.034
	Min	25.000	24.760	0.110	25.000	24.883	0.065	25.000	24.959	0.020	25.000	24.980	0.007	25.000	24.987	0.000
30	Max	30.130	29.890	0.370	30.052	29.935	0.169	30.033	29.980	0.074	30.021	29.993	0.041	30.021	30.000	0.034
	Min	30.000	29.760	0.110	30.000	29.883	0.065	30.000	29.959	0.020	30.000	29.980	0.007	30.000	29.987	0.000
40	Max	40.160	39.880	0.440	40.062	39.920	0.204	40.039	39.975	0.089	40.025	39.991	0.050	40.025	40.000	0.041
	Min	40.000	39.720	0.120	40.000	39.858	0.080	40.000	39.950	0.025	40.000	39.975	0.009	40.000	39.984	0.000
50	Max	50.160	49.870	0.450	50.062	49.920	0.204	50.039	49.975	0.089	50.025	49.991	0.050	50.025	50.000	0.041
	Min	50.000	49.710	0.130	50.000	49.858	0.080	50.000	49.950	0.025	50.000	49.975	0.009	50.000	49.984	0.00
60	Max	60.190	59.860	0.520	60.074	59.900	0.248	60.046	59.970	0.106	60.030	59.990	0.059	60.030	60.000	0.049
	Min	60.000	59.670	0.140	60.000	59.826	0.100	60.000	59.940	0.030	60.000	59.971	0.010	60.000	59.981	0.000
80	Max	80.190	79.850	0.530	80.074	79.900	0.248	80.046	79.970	0.106	80.030	79.990	0.059	80.030	80.000	0.049
	Min	80.000	79.660	0.150	80.000	79.826	0.100	80.000	79.940	0.030	80.000	79.971	0.010	80.000	79.981	0.000
100	Max	100.200	99.830	0.610	100.087	99.880	0.294	100.054	99.964	0.125	100.035	99.988	0.069	100.035	100.000	0.057
	Min	100.000	99.610	0.170	100.000	99.793	0.120	100.000	99.929	0.036	100.000	99.966	0.012	100.000	99.978	0.000
120	Max	120.220	119.820	0.620	120.087	119.880	0.294	120.054	119.964	0.125	120.035	119.988	0.069	120.035	120.000	0.057
	Min	120.000	119.600	0.180	120.000	119.793	0.120	120.000	119.929	0.036	120.000	119.966	0.012	120.000	119.978	0.000
160	Max	160.250	159.790	0.710	160.100	159.855	0.345	160.063	159.957	0.146	160.040	159.986	0.079	160.040	160.000	0.065
	Min	160.000	159.540	0.210	160.000	159.755	0.145	160.000	159.917	0.043	160.000	159.961	0.014	160.000	159.975	0.000
200	Max	200.290	199.760	0.820	200.115	199.830	0.400	200.072	199.950	0.168	200.046	199.985	0.090	200.046	200.000	0.075
	Min	200.000	199.470	0.240	200.000	199.715	0.170	200.000	199.904	0.050	200.000	199.956	0.015	200.000	199.971	0.000
250	Max	250.290	249.720	0.860	250.115	249.830	0.400	250.072	249.950	0.168	250.046	249.985	0.090	250.046	250.000	0.075
	Min	250.000	249.430	0.280	250.000	249.715	0.170	250.000	249.904	0.050	250.000	249.956	0.015	250.000	249.971	0.000
300	Max	300.320	299.670	0.970	300.130	299.810	0.450	300.081	299.944	0.189	300.052	299.983	0.101	300.052	300.000	0.084
	Min	300.000	299.350	0.330	300.000	299.680	0.190	300.000	299.892	0.056	300.000	299.951	0.017	300.000	299.968	0.000
400	Max	400.360	399.600	1.120	400.140	399.790	0.490	400.089	399.938	0.208	400.057	399.982	0.111	400.057	400.000	0.093
	Min	400.000	399.240	0.400	400.000	399.650	0.210	400.000	399.881	0.062	400.000	399.946	0.018	400.000	399.964	0.000
500	Max	500.400	499.520	1.280	500.155	499.770	0.540	500.097	499.932	0.228	500.063	499.980	0.123	500.063	500.000	0.103
	Min	500.000	499.120	0.480	500.000	499.615	0.230	500.000	499.869	0.068	500.000	499.940	0.020	500.000	499.960	0.000

Hole Basis Transition and Interference Fits (Metric)

Basic Size		LOCATIONAL TRANSITION			LOCATIONAL TRANSITION			LOCATIONAL INTERFERENCE			MEDIUM DRIVE			FORCE		
		Hole H7	Shaft k6	Fit	Hole H7	Shaft n6	Fit	Hole H7	Shaft p6	Fit	Hole H7	Shaft s6	Fit	Hole H7	Shaft u6	Fit
1	Max	1.010	1.006	0.010	1.010	1.010	0.006	1.010	1.012	0.004	1.010	1.020	-0.004	1.010	1.024	-0.008
	Min	1.000	1.000	-0.006	1.000	1.004	-0.010	1.000	1.014	-0.012	1.000	1.014	-0.020	1.000	1.018	-0.024
1.2	Max	1.210	1.206	0.010	1.210	1.210	0.006	1.210	1.212	0.004	1.210	1.220	-0.004	1.210	1.224	-0.008
	Min	1.200	1.200	-0.006	1.200	1.204	-0.010	1.200	1.206	-0.012	1.200	1.214	-0.020	1.200	1.218	-0.024
1.6	Max	1.610	1.606	0.010	1.610	1.610	0.006	1.610	1.612	0.004	1.610	1.620	-0.004	1.610	1.624	-0.008
	Min	1.600	1.600	-0.006	1.600	1.604	-0.010	1.600	1.606	-0.012	1.600	1.614	-0.020	1.600	1.618	-0.024
2	Max	2.010	2.006	0.010	2.010	2.010	0.006	2.010	2.010	0.004	2.101	2.020	-0.004	2.010	2.024	-0.008
	Min	2.000	2.000	-0.006	2.000	2.004	-0.010	2.000	2.006	-0.012	2.000	2.014	-0.020	2.000	2.018	-0.024
2.5	Max	2.510	2.506	0.010	2.510	2.510	0.006	2.510	2.512	0.004	2.510	2.520	-0.004	2.510	2.524	-0.008
	Min	2.500	2.500	-0.006	2.500	2.504	-0.010	2.500	2.506	-0.012	2.500	2.514	-0.020	2.500	2.518	-0.024
3	Max	3.010	3.006	0.010	3.010	3.010	0.006	3.010	3.012	0.004	3.010	3.020	-0.004	3.010	3.024	-0.008
	Min	3.000	3.000	-0.006	3.000	3.004	-0.010	3.000	3.006	-0.020	3.000	3.014	-0.020	3.000	3.018	-0.024
4	Max	4.012	4.009	0.011	4.012	4.016	0.004	4.012	4.020	0.000	4.012	4.027	-0.007	4.012	4.031	-0.011
	Min	4.000	4.001	-0.009	4.000	4.008	-0.016	4.000	4.012	-0.020	4.000	4.019	-0.027	4.000	4.023	-0.031
5	Max	5.012	5.009	0.011	5.012	5.016	0.004	5.012	5.020	0.000	5.012	5.027	-0.007	5.012	5.031	-0.011
	Min	5.000	5.001	-0.009	5.000	5.008	-0.016	5.000	5.012	-0.020	5.000	5.019	-0.027	5.000	5.023	-0.031
6	Max	6.012	6.009	0.011	6.012	6.016	0.004	6.012	6.020	0.000	6.012	6.027	-0.007	6.012	6.031	-0.011
	Min	6.000	6.001	-0.009	6.000	6.008	-0.016	6.000	6.012	-0.020	6.000	6.019	-0.027	6.000	6.023	-0.031
8	Max	8.015	8.010	0.014	8.015	8.019	0.005	8.015	8.024	0.000	8.015	8.032	-0.008	8.015	8.037	-0.013
	Min	8.000	8.001	-0.010	8.000	8.010	-0.019	8.000	8.015	-0.024	8.000	8.023	-0.032	8.000	8.028	-0.037
10	Max	10.015	10.010	0.014	10.015	10.019	0.005	10.015	10.024	0.000	10.015	10.032	-0.008	10.015	10.037	-0.013
	Min	10.000	10.001	-0.010	10.000	10.010	-0.019	10.000	10.015	-0.024	10.000	10.023	-0.032	10.000	10.028	-0.037
12	Max	12.018	12.012	0.017	12.018	12.023	0.006	12.018	12.029	0.000	12.018	12.039	-0.010	12.018	12.044	-0.015
	Min	12.000	12.001	-0.012	12.000	12.012	-0.023	12.000	12.018	-0.029	12.000	12.028	-0.039	12.000	12.033	-0.044
16	Max	16.018	16.012	0.017	16.018	16.023	0.006	16.018	16.029	0.000	16.018	16.039	-0.010	16.018	16.044	-0.015
	Min	16.000	16.001	-0.012	16.000	16.012	-0.023	16.000	16.018	-0.029	16.000	16.028	-0.039	16.000	16.033	-0.044
20	Max	20.021	20.015	0.019	20.021	20.028	0.006	20.021	20.035	-0.001	20.021	20.048	-0.014	20.021	20.054	-0.020
	Min	20.000	20.002	-0.015	20.000	20.015	-0.028	20.000	20.022	-0.035	20.000	20.035	-0.048	20.000	20.041	-0.054
25	Max	25.021	25.015	0.019	25.021	25.028	0.006	25.021	25.035	-0.001	25.021	25.048	-0.014	25.021	25.061	-0.027
	Min	25.000	25.002	-0.015	25.000	25.015	-0.028	25.000	25.022	-0.035	25.000	25.035	-0.048	25.000	25.048	-0.061
30	Max	30.021	30.015	0.019	30.021	30.028	0.006	30.021	30.035	-0.001	30.021	30.048	-0.014	30.021	30.061	-0.027
	Min	30.000	30.002	-0.015	30.000	30.015	-0.028	30.000	30.022	-0.035	30.000	30.035	-0.048	30.000	30.048	-0.061
40	Max	40.025	40.018	0.023	40.025	40.033	0.008	40.025	40.042	-0.001	40.025	40.059	-0.018	40.025	40.076	-0.035
	Min	40.000	40.002	-0.018	40.000	40.017	-0.033	40.000	40.026	-0.042	40.000	40.043	-0.059	40.000	40.060	-0.076
50	Max	50.025	50.018	0.023	50.025	50.033	0.008	50.025	50.042	-0.001	50.025	50.059	-0.018	50.025	50.086	-0.045
	Min	50.000	50.002	-0.018	50.000	50.017	-0.033	50.000	50.026	-0.042	50.000	50.043	-0.059	50.000	50.070	-0.086
60	Max	60.030	60.021	0.028	60.030	60.039	0.010	60.030	60.051	-0.002	60.030	60.072	-0.023	60.030	60.106	-0.057
	Min	60.000	60.002	-0.021	60.000	60.020	-0.039	60.000	60.032	-0.051	60.000	60.053	-0.072	60.000	60.087	-0.106
80	Max	80.030	80.021	0.028	80.030	80.039	0.010	80.030	80.051	-0.002	80.030	80.078	-0.029	80.030	80.121	-0.072
	Min	80.000	80.002	-0.021	80.000	80.020	-0.039	80.000	80.032	-0.051	80.000	80.059	-0.078	80.000	80.102	-0.121
100	Max	100.035	100.025	0.032	100.035	100.045	0.012	100.035	100.059	-0.002	100.035	100.093	-0.036	100.035	100.146	-0.089
	Min	100.000	100.003	-0.025	100.000	100.023	-0.045	100.000	100.037	-0.059	100.000	100.071	-0.093	100.000	100.124	-0.146
120	Max	120.035	120.025	0.032	120.035	120.045	0.012	120.035	120.059	-0.002	120.035	120.010	-0.044	120.035	120.166	-0.109
	Min	120.000	120.003	-0.025	120.000	120.023	-0.045	120.000	120.037	-0.059	120.000	120.079	-0.101	120.000	120.144	-0.166
160	Max	160.040	160.028	0.037	160.040	160.052	0.013	160.040	160.068	-0.003	160.040	160.125	-0.060	160.040	160.215	-0.150
	Min	160.000	160.003	-0.028	160.000	160.027	-0.052	160.000	160.043	-0.068	160.000	160.100	-0.125	160.000	160.190	-0.215
200	Max	200.046	200.033	0.042	200.046	200.060	0.015	200.046	200.079	-0.004	200.046	200.151	-0.076	200.046	200.265	-0.190
	Min	200.000	200.004	-0.033	200.000	200.031	-0.060	200.000	200.050	-0.079	200.000	200.122	-0.151	200.000	200.236	-0.265
250	Max	250.046	250.033	0.042	250.046	250.060	0.015	250.046	250.079	-0.004	250.046	250.169	-0.094	250.046	250.313	-0.238
	Min	250.000	250.004	-0.033	250.000	250.031	-0.060	250.000	250.050	-0.079	250.000	250.140	-0.169	250.000	250.284	-0.313
300	Max	300.052	300.036	0.048	300.052	300.066	0.018	300.052	300.088	-0.004	300.052	300.202	-0.118	300.052	300.382	-0.298
	Min	300.000	300.004	-0.036	300.000	300.034	-0.066	300.000	300.056	-0.088	300.000	300.170	-0.202	300.000	300.350	-0.382
400	Max	400.057	400.040	0.053	400.057	400.073	0.020	400.057	400.098	-0.005	400.057	400.244	-0.151	400.057	400.471	-0.378
	Min	400.000	400.004	-0.040	400.000	400.037	-0.073	400.000	400.062	-0.098	400.000	400.208	-0.244	400.000	400.435	-0.471
500	Max	500.063	500.045	0.058	500.063	500.080	0.023	500.063	500.108	-0.005	500.063	500.292	-0.189	500.063	500.580	-0.477
	Min	500.000	500.005	-0.045	500.000	500.040	-0.080	500.000	500.068	-0.108	500.000	500.252	-0.292	500.000	500.540	-0.580

Shaft Basis Clearance Fits (Metric)

Basic Size		LOOSE RUNNING			FREE RUNNING			CLOSE RUNNING			SLIDING			LOCATIONAL CLEARANCE		
		Hole C11	Shaft h11	Fit	Hole D9	Shaft h9	Fit	Hole F8	Shaft h7	Fit	Hole G7	Shaft h6	Fit	Hole H7	Shaft h6	Fit
1	Max	1.120	1.000	0.180	1.045	1.000	0.070	1.020	1.000	0.030	1.012	1.000	0.018	1.010	1.000	0.016
	Min	1.060	0.940	0.060	1.020	0.975	0.020	1.006	0.990	0.006	1.002	0.994	0.002	1.000	0.994	0.000
1.2	Max	1.320	1.200	0.180	1.245	1.200	0.070	1.220	1.200	0.030	1.212	1.200	0.018	1.210	1.200	0.016
	Min	1.260	1.140	0.060	1.220	1.175	0.020	1.206	1.190	0.006	1.202	1.194	0.002	1.200	1.194	0.000
1.6	Max	1.720	1.600	0.180	1.645	1.600	0.070	1.620	1.600	0.030	1.612	1.600	0.018	1.610	1.600	0.016
	Min	1.660	1.540	0.060	1.620	1.575	0.020	1.606	1.590	0.006	1.602	1.594	0.002	1.600	1.594	0.000
2	Max	2.120	2.000	0.180	2.045	2.000	0.070	2.020	2.000	0.030	2.012	2.000	0.018	2.010	2.000	0.016
	Min	2.060	1.940	0.060	2.020	1.975	0.020	2.006	1.990	0.006	2.002	1.994	0.002	2.000	1.994	0.000
2.5	Max	2.620	2.500	0.180	2.545	2.500	0.070	2.520	2.500	0.030	2.512	2.500	0.018	2.510	2.500	0.016
	Min	2.560	2.440	0.060	2.520	2.475	0.020	2.506	2.490	0.006	2.502	2.494	0.002	2.500	2.494	0.000
3	Max	3.120	3.000	0.180	3.045	3.000	0.070	3.020	3.000	0.030	3.012	3.000	0.080	3.010	3.000	0.016
	Min	3.060	2.940	0.060	3.020	2.975	0.020	3.006	2.990	0.006	3.002	2.994	0.002	3.000	2.994	0.000
4	Max	4.145	4.000	0.220	4.060	4.000	0.090	4.028	4.000	0.040	4.016	4.000	0.024	4.012	4.000	0.020
	Min	4.070	3.925	0.070	4.030	3.970	0.030	4.010	3.988	0.010	4.004	3.992	0.004	4.000	3.992	0.000
5	Max	5.145	5.000	0.220	5.060	5.000	0.090	5.028	5.000	0.040	5.016	5.000	0.024	5.012	5.000	0.020
	Min	5.070	4.925	0.070	5.030	4.970	0.030	5.010	4.988	0.010	5.004	4.992	0.004	5.000	4.992	0.000
6	Max	6.145	6.000	0.220	6.060	6.000	0.090	6.028	6.000	0.040	6.016	6.000	0.024	6.012	6.000	0.020
	Min	6.070	5.925	0.070	6.030	5.970	0.030	6.010	5.988	0.010	6.004	5.992	0.004	6.000	5.992	0.000
8	Max	8.170	8.000	0.260	8.076	8.000	0.112	8.035	8.000	0.050	8.020	8.000	0.029	8.015	8.000	0.024
	Min	8.080	7.910	0.080	8.040	7.964	0.040	8.013	7.985	0.013	8.005	7.991	0.005	8.000	7.991	0.000
10	Max	10.170	10.000	0.260	10.076	10.000	0.112	10.035	10.000	0.050	10.020	10.000	0.029	10.015	10.000	0.024
	Min	10.080	9.910	0.080	10.040	9.964	0.040	10.013	9.985	0.013	10.005	9.991	0.005	10.000	9.991	0.000
12	Max	12.205	12.000	0.315	12.093	12.000	0.136	12.043	12.000	0.061	12.024	12.000	0.035	12.018	12.000	0.029
	Min	12.095	11.890	0.095	12.050	11.957	0.050	12.016	11.982	0.016	12.006	11.989	0.006	12.000	11.989	0.000
16	Max	16.205	16.000	0.315	16.093	16.000	0.136	16.043	16.000	0.061	16.024	16.000	0.035	16.018	16.000	0.029
	Min	16.095	15.890	0.095	16.050	15.957	0.050	16.016	15.982	0.016	16.006	15.989	0.006	16.000	15.989	0.000
20	Max	20.240	20.000	0.370	20.117	20.000	0.169	20.053	20.000	0.074	20.028	20.000	0.041	20.021	20.000	0.034
	Min	20.110	19.870	0.110	20.065	19.948	0.065	20.020	19.979	0.020	20.007	19.987	0.007	20.000	19.987	0.000
25	Max	25.240	25.000	0.370	25.117	25.000	0.169	25.053	25.000	0.074	25.028	25.000	0.041	25.021	25.000	0.034
	Min	25.110	24.870	0.110	25.065	24.948	0.065	25.020	24.979	0.020	25.007	24.987	0.007	25.000	24.987	0.000
30	Max	30.240	30.000	0.370	30.117	30.000	0.169	30.053	30.000	0.074	30.028	30.000	0.041	30.021	30.000	0.034
	Min	30.110	29.870	0.110	30.065	29.948	0.065	30.020	29.979	0.020	30.007	29.987	0.007	30.000	29.987	0.000
40	Max	40.280	40.000	0.440	40.142	40.000	0.204	40.064	40.000	0.089	40.034	40.000	0.050	40.025	40.000	0.041
	Min	40.120	39.840	0.120	40.080	39.938	0.080	40.025	39.975	0.025	40.009	39.984	0.009	40.000	39.984	0.000
50	Max	50.290	50.000	0.450	50.142	50.000	0.204	50.064	50.000	0.089	50.034	50.000	0.050	50.025	50.000	0.041
	Min	50.130	49.840	0.130	50.080	49.938	0.080	50.025	49.975	0.025	50.009	49.984	0.009	50.000	49.984	0.000
60	Max	60.330	60.000	0.520	60.174	60.000	0.248	60.076	60.000	0.106	60.040	60.000	0.059	60.030	60.000	0.049
	Min	60.140	59.810	0.140	60.100	59.926	0.100	60.030	59.970	0.030	60.010	59.981	0.010	60.000	59.981	0.000
80	Max	80.340	80.000	0.530	80.174	80.000	0.248	80.076	80.000	0.106	80.040	80.000	0.059	80.030	80.000	0.049
	Min	80.150	79.810	0.150	80.100	79.926	0.100	80.030	79.970	0.030	80.010	79.981	0.010	80.000	79.981	0.000
100	Max	100.390	100.000	0.610	100.207	100.000	0.294	100.090	100.000	0.125	100.047	100.000	0.069	100.035	100.000	0.057
	Min	100.170	99.780	0.170	100.120	99.913	0.120	100.036	99.965	0.036	100.012	99.978	0.012	100.000	99.978	0.000
120	Max	120.400	120.000	0.620	120.207	120.000	0.294	120.090	120.000	0.125	120.047	120.000	0.069	120.035	120.000	0.057
	Min	120.180	119.780	0.180	120.120	119.913	0.120	120.036	119.965	0.036	120.012	119.978	0.012	120.000	119.978	0.000
160	Max	160.460	160.000	0.710	160.245	160.000	0.345	160.106	160.000	0.146	160.054	160.000	0.079	160.040	160.000	0.065
	Min	160.210	159.750	0.210	160.145	159.900	0.145	160.043	159.960	0.043	160.014	159.975	0.014	160.000	159.975	0.000
200	Max	200.530	200.000	0.820	200.285	200.000	0.400	200.122	200.000	0.168	200.061	200.000	0.090	200.046	200.000	0.075
	Min	200.240	199.710	0.240	200.170	199.885	0.170	200.050	199.954	0.050	200.015	199.971	0.015	200.000	199.971	0.000
250	Max	250.570	250.000	0.860	250.285	250.000	0.400	250.122	250.000	0.168	250.061	250.000	0.090	250.046	250.000	0.075
	Min	250.280	249.710	0.280	250.170	249.885	0.170	250.050	249.954	0.050	250.015	249.971	0.015	250.000	249.971	0.000
300	Max	300.650	300.000	0.970	300.320	300.000	0.450	300.137	300.000	0.189	300.069	300.000	0.101	300.052	300.000	0.084
	Min	300.330	299.680	0.330	300.190	299.870	0.190	300.056	299.948	0.056	300.017	299.968	0.017	300.000	299.968	0.000
400	Max	400.760	400.000	1.120	400.350	400.000	0.490	400.151	400.000	0.208	400.075	400.000	0.111	400.057	400.000	0.093
	Min	400.400	399.640	0.400	400.210	399.860	0.210	400.062	399.943	0.062	400.018	399.964	0.018	400.000	399.964	0.000
500	Max	500.880	500.000	1.280	500.385	500.000	0.540	500.165	500.000	0.228	500.083	500.000	0.123	500.063	500.000	0.103
	Min	500.480	499.600	0.480	500.230	499.845	0.230	500.068	499.937	0.068	500.020	499.960	0.020	500.000	499.960	0.000

Shaft Basis Transition and Interference Fits (Metric)

Basic Size		LOCATIONAL TRANSITION			LOCATIONAL TRANSITION			LOCATIONAL INTERFERENCE			MEDIUM DRIVE			FORCE		
		Hole K7	Shaft h6	Fit	Hole N7	Shaft h6	Fit	Hole P7	Shaft h6	Fit	Hole S7	Shaft h6	Fit	Hole U7	Shaft h6	Fit
1	Max	1.000	1.000	0.006	0.996	1.000	0.002	0.994	1.000	0.000	0.986	1.000	-0.008	0.982	1.000	-0.012
	Min	0.990	0.994	-0.010	0.986	0.994	-0.014	0.984	0.994	-0.016	0.976	0.994	-0.024	0.972	0.994	-0.028
1.2	Max	1.200	1.200	0.006	1.196	1.200	0.002	1.194	1.200	0.000	1.186	1.200	-0.008	1.182	1.200	-0.012
	Min	1.190	1.194	-0.010	1.186	1.194	-0.014	1.184	1.194	-0.016	1.176	1.194	-0.024	1.172	1.194	-0.028
1.6	Max	1.600	1.600	0.006	1.596	1.600	0.002	1.594	1.600	0.000	1.586	1.600	-0.008	1.582	1.600	-0.012
	Min	1.590	1.594	-0.010	1.586	1.594	-0.014	1.584	1.594	-0.016	1.576	1.594	-0.024	1.572	1.594	-0.028
2	Max	2.000	2.000	0.006	1.996	2.000	0.002	1.994	2.000	0.000	1.986	2.000	-0.008	1.982	2.000	-0.012
	Min	1.990	1.994	-0.010	1.986	1.994	-0.014	1.984	1.994	-0.016	1.976	1.994	-0.024	1.972	1.994	-0.028
2.5	Max	2.500	2.500	0.006	2.496	2.500	0.002	2.494	2.500	0.000	2.486	2.500	0.000	2.482	2.500	-0.012
	Min	2.490	2.494	-0.010	2.486	2.494	-0.014	2.484	2.494	-0.016	2.476	2.494	-0.024	2.472	2.494	-0.028
3	Max	3.000	3.000	0.006	2.996	3.000	0.002	2.994	3.000	0.000	2.986	3.000	-0.008	2.982	3.000	-0.012
	Min	2.990	2.994	-0.010	2.986	2.994	-0.014	2.984	2.994	-0.016	2.976	2.994	-0.024	2.972	2.994	-0.028
4	Max	4.003	4.000	0.011	3.996	4.000	0.004	3.992	4.000	0.000	3.985	4.000	-0.007	3.981	4.000	-0.011
	Min	3.991	3.992	-0.009	3.984	3.992	-0.016	3.980	3.992	-0.020	3.973	3.992	-0.027	3.969	3.992	-0.031
5	Max	5.003	5.000	0.011	4.996	5.000	0.004	4.992	5.000	0.000	4.985	5.000	-0.007	4.981	5.000	-0.011
	Min	4.991	4.992	-0.009	4.984	4.992	-0.016	4.980	4.992	-0.020	4.973	4.992	-0.027	4.969	4.992	-0.031
6	Max	6.003	6.000	0.011	5.996	6.000	0.004	5.992	6.000	0.000	5.985	6.000	-0.007	5.981	6.000	-0.011
	Min	5.991	5.992	-0.009	5.984	5.992	-0.016	5.980	5.992	-0.020	5.973	5.992	-0.027	5.969	5.992	-0.031
8	Max	8.005	8.000	0.014	7.996	8.000	0.005	7.991	8.000	0.000	7.983	8.000	-0.008	7.978	8.000	-0.013
	Min	7.990	7.991	-0.010	7.981	7.991	-0.019	7.976	7.991	-0.024	7.968	7.991	-0.032	7.963	7.991	-0.037
10	Max	10.005	10.000	0.014	9.996	10.000	0.005	9.991	10.000	0.000	9.983	10.000	-0.008	9.978	10.000	-0.013
	Min	9.990	9.991	-0.010	9.981	9.991	-0.019	9.976	9.991	-0.024	9.968	9.991	-0.032	9.963	9.991	-0.037
12	Max	12.006	12.000	0.017	11.995	12.000	0.006	11.989	12.000	0.000	11.979	12.000	-0.010	11.974	12.000	-0.015
	Min	11.988	11.989	-0.012	11.977	11.989	-0.023	11.971	11.989	-0.029	11.961	11.989	-0.039	11.956	11.989	-0.044
16	Max	16.006	16.000	0.017	15.995	16.000	0.006	15.989	16.000	0.000	15.979	16.000	-0.010	15.974	16.000	-0.015
	Min	15.988	15.989	-0.012	15.977	15.989	-0.023	15.971	15.989	-0.029	15.961	15.989	-0.039	15.956	15.989	-0.044
20	Max	20.006	20.000	0.019	19.993	20.000	0.006	19.986	20.000	-0.001	19.973	20.000	-0.014	19.967	20.000	-0.020
	Min	19.985	19.987	-0.015	19.972	19.987	-0.028	19.965	19.987	-0.035	19.952	19.987	-0.048	19.946	19.987	-0.054
25	Max	25.006	25.000	0.019	24.993	25.000	0.006	24.986	25.000	-0.001	24.973	25.000	-0.014	24.960	25.000	-0.027
	Min	24.985	24.987	-0.015	24.972	24.987	-0.028	24.965	24.987	-0.035	24.592	24.987	-0.048	24.939	24.987	-0.061
30	Max	30.006	30.000	0.019	29.993	30.000	0.006	29.986	30.000	-0.001	29.973	30.000	-0.014	29.960	30.000	-0.027
	Min	29.985	29.987	-0.015	29.972	29.987	-0.028	29.965	29.987	-0.035	29.952	29.987	-0.048	29.939	29.987	-0.061
40	Max	40.007	40.000	0.023	39.992	40.000	0.008	39.983	40.000	-0.001	39.966	40.000	-0.018	39.949	40.000	-0.035
	Min	39.982	39.984	-0.018	39.967	39.984	-0.033	39.958	39.984	-0.042	39.941	39.984	-0.059	39.924	39.984	-0.076
50	Max	50.007	50.000	0.023	49.992	50.000	0.008	49.983	50.000	-0.001	49.966	50.000	-0.018	49.939	50.000	-0.045
	Min	49.982	49.984	-0.018	49.967	49.984	-0.033	49.958	49.984	-0.042	49.941	49.984	-0.059	49.914	49.984	-0.086
60	Max	60.009	60.000	0.028	59.991	60.000	0.010	59.979	60.000	-0.002	59.958	60.000	-0.023	59.924	60.000	-0.057
	Min	59.979	59.981	-0.021	59.961	59.981	-0.039	59.949	59.981	-0.051	59.928	59.981	-0.072	59.894	59.981	-0.106
80	Max	80.009	80.000	0.028	79.991	80.000	0.010	79.979	80.000	-0.002	79.952	80.000	-0.029	79.909	80.000	-0.072
	Min	79.979	79.981	-0.021	79.961	79.981	-0.039	79.949	79.981	-0.051	79.922	79.981	-0.078	79.879	79.981	-0.121
100	Max	100.010	100.000	0.032	99.990	100.000	0.012	99.976	100.000	-0.002	99.942	100.000	-0.036	99.889	100.000	-0.089
	Min	99.975	99.978	-0.025	99.955	99.978	-0.045	99.941	99.978	-0.059	99.907	99.978	-0.093	99.854	99.978	-0.146
120	Max	120.010	120.000	0.032	119.990	120.000	0.012	119.976	120.000	-0.002	119.934	120.000	-0.044	119.869	120.000	-0.109
	Min	119.975	119.978	-0.025	119.955	119.978	-0.045	119.941	119.978	-0.059	119.899	119.978	-0.101	119.834	119.978	-0.166
160	Max	160.012	160.000	0.037	159.988	160.000	0.013	159.972	160.000	-0.003	159.915	160.000	-0.060	159.825	160.000	-0.150
	Min	159.972	159.975	-0.028	159.948	159.975	-0.052	159.932	159.975	-0.068	159.875	159.975	-0.125	159.785	159.975	-0.215
200	Max	200.013	200.000	0.042	199.986	200.000	0.015	199.967	200.000	-0.004	199.895	200.000	-0.076	199.781	200.000	-0.190
	Min	199.967	199.971	-0.033	199.940	199.971	-0.060	199.921	199.971	-0.079	199.849	199.971	-0.151	199.735	199.971	-0.265
250	Max	250.013	250.000	0.042	249.986	250.000	0.015	249.967	250.000	-0.004	249.877	250.000	-0.094	249.733	250.000	-0.238
	Min	249.967	249.971	-0.033	249.940	249.971	-0.060	249.921	249.971	-0.079	249.831	249.971	-0.169	249.687	249.971	-0.313
300	Max	300.016	300.000	0.048	299.986	300.000	0.018	299.964	300.000	-0.004	299.850	300.000	-0.118	299.670	300.000	-0.298
	Min	299.964	299.968	-0.036	299.934	299.968	-0.066	299.912	299.968	-0.088	299.798	299.968	-0.202	299.618	299.968	-0.382
400	Max	400.017	400.000	0.053	399.984	400.000	0.020	399.959	400.000	-0.005	399.813	400.000	-0.151	399.586	400.000	-0.378
	Min	399.960	399.964	-0.040	399.927	399.964	-0.073	399.902	399.964	-0.098	399.756	399.964	-0.244	399.529	399.964	-0.471
500	Max	500.018	500.000	0.058	499.983	500.000	0.023	499.955	500.000	-0.005	499.771	500.000	-0.189	499.483	500.000	-0.477
	Min	499.955	499.960	-0.045	499.920	499.960	-0.080	499.892	499.960	-0.108	499.708	499.960	-0.292	499.420	499.960	-0.580

AUTHOR:	SCALE:
TEAM: SECTION:	DATE:

AUTHOR:	SCALE:
TEAM: SECTION:	DATE:

AUTHOR:	SCALE:
TEAM: SECTION:	DATE:

AUTHOR:	SCALE:
TEAM: SECTION:	DATE:

AUTHOR: SCALE:

TEAM: SECTION: DATE:

AUTHOR: .... SCALE: ....

TEAM: .... SECTION: .... DATE: ....

AUTHOR:	SCALE:
TEAM: SECTION:	DATE:

AUTHOR:	SCALE:
TEAM: SECTION:	DATE:

AUTHOR:	SCALE:
TEAM: SECTION:	DATE:

Index

D

E

F

N

O

P

R

S